中国国家大剧院
建筑设计国际竞赛方案集

A COLLECTION OF DESIGN SCHEMES FOR THE INTERNATIONAL ARCHITECTURAL COMPETITION OF THE NATIONAL GRAND THEATER P.R.CHINA

中国建筑工业出版社
CHINA ARCHITECTURE & BUILDING PRESS

《中国国家大剧院国际竞赛方案集》 编委会领导小组

万嗣铨　艾青春　姚　兵
王争鸣　张永嘉　周庆琳
金志舜

主　　编　周庆琳

工作人员　迟　鸣　杨　帆　于艳梅

目 录
CONTENTS

前　言

中国最高艺术表演中心——国家大剧院的建设，从50年代起，几经论证和策划，历时近四十年。1997年中国政府决定由政府投资，在北京天安门广场人民大会堂西侧兴建国家大剧院，并为此成立了国家大剧院工程业主委员会，该委员会在国家大剧院建设领导小组领导下开展工作。为了能够将本工程建成国际一流的艺术殿堂，中国政府决定在国内外广泛征求设计方案。为此，国家大剧院工程业主委员会邀请了国内外一些著名的设计单位参加方案竞赛。

由于本工程性质和位置的特殊以及工程本身的复杂，决定了它是一件极具难度和挑战的工作。在一年零四个月的时间里，经过两轮竞赛和三次修改，于1999年7月22日由中央审定，确定了法国巴黎机场公司设计，清华大学配合的设计方案为实施方案。

□第一轮竞赛

国家大剧院工程业主委员会于1998年4月13日共邀请了17家国内外建筑设计单位参加设计竞赛，其中国内11家(含香港特别行政区3家)国外6家。此外还有19家自愿参加的单位，其中国内5家，国外14家。经1998年4月13日至7月13日共三个月的运作时间，截止7月13日收到36家设计单位交来44个方案，其中国内24个(含香港特别行政区4个)国外20个。7月14日~23日技术委员会对44个方案进行了技术审查，7月27日~31日由评委会进行了评选。

评委会由11人组成，其中国内评委8名(含香港特别行政区1名)国外评委3名。根据竞赛的规定，评委会应评出三个方案上报领导小组。评委们对44个方案进行了认真的研究，充分的讨论，认为在全部竞赛方案中，还没有一个方案能较综合地、圆满地、高标准地达到设计任务书提出的要求，无法提出3个可以直接上报领导小组的方案。评委会根据竞赛文件中"如条件不具备，可以缺额"的规定，以无记名投票方式选出五个得票过半数的方案并建议业主委员会再进行一次新一轮设计创作。这5个方案是(按编号顺序)：

- ·101号　法国巴黎机场公司提供
- ·106号　英国塔瑞·法若建筑师事务所提供
- ·201号　日本矶崎新建筑师株式会社提供
- ·205号　中国建设部建筑设计院提供
- ·507号　德国国际建筑设计公司提供

根据评委会建议，经领导小组批准业主委员会于8月24日请参赛方进行了第二轮设计创作。

□第二轮竞赛

此次送选的方案共14个，除上一轮评委会推荐的5个方案外，另有由业主委员会邀请参赛的国内(含香港)4个设计单位，以及曾参与上一轮竞赛，并自愿参加本轮竞赛的国外5个设计单位参加竞赛。评委会11位成员除有个别变动外基本上与第一轮相同，于1998年11月14日至11月17日进行了评审。

评选期间，业主委员会负责人对评选方式作出新的建议。评委们根据实际情况，对评选方法作出如下的决定：

1．从第一轮所遴选的5个方案中，选出两名。

2．从第一轮竞赛后业主委员会所邀请的四个国内参赛方案中，遴选出一名。

3．从第一轮竞赛后自愿参加第二轮竞赛的方案中，遴选出一名。

4．以上投票所得的方案提供领导小组及有关方面参考。

经过评委会无记名投票，结果如下：

1．从第一轮所遴选的5个方案中，选出：

1号方案，法国巴黎机场公司

2号方案，日本矶崎新建筑师株式会社

(另：5号方案，英国塔瑞·法若建筑师事务所，以两票之差逊于上述两个方案。)

2．从国内受邀单位中，选出：

6号方案，北京市建筑设计研究院

8号方案，清华大学

(以上两个方案因票数相同，一并列入，提供领导小组参考。)

3、从参与第一轮竞赛并自愿参加第二轮竞赛的5个方案中，选出：

12号方案，奥地利汉斯·豪莱建筑师事务所。

绝大多数评委认为，上述所列举的方案有些虽然已达到比较高的水准，但对特定的地段条件以及其他因素来讲，这些方案均不够完美，或多或少存在不同程度的问题，有的有较严重的缺陷，提请领导小组和决策人慎重考虑。

领导小组认真分析研究了评委会的意见，决定由法国巴黎机场公司与清华大学合作，英国塔瑞·法若建筑师事务所与北京市建筑设计研究院合作，对第二轮的方案进行一次修改，同时领导小组又根据聘请的专家组中大多数专家意见，决定由加拿大卡洛斯建筑师事务所与建设部建筑设计院合作，对其方案一起进行修改。本次修改自1998年12月15日至1999年1月31日进行。

□第二轮第一次修改

共收到四个修改方案，即(按英文字母顺序排列)：

- ·1号　法国巴黎机场公司＋清华大学提供
- ·2号　加拿大卡洛斯建筑师事务所＋建筑部建筑设计院提供
- ·3号　英国塔瑞·法若建筑师事务所＋北京市建筑设计研究院提供
- ·4号　清华大学＋法国巴黎机场公司提供

领导小组聘请专家一组、二组成员及上两轮的部分国内评委对上述修改方案进行讨论评议，大多数专家认为本次修改方案又有了很大的改进和提高，但仍不够满意，还需继续修改。

同时经部分参赛设计人员和部分专家的建议，经领导小组及北京市规划部门同意，将用地位置作了调整，即整个用地向南移70m，将原南侧绿地移至北侧，因此也需修改方案。从1999年3月2日至5月4日进行了第二次修改。

□第二轮第二次修改

参加本次方案修改的单位有(按英文字母顺序排列)：

·1号　法国巴黎机场公司设计、清华大学配合

·2号　加拿大卡洛斯建筑师事务所与建设部建筑设计院合作设计

·3号　英国塔瑞·法若建筑师事务所与北京市建筑设计研究院合作设计

·4号　清华大学设计、法国巴黎机场公司配合

对上述方案国家大剧院建设领导小组及专家组、工程业主委员会、艺术委员会成员听取了4位主设计师的介绍，进行评论。大多数专家认为法国巴黎机场公司提交的方案构思独特，造型新颖，很有创意。该方案还对天安门广场地区整体规划提出了大胆的设想，得到了与会者的赞同。如果这一设想得以实现，将会大大改善天安门地域的环境，领导小组采纳了这个意见。经北京市规划部门同意，业主委员会又将设计条件作了调整：扩大了绿地范围，并将大剧院用地再次南移，使大剧院的东西轴线同人大会堂东西轴线对齐。根据修改的条件请参赛单位对方案再作一次调整。根据专家组讨论的意见，本次仅邀请了3个单位参加调整修改。

□第二轮竞赛第三次修改

参加单位有：

·法国巴黎机场公司设计、清华大学配合

·英国塔瑞·法若建筑师事务所与北京市建筑设计研究院合作设计

·清华大学设计、法国巴黎机场公司配合

7月上、中旬业主委员会先后邀请了全国和北京市部分人大代表和政协委员进行座谈，征求意见，多数人赞成法国巴黎机场公司提交的方案。领导小组在认真听取、分析各方面意见的基础上，经过深入研究后一致推荐法国巴黎机场公司方案，并决定连同英国法瑞·塔若建筑师事务所和清华大学方案一同上报中央审定。为时一年零四个月的方案竞赛活动到此结束。

Foreword

Ever since the 1950's,the National Grand Theater,the top performing art center of the People's Republic of China,has undergone over 40 years of planning and evaluation.In 1997,the Chinese government decided to build the Theater at the west side of the People's Great Hall by the side of Tian'anmen Square;and consequently founded the Proprietor Committee of the National Grand Theater that has been developing the project under the Leader Team.In order to achieve a first class international performance palace,the Chinese government decided to collect design schemes both from abroad and in China.For this purpose the Proprietor Committee invited some well-known architects to take part in a Scheme Design Competition.

Due to the unique characteristics of the project and the site,as well as the complication of the building,to achieve a satisfactory design is extremely difficult and challenging.Two rounds of competition and three times of modification have been done over one year and four mouths.On July 22,1999,the central government selected the design of France ADP(Aeroports de Paris) which was assisted by China Tsinghua University ,as the final winner scheme.

· The First round competition

The Proprietor Committee of the National Grand Theater invited 17 design companies on April 13,1998 from China and abroad which include 11 Chinese (with three Hong Kong companies) and 6 foreign companies.Apart from the invited there were 19 more volunteers,with 5 from China and 14 from overseas.After three months work 36 companies submitted 44schemes on July 13,among them there were 24 Chinese (including four from Hong Kong) and 20 foreign entries.The Technical Committee assessed the 44 entries from July 14 to 23,1998 and the Jury evaluated them from July 27 to 31,1998.

The Jury comprises of 11 members,with 8 local(including one from Hong Kong)and 3 overseas.The Rules of the competition requires the Jury to choose 3 works from all the entries,and thereafter the Leader

Team shall further decide one winner out of the three.After careful study and full discussion over the 44 entries produced in the first round of competition,the Jury concluded with regret that non of the works meet the required standard comprehensively and satisfactorily.The Jury found that it is not in a position to provide the Lead Team with the required three entries.Pursuant to the section 8 of the Document of the Design Scheme Competition of the National Grand Theater:"should the requirements can not be satisfied,the Jury may leave the position vacant."and by an anonymous voting,the Jury instead selected five entries which received more than half of the votes as the recommended schemes.The Jury suggested the Proprietor Committee to consider those five entries as candidates for the second round of competition.

The five selected schemes were:

No.101 France ADP (Aeroports de Paris)

No.106UK Terry Farrell & Partners

No.201 Japan Arata Isozaki & Associates

No.205 China Architecture Design Institute Ministry of Construction

No.507 Germany HPP International Planungsgesellschaft mbH

According to the suggestion of the Jury the Leader Team approved that the Proprietor Committee held the second round design scheme competition from August 24 to November 10,1998.

· The second round competition

There were 14 schemes submitted;besides the 5 designers recommended by the Jury from the first round competition,there were 4 invited local designers (including one from Hong Kong) and 5 foreign volunteers who had participated in the first competition,Except some minor changes,the 11 members of the Jury were the same as the first round competition;and they were evaluating the entries from November 14 to 17,1998.

In the course of the evaluation,the Chairman of the Proprietor Committee made a new suggestion regarding the selecting method.Based on actual situation,the Jury members agreed to make the following changes to the selecting method.

A.To select 2 out of the 5 schemes recommended from the first round competition.

B.To select 1 scheme from the 4 participating Chinese designers invited to join in the second round by the Proprietor Committee.

C.To select 1 scheme from the voluntary participants.

D.Schemes selected through voting should be provided to the Leader Team and relevant departments for consideration.

The following was the result of the anonymous voting by the Jury.

A.From among the 5 designers recommended in the first round competition,the following ones were selected.

No.1 France ADP(Aeroports de Paris)

No.2 Japan Arata Isozaki &Associates

(Next to the first 2 selections was No.5 by Terry Farrell & Partners of UK due to the lack of 2 votes.)

B.Among the schemes of the invited Chinese designers,the following were selected.

No.6 China Beijing Institute of Architectural Design & Research

No.8 China Tsinghua University

(Since the 2 schemes got the same votes,both schemes were submitted for the consideration of the Leader Team.)

C. From the 5 voluntary schemes,No.12 by Hans Hollein + Heinz Neumann Design Group of Austria was selected.

The overwhelming majority of the judges concluded that among the above schemes,some achieved very high level of design quality.However, because of this specific site and other related factors,non of these schemes was perfect.Each had its own defects,some with quite serious ones,so the Jury reminded the Leader Team and decision makers to carefully consider the situation.

Having reviewed fully the suggestions by the Jury,the Leader Team recommended and made the following arrangement,which enable the foreign front-rummers get the needed cooperation from a domestic China counterparts in amending and improve their works:the France ADP (Aeroports de Paris) and Tsinghua University of China;the Terry Farrell & Partners of UK and Beijing Institute of Architectural Design & Research.Also by the majority opinion of the Expert Group,the Leader Team recommended and arranged the cooperation between Carlos Ott &Associates of Canada and the Architecture Design Institute Ministry of Construction of China,in joint effort to improve the Canadian entry.This phrase of modification was from Dec.15,1998 to Jan.31,1999.

· The first modification on the 2nd round scheme design competition

There were totally 4 modified schemes received.(In alphabetic order)

No.1 France:ADP (Aeroports de Paris)+China:Tsinghua University

No.2 Canada:Carlos Ott & Associates + China: Architecture Design Institute Ministry

No.3 UK:Terry Farrell & Partners +China:Beijing Institute of Architectural Design &Research

No.4 China: Tsinghua University + France:ADP(Aeroports de Paris)

The Leader Team invited the two Expert Groups and local members of the jury of the National Grand Theater to evaluate the above modified schemes.Most experts considered them were impressively improved compared with last time but still not fully satisfying.Meanwhile on the suggestion of some of the designers amd experts,with the approval of the Leader Team as well as the Municipal Planning Department the construction site was moved 70 meters southward,with the greenery patches shifted from the south-side to the north-side of the site.The second improvement of the schemes were made from March 2 to May 4,1999.

· The second modification on the 2nd round scheme design competition

Participants of this stage:(In alphabetic order)

No.1 Designer: France ADP(Aeroports de Paris)

Cooperator:China Tsinghua University

No.2 Designer:Canada Carlos Ott & Associates

Cooperator:China Architecture Design Institute Ministry

No.3 Designer:UK Terry Farrell & Partners

Cooperator:China Beijing Institute of Architectural Design & Research

No.4 Designer:China Tsinghua University

Cooperator:France ADP (Aeroports de Paris)

For the above schemes,the Leader Team,the Experts Group,the Propriety Committee and Art Consultants Committee listened the introduction of the four main designers.After discussion,most experts considered the scheme provided by France ADP reflects an unique concept and bccn novelly shaped,and creative.The scheme also proposed a bold plan for Tian'anmen Square area which was appreciated by the meeting presence.If this propose could be realized the surround of Tian'anmen Square would be significantly improved.The Leader Team supported this propose.On the agreement of the Beijing Municipal Planning Department,the Propriety Committee further adjusted the requirement of the design:by enlarging the greenery patches and moving the site southward again,so the east-west axle of the Theater and the People's Great Hall resulted at the same line.According to the adjusted requirement,the designers made another modification to their schemes.Based on the experts' opinion the following companies were invited to take part in the modification of this stage.

· The third modification on the 2nd round scheme design competition

Participants of this stage:(In alphabetic order)

No.1 Designer:France ADP (Aeroports de Paris)

Cooperator:China Tsinghua University

No.2 Designer:UK Terry Farrell & Partners

Cooperator:China Beijing Institute of Architectural Design & Research

No.3Designer:China Tsinghua University

Cooperator:France ADP(Aeroports de Paris)

In the early July,Propriety Committee invited some of the deputies of the National People's Congress and the Political Consultative Conference from country and Beijing city to have an informal discussion to ask for their opinion,and most of the deputies supported the design of France ADP (Aeroports de Paris).Having fully heard and analyzed all sorts of suggestions and thoroughly studied,the Leader Team unanimously recommend French design together with the schemes from UK Terry Farrell& Partners and China Tsinghua University to be submitted for the central government to examine and finalize.By now, the Scheme Design Competition that lasted for one year and four mouths finally come to an end.

鸣　谢

　　本次中国《国家大剧院》设计方案的竞赛活动，得到了国内外众多的单位和个人热情的支持与关注。从 1998 年 4 月 13 日至 1999 年 7 月，在这一年零四个月的时间内，经过了二轮竞赛三次修改，共收集到了 69 个方案。对这些珍贵的作品进行了认真的分析和研究，深感这里凝聚着作者的汗水和智慧，我们为他们所付出的艰辛劳动与认真负责的态度所感动，在他们的作品正式展现在我们面前的时候，再一次向他们表示衷心地谢意！

THANKS

　　The Scheme Design Competition of the National Grand Theater of China had been blessed by enormous warmhearted support and attention from many designing offices and individuals,both domestic and overseas.Within the time of tone year and four months during which two round of competitions and three times of modification were conducted,we received a total of 69 schemes.Upon serious analysis and study of those works,we are exceedingly impressed by the hard work and earnest attitude devoted by the designers.Now.When those great works finally unveiled in front of our eyes,we once again express our hearty thanks.

建设领导小组名单

组 长	贾庆林	中共中央政治局委员
		中共北京市委书记
组 员	胡光宝	中共中央办公厅常务副主任
	何椿霖	全国人大常委会秘书长
	马 凯	国务院副秘书长
	曾培炎	国家发展计划委员会主任
	刘忠德	全国政协教科文卫体委员会主任
	孙家正	文化部部长
	俞正声	建设部部长
	张佑才	财政部副部长
	刘 淇	北京市市长

工程业主委员会名单

主 席	万嗣铨	北京市政协副主席
副主席	艾青春	文化部副部长
	姚 兵	建设部总工程师
委 员	王争鸣	中共北京市委城建工委副书记
	张永嘉	文化部文化设施建设管理中心主任
	周庆琳	建设部建筑设计院总建筑师

专家一组名单

组　长　吴良镛　科学院院士　　工程院院士
　　　　　　　　清华大学建筑学院教授

副组长　马国馨　工程院院士
　　　　　　　　北京市建筑设计研究院副总建筑师
　　　　崔　恺　建设部建筑设计院副院长兼总建筑师

组　员　平永泉　北京市城市规划管理局局长
　　　　　　　　高级建筑师
　　　　齐　康　科学院院士
　　　　　　　　东南大学建筑系教授
　　　　邢同和　上海现代设计(集团)总建筑师
　　　　何　弢　香港建筑师学会会长
　　　　何镜堂　华南理工大学建筑学院院长
　　　　　　　　工程设计大师
　　　　许宏庄　原文化部计财司总工程师
　　　　宣祥鎏　原首都规划建设委员会副主任
　　　　窦以德　中国建筑学会秘书长
　　　　戴复东　同济大学建筑系教授

专家二组名单

组　长　李　畅　原中央戏曲学院舞美系副主任
　　　　　　　　教授

副组长　王世全　中国交响乐团总音响师
组　员　于建平　国防科工委工程设计研究院总工程师
　　　　王炳麟　清华大学建筑学院教授
　　　　李布白　中国艺术科学技术研究所高级工程师
　　　　吴雁泽　中国音乐家协会党组书记
　　　　　　　　一级演员
　　　　陈　治　原中国青年艺术剧院灯光组组长
　　　　项端祈　北京市建筑设计研究院教授级高工
　　　　骆学聪　广播电影电视部设计院副总工程师
　　　　段惠文　北京有色冶金研究设计总院副总工程师

第一轮竞赛评委会名单

主 席　吴良镛　　科学院院士　　工程院院士
　　　　　　　　清华大学建筑学院教授
委 员　里卡杜·包费尔　　世界著名建筑师(西班牙)
　　　　　　　　包费尔建筑设计事务所所
　　　　　　　　长
　　　　芦原义信　世界著名建筑师(日本)
　　　　　　　　芦原建筑设计研究所所长、原日
　　　　　　　　本建筑学会会长
　　　　张锦秋　　工程院院士
　　　　　　　　工程设计大师
　　　　　　　　中国建筑西北设计研究院总建筑师
　　　　何镜堂　　华南理工大学建筑学院院长
　　　　　　　　工程设计大师
　　　　周干峙　　科学院院士工程院院士
　　　　阿瑟·爱里克森　　世界著名建筑师(加拿大)
　　　　宣祥鎏　　原首都规划建设委员会副主任
　　　　彭一刚　　工程院院士
　　　　　　　　天津大学教授
　　　　傅熹年　　工程院院士
　　　　　　　　中国建筑技术研究院教授级高级建
　　　　　　　　筑师
　　　　潘祖尧　　原香港建筑师学会会长

第二轮竞赛评委会名单

主 席　吴良镛　　科学院院士　　工程院院士
　　　　　　　　清华大学建筑学院教授
委 员　齐 康　　科学院院士
　　　　　　　　东南大学建筑系教授
　　　　里卡杜·包费尔　　世界著名建筑师(西班牙)
　　　　　　　　包费尔建筑设计事务所所
　　　　　　　　长
　　　　张锦秋　　工程院院士
　　　　　　　　工程设计大师
　　　　　　　　中国建筑西北设计研究院总建筑师
　　　　何镜堂　　华南理工大学建筑学院院长
　　　　　　　　工程设计大师
　　　　周干峙　　科学院院士　　工程院院士
　　　　阿瑟·爱里克森　　世界著名建筑师(加拿大)
　　　　宣祥鎏　　原首都规划建设委员会副主任
　　　　傅熹年　　工程院院士
　　　　　　　　中国建筑技术研究院教授级高级建
　　　　　　　　筑师
　　　　潘祖尧　　原香港建筑师学会会长
　　　　戴复东　　同济大学建筑系教授

中国国家大剧院
建筑设计方案竞赛文件

一、前言

中国国家大剧院是中国最高表演艺术中心。自50年代起，几经论证和策划，历时近四十年。在中国改革开放、经济蓬勃发展的今天，中国政府决定投资在首都北京天安门广场区段兴建国家大剧院，它将成为弘扬民族文化，反映时代精神，汇集世界现代建筑艺术与科学技术于一身，贡献人类表演艺术事业发展的宏伟巨作。为此，我们热忱邀请关心中国现代化建设，富有创新精神及丰富设计经验的国际、国内设计单位参与这一意义深远的设计方案竞赛，将国家大剧院真正建设成国际一流的艺术殿堂。中国人民将由衷地感谢你们的友谊和贡献。

本次竞赛活动在国家大剧院建设领导小组监督下按有关行政部门制定的规定进行，以保证竞赛活动的合法和公正。

二、工程概况

1．甲方：国家大剧院业主委员会

2．项目：国家大剧院工程总建筑面积12万 m²。内容包括一个2500座的歌剧院，一个2000座的音乐厅，一个1200座的话剧院和一个300～500座的小剧院和其他附属设施(详见设计任务书)。

3．地点：国家大剧院基地位于北京长安街，天安门广场及人民大会堂西侧，西长安街以南，石碑胡同以东，东绒线胡同以北，人民大会堂西路以西。基地东西224～244m，南北向约166m，总面积约3.89公顷。

4．依据：

(1)根据政府批准的国家大剧院可行性研究报告而编制的《国家大剧院设计任务书》。

(2)国家大剧院建筑设计方案竞赛文件附件。

三、设计成果要求

1．设计报告书

(1)说明内容要求(材料装订按以下顺序排列)

A．建筑总体布局说明

B．单体建筑方案构思说明

C．声学设计方案说明

D．结构选型方案说明

E．消防安全系统说明

F．主要内外装饰材料说明

(2)技术经济指标　格式见表1

(3)各观众厅数据　格式见表2

(4)造价估算(仅作参考)　格式见表3

以上文字部分用 A3(297mm × 420mm)纸打印。

2．图纸及模型

(1)总平面图　1：500

(2)交通分析图　1：500

(3)绿化及环境设计图　1：500

(4)建筑各层平面图、剖面图及立面图　1：300

(5)各观演厅，1：100声线、视线分析图及1：100舞台机械布置方案

(6)北侧、南侧透视图各一张

(7)夜景透视图一张

(8)公共大厅、歌剧院观众厅透视图各一张

(9)1／300单体模型一个，底盘尺寸1200mm × 1200mm，要求色彩反映实际效果，1／1000体量模型一个，要求全白色，底盘尺寸为1／1000的用地红线范围。(总体环境模型由甲方制作)。

(10)图纸标注尺寸均以毫米为单位，标高以米为单位。以上图纸提供一份，均应贴裱在1000mm × 700mm的轻质板上、另外提供上述报告书和图纸文件的A3(297mm × 420mm)缩印本16套。

四、报送竞赛文件的规定

1．为保证方案竞赛的公正性,各参赛单位均以匿名方式报送竞赛方案并不得在图纸及文件上表示表明身份的标记。评选时不需要参赛方向评委介绍方案。

2．在正本(原图)首面背后右下角标明作者名称并用不透明纸封住。

3. 为保证按时进行评选，各参赛方届时应派人来北京报送方案。

五、竞赛时间安排

1. 发送竞赛文件和现场踏勘时间

1998年4月13日召开竞赛文件发布会，时间地点另行通知。

2. 答疑时间

1998年4月30日前提问，提问方式采用附件中的质疑表（表4），以传真方式寄给甲方，过时不再解答。5月15日甲方以书面形式快递给各参赛方。

3. 提交方案时间

1998年7月13日下午14时前提交到业主委员会指定地点。14：30举行开标仪式，同时宣布评委会名单。

4. 技术审查时间

1998年7月14日～7月23日由竞赛技术委员会对参赛方案进行规划、建筑、声学、舞台工艺、消防等方面的技术审查。

5. 评选时间

1998年7月24日～7月29日

6. 8月3日由业主委员会以书面方式向各参赛单位发布评选结果。

六、参赛单位的邀请

1. 参赛单位应具有丰富的相关工程设计经验和相应设计资质。

2. 邀请参赛单位数量：国内、国外共计15～20家。

七、评委会组成

1. 评委会由11名国内、外著名建筑及剧场设计专家组成，名单在评选前公布。

2. 由若干名规划、建筑、声学、舞台工艺、消防等专家组成一个技术委员会对参赛方案进行初审。技术委员会对评选委员会提供初审咨询，但不参与评审和投票。

八、评选办法

1. 对报送方案先由技术委员会进行初审，并将初审意见报送评委会，对初审的意见由评委会进行处理。

2. 评委会委员在充分讨论的基础上，以无记名投票方式，以简单多数产生三个提名方案。如条件不具备时可以缺额。

3. 评审工作结束后，由评委会写出"评审报告"，并由全体评委签字后交给业主委员会。

4. 全部参赛方案在评委会评选工作结束后，向领导小组汇报前公开展览。

5. 业主委员会将"评审报告"及三个提名方案向国家大剧院建设领导小组汇报，被提名方可以自愿参加介绍方案，最终由领导小组确定中选方案。

6. 如有未被邀请单位自愿参加竞赛，其方案以同等条件参加评选，如被提名可同样获应得奖金，未被提名则不支付成本金。

九、评选原则

1. 设计方案必须符合竞赛文件的规定，否则视为不合格作品。

2. 充分体现作为国家最高艺术殿堂的特征，满足使用功能要求。

3. 充分地考虑到作为中华人民共和国政治文化中心地段的特定地理环境的要求。

4. 技术先进、经济合理、有充分的可实施性。

5. 各项技术经济指标计算准确、合理。

十、费用

1. 被邀请参加竞赛单位在领取竞赛文件时应提供有法人签字的有效证件及1000美元(国内单位8000元人民币)的保证金，待评审结束后退回。如未按规定报送文件者保证金不退。

2. 提名方案每个获奖金8万美元，中选方案的奖金在设计费中扣除。

3. 未获提名的方案，凡符合设计竞赛文件规定者，每个邀请单位可获成本费3万美元。国外设计单位另加1万美元差旅费，所有被邀请单位来京的差旅、食宿费自理。

4. 在中国境内发生的税金由参赛单位自理。

5. 奖金和成本费外方支付美元，中方支付当日牌价的人民币。

十一、版权

1.甲方对所征集的方案，有权展览、印刷、出版等。

2.所有报送的参赛文件评选后均不退回。

十二、关于中选方案实施的规定

1.中选单位有义务按业主意见作必要的修改，并应负责完成建筑、结构、设备、室内、声学、环境等设计内容，设计内容必须符合中国现行各项设计规范，施工图设计深度必须达到中国现行的民用建筑施工图设计深度的要求。

2.如外方方案中选，根据中国有关规定，必须与国内一家甲级设计单位进行合作设计。

3.完成全部设计的设计费，外方为建安投资(不包括舞台设备和特殊室内装修)的5%(包括同国内设计单位合作设计的费用)。

国内设计单位的设计费按国内规定办。

十三、竞赛文件用语：中文、英文

所有竞赛文件、说明、图纸均为中、英文对照，如中文文本与英文文本有不一致之处，以中文文本为准。

十四、参加本次方案竞赛者均视为承认本文件的所有条款。

甲方地址：北京市西城区石碑胡同 4 号

邮　　编：100031

传　　真：(010)66064707

电　　话：(010)66064705

联 系 人：杨　帆

The Document of the Design Scheme Competition of the National Grand Theater,China

1.PREFACE

The National Grand Theater will be the highest performing arts center of China.Since the 1950's,it has undergone a long process of planning and presentations.Today with the successful reform and open-policy and rapid growth of the economy in China,the Chinese government has decided to erect the National Grand Theater in the area of the Tian'anmen Square of Beijing.It will be a monumental work to carry the Chinese cultural forward,to reflect the spirit of the time,to collect modern architectural arts and high technologies in the world,and to contribute to the development of human performing arts.We would like to cordially invite both local and overseas architects,who are concerned about the modernization of China,who have originality of thought and have rich design experience,to participate in this significant design scheme competition so as to create the National Grand Theater as one of the best art's palaces in the world.The Chinese people would like to thank you sincerely for your friendship and contribution to China in advance.

The competition will be carried out under the supervision of the Project Leader Team of the National Grand Theater and in conformity with the regulations by the related authorities,in order to secure legality and justice in the competition.

2.OUTLINE OF THE PROJECT

2.1 Client

The client of this project is the Proprietor Committee of the National Grand Theater.

2.2 Project

The overall construction area of the National Grand Theater will be 120,000 square meters.It comprises an opera house of 2,500 seats,a concert hall of 2,000 seats,a theater for modem drama of 1,200 seats,a mini theater of 300-500 seats and other facilities.(The details can be referred to in the Design Program.)

2.3 Location of Site

The site of the National Grand Theater is located on the Chang'an Street,Beijing,to the west of the Tian'anmen Square and the Great Hall of the People,south of west Chang'an Street,east of Shibeihutong,north of East Rongxian-hutong and west of the Great Hall of the People.The length of the Site from east to west is 224-244 meters,and the length from north to south is about 166 meters.The total area of the site covers 3.89ha.

2.4 Design Instructions

The Design Instructions are the following:

1)The Design Program of the National Grand Theater is approved by the Chinese Government and based on the Feasibility Study of the National Grand Theater.

2)Appendices of the Document of the Design Scheme Competition of the National Grand Theater.

3.REQUIRED DESIGN DOCUMENTS

3.1 Design Reports

1)Contents of instruction:(The arrangement of documents should be as follows.)

A.Instructions of overall architectural layout

B.Instructions of conception of each single architectural scheme.

C.Instructions of acoustic designs.

D.Instructions of structural selection scheme

E.Instructions of fire fighting and safety system

F.Instructions of main material for exterior and interior decoration

2)Technical and economic Indicators(Use Form 1)

3)Room program of auditorium halls(Use Form 2)

4)Cost evaluation(Only to be used as reference.Use Form 3)

The text above should be printed on A3 paper.(297mm × 420mm).

3.2 Drawings and Models

1)An overall plan(1 ： 500)

2)A traffic plan(1 ： 500)

3)A plan of green space and surroundings(1 ： 500)

4)Plans of floors,sections and elevations(1 ： 300)

5)Acoustic and visual analysis drawings for each auditorium hall (1：100)and layout drawings of equipments for each performing hall(1：100)

6)A perspective drawing from north and a perspective drawing from south.

7)A perspective drawing of night scene.

8)A perspective drawing of lobby and a perspective drawing of opera house.

9)A scale model(1：300).The base of this model should be 1200mm × 1200mm in size.The color of the model must present real effect.

Another scale model(1：1000).This model must be colored totally white.The base of the model should be sized on the construction site of 1：1000.(A model of the surroundings has been made by the Client.)

10)The dimention unit showings in the drawings are in mm,except that the elevations are in m.Entrants should submit one set of drawings of each items above.Each drawing should be kept close to a light board.In addition,submitted documents must include 16 copies of the reports and the drawings reduced to A3-size(297mm × 420mm).

4.INSTRUCTIONS FOR SUBMISSION OF DOCUMENTS

4.1 In order to secure justice in the competition,entries will be anonymous.Any symbol that could identify the entrants should not be marked on all drawings and documents.In the evaluation of entries,it will be not necessary that entrants explain their design scheme to the Jury.

4.2 Entrants should write their name at the bottom right corner on the back of the original drawings,and seal the name with a opaque paper.

4.3 In order to ensure that evaluation can be carried out on schedule,entrants should dispatch staffs to delivery the documents of the design scheme to Beijing.

5.SCHEDULE OF COMPETITION

5.1 Distribution of Competition Document and Visit to Site

The Release Meeting of the Design Scheme Competition of the National Grand Theater will be held on April 13-14,1998.During the Release Meeting entrants will visit the construction site.The appointed time and address will be told next time.

5.2 Queries Time

Queries must be submitted to the Proprietor Committee by April 30,1998 by facsimile.Entrants are required to use the Queries Form(Form 4)of Appendices.They will be answered in writing on May 15,1998 by express mail.Queries received after the appointed time will not be answered.

5.3 Submission Time of Entries

All entries must be dispatched to the address appointed by the Proprietor Committee by 14：00 p.m. on July 13,1998.The ceremony of opening the entries will be started at 14：30 on the same day.Mean while,the name list of the Jury will be announced.

5.4 Technical Assessment Time

The technical assessment of planning,architecture,acoustics,technique of performing stage and fire fighting system for each entry will be carried out by the Technical Committee of the Competition from July 14 to July 23,1998.

5.5 Evaluation Time

Evaluation of the entries will be held from July 24 to July 29,1998 by the Jury.

5.6 The results of Evaluation of the entries will be announced on August 3 to the entrants by the Proprietor Committee in writing.

6.INVITATION OF ENTRANTS

6.1 Entrants should have rich experience in designing related projects and competent qualifications for architectural design.

6.2 15-20 local and overseas entrants will be invited to participate in the Competition.

7.THE COMPONENTS OF JURY

7.1 The Jury comprises 11 local and overseas members who are well—known experts on architecture or theater design.

7.2 A technical committee composed of experts on planning, architecture,acoustics,technique of performing stage and fire fighting system will make a preliminary examination of the entries.The technical committee advises the Jury of the results of the preliminary examination,but do not take part in judging or have the right to vote.

8.PROCESS OF EVALUATION

8.1 All entries will undergo a preliminarily examination by the technical committee.The results of the examination will be reported to the Jury and dealt with by the Jury.

8.2 The Jury will select three nominated design schemes from entries through a secrete vote after full discussion.If the requirements can not be satisfied,the Jury may leave the positions vacant.

8.3 After the evaluation,the Jury will make an assessment report with the signatures of all members of the Jury and present it to the Proprietor Committee.

8.4 After the evaluation all entries will be put on exhibition before being reported to the Leader Team.

8.5 The Proprietor Committee will present the assessment report and three nominated design schemes to the Leader Team of the National Grand Theater.The entrants nominated have an opportunity to give a presentation of their desigh schemes to the Leader Team.Finally,the wining design scheme will be decided by the Leader Team.

8.6 If design companies or firms which have not been invited voluntarily participate in the competition,their entries will be evaluated on an equal basis with the entries formally invited.If any voluntary entry is selected as one of the nominated design schemes through the evaluation,it will also win a prize of 80,000 US dollars.Voluntary entries which are not selected as the nominated design scheme through the evaluation,will not be payed any design cost.

9.PRINCIPLES OF EVALUATION

9.1 Every entry must satisfy the requirements of the competition documents,otherwise it will be considered as an unqualified entry.

9.2 Design should sufficiently embody its characteristics as a sanctuary of the superb art of the country,and should satisfy performance needs and requirements of theater functions.

9.3 Design should sufficiently consider the characteristics of specific location of the site which is the political and cultural center of the People's Republic of China.

9.4 The Theater should be fitted with advanced technologies.Design should be economical,rational and fully feasible.

9.5 All of the economical and technological indictors of the design should be calculated correctly and reasonably.

10.PRIZE AND COST

10.1 Entrants invited to the Competition are required to bring valid certification of corporative registration and leave a deposit of 1,000 US dollars(the local entrants are required to deposit 8,000 RMB)when collecting the competition program and documents.The deposit will be refunded to entrants after evaluation.Any entrant who does not present his entry documents according to the requirements will not be refunded the deposit.

10.2 Each nominated entry will win a prize of 80,000 US dollars.If it is selected as winning design scheme,the prize will be included in the design payment.

10.3 All non-nominated entries which are invited entrants and satisfy the regulations of the Competition will be paid 30,000 US dollars to cover the design cost.The foreign companies invited will receive another payment of 10,000 US dollars to cover the cost of travel to Beijing and the cost of staying in Beijing.

10.4 Any tax charged to entrants in China should be paid by themselves.

10.5 The prizes and payments for covering design cost will be paid to foreign entrants with US dollars and to local entrants with RMB by the same day's exchange rate.

11.COPYRIGHT

11.1 The Proprietor Committee has a right to publish and display the entries and so on.

11.2 All entries will not be returned after the evaluation.

12.REQUIREMENTS TO WINNING DESIGN COMPANY

12.1 The winning design company has obligations to make necessary amendments of the design scheme according to the requirements of the client and must complete the architectural design,structural design,equipment design,interior design,acoustic design,environmental design ect.The design must meet the requirements of all current standards,regulations and codes for architectural design of the people's Republic of China.The Construction drawings must meet the requirements of civil building in force in China.

12.2 If a foreign entry is selected as winning design scheme,the design company will be required,in accordance with Chinese law and regulations,to work with a Chinese architectural design company which is a licensed first class design company in China.

12.3 The total amount of design payment to the foreign winning company which completed the all design,including the cost of local cooperative desigh,will be 5% of the construction in vestment except the investment of stage equipments and the special interior decoration.In the case of local winning company,the total amount of design payment will be decided in accordance with local regulations.

13.COMPETITION LANGUAGE:CHINESE AND ENGLISH

All competition documents and drawings submitted by entrants must be written in both Chinese and English.The Chinese documentation will be standard language when there is any difference between Chinese and English in thc documentation.

14. All entrants of the Competition will be regarded as acknowledging all articles of this document.

Information:

Ms.Yang,Fan

Proprietor Committee of the National Grand Theater

4 Shibeihutong,Xicheng District

Beijing 100031

China

Tel:+86-10-6606-4705

Fax:+86-10-6606-4707

中国国家大剧院建筑设计方案竞赛文件附件

国家大剧院规划设计条件

一、用地范围

北至西长安街南红线，东距人民大会堂西侧路西红线25m处，南至大剧院南侧路北红线，西至石碑胡同东红线。南北长166m，东西宽：北部224m，南部约244.9m，规划用地约3.89公顷。用地东侧有25m宽的城市地下管廊，地上可作停车、绿化、循环路使用，但不设围墙，允许市政部门进入检修。(见附图一)

二、交通组织

1. 西长安街红线宽120m，规划道路断面为一幅路，面宽50m。

石碑胡同红线宽30m，规划为一幅路，面宽16m。

大剧院南侧路红线宽30m，规划为一幅路，面宽16m。

四周道路都为机动车路。

2. 在用地四边的适当位置开设机动车出入口，且应在用地内留有通畅的双向车道循环路。东侧可利用市政管廊作循环道路。由于长安街上有地铁和过街人行道的出入口以及公交路线的停车站，故北入口的设计需考虑和上述这些设施的关系，减少相互干扰。

3. 人流出入口可在四边设置。

三、停车

需按不少于10辆／百座的指标配置机动车停车位，并且地面应考虑适量的大车车位。

四、退红线距离

1. 主体建筑北退红线20m。南边以内部循环路为退线距离，西边由于石碑胡同只拓宽了半幅红线15m，交通较紧张，建议西侧多退红线以缓解近期交通拥挤状况。剧院的公共出入口应多退红线，形成人流集散广场。

2. 地下建筑退用地红线距离不得小于5m，以满足户线及附属设施的需要。

五、市政

1. 在四周城市道路下设有上水、雨污水、煤气、电力、电信、供热等管线，已为本工程做好市政规划，待本工程进入施工设计阶段，再提供具体方案。根据市政规划要求在本用地内需设置热力交换站、变电室及2000门模块局等设施。

2. 在本用地的北侧，临西长安街西边有地下人行出入口，东边有地铁出入口及二者之间的地下连通道。国家大剧院应设地下出入口与地铁相通。

3. 在本用地的东北、西北角已建成地铁风口各一处，其处置办法详见附图。

六、建筑高度

沿西长安街的建筑高度按30m控制，局部根据功能的需要可适当提高，最高不得超过45m。

七、绿化

1. 用地内需适当布置绿地。

2. 南侧绿地是人民大会堂、国家剧院及南侧规划的大型公共建筑共同使用的城市绿化广场，大剧院的设计要充分利用这个环境。

八、城市设计要求

1. 应在建筑的体量、形式、色彩等方面与天安门广场的建筑群及东侧的人民大会堂相协调。

2. 在建筑处理方面需突出自身的特色和文化氛围，使其成为首都北京跨世纪的标志建筑。

3. 建筑风格应体现时代精神和民族传统。

天安门广场规划方案简介

　　天安门广场位于北京城市中心，也是南北中轴线和东西长安街的中点。天安门广场是首都政治、文化中心的象征。

　　天安门广场北端是天安门及紫禁城，南端为正阳门，人民英雄纪念碑和毛主席纪念堂位于广场中部和南部，历史博物馆和人民大会堂分列广场东西，广场总占地面积约44公顷。

　　按照天安门广场规划方案，广场东南及西南侧将建设国家艺术宫东馆和西馆，历史博物馆东侧和人民大会堂西侧将建设公安部大楼和国家大剧院，这些新建筑将与传统建筑共同构成广场严整、庄重、对称的格局。

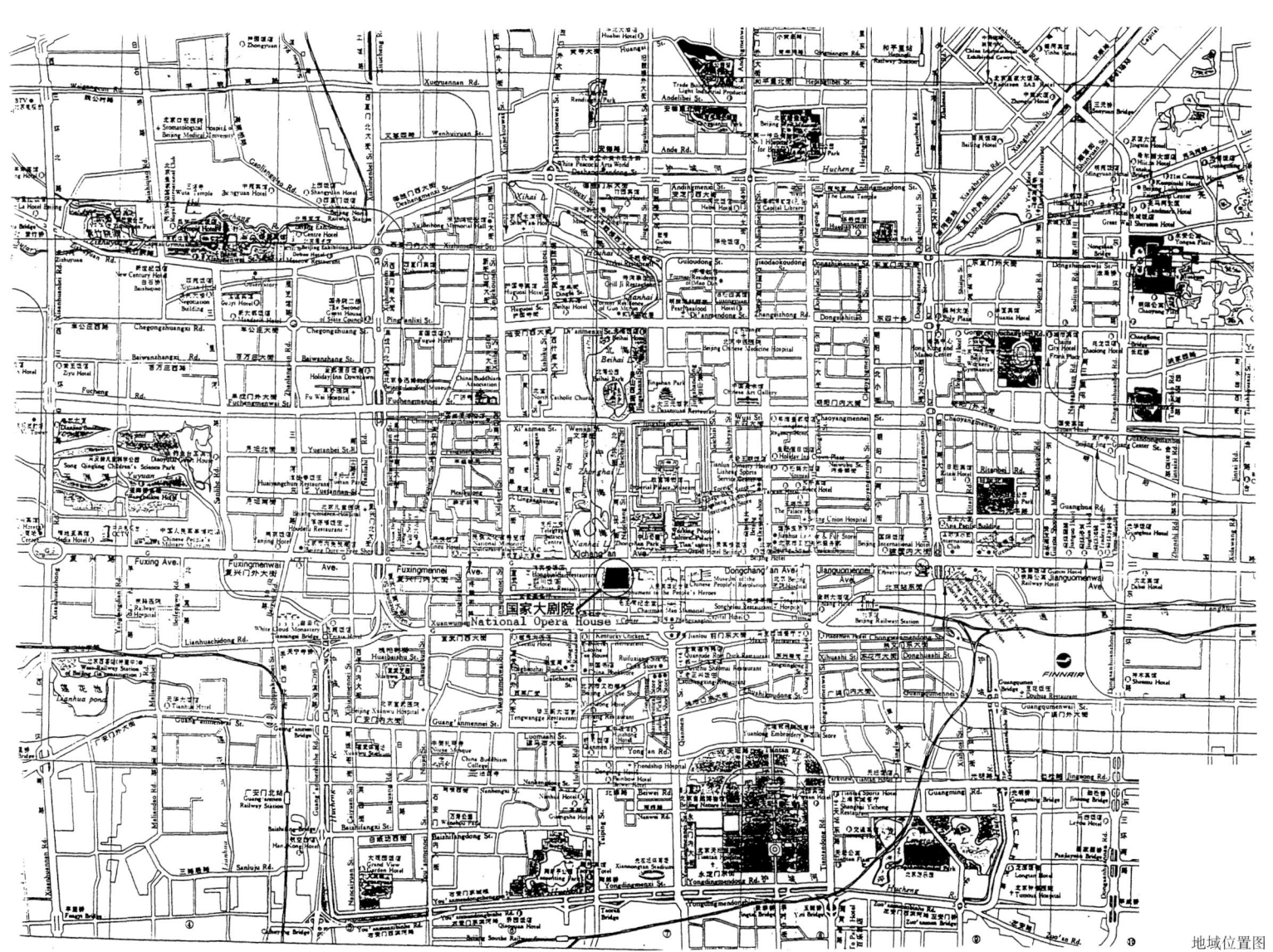

地域位置图

19

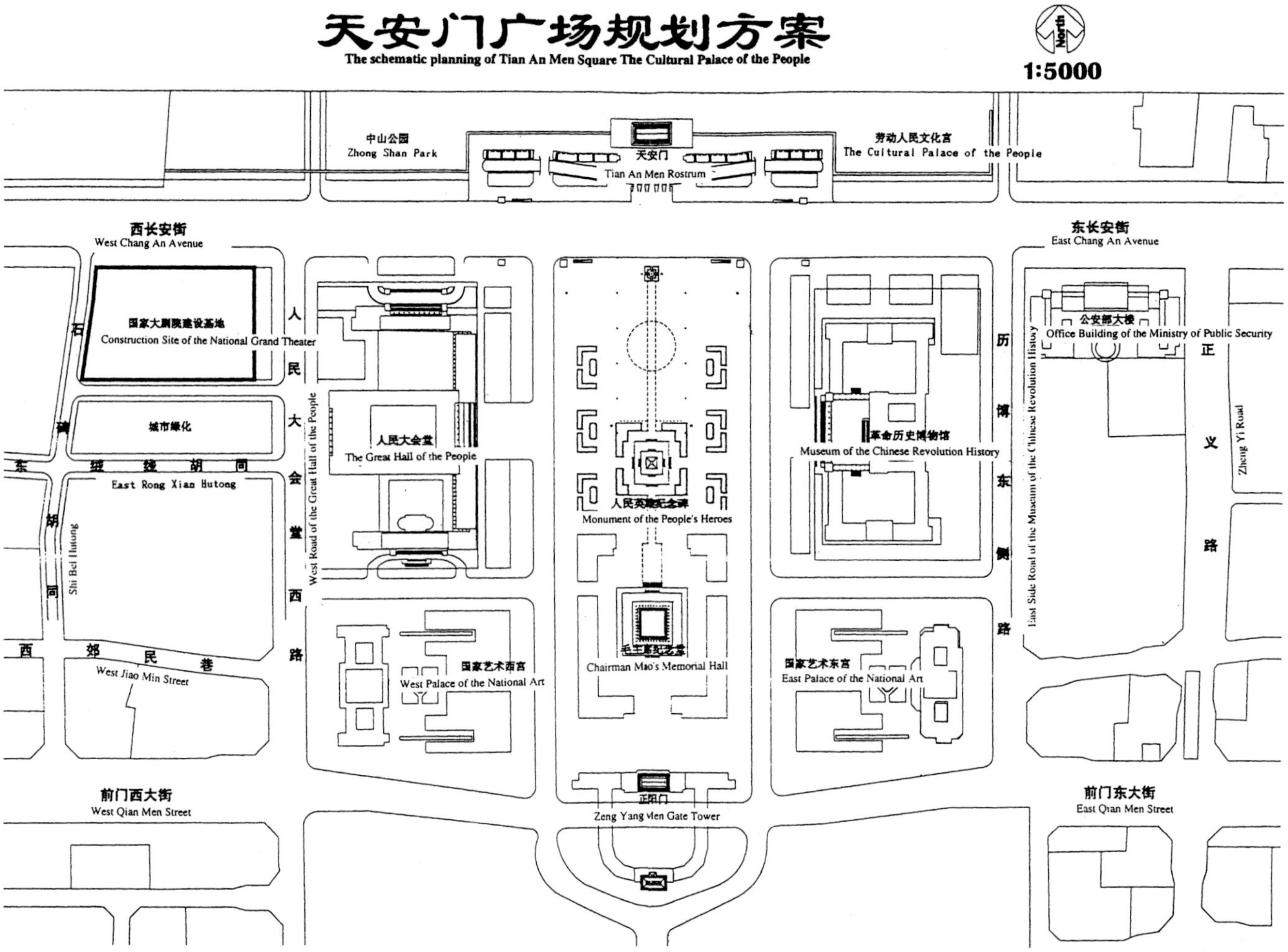

天安门广场规划方案
The schematic planning of Tian An Men Square The Cultural Palace of the People

1:5000

中山公园
Zhong Shan Park

天安门
Tian An Men Rostrum

劳动人民文化宫
The Cultural Palace of the People

西长安街
West Chang An Avenue

东长安街
East Chang An Avenue

国家大剧院建设基地
Construction Site of the National Grand Theater

公安部大楼
Office Building of the Ministry of Public Security

人民大会堂西路
West Road of the Great Hall of the People

城市绿化

人民大会堂
The Great Hall of the People

革命历史博物馆
Museum of the Chinese Revolution History

历史博物馆东侧路
East Side Road of the Museum of the Chinese Revolution History

东绒线胡同
East Rong Xian Hutong

石碑胡同

师北胡同
Shi Bei Hutung

人民英雄纪念碑
Monument of the People's Heroes

正义路
Zheng Yi Road

西郊民巷
West Jiao Min Street

国家艺术西宫
West Palace of the National Art

毛主席纪念堂
Chairman Mao's Memorial Hall

国家艺术东宫
East Palace of the National Art

前门西大街
West Qian Men Street

正阳门
Zeng Yang Men Gate Tower

前门东大街
East Qian Men Street

天安门广场规划方案图

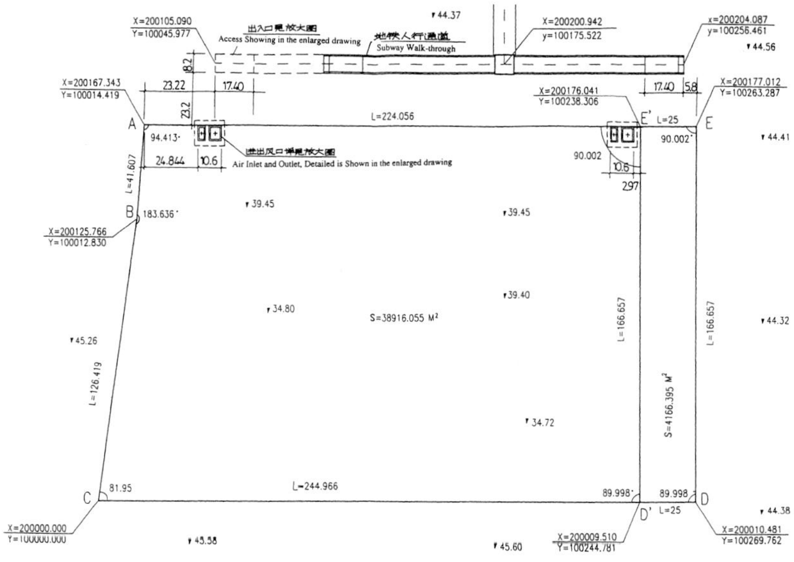

出入口见放大图
Access Showing in the enlarged drawing

地铁人行通道
Subway Walk-through

进出风口详见放大图
Air Inlet and Outlet, Detailed is Shown in the enlarged drawing

X=200105.090
Y=100045.977

X=200200.942
y=100175.522

X=200204.087
y=100256.461

▼44.37

▼44.56

X=200167.343
Y=100014.419

23.22   17.40

X=200176.041
Y=100238.306

17.40   5.8

X=200177.012
Y=100263.287

8.2

23.2

L=224.056

L=25

A   94.413'

90.002

90.002

E'   E

24.844   10.6

▼44.41

0.5

2.97

B   183.636'

▼39.45

▼39.45

L=41.607

X=200125.766
Y=100012.830

▼34.80

S=38916.055 M²

L=166.657

L=166.657

▼44.32

▼45.26

L=126.419

▼34.72

S=4166.395 M²

C   81.95

L=244.966

89.998

89.998   D

D'   L=25

▼44.38

X=200000.000
Y=100000.000

▼45.58

▼45.60

X=200009.510
Y=100244.781

X=200010.481
Y=100269.762

用地范围图

地铁排、进风口设计要求

地铁排、进风口可按三种型式处理，主要要求如下：

1. 单建下沉式，进风、排风风口分建

一般设在绿地内，进风口与排风口相距10m，进风口下口距地面不小于1m，风口采用格栅形式，进风口与排风口面积各为20m²。

风口四周须设栏杆，栏杆距风口不小于3m，并考虑通风设备维修进、出吊装设备的工作场地及道路条件。

2. 单建地面式，进、排风口合建

进风口格栅下口距地面的高度应大于2m，排风口下口比进风口上口高出5m，进风口与排风口的格栅面积各为20m²。围墙须考虑通风设备维修进、出吊装设备的工作场地及道路条件。

3. 与建筑结合

进风与排风道的净面积各为16m²，进风口与排风口须设格栅，面积各为20m²。进风口与排风口高差不小于5m。

风道平面型式要平顺，如需改变方向时，转角应大于90°，且应加导流措施。

风道不宜过长，风道加长后须经原设计单位检算，地铁总公司认可。

注：地铁风口单建型还可根据进、排风口分建或合建的条件，组合其他方案，但必须满足以上所提要求。

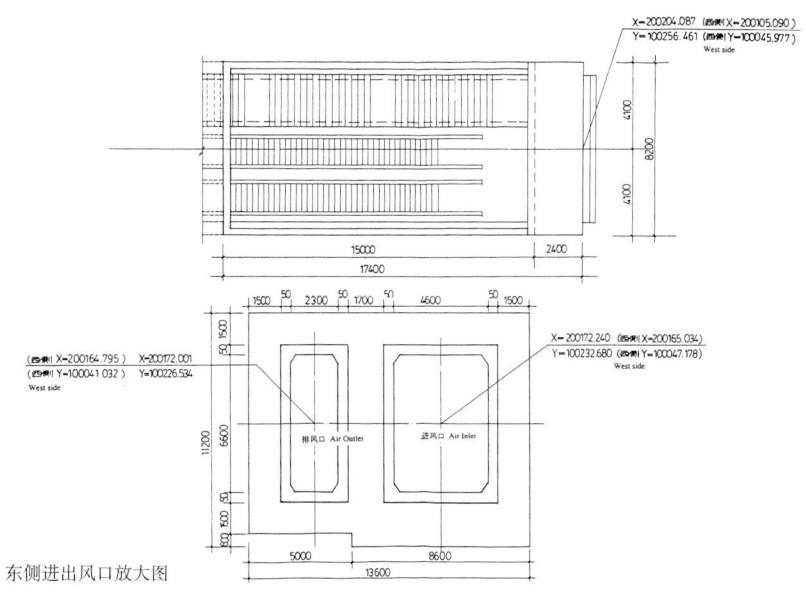

东侧进出风口放大图

国家大剧院建筑场地岩土工程背景资料

一、自然地理、气象及气候条件

拟建场区所处的北京地区具有暖温带半湿润半干旱大陆型季风气候的特点。北京平原地区平均气温在11℃～12℃之间，每年1月份的平均气温最低，平均在－4℃～－5℃，7月平均气温最高，平原地区为25℃～26℃，无霜期平原地区在190天以上。北京大部分地区年降水量在550mm～660mm之间。夏季降水量占全年的70%以上；冬季降水一般只有10mm左右。北京地区的风向有明显的季节变化，冬季以北和西北风为主，夏季多偏南风，春、秋为南北向转换季节，年平均风速为2m／s～3m／s，其中1月份最大风速可达24m／s。

二、拟建场区工程地质条件

1.地质环境背景

《国家大剧院》工程拟建场地处于永定河冲洪积扇中部，地质土质以粘性土、粉土与砂卵石土交互层为主，本场区第四纪地层厚度(相当于基岩埋深)约在80m左右。

北京地区主体地质构造为早第三纪前的断裂及其控制的断块构造。主要有三组断裂带，主干断裂带为北北东向，其次北东向和北西向断裂带。北京市区的地震基本烈度(50年超越概率为10%)为8度。本工程距上述断裂带的最近距离约10km左右。

2.拟建场区和周围地区古河道的分布及变迁史

拟建场区地貌单元宏观上位于永定河冲洪积扇中部，根据资料，历史上曾有2条湮废的古河道贯穿场区，其中"三海大河"古河道自西北向东南穿越本场区，据已有资料，该古河道深约10m～15m，河底标高30m～35m；元大都南护城河古河道沿元大都南城墙外护城河东西向穿越人民大会堂北侧和拟建场区北部，该段护城河在明初填塞，古河道深约7m～8m。

3.现状地形与地物

拟建场地地形基本平坦，现状自然地面标高一般在44.66m～45.89m之间。80年代中后期，该场地曾拟做为某办公楼用地，旧有房屋均已拆除，基坑已挖至原自然地面以下约12m。

4.地层土质及其工程特性概述

原拟建某办公楼建筑场区范围自然地面下33m内的地层按成因年代分为人工堆积层、新近沉积层、第四纪沉积层。在综合拟建场区及周边地层资料基础上，对地层情况分述如下：

人工堆积层主要有粉质粘土填土、房渣土、杂填土层等，厚度约在自然地面下4.00m～8.00m。

新近沉积层为古河道所形成的沉积土层，由粘性土、砂类土层组成，局部有黑色及灰色含有机质的土层，厚度3.00m～10.00m，层底深度10.00m～15.00m。

第四纪沉积层包括厚度在1.00m～5.00m的砂卵石层，厚度在3.50m～4.50m的粘质粉土、砂质粉土、重粉质粘土、粘土层土，和卵石、细砂层、粘质粉土、砂质粉土层。

三、拟建场区水文地质条件

1.水文地质背景

根据《北京市区浅层地下水位动态规律研究》成果，在北京市规划市区，对建筑工程有影响的浅层地下水分为3个区、7

个亚区，该工程场区位于北京市区工程水文地质分区的II$_b$亚区。从宏观分析看，目前场区浅层地下水主要有上层滞水、层间潜水及潜水—承压水。下图为1984年1月底勘探时的地下水赋存情况。

第一层上层滞水赋存于填土及新近沉积之粉质粘土层中，主要受大气降水补给，并以蒸发及地下径流为主要排泄方式。1984年在该场地进行初勘时，第二层层间潜水位于标高29.00m～29.84m

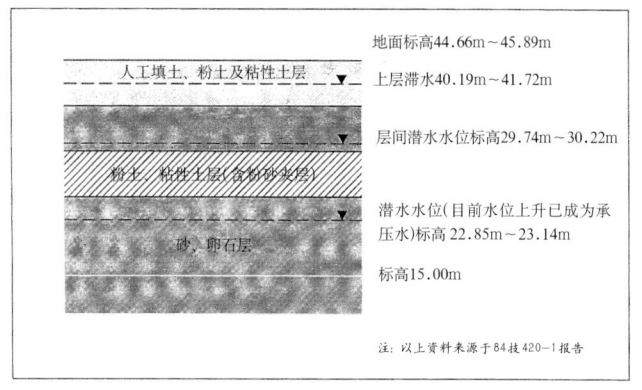

地面标高44.66m～45.89m

上层滞水40.19m～41.72m

层间潜水水位标高29.74m～30.22m

潜水水位(目前水位上升已成为承压水)标高22.85m～23.14m

标高15.00m

注：以上资料来源于84技420-1报告

工程剖面示意图

以上的砂、卵石层中，主要接受"天窗"泄漏及地下径流补给，并以地下径流为主要排泄方式。第三层水位于标高24.65m～25.82m以下的砂、卵石中，1984年勘探时水位在标高22.85m～23.14m，无承压性。随着水位的逐渐回升，到1991年底水位已到达砂、卵石底板，之后受官厅水库放水等因素影响水位继续上升，具有承压性。这层水主要接受地下径流补给，并以地下径流为主要排泄方式。

该工程场区曾于1959年达到近50年的最高地下水位，标高为41m～42m。近年来，场区层间潜水呈上升趋势，近3～5年最高水位标高在31.00m左右。场区第3层承压水受永定河放水及地铁施工降水影响幅度较大，如将来的基础设计施工工作涉及该层地下水的影响，则可作为拟建建筑物详勘时的工作重点之一。

四、建筑场地抗震设计条件

1.地震基本烈度

根据中华人民共和国国家地震局所颁发的《中国地震烈度区划图(1990)》和中华人民共和国国家标准《建筑抗震设计规范》(GBJ11—89)，北京市市区的地震基本烈度为8度。在进行抗震设计时，北京地区只考虑近震的地震影响。

2.场地土类型、建筑场地类别

工程拟建场区第四纪覆盖层的厚度d_{ov}约在80m左右。目前该拟建场区尚无实测剪切波速资料(前期勘察时，规范无要求)作为直接判定地震场地类别的依据，根据对于北京市区地震场地类别区划的研究成果资料，拟建场区地震场地类别可暂按II类考虑。

3.地基土层的地震液化判定

根据目前所掌握资料的初步分析，在地震烈度为8度，地下水上升至新近沉积土层顶板时，新近沉积层上部的砂层存在液化的可能。

表1：技术经济指标

基地面积	总建筑面积	建筑高度	建筑覆盖率
	地上： 地下：		
容积率	绿化覆盖率	停车数量	自行车停车数量
		地上： 地下：	

表3：投资估算

土建造价 （含土建设备）	
舞台设备造价 （音响、舞台、设备、灯光）	
装修造价	

表2：各观众厅数据

		歌剧院	音乐厅	戏剧院	小剧场
观众厅面积(m²)					
观众厅体积(m²)					
座席数	总座席数				
	其中：池座				
	楼座				
	休息厅面积(m²)				
舞台尺寸(m²)	主舞台				
	左右侧台				
	后舞台				
	台仓				
	升降乐池				
最大视距(m)					
最大俯角(°)					

表4：质疑表

中国国家大剧院工程方案竞赛质疑		
质疑单位：		
质疑单位传真：	提问时间	
问题：		

The Document of the Design Scheme Competition of the National Grand Theater,China Annex

Terms for the Planning & Design of the "National Grand Theater"

Ⅰ.Site

In the north:to the south property line on Xi Chang An Avenue;In the east:25 m from the west property line on the west side road of the People's Great Hall;In the south:to the north property line on the south side road of the National Grand Theater;In the west:to the east property line of the Shi Bei Hutong.It is 166 m long from the north to the south.In the north part,it is 224 m wide from east to west and in the south part,244.9m wide.The site area in planning is about 3.89 hectares.There is a 25 m-wide urban underground pipeline tunnel in the east side of the site,while its ground surface may be used as parking,green area and circulating road,but without enclosing wall.The Municipal Administration staves are allowed to go in to examine and repair.(See attached drawing 1)

Ⅱ.Access:

1.The width between the property lines on the Xi Chang An Avenue is 120m.The road section planned is of a road without separator and its width is 50m.

The width between the Shi Bei Hutong property lines is 30 m.The road section planned is of a road without separator and its width is 16m.

The width between the property lines on the south side road of the Theater is 30m,the road section planned is of a road without separator and its width is 16m.

The roads surrounding the theater are all motor ways.

2.The access for motor vehicles should be set up at the appropriate locations surrounding the site,and the clear circulation road with two-way traffic lane should be constructed within the site.On the east side,the municipal pipeline tunnel may be utilized as the circulation road.As there are accesses for the subway and the pedestrian crosswalke as well as the parking lot for public transit lines on Chang An Avenue,the relation with the above-mentioned facilities should be considered in the design of the north entry as to reduce mutual interference.

3.Accesses for pedestrians may be set up on the four sides of the theater.

Ⅲ.Parking:

Parking stalls for motor vehicles should be arranged at a quota of not less than 10 cars/100 seats,and stalls for large vehicles on the ground should be considered.

Ⅳ.Receding distance from property line

1.The main buildings should recede 20 m from the property line on the north side.On the south side,the receding distance is the width of the internal circulation road.On the west side,due to the fact that Shi Bei Hutong has been widened only 15 m,the traffic is rather congested.It is suggested that on the west side the distance of receding from the property line should be larger in order to alleviate the recent traffic congestion.The distance of receding from the property line at the common access of the theater should be larger so as to form a square for gathering and scattering.

2.The receding distance from property line of the site for underground structures should not be less than 5 m so as to meet the needs for arrangement of household lines and the accessory installations.

Ⅴ.Municipal works

1.Pipe lines for water supply,storm and sanitary drainage,coal gas,electric power,communication,and heating,etc.have already been planned under the urban roads surrounding the site of the project in the municipal plan for urban facilities.The specific plan will be provided when the project begins its design and construction phase.Installations for heating power exchange station,transformer room and a telephone modulus with 2000 lines,etc.should be provided according to the requirement of the municipal plan for urban facilities.

2.On the north side of the site,an access for the underground walkthrough near the west side of Xi Chang An Avenue,an access for the subway on the east side and a linking passage between them are available.An underground access for the theater should be provided to connect with the subway.

3.The measures dealing with the air inlets and outlets at two places for the subway at the northeast and northwest corners of the site are shown in the attached drawings.

Ⅵ.Building height limitation

The height of the buildings along the Xi Chang An Avenue should be limited to 30 m.The height of parts of them may be higher,but not higher than 45 m according to their function needs.

Ⅶ.Green space

1.Green space should be arranged properly within the site.

2.The green space on the south side which is an urban green square shared by the People's Great Hall,the National Grand Theater and the planned large-sized public buildings should be considered in the design.

Ⅷ.Requirements of urban design

1.The mass,configuration,and color of the buildings should be in harmony with the architectural complex of the Tian An Men Square and the People's Great Hall on its east side.

2.In the architectural design,the characteristics of the buildings themselves and the cultural atmosphere should be predominant so as to become a transcentury land mark building in the capital,Beijing.

3.The building style should embody the spirit of the era and the national tradition.

Brief Introduction to the Planning Scheme
of the Tian An Men Square

The Tian An Men square is located at the urban center of Beijing city,and also on the central point of the south-north central axis and the Chang An Avenue.TheTian An Men Square is the symbol of the political and cultural center of the capital.

The north end of the Tian An Men square is the rostrum of Tian An Men and the Forbidden City and the south end is the rostrum of Zheng Yang Men.The parcel area of the square is about 44 hectares with the Monument to the People's Heroes at the middle part,Chairman Mao's Memorial Hall at the south part,the Historical Museum at its east and the People's Great Hall at its west..

According to the planning scheme of the Tian An Men square,the East Hall and the West Hall of the National Art Palace will be built at the southeast and southwest sides of the square and the Office building of the Ministry of Public Security and the National Grand Theater will be constructed at the east side of the Historical Museum and at the west side of the People's Great Hall respectively.An orderly,solemn and symmetrical layout of the square will be constituted by these new buildings together with the traditional structures.

The Design Requirements for Air Inlet
and Outlet of the Subway

The air inlet and outlet of the subway may be designed in the following three types and the main requirements are as follows:

1.Separating and sunken type with separated air inlet and outlet

Generally they should be built within the green area.The distance between the air inlet and outlet is 10 m.The distance between the lower part of the air inlet and the ground surface is not less than 1 m.The grid type is adopted for the air inlet and outlet.The areas of both the inlet and the outlet are 20m².

The railings should be set up around the air inlet and outlet.The distance from railings to the air inlet and outlet is not less than 3 m.The working area and road conditions should be considered for access of hoisting installations for maintenance and repair of ventilation equipment.

2.Separating and ground type with communal air inlet and outlet

The height from the bottom of the air inlet grid to the ground surface should be more than 2 m.The bottom of the air outlet is 5 m higher than the top of the air inlet.The areas of the grids for both the air inlet and the air outlet are 20m².The working area and road conditions should be considered for access of hoisting installations for maintenance and repair of ventilation equipment in the design of enclosing wall.

3.Communal type and Integrated with buildings

The net areas of the air intake and air exhaust duct are both 16m².The grids should be installed for air inlet and outlet,its area being 20m² each.The height difference between the air inlet and outlet should be not less than 5m.

The surface of the air duct should be smooth,and if direction is to be changed the anger of rotation should be more than 90°, and devices should be added to guide the air current.

The air duct should not be too long.If it is to be extended,examination and calculation must be carried out by the original design company and approved by the General Subway Corporation.

Note:For the subway air inlet and outlet with separating type,the schemes other than the above mentioned can be produced according to the conditions of separating type of communal type for the air inlet and outlet,but in any case the above mentioned requirements must be fulfilled.

Background Information Regarding the Geotechnical Engineering of the Construction Site of the National Grand Theater

I .Physical geographical,meteorological and climate conditions

The Beijing area where the construction site of the proposed project is located has the features of warm temperate zone,semi-wet and semi-dry continental monsoon climate.The annual average temperature in the plain area of Beijing is between 11℃ and 12℃.The monthly average temperature of January is the lowest each year,and between-4℃ and -5℃ in the plain area.The monthly average temperature of July is the highest,and between 25℃ and 26℃ in the plain area.The frostless season in the plain area is above 190 days.The annual precipitation in most districts of Beijing is between 550mm～660mm.The precipitation in summer amounts to over 70% of that in the whole year and generally in winter the precipitation is only about 10mm.The wind direction in Beijing area has apparent changes in different seasons.In winter there are mostly north and northwest winds,and in summer mostly wind by south.The spring and autumn are the seasons in which the south wind and the north wind change their directions.The annual average wind speed is 2m/s～3 m/s,and in January it may reach to a maximum of 24m/s.

II .Engineering geological conditions of the proposed construction site

1.The background of the geological condition

The proposed construction site of the "National Grand Theater"is located in the middle part of the alluvial-diluvial fan of the Yong Ding River.The stratum is mainly composed of cohesive soil,silty soil,and the alternated layers of sand and pebble gravel.The thickness of the stratum of the Quaternary period(equivalent to the embedded depth of bedrock)is about 80m.

The main geologic structure of Beijing area are the faults and rupture blocks bounded by the faults before the early Tertiary period.There are mainly three groups of fault zones:1/the main fault zone with NNE direction,2/the fault zone with NE direction,and 3/the fault zone with NW direction.The basic intensity in the urban area of the city of Beijing is Ⅷ (The exceedance probability of 50 years is 10%).The nearest distance between the construction site of this project and the above-said fault zones is about 10km.

2.The distribution and the history of vicissitudes of the abandoned channels under the proposed construction site and its surrounding area

The geomorphological unit of the proposed construction site is located in the middle part of Yong Ding River's alluvial-diluvial fan in macro-view.According to the data of the Beijing Geotechnical Institute,in history,there were two abandoned channels running through the site.The abandoned channel of"San Hai Da He"passed through the site from the NW to the SE.The depth of the channel is about 10 m～15 m and the level of the its bottom is about 30 m～35 m.The abandoned channel of the south moat of the Yuan Dynasty's capital ran along the moat outside the south city wall of the Yuan Dynasty's capital with EW direction through the north side of the People's Great Hall and the north part of the site.This channel section of the moat was filled up in the early Ming Dynasty with the depth of 7m～8m.

3.Present topography and ground objects

The topography of the proposed site is basically smooth and its natural ground elevation is generally between 44.66m～45.89m.In the middle and late years of the 1980s,the site was intended to be used as the site for an office building.The original buildings within the site have all been demolished.The foundation pit was dug 12m below the original natural ground level.

4.The profile of the stratum and soil engineering characteristics

The stratum within 33m below the natural ground level in the proposed site may be divided into the following three stratums based on their genetic era and the data related to the stratum of the site and its surrounding area.

*The artificial filling stratum:It is mainly composed of layers of silty clay fill and miscellaneous fill.The thickness of the stratum is 4.00m～8.00m below the natural ground level.

*Newly sediment deposit stratum:The stratum was formed by abandoned channels and consists of the layers of clay,sandy soil and the layer of black and gray soil with organic matter locally.The thickness of the stratum is 3.00 m～10.00 m and the depth of the bottom of the stratum is 10.00m～15.00m.

*The Quaternary period deposit stratum:This stratum includes the layers of sand pebble with a thickness of 1.00m～5.00m and the layers

of clay silt,sandy silt,heavy silty clay and clay with a thickness of 3.50 m～4.50 m as well as the layers of pebble and fine sand,and clay silt and sand silt.

The hydrogeological conditions of the proposed site

The hydrogeology background

According to the results of "Research on the Dynamic Behavior of Shallow Water Level in the Urban Area of Beijing City",the shallow water effected to the construction projects in the planned urban area in the city of Beijing can be zoned into three zones and seven sub-zones.The site of the project is located in the Ⅱ b sub－zone in the Engineering Hydrogeological Zoning of the Urban Area of Beijing City.From the macroscopic analysis,at present the shallow water of the site mainly includes the upper layer unconfined water,inter-layer unconfined water and confined water.

The following profile sketch shows the ground water storage investigated at the end of January,1984.

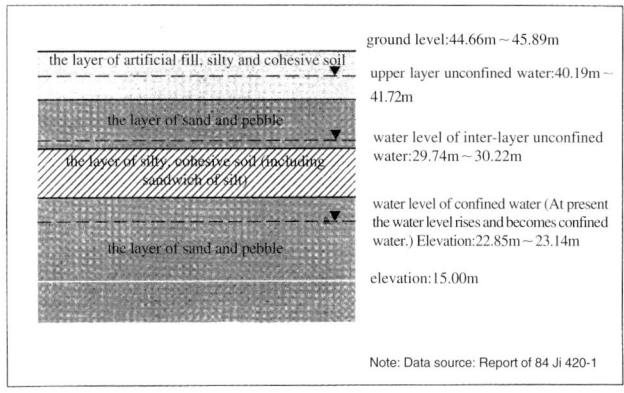

ground level:44.66m～45.89m

the layer of artificial fill, silty and cohesive soil

upper layer unconfined water:40.19m～41.72m

the layer of sand and pebble

the layer of silty cohesive soil (including sandwich of silt)

water level of inter-layer unconfined water:29.74m～30.22m

water level of confined water (At present the water level rises and becomes confined water.) Elevation:22.85m～23.14m

the layer of sand and pebble

elevation:15.00m

Note: Data source: Report of 84 Ji 420-1

PROFILE SKETCH

The upper layer unconfined water of the lst layer stores in the layers of fill and newly deposit silty clay and mainly supplied by the rain water and discharged by evaporation and underground runoff.In 1984,when the site was initially investigated,the inter-layer unconfined water of the second layer stores in the layers of sand and pebble above the level of 29.00m°´29.84m and mainly supplied by water from funnel in the stratum and underground runoff and discharged by the underground runoff.The water of the third layer was in the layers of sand and pebble with an elevation of below 24.65m°´25.82m.In 1984,the unconfined water level was 22.85m°´23.14m.With the gradual rise of the water level,by

the end of 1991,it reached the bottom of the layer of sand and pebble stone.Later it continuously rose due to the effect of the discharge of the Guan Ting Reservoir and then becomes confined water.The water in this layer was mainly supplied by underground runoff and discharged in the way of underground runoff.

In 1959,the water level in the site once reached the highest for nearly 50 years,that is 41m～42m.In recent years,the inter-layer unconfined water level has a tendency to rise and in the recent 3～5 years,its highest level has been about 31.00m.The confined water level in the third layer is largely affected by the discharge of Yong Ding River and the lowering water level due to the construction of the subway.In the future,if the work of foundation design and construction may involve the effect of ground water in this layer,then the attention should be paid at the final investigation of the proposed building sites.

Ⅳ.Terms of seismic design for the construction site

1.Basic intensity

The basic intensity in the urban ared of Beijing city is Ⅷ based on the "China Seismic Intensity Zoning Map"issued by the State Seismology Bureau of the People's Republic of China and The National Standard of the People's Republic of China-"The Code for Seismic Design of Buildings"(GBJ 11-89).For the Beijing area,only the effect of near-earthquake should be considered in the seismic desigh.

2.Site soil type and construction site category

The thickness d_{ov} of the overlaying layer of the Quaternary period in the proposed construction site of the project is about 80 m.At present time,there is no any data on experimental shear wave velocity in the proposed site for direct determination of the site category.The construction site category Ⅱ can be considered temporarily for the proposed construction site based on the research results on construction site category zoning in urban area of Beijing city conducted by the Beijing Geotechnical Institute.

3.Seismic liquefaction discrimination of the subsoil

According to the preliminary analysis of the data available at present time,when the earthquake intensity is Ⅷ and the ground water raises to the top of newly sediment deposit layer,the sand layer in the upper part of the newly sediment deposit has a potential of liquefaction.

Table1: Technical Economic Index

Site Area	Total Floor Area	Building Height	Building Coverage Ratio
	Above Ground Under Ground		
Plot Ratio	Green Coverage Ratio	Number of Stalls	Number of Bicycle Stalls
		Above Ground Under Ground	

Table3: Investment Estimation

Construction Cost (Including Fixed Equipment)	
Stage Installation Cost (Sound, Stage, Apparatus, Light Fixtures)	
Decoration Cost	

Table2: Data For Auditoriums

		Opera House	Concert Hall	Theater	Mini Theater
	Floor Area of Auditorium				
	Volume of Auditorium				
Seating Capacity	Seating Capacity(Seats)				
	Among Them:Auditorium				
	Floor Seat				
	Lounge Floor Area				
Stage Dimentions	Main Stage				
	Left and Right Side Stages				
	Back Stage				
	Under Stage Storage				
	Elevating Orchestra Pit				
	Max. Visual Distance				
	Max. Angle of Depression				

Table4: Table For Question

Questions to the Architectural Design Scheme Competition **of the National Grand Theater, China**		
Putting Question Organization:		
Fax of the Organization:	Time of Putting Question:	
Questions:		

设计任务书

中国国家大剧院是中国最高表演艺术中心，是弘扬中华民族文化，促进精神文明建设，展现音乐戏剧水平，推动国际文化交流的重要场所。国家大剧院应满足各种表演艺术形式演出的需要，功能齐全，视听条件优良，技术先进可行，设备完善，经济合理。

由于剧院建设地点处于中华人民共和国首都政治文化中心——天安门广场一侧的特定条件，建筑造型应与周围建筑谐调，形成广场建筑群与相交接的长安街的有机组成部分，在充分弘扬城市整体美的前提下，剧院建筑本身体现庄重、典雅的艺术表现力和鲜明的人民性、时代性。建成之后将为北京这一历史名城增添新的光彩。最终设计应满足中华人民共和国各项有关规范的规定。

一、建设地点

基地位于北京长安街，天安门广场及人民大会堂西侧，长安街以南，石碑胡同以东，东绒线胡同以北，人民大会堂西路以西。基地东西向长224～245m，南北向约166m，总面积约3.89公顷。

二、建筑规模及投资

建筑总面积为120000m²(允许上下浮动10%)，工程投资约20亿元人民币。

三、工期

全部工程于2001年底完成。
1998年完成方案初步设计。
1999年春季开工。
2001年底完工。
2002年1～6月调试。

四、设计内容、要求、面积分配

1.各部门内容及面积总表

名　称	建筑面积(m²)	系数	使用面积(m²)	摘　要
一、歌剧院	22529	0.65	14644	
1.前厅和休息厅			3500	前厅和休息厅、贵宾室、急救室、服务员室等
2.观众厅			2000	观众厅
3.舞台、乐池、台仓			4538	主台、左侧台、右侧台、后舞台、乐池、台仓等
4.后台业务用房			4156	化妆室、休息室、换装室、候演室、道具间、办公室、布景组装场、服装鞋帽间、乐器库、舞台用品库等
5.演出技术用房			450	总监室、导演室、调光室、调音室、放映室、转播室、灯光室、设备间等
二、音乐厅	11987	0.65	7792	
1.前厅和休息厅			2800	前厅和休息厅、贵宾室、急救室、服务员室等
2.观众厅			1600	观众厅
3演奏台、管风琴室			1156	演奏台、左右侧台、管风琴室、台仓等
4.后台业务用房			2036	化妆室、休息室、候演室、办公室、乐器库、演出用品库
5.演出业务用房			200	总监室、调光室、调音室、转播室、灯光室、设备间等
三、戏剧场	12006	0.65	7804	
1.前厅和休息厅			1680	前厅和休息厅、贵宾室、急救室、服务员室等
2.观众厅			960	观众厅
3.舞台、乐池、台仓			2218	主台、左右侧台、后舞台、升降大台唇兼乐池、台仓等
4.后台业务用房			2556	化妆室、休息室、换装室、候演室、道具间、办公室、布景组装场、服装鞋帽间、乐器库、舞台用品库等
5.演出技术用房			390	总监室、导演室、调光室、调音室、放映室、转播室、同声传译室、灯光室、设备间等
四、小剧场	7427	0.65	4828	
1.前厅和休息厅			1550	前厅和休息厅、贵宾室、贵宾活动室、急救室、服务员室

2.观众厅与舞台			1900	观众厅、舞台、台仓等
3.后台业务用房			1078	化妆室、休息室、换装室、候演室、道具间、办公室、服装鞋帽间、乐器库、舞台用品库等
4.演出技术用房			300	总监室、导演室、调光室、调音室、放映室、转播室、同声传译室、灯光室、设备间等
五、公共剧务用房	11169	0.65	7260	
1.排练厅			2073	大、中、小排练厅、合唱排练厅、乐队分部排练厅等
2.练习琴房			864	练习琴房
3.售票广告			350	票务中心，宣传广告
4.制作修理间			1480	绘景间，布景、道具、服装、鞋帽、幻灯制作、修理
5.布景、材料库			1255	布景、材料、物品等库房
6.音像制作室			1238	录音、录像制作
六、通用设备用房	13100	0.65	8515	
1.通用设备机房			7055	
2.维修值班室			960	
3.维修品库			500	
七、行政、业务、后勤用房	9030	0.65	5870	
1.行政业务管理用房			3200	办公室、会议室、计算机房等
2.后勤用房			2670	传达、警卫、食堂、浴室、环卫、车队、医务、仓库等
八、服务配套设施	10369	0.65	6740	
1.艺术展厅			1740	艺术展厅、陈列室、业务用房
2.表演艺术交流部			1800	阅览室、视听室、书库、研究室、交谊厅等
3.艺术商店			1650	商店、银行、邮局、仓库等
4.快餐及咖啡厅			1550	快餐、自助餐、咖啡厅
九、地下停车场	21000	0.65	13650	停放小轿车500辆，非机动车1000辆
十、人防工程	538	0.65	350	人防设备用房
总　　计	119155		77453	

2.国家大剧院功能关系图

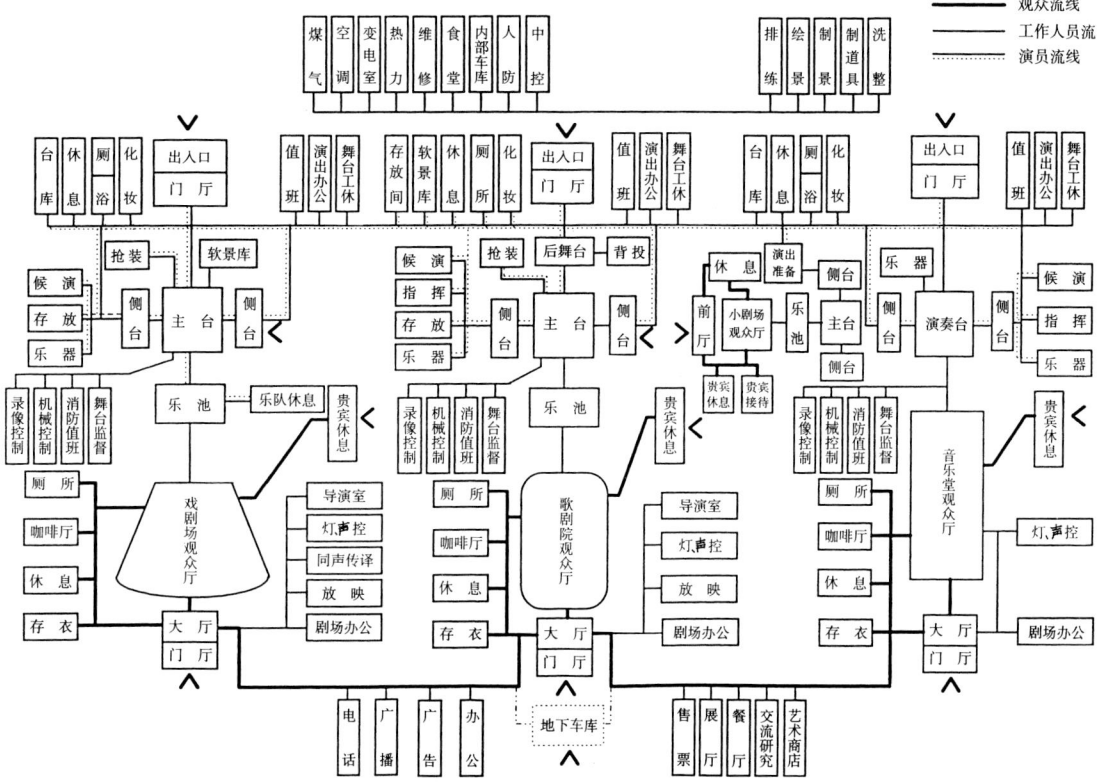

五、电气的要求

(1)本工程供电的高压电源采用 10kV。

(2)高压电源应选用二回路同时供电。

(3)本工程应设置应急电源设备。

(4)照明电源与空调电源应由不同变压器供电。

(5)舞台照明电源供电指标采用 280～360VA／座。并考虑舞台机械设备及电视转播用电。

(6)电讯、音响广播、消防自控、闭路电视、安保、电脑管理、电视转播室等弱电系统电源应与舞台照明电源分开,由不同变压器供电。

六、空调的要求

(1)空调设计标准：

a.整个建筑按全空调设计。

b.设计应依据《剧场建筑设计规范》的有关规定。

c.全楼的空调要采用 DDC(直接数字自动控制)控制系统。

d.观众厅、舞台要有防排烟设计。

(2)空调设计要求：

a.系统设计要考虑节能、全年运行。

b.系统设计应确保观众厅的音响效果,采取必要的隔声、吸声、消声处理。

c.舞台空调设计应给予重视。

d.观众厅的气流组织设计要有依据,应尽量不影响观众厅音响效果。

(3)冷源要求采用电动式制冷机。

七、给排水及消防设计的要求

(1)一般条件：

a.为保证剧院的生活和消防用水,要求在剧院周围设置环形供水管网,并自附近市政给水管引入二条户管,与该环网相接,要求每条户管均能供应剧院全部需水量。

b.剧院排出的生活污水须经化粪池处理后排入市政污水管道。

c.剧院空调冷却水须经冷却塔冷却后循环使用。

(2)剧院须设有下列消防灭火系统：

a.消火栓系统。

b.自动喷淋灭火系统。

c.水幕系统。

d.雨淋系统。

e.泡沫、水喷淋系统。

f.气体灭火系统。

(3)剧院须设有下列给水排水系统：

a.给水系统。

b.热水供应系统。

c.生活污水系统。

d.雨水系统。

e.循环冷却水系统。

八、热力及燃气供应

本工程的热源为城市高温水,热网检修期间用电或燃油锅炉供应。

厨房热源采用城市低压煤气。

九、楼宇自控的要求

(1)对空调系统进行自动调节。

(2)对喷淋泵、消防泵、生活泵、排污泵等运行状态进行监测并可对部分泵进行控制。

(3)应设火灾自动报警系统、安全系统、车库管理系统、卫星电视系统、通讯系统。

The National Grand Theater is the highest performing art center of China,an important place for developing the Chinese national culture,promoting cultural civilization,presenting the high levels of music and play,and pushing the international culture exchange forward.With complete functions,excellent visual and auditory conditions,advanced technologies,complete sets of equipment,and economical and rational design.the National Grand Theater should meet the performance needs of various forms of acting arts.

For the specific requirements of the construction site where is in the political and cultural center area of the People's Republic of China,the architecture should be harmonious with the surrounding buildings,create a space as an organic part of the Tian'an Men Square and the Chang'an Street,present solemmn and elegant art performing by the building itself,and show distinctive affinity of the people and the trend of the time with the presupposion of developing the beauties of the entire city.After construction,the Theater will add a new lustre to the histarical culture city of Beijing.

Ⅰ.Location:

The base is situated at Chang'an Avenue,Beijing,west side of Tian An Men Square and the People's Great Hall.South of west Chang'an Avenue,east of Shi Bei hutong,north of east Rong Xian hutong,west of the People's Great Hall Xi Lu.

The dimensions of the base are 224-245m long from east to west and 166m from south to north.The total area of the base is about 3.89 hectares.

Ⅱ.project scale and investment:

The total floor area is 120,000m^2 with possible floating of 10%.

The total investment of the project is about 2 billion yuan RMB

Ⅲ.Construction Period:

The entire project will be completed by the end of 2001

1998:completion of scheme and preliminary design

In the spring of 1999:commencement of work

By the end of 2001:final completion

January to June of 2002:debugging and running-in

Ⅳ.Design contents,requirements,and floor area distribution

Ⅰ.Master Table of Floor Area Distribution for Components

Name	Floor area (m^2)	Coefficient	Usable floorarea (m^2)	Briefing
Ⅰ.Opera house	22529	0.65	14644	
1.entrance lobby & lounge			3500	entrance lobby & lounge,VIP room,first aid room,attendant room,etc.
2.auditorium			2000	auditorium
3.stage,orchestra pit,under stage storage			4538	main stage left and right side,back stage,orchestra pit,under stage storage,etc.
4.premises for rear stage vocation			4156	dressing room,lounge,room for changing costume,waiting room for performance,props room,office,scenery assembly area,room for costume,shoes & caps,store for musical instruments,store for stage articles,etc.
5.premises for performance techniques			450	chief inspector room,director room,dimmer room,sound regulation room,projection room,relay room,light-control room,equipment room,etc.
Ⅱ.Concert hall	11987	0.65	7792	
1.entrance lobby & lounge			2800	entrance lobby & lounge,VIP room,first aid room,attendant room,etc.
2.auditorium			1600	auditorium
3.orchestra stand,pipe organ room			1156	orchestra stand,left & right side stages,pipe organ room,under stage storage,etc.
4.premises for rear stage vocation			2036	dressing room,lounge,waiting room for performance,office,store for musical instruments,storage for performance articles
5.premises for performance vocation			200	chief inspector room,dimmer room,sound regulation room,relay room,light-control room,equipment room,etc.
Ⅲ.theater	12006	0.65	7804	
1.entrance lobby & lounge			1680	entrancelobby & lounge,VIP room,first aid room,attendant room,etc.
2.auditorium			960	auditorium
3.stage,orchestra pit,under stage storage			2218	main stage,left & right side stages,back stage,lifting large forestage (also used for orchestra pit),under stage storage,etc.

4.premises for rear stage vocation			2556	dressing room,lounge,room for changing costume,waiting room for performance,props room, office,scenery assembly area,room for costumes,shoes & caps,store for musical instruments, store for stage articles,etc.
5.premises for performance techniques			390	chief inspector room,director room,dimmer room,sound regulation room,projection room,relay room, simultaneous interpretation room,light-control room,equipment room,etc.
IV.Mini theater	7427	0.65	4828	
1.entrance lobby & lounge			1550	entrance lobby & 1ounge,VIP room,room for VIP's activity,first aid room,attendant room
2.auditorium & stage			1900	auditorium,stage,under stage storage,etc.
3.premises for rear stage vocation			1078	dressing room,lounge,room for changing costume,waiting room for performance,props room, office,room for costumes,shoes & caps,store for musical instruments,store for stage articles,etc.
4.premises for performance techniques			300	chief inspector room,director room,dimmer room,sound regulation room,projection room,relay room, simultaneous interpretation room,light-control room,equipment room,etc.
V .Premises for public stage management	11169	0.65	7260	
1.rehearsal hall			2073	large,medium and small sized rehearsal halls,rehearsal hall for chorus,rehearsal hall for orchestra,etc.
2.practice room for piano,etc.			864	practice room for piano,etc.
3.booking and advertising			350	ticket center,advertisement
4.making & repairing workshop			1480	painting scenery room;making & repairing scenery,props,costumes,shoes,caps,and slides.
5.stores for scenery and material			1255	Stores for scenery,materials,and articles.
6.workshop for produce of recording and video			1238	producing recording and video.
VI.Premises for general purpose equipment	13100	0.65	8515	
1.General-purpose equipment room			7055	
2.Room on maintenance duty			960	
3.store for maintenance articles			500	
VII.premises for administration, vocation and logistics	9030	0.65	5870	
1.premises for administration and vocation management			3200	office,meeting room,computer room,etc.
2.premises for logistics			2670	reception office,guard room,dining hall,bathroom,general sanitation room,driver room,clinic,storage,etc.
VIII.Premises for service necessary facilities	10369	0.65	6740	
1.art exhibition hall			1740	art exhibition hall,display, room,vocational room
2.department for performance art exchange			1800	reading room,audio-visual room,stack room,study room,communication hall,etc.
3.art shop			1650	shop,bank,post office,storage etc.
4.snack bar & cafe			1550	snack,buffet,cafe
IX.Underground parking area	21000	0.65	13650	500 car stalls,1000 non-motor driven vehicle stalls
X.Civil air defense engineering	538	0.65	350	premises for civil air defense equipment
Total	119155		77453	

2.Function Relationshlp of the National Grand Theater

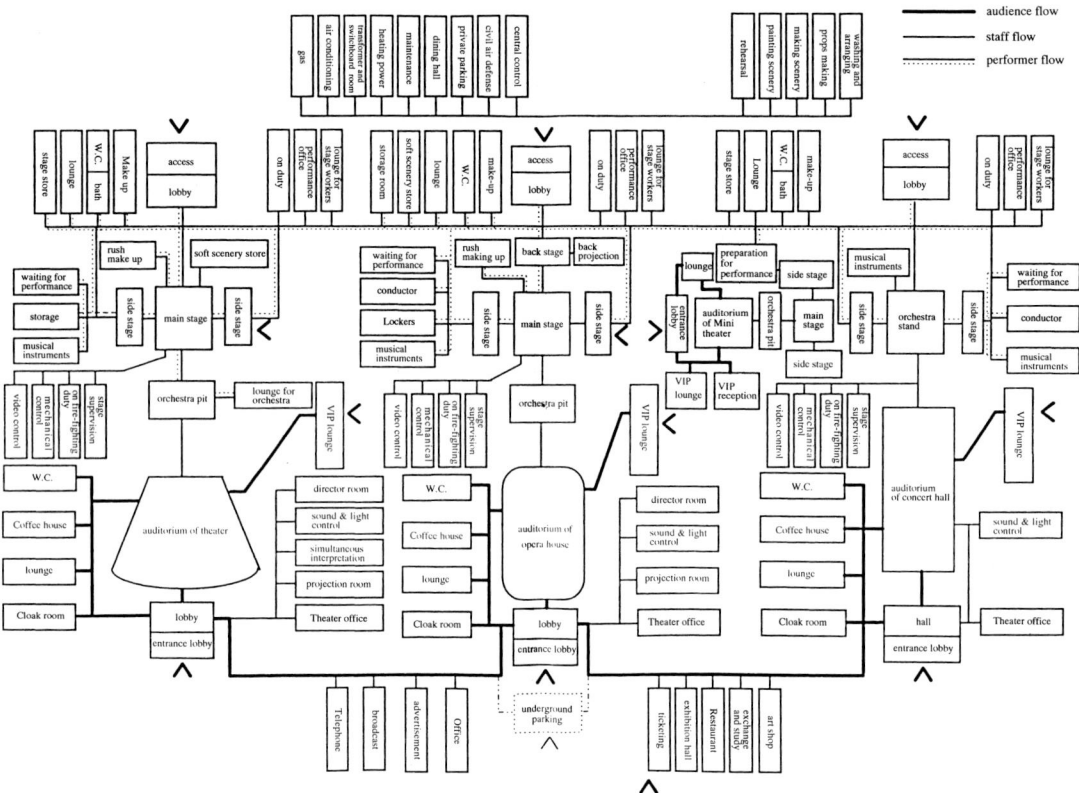

Ⅴ.Requirements for electrical design

(1)The high-voltage power supply with 10 kV is adopted for this project.

(2)The high-voltage power should be supplied simultaneously in two circuits.

(3)Emergency power source equipment should be installed for this project.

(4)Power source for lighting and air conditioning should be supplied by different transformers.

(5)The quota of power supply for stage illumination should be 280~360VA/seat.The power supply for machines and equipment of the stages and TV relay should be considered.

(6)Power source for weak current system(such as telecommunication,sound broadcast,automatic fire-fighting,closed circuit TV,security,computer management and television relay room,etc.)should be separated from that for stage illumination and supplied by separated transformers.

Ⅵ.Requirements for air conditioning design

(1)Standard for air conditioning design

a.Air conditioning should be designed for all of the premises of the theater.

b.The design should conform with the related articles and clauses of the "Design Code for Theaters".

c.The DDC(direct,digital and automatic control)control system should be used for the air conditioning of the entire theater.

d.The smoke prevention and evacuation design of the auditoriums and the stages,should be available.

(2)Design requirements of air conditioning

a.Energy saving and whole year operation should be considered in system design.

b.The acoustics of the auditorium should be ensured in system design with necessary measures for sound insulation,sound absorption and sound elimination.

c.Pay attention to air conditioning design for the stage.

d.The air flow design for the auditorium should be based on enough data,and not affected the acoustics of the auditorium as much as possible.

(3)It is required to use electric refrigerating machine for refrigeration.

Ⅶ.Requirements for water supply,drainage and fire fighting design

(1)General conditions:

a.In order to ensure water supply for domestic use and fire fighting, it is required to set up a circular water supply network surrounding the theater and two pipe lines for households from the nearby municipal water pipe network which should be connected with the circular network.It is also required that each such pipe line should supply enough of water for the theater.

b.Domestic sewage from the theater must be drained into the municipal sewage conduit after being treated by septic tanks.

c.The water used for air conditioning of the theater must be reused after being cooled by cooling tower.

(2)The following fire-fighting and fire extinguishing systems should be set up in the theater

a.Fire hydrant system

b.Automatic sprinkler system

c.Water curtain system

d.Deluge system

e.Foam,water sprinkling system

f.Gas fire-extinguishing system

(3)The following water supply and drainage systems must be set up in the theater:

a.Water supply system

b.Hot water supply system

c.Domestic sewage system

d.Rainwater system

e.Circulating cooling water system

Ⅷ.Heating power and gas supply

The heat source for this project is the city high temperature water.During the period of examination and repair of the heating network,it may be supplied by electricity or oil burning boiler.

The heat source for kitchen may be supplied by city low pressure coal gas.

Ⅸ.Requirements for building intelligence design

(1)Automatic control for air conditioning system is provided.

(2)The monitoring and measurement for operation of sprinkling pump,fire-fighting pump,domestic water pump,drainage pump,etc.should be realized and be controlled for some of the pumps.

(3)Automatic fire alarm system,security system,garage management system,satellite television system and communication system should be set up.

问 题 解 答

一、规划方面的问题

● 为什么北侧建筑物退红线20m?主体建筑以外的室外大台阶能否占用此20m范围?其他方向退红线有何要求?

答：建筑物在北侧退后20m是长安街整体规划的要求。此20m可作为北侧入口广场、绿化和内部循环路用。主体外的台阶、花池等可占用此20m范围。其他方向退红线的要求见附件"国家大剧院规划设计条件"第四条。

● 在满足基底退红线的情况下，上部挑出建筑红线是否可以?

答：因长安街规划是整体的，沿街建筑及天安门广场两侧建筑都退20m，挑出过多，会使整体形象受影响，因此不宜多挑。

● 长安街道路红线宽120m，其中道路50m，剩余的南侧35m的用途?是否可同建筑一起考虑?

答：剩余的35m可用作人行道、绿化、地下管线、地铁出入口等，此地的用途可以同建筑设计一起考虑。但不作硬性规定。

● 是否可以把场地南边的花园紧靠剧场，作为场地的一部分?具体说是否可以用空中通道或地下通道方式把花园和剧院相连接?

答：不可把南边花园紧靠剧场，中间有30m的红线间距，此绿地是供本剧院、人民大会堂及南边的公共建筑共用的城市绿地。

● 国家大剧院基地西侧绿化情况如何?

答：基地西侧将作为政府办公用地，控制高度在18m，绿化面积比较多，目前尚未作具体规划。

● 本地块南侧60m×280m绿地的南侧，将作为"公众用途"，请问"公众用途"是指什么?若有建筑物，是否有高度限制?

答：南侧将建设大型办公公共建筑。高度限制为30m。

● 请明确一下场地东侧道路的宽度。

答：30m。

● 可否在控制高度30m与45m之间指出明确界限?

答：该建筑物的主要部分特别是北侧高度应控制在30m以下，少部分如舞台升起部分高度应控制在45m以下。

● 在规划中有无自周边道路到建筑物的斜线限制?

答：没有。

● 在规划要求上，需否控制建筑覆盖率指标以满足绿化面积?覆盖率的计算方法是否以建筑物各部分屋顶投影面积计算?

答：本工程对绿化面积没有作具体规定。绿化面积大，应算作方案的优点。覆盖率的计算应是建筑物基底面积与用地面积之比。

二、交通方面的问题

● 四边道路的功能特点是什么?

答：北侧西长安街为城市主干道，组织行人交通为主，东侧人大会堂西路为城市次干道，石碑胡同、东绒线胡同为城市支路，以组织机动车交通为主，剧院南路为城市支路以行人和车辆并重。

● 人民大会堂西侧路是否有红绿灯控制?

答：人民大会堂西侧路与西长安街的路口有红绿灯控制，而与东绒线胡同路口近期无红绿灯控制。

● 交通分析图是否可以认为就是行人与车辆流向用地区域及建筑物的动态线?

答：应是行人与车辆流入及流出用地区域及建筑物的动态线。

● 可否提供用地附近西长安街一带地铁交通网络图纸?

答：在用地附近仅在长安街下有一条地铁线路。

● 明确行人运动情况，特别是从"紫禁城"方向来的行人情况。

答：四个方向均可能有行人进出。从"紫禁城"方向来的行人通过西长安街上的地下通道过街，再进入剧院。

● 请确认停车数为：地下500车位，地面50辆小车位、20辆大客车位。

答：是的。欢迎再多设车位。

● 地下车库需要几个出入口?平时是否可向社会开放?

答：根据中国《汽车库建筑设计防火规范》停车库的汽车疏散出口不应少于两个，本工程宜多于两个。地下停车库平时可以考虑向社会开放。

● 地下车库出入口的位置与道路之间的距离有何要求?

答：地下车库出入口的位置要结合四周道路的特点设置，与人流出入口分开，地下车库出入口不要放在长安街方向上。建议出口坡道外的水平道路部分的长度除满足转弯半径的要求外，最好不要小于一个半车的长度，并与建筑用地内的环路连接。

● 如果多设停车位是否影响交通?

答：不会影响。

● 自行车交通如何处理?每辆车所占面积是多少?

答：自行车交通从建筑东、西两侧进入，四周城市道路均设有非机动车道。自行车均停放在地下非机动车库，面积按1.8m²/辆计算。

● 大剧院的主入口希望在那个方向?

答：可以结合人流、道路特点及城市规划的要求由建筑师自己来决定。

● 北入口为礼仪出入口是否允许车辆穿行?贵宾是否需要单独设出入口?

答：北入口允许少量的贵宾车辆出入，但要与行人路线分开。需要设置供贵宾用的单独出入口。

● 场地周边的机动车的交通流量是多少?

答：西长安街高峰时期双向机动车每小时 6000 辆。

三、地铁方面的问题

● 地铁人行通道的位置是否可以改变?标高如何?如何与剧院建筑结合?建筑红线与地铁人行通道之间是否有城市管道?

答："附件"提供的地铁人行通道的位置不可改变。根据本次竞赛的要求，从地铁的地下人行通道可直接进入剧院内，供观众进入和疏散。进入剧院通道的位置由设计人员定。建筑红线与地铁人行通道之间有城市管道。地铁人行通道底标高为 32.776m，顶标高为 37.776m，在此范围内与城市管道无矛盾。

● 剧院通往地铁人行通道的出入口是否可以设置公用大厅?

答：设在建筑用地红线以内，由设计人确定；设在城市道路红线内，需同有关部门协商。

● 场地西北侧地下通道与场地西北侧地铁通道是一个吗?

答：是一个。

● 地铁风口是否可以横向、纵向移动?

答：风口可以在国家大剧院的红线范围内移动，但要满足"附件"中"地铁排、进风口设计要求"第 3 条的规定。

● "进风口和排风口之间的距离是 10m"。这里 10m 指的是垂直距离还是水平距离?

答：指的是水平距离。

● 地铁的振动情况如何?

答：在本地段范围内的地铁轨道未作特殊减振措施，因目前地铁正在施工，噪声尚无法测到。

四、建筑方面的问题

● 本次方案竞赛的面积如何计算?

答：①按轴线尺寸计算面积；②有柱、有顶盖部分要计算面积；③层高在 2.2m 以下的管道层可不计算面积。

● 地下车库是否算建筑面积?

答：地下车库要计算面积。

● 有柱、有盖没有墙的部分是否算建筑面积?

答：要计算建筑面积。

● 在计算地面上的建筑面积时，是否包括所有外墙及围墙的面积?

答：计算面积按轴线尺寸。围墙不计算面积。

● 建筑面积浮动 10% 是如何考虑的?

答：为了得到一个优秀的方案，我们尽量为建筑师创造一个宽松的创作条件。

● 建筑利用系数 0.65 是否可以认为就是建筑面积减去楼梯间、走廊、管理人员用厕所等的面积?

答：建筑利用系数是指使用面积占相应部分的建筑面积的百分比。

● 请确认人防面积是 3% 还是 0.3%；是 350m² 还是 3500m²?有何具体要求?

答：是 350m²。留出 350m² 面积即可。

● 只要在 10% 的范围内，我们是否可以在地面和地下增加更多的零售商店区域，以增加更多的文化和娱乐面积?

答：可以。

● 东、西侧退红线不清楚，请明确。

答：西侧退红线见"附件"规划设计条件四。最少应能满足安排下双向车道。东侧建筑线不能越过 25m 宽的地下管廊区。

● 请明确退红线和地下室退红线的要求。

答：地上建筑退红线见"附件"规划设计条件四及本次回答问题。地下建筑四边均退红线最小 5m。

● 环路是否属于场地设计范围?

答：是的。

● 环路是否可以位于退进线以外、红线以内?

答：是的。

● 建筑表现图是否可以用色彩?可否提交方案的录像等资料?

答：建筑表现图可以用色彩。除去文件规定的必须上交的设计成果外，可以附加说明设计意图的其他资料。

● 公共大厅透视图应作哪一部分?

答：设计者认为最精彩的部分。

● 透视图是否可以采用电脑画的?

答：可以。

● 要求的"总平面图 1∶500"是否就是总体配置图?

答：是的。

● 声线和视线分析图要作到哪种深度?小剧院是否也要作声线、视线分析图。

答：要求从平面和剖面上表示声线的分布情况和视线的各种角度，同时还要表明地面升起标准。

小剧场不要求作声线和视线分析图。

● 地铁出入口至大剧院是否考虑无障碍设计?

答：根据规范要求本工程均要考虑执行中国《方便残疾人使用的城市道路和建筑物设计规范》。

● 剧场楼梯的宽度，每级台阶的高度、宽度以及楼梯的级数有何规定？

答：踏步宽度不应小于0.28m，踏步高度不应大于0.16m，连续踏步不超过18级。超过18级时，每增加一级，踏步宽加大0.01m，踏步高度作相应降低，但最多不得超过22级。

● 大剧院内部是否可以考虑雕塑、壁画及艺术装饰？

答：可以。

● 大剧院内部人员按500人考虑是否够？

答：此500人仅只是内部的管理人员，不包括剧团和演员的人数。

● 可否提供较详尽的地形图(或红线范围南北剖面)，以便在计算土建造价时能较准确确定土方量。

答：本次竞赛在计算土方量时可不考虑现状，均按平地计算。

● 在"说明内容要求"一项中是否可以附上工法建议？

答：可以，但不作规定。

● 该地块是否作过地质勘察？若有，是否可以提供有关勘察报告？

答：作过简单的勘察。见"附件"建筑场地岩土工程背景资料。

● "在正本(原图)首面背后右下角标明作者名称"指的是设计报告书还是图纸？

答：是第一页图纸，准确地说应是贴在贴图纸的轻质板背后的右下角。

● 设计报告书是否需要装订？如果需要装订，A3纸是竖向还是横向装订？要求提供16套缩印本是否是将图纸及报告书装订在一起？

答：A3大小的设计报告书同A3大小的图纸缩印本装订在一起。横向装订。

● 请明确投资估算中土建造价、舞台设备造价、装修造价的具体内容。

答：土建造价包括：全部土建工程，土建设备工程(如上下水、空调、电器等设备)，一般装修工程，各观众厅及门厅、休息厅的装饰工程，电梯、观众厅的固定坐椅，室外红线内的道路、绿化，室外管线工程。

舞台设备造价包括：音响设备、灯光设备、舞台设备(如各种机械舞台、吊杆、各种幕布、管风琴等)。

装修造价包括：特殊装修工程(如贵宾休息室、公共餐厅等)、公共艺术品(如雕塑、绘画、特殊装饰)。

说明：所有活动家具如餐厅桌椅、沙发、演奏用谱架、练习用钢琴均不包括在本次造价之内。

五、消防方面的问题

● 本工程执行中国《高层民用建筑设计防火规范》还是执行中国《建筑设计防火规范》？

答：本工程执行中国《建筑设计防火规范》。

● 本工程建筑面宽大于150m，是否要考虑消防通道？

答：本工程较为方正，建筑主立面可不考虑消防车穿行，如有内院，应考虑消防车辆进入。

● 消防人员进入通道及云梯作业面有何管理要求？如夜晚所有场馆全部关闭时，是否要求保证消防人员通道畅通？

答：消防人员在火灾情况下应能随时进入建筑物，云梯作业面不应布置在妨碍登高消防车操作的绿化、架空线处。

● 防火分区的最大面积？有无超过该面积的特例？

答：带有自动灭火设备时，防火分区为5000m²，不得超过。

● 排烟分区的最大面积是多少？有无超过特例？

答：500m²，不得超过。

● 舞台与观众厅之间有无必要设置防火隔声幕？

答：需要设置。

● 是否允许把剧场安排在地下？

答：小剧场可安排在地下一层。

六、工艺、设备方面的问题

● 三个剧场排列方式是否按工艺图的方式排列？

答：可不按工艺图排列，工艺图是功能示意图，供建筑设计时参考。

● 戏剧场的京剧乐队位置是在侧面还是在乐池？

答：在侧面和在乐池都需要。

● 音乐厅的演奏台是否包括了合唱队的位置？音乐厅的侧台是如何考虑的？

答：音乐厅的演奏台包括合唱队的位置，音乐厅的侧台考虑能够放置大型乐器、乐谱架及台板等。

● 戏剧场的后舞台是否需要？

答：戏剧场的后舞台是需要的，任务书上没有限定具体尺寸，设计者可根据自己的舞台设计方案，确定其尺寸。

● 设计任务书英文部分，主舞台描述中有一引号中的内容不清楚，应是什么意思？

答：是中国字的"品"字。

● 歌剧院、音乐厅演出时是自然声还是电声？

答：按自然声设计，应备有电声系统。

● 管风琴练习室里配置的管风琴是声学式还是电气式？

答：是声学式。

● 观众厅的最远视距的计算方法如何统一？

答：观众厅最远视距按水平投影距离计算。

● 空调是否按剧场分系统控制？

答：是的。

● "雨淋系统"是否为开放型喷淋系统？

答：是的。

● 是否要设生活及消防储水池?容量多少?

答：可以考虑设生活及消防储水池，生活水池150m³，消防水池300m³。

● 要求设化粪池，请明确其污水排放标准?

答：①水温不高于40℃；②不腐蚀管道，pH值为6～10；③不阻塞管道；④不致产生易燃、爆炸和有毒气体；⑤不伤害养护工作人员；⑥不影响污水的利用、处理和排放；⑦对伤寒、痢疾、炭疽、结核、肝炎等病原体必须严格消毒灭除；⑧放射性物质应严格按照《放射性防护规定》执行。

七、其他方面问题

● 设计文件提出的"人民性"的含义是什么?群众是否参加投票?是仅考虑北京的群众还是全国的群众?

答：人民性是考虑到人民群众的需要。报送的方案将在北京公开展出，参观的人不仅是北京群众还可能有外地的甚至国外的群众，我们将征求他们的意见。最后由评委会决定三个提名方案。

● 谁来决定三个提名方案中的中选方案?

答：三个提名方案报送国家大剧院建设领导小组，最终由领导小组确定中选方案。

● 评委会有多少人组成?有多少外国人?

答：见竞赛文件第七条。

● 参赛单位是否只允许提供一个建筑设计方案?

答：可以提供一个以上设计方案，但业主委员会支付的成本费不增加。

● 可否提供关于提交竞赛方案的具体地点?

答：在报送方案前我们将通知具体报送地点。

● 图纸大小与数量有什么要求?

答：见竞赛文件第三条。

● 请确认有关提交图纸所用轻质板的尺寸。

答：轻质板尺寸为1000mm×700mm。

● 是否可以考虑采用进口材料、技术和设备?

答：可以。

● 每张图都要附在轻质板上是什么意思?轻质板是指什么?

答：为了保证图纸的平整，要求把提供的大图贴在一张任何材质的板子上。

● 该剧院、音乐厅是否有常驻剧团、乐队，或只是供到访的剧团、乐队使用?

答：不考虑常驻剧团、乐队。

● 参赛的设计单位为了组成必要的技术和设计专家组，是否可以从参加竞赛的其他设计单位中选择一家工程公司?

答：可以。

● 税收情况如何?

答：境外单位按收入的16%在中国交所得税。

Questions and Answers

1.QUESTIONS AND ANSWERS OF PLANING

Q:Why should buildings recede 20 meters from the north property line?

Can outdoor flights of steps to buildings be built in this 20 meters-wide area?Is there any requirement about receding distance from the property lines on other sides?

A.The requirement of receding 20 meters from the north property line is according to the Chang An Avenue Planning.This 20 meters-wide area can be used as open space of north entrance,green space and inner circulation road of the side.Outdoor flights of steps,flower beds and so on can be built in this area.The requirements about receding distance from the property lines on other sides can be referred to in Terms for the Planning & Design of the National Grand Theater,article,4,Annex.

Q:Is it allowed that the overhang goes beyond the property line in case of satisfying the requirements for receding distance from property line at the base?

A.Since all of the buildings along the Chang An Avenue and at both sides of the Tian An Men Square should recede 20 m from the property line in the Integrated Planning of Chang An Avenue,far beyond the property line of the overhang may make an impact on configuration as a whole.Therefore,far beyond the property line of the overhang should not be considered.

Q:The width between the two property lines of the Chang An Avenue is 120 meters.The road section is 50 meters wide.What is use of the left 35 meters-wide area on the south side?Can it be considered with design of buildings?

A:The left 35 meters-wide area can be used as pedestrian road,green spaces,underground pipelines,entrances of subway,etc.The uses of this area can be considered with design of buildings,but will not be specifically required.

Q:Can the green space at south side of the site be closely linked up with the theater and be recognized as a part of the site?Particularly,may the green space be connected with the theater by hanging or underground passage?

A:The green space at south side of the site may not be closely linked up with the theater.The green space and the theater are separated by a space of 30 m in between.The green space is for public use including pedestrians from the theater,the great hall of the people,and the public buildings at south side.

Q:What is the green plan to the west of the site?

A:The west of the site will be used as governmental office area.The building height limitation is 18 meters.There will be much green space in the area,but the plan has not been made yet.

Q:The south area to the green space of 60m × 280m to south of the site will be planed as public use.What does the public use refer to? Is there any building height limitation if something will be built in the area?

A:It is planned to build large-scale public buildings and office buildings in the south area.The building height limitation is 30 meters in the area.

Q:Please give the road width to east of the site.

A:It is 30 meters.

Q:Is it possible to draw a clear line of demarcation between 30m and 45m of the height limitation

A:The height of the main part of the theater,especially the building height at north side,shall be no more than 30m.The height of some part of the theater,such as the height of stage rising part,shall be no more than 45m.

Q:Is there any oblique line limitation from around roads to the Theater building in the planing requirements?

A:No,there is not.

Q:As planning requirements,is it necessary to control the building coverage in order to meet area of green space?Is the building coverage should be calculated by area of roof projections?

A:There is no specific requirement on area of green space in this project.Having larger green space will be considered as a strong point of the design scheme.The building coverage should be the ratio of area of the building base to area of the site.

2.QUESTIONS AND ANSWERS OF THE ACCESS

Q:What are the function characteristics of the roads surrounding the construction site?

A:The West Chang An Avenue at north side is an urban main truck road and arranged mainly for pedestrians.The West Road of the Great Hall of the People at east side is an urban secondary truck road and both Shi Bei Hutong at west side and East Rong Xian Hutong at south side are urban branch roads,and they are mainly arranged for vehicles.The road at south of the theater is an urban branch road and arranged for both pedestrians and vehicles.

Q:Is there traffic signal control at West Road of the Great Hall of the People?

A:There is traffic signal control at the crossing of the West Road of the Great Hall of the People and West Chang An Avenue.There is no traffic signal control at the crossing of the West Road of the Great Hall of the People and East Rong Xian Hutong in the near future.

Q:Do the flowing lines of the pedestrians and vehicles flowed to the project site and into the buildings identify with the traffic analysis drawing?

A:The traffic analysis drawing should be the flowing lines of the pedestrians and vehicles flowed from and to the project site and the buildings.

Q:Could provide a map of subway traffic network nearby the project site along the West Chang An Avenue?

A:Nearby the project site there is only one subway line under the Chang An Avenue.

Q:Please make clear the pedestrian movement,especially the pedestrians coming from the "Forbidden City"direction.

A:Pedestrians may come from four directions.The pedestrians coming from the "Forbidden City"direction enter the theater through underground passage at West Chang An Avenue.

Q:Please confirm the number of stalls.Are they 500 stalls underground,50 car stalls and 20 bus stalls on the ground?

A:Yes. Setting up more stalls are welcomed.

Q:How many accesses are needed for underground parking?May they provide service to the society in normal times?

A:The number of vehicle escape for the parking shall not be less than two according to the China's"Fire Fighting Code for Design of Garage".The number of vehicle escape for the parking should be more than two for this project.Proving service to the outside may be considered in normal times.

Q:What is the requirements of the distance between access to underground parking and the roads?

A:The access of underground parking should not be set up in the direction of Chang An Avenue,and should be separated from the access of pedestrians in consideration of function characteristics of surrounding roads.It is suggested that the length of horizontal part of the road outside the exit ramp should not be less than one and half length of the vehicle except satisfying the requirements of turning radius.The horizontal part of the road outside the exit ramp should be connected with ring road in the construction site of the project.

Q:If more stalls are set up,may the traffic be impacted?

A:It may not be impacted.

Q:How to deal with the bicycle path?What is the parking area per one bicycle?

A:The bicycles enter from both east and west sides of the theater.Nonmotor vehicle path are available on the surrounding urban roads.The bicycles are placed in the underground non-motor vehicle parking area.The parking area for bicycles may be calculated by 1.8m²/bicycle.

Q:In which direction the main entrance of the theater is expected to be set up?

A:It may be decided by architect in consideration of pedestrian flow,road function characteristics,and urban planning requirements.

Q:Should the etiquette access at north entrance be allowed to vehicle pass through?Should the access for VIP be separated?

A:Some vehicles for VIP are allowed to pass through the north entrance, and the path for vehicles should be separated from the pedestrian path.It is needed to set up separated access for VIP.

Q:What is the traffic flow of vehicles on the roads surrounding the construction site?

A:The peak traffic of two ways for West Chang An Avenue is 6000 vehicles per hour.

3.QUESTIONS AND ANSWERS OF THE SUBWAY

Q:Is it possible to change the location of the subway walk-through? What is its elevation?How to integrate it with the theater?Are there any urban underground pipelines between the property line and the subway walk-through?

A:The location of the subway walk-through specified in the Annex can not be changed.According to the requirements of the competition,the audiences may be directly entered into and conveniently evacuated from the theater through subway underground walk-through.The location of the entryway of the theater may be decided by the designer.There are underground pipelines between the property line and the subway walk-through.Fortunately,the bottom and top elevations of subway walk-through are 32.776m and 37.776m respectively,and there are no urban pipelines within the elevation between the top and bottom level of subway walk-through.

Q:Can the public hall be set up at the access to subway walk-through from the theater?

A:The designer may decide when it is set up within the property line of the project site.Negotiation with departments concerned is necessary when it is located within the property lines of an urban road.

Q:Does the underground passage at the north-west side of the site identify with the subway passage at the north-west side of the site?

A:Yes.They are identical.

Q:Is it possible to relocate the air inlet and outlet of the subway in horizontal and longitudinal directions?

A:The air inlet and outlet of the subway may be relocated within the property lines of the National Grand Theater.However,the article 3 of "The Design Requirements for Air Inlet and Outlet of the Subway" in the Annex should be satisfied.

Q:What the 10 m in "The distance between the air inlet and outlet is 10 m " means?Does it mean vertical distance or horizontal one?

A:It means horizontal distance.

Q:What is the vibration performance of the subway?

A:Special vibration isolation measures are not taken for the rails of subway in surrounding area of the project.The data of noise are not available since the subway is under construction.

4.QUESTIONS AND ANSWERS OF THE ARCHITECTURE

Q:How to calculate floor area in the design scheme competition?

A:a/The floor area should be calculated by axes dimension lines;

b/The floor area of the space with columns and roofs should be considered;

c/The floor area can be neglected for the story with height less than 2.2m and used for pipeline installation.

Q:Should the floor area of the underground parking be considered?

A:It should be considered.

Q:should space with columns,roof,and floor and without walls be considered as floor area?

A:Yes,It should be.

Q:Is the area of exterior wall and enclosing wall included in floor area of the buildings above the ground?

A:The area is calculated by axes dimension lines.The area of enclosing wall is neglected.

Q:How to consider 10% for floor area floating?

A:We hope to create a tolerant condition of architectural creation for architects as far as possible in order to get an excellent scheme.

Q:May the "building utilization factor:0.65" in the Design Program be recognized as the floor area minus the area of staircase,corridor,and toilet for management staff etc.?

A:The building utilization factor is the ratio between usable floor area and corresponding building floor area.

Q:Please confirm that the area for civil air defense is 3% or 0.3%?Is 350m^2 or 3500m^2?Is there any specific requirement?

A:The area for civil defense is 350 m^2.That means an area of 350 m^2 should be preserved.

Q:May we increase retail shop area on the ground and underground in order to get more area for cultural and recreational activities provided within the scope of 10%?

A:Yes.You may do.

Q:The receding lines at east side and west side are not clear,please make a definite answer.

A:See IV of the Terms for the Planning and Design of the "National Grand Theater" in Annex for the receding line at west side.At least,the requirement for two way traffic lane should be satisfied.The building line at east side should not be beyond underground pipe gallery with width of 25m.

Q:Please definite the requirements for receding line and basement receding line.

A:See IV of the Terms for the Planning and Design of the "National Grand Theater" in the Annex and the present questions and answers for the receding line of buildings on the ground.The underground part of the buildings and structures should recede 5 m from the property line at their four sides.

Q:Should the circulation road be considered in the site design?

A:Yes.It should be.

Q:Can the circulation road be arranged within the property line and outside the receding line?

A:Yes.

Q:Can the color architectural perspective drawings and the data including video tape of the design scheme be accepted?

A:The architectural perspective drawings can be colored.The additional data illustrating design intention can be attached except the design documents necessary to provide according to the document of the design scheme competition.

Q:Which part of the public hall shall be included in its perspective drawing?

A:The most interesting part recognized by the designer.

Q:Is the perspective drawings made by computer accepted?

A:Yes.Accepted.

Q:Does the required "General Plan 1：500" identify with General Layout Plan?

A:Yes.

Q:To which extent the acoustic and visual analysis drawings should reach? Is the acoustic and visual analysis needed for mini theater?

A:It is required to show acoustic distribution and various visual angles as well as the ground ascending standard.The acoustic and visual analysis drawings are not needed for the mini theater.

Q:Should the barrier-free design be considered for access to grand theater from subway entry and exit?

A:The China's "Design Code for Urban Roads and Buildings Convenient to Disabled Person" should be followed for this project.

Q:Are there specifications concerning width of stairs,riser height,tread run,and number of steps?

A:The tread run should be not less than 0.28m.The riser height should be not more than 0.16 m.The number of continuous steps should be not more than 22.In case of 18 and more,the tread run should increase 0.01m by every step,and the riser height should decreased correspondingly.

Q:May the sculptures,frescoes,and artistic decorations be set up in the grand theater?

A:Yes,it is allowed.

Q:Are 500 staffs enough for the grand theater?

A:The 500 persons are the management staffs,and the troupe staffs and performers are not included.

Q:Could provide topographic details (or north-south section within the property line) in order to precisely determine earth volume in the calculating construction cost?

A:In calculation of earth volume in this design scheme competition,the flat ground instead of the present condition should be considered.

Q:Does the construction methold allow to be attached to the "Contents of instruction" in the Required Design Documents of the Document of the Design Scheme Competition?

A:It is allowed,but it is not necessarily required.

Q:Has the geological examination been done for this site?If so,is it possible to provide geological investigation report concerned?

A:The simple geological investigation has been done.See "Background Information Regarding the Geological Engineering of the Construction Site of the National Grand Theater" in the Annex.

Q:Does the statement "Entrants should write their name at the bottom right corner on the back of the original drawings" refer to design reports or drawings?

A:It refer to the first page of drawings.Precisely,entrants should stick their name on the bottom right corner on the back of the light board with drawing pasted.

Q:Does the design report need to bind up all documents concerned into one volume?If it is needed,does it bind up in vertical or horizontal way for A3-size?Do the required 16 reduced copies need to bind up drawings and reports into one volume?

A:It needs to bind up the design report in A3-size and the drawings reduced to A3-size into one volume in horizontal way.

Q:Please make clear concrete content of construction cost,stage facility cost,and finishing and decoration cost in the investment estimation.

A:The construction cost includes costs of:construction works;facility works(such as work for water supply and sewage,air conditioning,and electrical equipment etc.);ordinary finish works;decoration works for all of the auditoriums,entrance lobby,and lounges;lift work;fixed seats in the auditoriums,roads and greening work within the outdoor property line and outdoor pipe line works.

The stage facility cost includes costs of: sound equipment,lighting equipment,stage equipment(such as various type of mechanical stages, hangers,curtains,pipe organ,etc.)

The finishing and decoration cost includes costs of:special decoration work(such as VIP lounges,public dinning hall,etc.),public arts(such as sculptures,paintings,special decorations).

Note;All of the traveling furniture,such as tables and chairs in dinning hall,sofa,music score stand for playing a music instrument use,piano for practice,are not included in the construction cost of this project.

5.QUESTIONS AND ANSWERS OF THE FIRE FIGHTING

Q:Should the China's "Fire Fighting Code for High Rise Civil Building Design" or the China's "Fire Fighting Code for Building Design" be followed?

A:The China's "Fire Fighting Code for Building Design" should be followed for this project.

Q:The width of building covering area of this project is more than 150 m,should the fire fighting access be considered?

A:Pass through a main facade of the buildings for a fire truck may not be considered because the site of this project is upright and foursquare.The fire truck should be allowed to enter the buildings in the inner compound.

Q:Are there any management requirement of fire fighting access for fire brigade and work area for scaling ladder?For example,during the night when the theater closed,is it needed to guarantee the fire fighting access be unblocked?

A:The fire brigade shall be allowed to enter buildings at any time in case of fire.The working face for scaling ladder should not be arranged at the green space or place with fly-over lines that hinder the lift fire truck from operation.

Q:What is the maximum area of fire compartment?Is there any special case of beyond the maximum area?

A:The maximum area of fire compartment is $5000m^2$ when the automatic fire-extinguishing equipment is available.To go beyond the limit is not allowed.

Q:What is the maximum area of smoke evacuation compartment?Is there any special case of beyond the maximum area?

A:The maximum area of smoke evacuation compartment is $500m^2$.To go beyond the limit is not allowed.

Q:Is it needed to set up fire and sound insulation curtain between the stage and the auditorium?

A:It is nceded to set up.

Q:Can the theater be arranged underground?

A:The mini theater can be arranged underground in first story.

6.QUESTIONS AND ANSWERS OF TECHNIQUE AND EQUIPMENT

Q:Is the arrangement pattern of the three theaters determined on the basis of process chart?

A:The process chart is a function schematic diagram and provided for reference in building design.Therefore,the arrangement pattern of the three theaters may not be determined on the basis of process chart.

Q:Is the location of the Beijing opera band in the theater situated at the side or in the orchestra pit?

A:Both at the side and in the orchestra pit.

Q:Is the space for chorus included in the instrument performance stage in the concert hall?How to consider the side stage in the concert hall?

A:The space for chorus is included in the instrument performance stage.The large-scale music instruments,music score stands,and platform etc.may be placed on the side stage in the concert hall.

Q:Is the back stage in the theater needed?

A:The back stage in the theater is needed.Its concrete dimensions are not specified in the Terms of Design and may be determined by the designer on the basis of their stage design scheme.

Q:What the content in quotation marks(in second line on P.41 of the Terms of Design in English) means?

A:It is a Chinese character "品".

Q:Is the natural sound system or electroacoustic system adopted during the performance in the opera house and concert hall?

A:Designed by natural sound system and the electroacoustic system is available.

Q:Is the pipe organ placed in the pipe organ practice room acoustics type or electric type?

A:It is acoustics type.

Q:How to unify the method for calculating furthest sight distance in auditorium?

A:The furthest sight distance in auditorium is calculated by horizontal projection distance.

Q:Is the air conditioning controlled by individual theater system?
A:Yes.

Q:Is the "deluge system" an open type spray system?
A:Yes.

Q:Is it needed to set up water tank for life and fire fighting?What is the volume?

A:May consider to set up water tank for life and fire fighting,and $150m^3$ for domestic water tank,$300m^3$ for fire fighting water tank.

Q:Please make clear the sewage discharge standard for the septic tank?

A:a/The sewage temperature is not higher than 40℃;

　b/Not make corrosion to pipeline,the value of PH is 6-10;

　c/Not clog pipeline;

　d/Not produce inflammable,explosive and toxic gas;

　e/Not harmful to the health of the maintenance workers;

　f/Not impact on the utilization,treatment,and discharge of the sewage;

　g/Pathogen of typhoid,dysentery,anthrax,tuberculosis,hepatitis etc. shall be strictly removed through disinfection.

　h/Radioactive substance shall be strictly treated by the requirements specified in the "Specification for Protection of Radioactivity".

7.OTHER QUESTIONS AND ANSWERS

Q:What is the meaning of "distinctive affinity of people" presented in the Design Program? Will people participate in the vote?Are they only the citizens in Beijing or people all over China?

A:"Distinctive affinity of people" is a concept considering people's needs.All entries will be put on an exhibition in Beijing.Visitors are not only citizens in Beijing but also people all over China and foreigners.Their opinions will be collected.Finally,three nominated design schemes will be selected by the Jury.

Q:Who will decide the winning design scheme from the three nominated design schemes?

A:Three nominated design schemes will be reported to the Leader Team of the National Grand Theater.Finally,a winning design scheme will be decided by the Leader Team.

Q:How many members are there in the Jury?How many foreign members are there in the Jury?

A:See the Document of the Competition,article,7.

Q:Is each entrant required to submit only one design scheme?

A:Each entrant can submit more than one design schemes.But payment to cover design cost from the Proprietor Committee will not be added .

Q:Please inform the detail address where entrants will go to submit their entries.

A:The detail information of the address will be announced before entrants submit their entries.

Q:What are the requirements for size and quantity of drawings?

A:See the Document of the Competition,article 3.

Q:What is the size of light boards for pasting drawings on?

A:The size of light boards is 1000mm × 700mm.

Q:Is it possible to use imported materials and equipment?

A:Yes,it is.

Q:What is meant by the statement,"Each drawing should be kept close to a light board",and what does light board refer to?

A:In order to keep drawings neat,it is required to paste each drawing on a light board which may be made of any materials.

Q:Is there any residential troupe or residential orchestra in the Theater?Or is the theater only used for visiting troupes and visiting orchestras?

A:Any residential troupe or residential orchestra will not be considered in the Theater.

Q:In forming a group of necessary technical and design consultants,is it possible that an entrant choose an engineering firm already participating to the competition within another group?

A:Yes,it is.

Q:What about taxes?

A:Foreign entrants should pay Income Tax of 16% of their income gained in China.

第一轮竞赛评语

一、中国"国家大剧院"是中国最高表演艺术中心，建筑面积12万平方米，占地3.89公顷。从1998年4月13日～7月13日公布建筑设计方案邀请竞赛以来，受到国内外建筑界、文化界、政治界和社会广泛的重视，表现出极大的兴趣，并积极踊跃参加。全世界都在注视着它，期望着一个能够展现新中国的面貌，反映建筑未来发展方向的高水准的方案问世。参赛者有36家设计单位，送来方案44个，包括来自国外设计方案20个，国内设计方案24个(含香港特别行政区方案4个)。

二、这是一次全国及国外建筑界广为关心的设计评选活动。演出建筑本身不仅集四种观演场厅于一身，内容繁多，且它位于中国重要的政治文化中心北京，天安门广场建筑群与长安街交汇处，北临大街，南面公共绿地，东与人民大会堂为邻。设计造型上如何与周围环境结合，而又"和而不同"富有鲜明的个性，表现出既具有民族文化特色又有时代精神；既具备庄重典雅而又亲切宜人；既具有开放性，便于群众交往，又利于运营管理；既能选用先进的技术，又能保证建设与长时间使用的经济合理性等等，总之，综合地满足设计任务书上的基本要求，是这一设计必须满足的原则。

三、全体评委认为就送审方案图纸、说明书、模型等，一般都认真细致地做了多方面的探索，说明参赛者在短暂的时间内及繁杂的设计要求下，作了极大的努力，付出了极大的劳动。但必须说明，方案水平参差不齐。经过对其设计构想、功能组织、流线处理、技术手段与外部环境的联系及作为"艺术殿堂"建筑形象的塑造等种种因素，反复充分讨论，评委们不能不遗憾地认为，在全部竞赛方案中，还没有一个方案能较综合地、圆满地、高标准地达到设计任务书提出的要求。我们推举不出不做较大修改即可作为实施的三个提名方案。

四、尽管如此，评委们经过深入分析研究，多次以无记名投票方式进行预选，最后正式投票(无一方案获全票)，选出得票过半数者五个方案，这些方案代表了各种建筑流派的不同思路，有的很有独创性，但均有很多需要改进提高的地方，有的虽然构思巧妙，但作为设计方案深度不够，有些方案对中华民族文化传统以及与天安门广场特定地段的相互关系考虑得很不够(这五个方案的优缺点见附页)，但我们相信这些方案的作者都有能力、有潜力把方案做好。

五、鉴于上述情况，评委们本着对建筑事业负责的精神，根据中国国家大剧院建筑设计方案竞赛办法第八条规定，方案提名"如条件不具备，可以缺额"，因此不能按照原有愿望，提出三个方案交领导小组确定中选。我们建议业主委员会将上述经过投票产生的五个方案，作为参加单位，给以充裕的时间，再进行一次新的一轮设计创作(并付以酬劳)，保持原有方案的独特的优点，努力修正改进存在的缺点，以期产生为全社会所喜爱的、为世界所瞩目的新的建筑精品问世。

中国国家大剧院建筑设计方案评审委员会
1998年7月31日

【附录】对五个方案优缺点的评论

—101方案

这是一个简洁的建筑，同时照顾南北两面，设计整体性强，南北有水池环绕，并有观赏台，可以极目远眺，建筑造型很有个性。

缺点：造型过于严整，像纪念性建筑，平面以大剧场为中心，在南北主入口处，空间显得局促，且人流在大剧院四周来往，交通组织欠佳，空间单调无变化，深色石头过于沉闷压抑，难以与周围环境协调，应进行修改。

—106方案

这是一个浪漫与理性相结合的设计，北部大厅具有很大的通透性，使整个大厅亦成为观众表演的舞台，这一设计概念富有新意。

缺点：在建筑上，对中国民族文化毫无反映，由于建筑只向长安街一面疏散，交通入口难以出入，并且这仅是一个概念性的设计，各方面仍需要进一步作具体的后续工作，本设计未注意建筑南面与公共绿地相结合，入口不明确，全部采用玻璃，这在北京冬季寒冷、夏日炎热的季节较长的地区，能源的消耗很大。

—201方案

剧场的构成灵活而有机，内部公共空间变化丰富，设计者利用柱廊、红墙等以增强与周围建筑环境的联系，在许多方面匠心独运，采用壳膜结构，有创造性。

但建筑屋顶造型过于独特，难以与广场环境协调，因这一方案最大的特点在屋顶，恐难以作较大的改进。

—205方案

设计者企图以严整的平面，开敞的柱廊，与人民大会堂封闭的实墙面既协调又对比，利用柱廊内开放活动空间，作为市民的公共活动场所，此外，还强调了中轴线的运用等等，说明作者在尝试中国传统与现代化相结合的追求。

缺点：柱廊漫长，显得单调，柱廊与屋顶曲线难以协调一致，"古琴"的含义实际上人们难以理解，建议在简化建筑词汇的同时简化建筑材料，追求整体协调。

—507方案

平面组合分区明确，空间通透，条理清晰，结构合理，具有很强的秩序感和合理的内容空间组织。

缺点：缺乏文化演出建筑的特性，缺少人情味，未反映中国民族文化特色，与周围环境不够协调，希望能进行改进。

中国国家大剧院工程设计方案评委会

职务 Position	姓名 Name	签字 Sign
主席 Chairman	吴良镛 Wu Liangyong	
委员 Member	阿瑟·爱里克森 Arthur Erickson	
委员 Member	傅熹年 Fu Xinian	
委员 Member	何镜堂 He Jingtang	
委员 Member	潘祖尧 Pan Zuyao	
委员 Member	彭一刚 Peng Yigang	
委员 Member	里卡杜·包费尔 Ricardo Bofill	
委员 Member	宣祥鎏 Xuan Xiangliu	
委员 Member	芦原义信 Yoshinobu Ashihara	
委员 Member	张锦秋 Zhang Jinqiu	
委员 Member	周干峙 Zhou Ganzhi	

Report of Jury on First Round Scheme Design Competition

1.The China "National Grand Theatre" is going to be the most important performing art centre in China.It is going to have an area of 120,000 sq.m.and occupies 3.89 hectare of land.Since the announcement of the invited architectural design competition between April 13th and July 13th,it has attracted immense attention and interest from the architectural profession at home and abroad,the cultural profession,the political profession and also from society at large.The whole sections of the society are involved in this competition.The world is watching and expecting a design scheme reflecting the face of new China and the direction of the development of the architectural world.There are 36 firms/design institutes participated in this competition.44 schemes are submitted.Among the participating competitors,there are 20 international entries,24 national entries which included 4 entries from the Hong Kong Special Administrative Region.

2.The judging of this competition attracted attention from architecture society at home and abroad.The National Theatre complex consists of four theatres of different uses,functionally very complicated.It is located in the cultural and political center of China-Beijing,at the intersection of Tian'anmen Square and Chang'an Street.There is a major traffic artery to the north of the site ,a park to the south,and the Great Hall of People to its east.The design for this site should reflect the following principles:the architectural image should be in concert with the surrounding environment;it should reflect the Chinese philosophy of "unity within difference";it should have the distinctive character of an arts facility and reflect "the spirit of the time";it should sit gracefully on the site and be easily approached by the general public;it should possess the quality of openness,at the same time be easy for management;it should reflect advanced technology and be feasible in construction and future operation.The scheme should satisfy basic professional requirements put forward in the program.

3.From the submitted design drawings,descriptive brochures and models,the Jury decided that the designers have made great efforts to explore various design directions.The participants made their endeavor in a short period of time through laborious efforts.We have to point out that the levels of finish in these submissions are different.Jury members discussed in detail each project's design concept,planning layout,circulation arrangement and technical strategy,exterior and environmental design,and its architectural image as the "Temple of Performing Arts".This jury has concluded with regrets that none of the submitted schemes satisfies all of the requirements described in the Program.The Jury cannot recommend three schemes without serious changes to be made on the scheme.

4.However,the Jury carefully analyzed all of the schemes and test anonymously voted many times before the final vote was cast (there was no scheme that received unanimous vote).In the final vote,five schemes received more than half of the votes and are selected as the recommended schemes.These schemes represented design concept of different schools of thought in architectural field.Some of the schemes are distinct in concept,but require further development and improvement of the concept.Others are uniquely conceived but were not developed to the level of detail required.Still others dit not show enough consideration to the Chinese cultural tradition and the relationship with this unique site near Tian'anmen Square (Jury's comment on these schemes are attached to this document).However we as Jury have the confidence that the designers of this project have the capability to develop their scheme further.

5.The Jury as responsible professionals deliberated that in accordance with the "Document of the Design Scheme Competition of the National Grand Theatre,China",Section No.8"······if the requirements can not be satisfied,the jury may leave the position vacant." we therefore decided that we could not select three schemes for the Leaders Committee to chose from as intended in the Program.We suggest that the client committee consider the five recommended schemes as first choice candidates,and give them more time to have a second round of design and pay them accordingly.(The above five schemes were selected through anonymous voting.)By retaining their original concepts yet revising their shortcomings,we hope the final submission will produce a work of excellence that meets the expectation of the society ,and is recognized by the world.

Jury of China "National Grand Theatre"
Architectural Design Competition
July 31st,1998

Appendix:Jury's Comment on the Five Selected Schemes

Scheme No.101

This is a very simple design with consideration to both the south and the north elevation.The overall design has a strong unified character.There are reflecting pools on the north and the south side.There is a spectator's gallery under the cone on the roof.The building shape is strong in character.

Shortcoming:The shape of the building is too monumental.In plan the opera house fills the center restricting circulation from north to south without an adequate entrance lobby.There is a lack of change in the public spaces.The dark stone facade is too ominous and depressing and very difficult to harmonize with the environment therefore should be revised.

Scheme No.106

This is a scheme that combines romance and rationalism.The atrium on the north is large and transparent which becomes a huge stage for theatergoers to perform

Shortcoming:the disbursement of the audience is on the north side only,circulation provisions are restrictive.However the scheme is at a schematic level,and requires further development.The entrance is not clearly defined with the large area of glass.In the climatic conditions of Beijing,the operation cost may be exorbitant.

Scheme No.201

The planning of the complex is free and organic.The internal public space is very rich.The roof is made of an compression shell structure.The architect has used the architectural metaphors such as the colonnade,and the red surrounding wall to enhance the connection with the unique surroundings.There are many creative design concepts in this scheme;however,the shape of the roof is too foreign to be in concert with the environment.Therefore it is not suitable for this specific site.The scheme itself is very hard to revise.

Scheme No.205

The designer attempted to use an open colonnade to unify his assembly of forms and to create a contrast to the solid walls of the Great Hall of the People.In planning it also provided a south-north axis through the site and an open public space.The designer tried to explore Chinese tradition and modernity in architectural design.

Shortcoming:the colonnade is too long and banal;the colonnade and the distorted roofline created a very contradictory relationship.

Scheme No.507

This scheme is very clear in the functional layout.It presents a strong sense of order and has a meaningful link between north and south which forms an interior public space in the complex.

Shortcomings:It lacks the characteristic of a performing arts facility and a sense of human feelings.It is not in harmony with the surrounding environment.

第二轮竞赛评语

一、"国家大剧院"第二次评审会于1998年11月14日至11月17日在北京召开，根据"国家大剧院"领导小组的意见，原则上仍聘请第一轮评选专家继续担任评审委员，另增加上海、南京专家各一人。计国内专家9人（包括香港评委1人），国外专家2人，共11人。此次送选方案共14个，除上一轮遴选的5个方案外，另有由业主委员会邀请参赛的4个方案，以及曾参与上轮竞赛、并志愿参加本轮竞赛的国外方案5个。会议起始两天，聆听各参赛单位主要设计人介绍方案情况，并回答有关问题。评委会对参赛者认真准备的报告表示满意。

二、评委们认为，经过第二轮近三个月时间的努力，各参赛方案的设计水平均有不同程度的改进和提高。这说明，进行第二轮评比的决定是正确的。

三、经过对参赛方案的认真讨论、反复交换意见，评委们认为，第一次报告书中所阐明的下列论点："设计造型上如何与周围环境结合，而又'和而不同'，富有鲜明的个性，表现出既具有民族文化特色又有时代精神；既具备庄重典雅而又亲切宜人；既具有开放性，便于群众交往，又利于运营管理；既能选用先进技术，又能保证建设与长时间使用的经济合理性等等"是正确的。本次评选坚持了前述原则，经过仔细研讨，对于这次任务的严肃性和重要性又加深了认识。

四、会议期间，业主委员会负责人对评选方式作出新的建议。评委们根据实际情况，对评选方法作出如下的决定：

1.从第一轮所遴选的五个方案中，选出两名。

2.从第一轮竞赛后业主委员会所邀请的四个国内参赛方案中，遴选出一名。

3.参加第一轮竞赛后，自愿并自费参加第二轮竞赛的方案中，遴选出一名。

4.以上投票所得的方案提供领导小组及有关方面参考。

五、经过评委会无记名投票，结果如下：

1.第一轮所遴选的五个方案中，选出：

1号方案，法国巴黎机场设计公司。

2号方案，日本矶崎新建筑师事务所。

（另：5号方案英国 TERRY FARRELL & PARTNER 事务所，以两票之差逊于上述两方案。）

2.国内受邀设计单位中，选出：

6号方案，北京市建筑设计研究院。

8号方案，清华大学。

（因票数相同，一并列入，提供领导小组参考）。

3.参与第一轮竞赛并自愿自费参加第二轮竞赛的五个方案中，选出：

12号方案，奥地利汉斯·豪莱建筑师事务所。

六、评委们反复、认真地讨论了上述方案，对其优点、缺点及不足之处主要有如下考虑：

1.1号方案：基本保持了上一轮方案的设计原则；兼顾南北两个立面；建筑造型有个性、整体性强；顶部设有观景平台，可以纵目远眺。本次方案在材料选择、色彩运用、室内空间处理等方面淡化了原方案的造型过于肃穆的问题，但整体看来，仍具有严整的纪念性建筑特征。建筑独立性过强，难与周围环境相谐调。另外由于歌剧院处于正中，南北两出入口处空间仍显局促。建筑形象缺乏剧院建筑特色。

2.2号方案：本方案基本上保持了原有方案的特点，如几个不同剧场的构成灵活而有机，内部公共空间变化丰富，外部利用柱廊、红墙以增强与周围建筑环境的联系，在许多方面匠心独运。但屋顶造型虽然做了一些调整、推敲，如屋面材料改用石板瓦等，但仍难以与周围环境相谐调。

（5号方案：原方案北部大厅具有很大的通透性，使其亦可成为观众表演的舞台，这一设计概念富有新意。新方案入口大厅做了一定改进；主轴线上将剧场架高，以沟通北入口与南部公园的联系。设计人对中国文化的研究探讨也作了一定努力。然而该方案失去了原有设计的比较简洁的特色，北面大片玻璃处理流于商业化。）

3.6号方案：整体方案在功能布局、空间处理上都做了新的努力。不足之处在于：平面布局及北入口至公园的通道组织与上一轮其他参赛方案有雷同之处，建筑造型四面尚缺乏统一处理。

4.8号方案：在利用地下空间的基础上，剧院东部开辟了大片绿化广场，试图适应观演、休闲等多功能用途。在建筑构图的探讨上，对于吸取民族文化传统作了新的尝试。但该方案的缺点是比例、尺度、色彩欠当，消防、构架等问题的处理尚需认真推敲。

5.12号方案：平面布局上，采用圆形大厅将三个主要演出厅有机地联系起来，空间组织灵活，外部形象力求规整，以符合地段需要。缺点方面：建筑外部造型过于琐碎、尺度欠当、不统一。屋顶的处理仍应推敲。

七、绝大多数评委认为，上述所列举的方案有些虽然已达到比较高的水准，但对特定的地段条件以及其他因素来讲，这些方案均不够完美，或多或少存在不同程度的问题，甚至有较严重的缺陷，对此宜提请领导小组和决策人慎重考虑。

八、评委们认为，方案的遴选对整个建筑的成功与否固然起着决定性作用，但基于上述所遴选方案的种种不成熟性，亟需在广泛征求意见的基础上，结合任务书研究的深入，以足够的时间进行深化和完善。

九、整个设计竞赛从始至终得到社会各界的广泛关注，国际驰名建筑师的参加，中外建筑师的交流，对增进中外建筑界的互相了解、提高中国建筑设计水平饶有意义。对于国外建筑师的一些积极建议，我们将向有关方面另行转达。

十、评委们四天来以极大的热情参与方案的评审工作，取得了上述的结论，但由于时间所限，在某些问题上，如对第2、8方案，尚未取得一致意见。评委个人如认为有必要，可在本评议书后另加附页，提出己见。

中国国家大剧院建筑设计方案评审委员会

1998年11月17日·北京·中国历史博物馆

中国国家大剧院设计方案第二轮评委会

职务 Position	姓名 Name	签字 Sign
主席 Chairman	吴良镛 Wu Liangyong	
委员 Member	包费尔，里卡杜 Ricardo Bofill	
委员 Member	齐　康 Qi Kang	
委员 Member	何镜堂 He Jingtang	
委员 Member	周干峙 Zhou Ganzhi	
委员 Member	张锦秋 Zhang Jinqiu	
委员 Member	宣祥鎏 Xuan Xiangliu	
委员 Member	爱里克森，阿瑟 Arthur Erickson	
委员 Member	傅熹年 Fu Xinian	
委员 Member	潘祖尧 Pan Zuyao	
委员 Member	戴复东 Dai Fudong	

Report of Jury on 2nd Round Design Competition

1.The second round evaluation by the jury of the National Grand Theatre was held in Beijing between November 14 and November 17,1998.According to the recommendation of the Leading Committee of the Construction of the National Grand Theatre,the members of the jury for the first round competition were invited back to continue judging the second round submission Added to this jury were 2 experts from Shanghai and Nanjing.Altogether there were 9 judges from China(including 1 from Hong Kong)and 2 experts from abroad.There were 14 schemes submitted for the second round.Besides the 5 schemes selected from the first round competition,9 more schemes from the first submission were invited or volunteered to participate in the second round.Two days before the jury meeting started,principal designers from each Participating firm or design institute had made presentations to the jury about their schemes and answered inquiries.The jury was satisfied with the presentations of the schemes' well prepared.

2.The judges concluded that through their efforts of about 3 months for the second round submission,the participating designers exhibited improvement of various levels in their schemes.This indicated that the decision to go on a second round evaluation was correct.

3.After careful discussion and repeated exchange of opinions,the judges reaffirmed the design principles put forward at the first round jury meeting.The design principles were as follows.The design schemes should have harmonious relationship with the surrounding environment,be "in harmony but different", have strong identity and reflect both the national cultural character and the spirit of the time;be graceful and also approachable;have the sense of openness,be convenient for the public to communicate with each other and also be easy to run and manage;be able to utilize advanced technology and feasible to construct and be used economically for an extended period of time,etc.The second round of selection process followed the above principles.Through careful discussion,the judges developed deeper understanding and appreciation of the significance of the above principles.

4.In the course of the jury meeting,the Chairman of Proprietor Committee made a suggestion regarding the selection method.Based on actual situation,the jury members agreed to make the following changes to the selection method.

A.To select 2 out of the 5 schemes selected from the first round competition.

B.To select 1 scheme from the 4 participating Chinese design institutes recommended to join in the second round evaluation by the Proprietor Committee.

C.To select 1 scheme from the voluntary participants.These voluntary participants attended the second round on their own expenses.

D.Schemes selected through voting should be provided to the Leading Committee for consideration.

5.The following was the result of the anonymous voting by the jury members.

A.From among the 5 schemes selected in the first round competition,the following ones were selected.

Scheme No.1 by ADP of France.

Scheme No.2 by Arata Isozaki & Associates of Japan.

(Next to the first 2 selections was No.5 by Terry Farrell & Parthers of Great Britain due to the lack of 2 votes.)

B.Among the schemes of the invited Chinese design institutes,the following schemes were selected.

Scheme No.6 by Beijing Institute of Architectural Design & Research.

Scheme No.8 by Tsinghua University.

(These 2 schemes received the same number of votes.Both schemes were included for the consideration of the Leading Committee.)

C.From the 5 voluntary schemes,scheme No.12 by Hans Hollein+Heinz Neumann Design Group of Austria was selected.

6.The jury members repeatedly discussed all of the above schemes.They arrived at the following evaluation.

A.Scheme No.1:It retained its original design principle.The scheme considered the elevation of both the North and the South.The architectural image had characters and sense of unity.A spectators' platform was located at the roof level.In this second round,the architect responded to the seriousness of its appearance by means of material selection,use of color,and interior spatial treatment.However,the overall image still had the problem of resembling memorial architecture.Also,the centrally located opera house made entrance spaces on the south and north sides too narrow.The building image was too strong,hard to be in harmony with the surrounding environment.the building image lacked the character for a grand theatre.

B.Scheme No.2:It retained its basic characteristics of the first submission,such as the free and organic composition of various venues,the free-flowing interior public space arrangement,the use of colonnade and red wall on the exterior to enhance a dialogue with the surrounding building environment.In various areas,this design exhibited masterful skills.However,the roof was still difficult to relate with the surroundings, although the architect made an effort to redesign it,e.g.changing the material of the roof to slate finish.

(In scheme No.5,the previous entrance lobby was very transparent,making it a stage for the public,This concept was very creative.In the revised scheme,improvement was made to this lobby design:the opera house on the main axis was elevated to create a connection to the landscaped garden to the south of the site.The architect made an effort in studying the Chinese culture.However the new scheme lost the simplicity and purity of the original submission.Large glass windows to the north facade tended to be too commercial in character.)

C.Scheme No.6:The design institute made a new effort in functional layout and spatial treatment.The scheme's shortcoming laid in the fact that the planning layout and the concourse between north entrance and the landscaped park to the north were similar to other schemes in the first round submission.The four facades lacked unity in the design.

D.Scheme No.8:On the basis of making use of the underground space,the scheme opened up the eastern part of the site for a large open landscaped plaza,trying to fulfil leisure use and outdoor performance.In the pursuit of architectural composition,the institute tried to integrate national cultural heritage into the design.However,the proportion,the scale and the color were inappropriate,fire code issues and the use of frame work element needed to be carefully studied.

E.Scheme No.12:on plan,three major performance spaces were connected organically by a rotunda and the spatial arrangement was organic.The exterior image was redesigned to fit into the surroundings.Its shortcomings:the building exterior massing was too complicated and the sense of scale inappropriate,and it lacked the sense of unity.The de-

sign of the roof should be reconsidered.

7.The overwhelming majority of the judges concluded that among the above schemes,some achieved very high level of design quality.However,because of this specific site and other related factors,none of the these schemes was perfect.Every one had its own defects,even quite serious ones.For this reason,we would like to remind the Leading Committee to consider these schemes carefully.

8.The jury members believed that the selection was very important to the success of the project,but it was difficult to choose an appropriate scheme for the site.However,based on the incomplete nature of all of the schemes,it was necessary to allow more time to further collect suggestions and study the program.

9.The competition attracted attention from all segments of the society from beginning.The participation of the internationally renowned architects and frims in the competition and the exchange of ideas between Chinese and foreign architects contributed significantly to the mutual understating between professionals,and helped improve Chinese architects' design qualities.We would present Positive suggestions from foreign architects in another report to the concerned parties.

10.The jury members worked enthusiastically in the past 4 days during the judging of the second round submissions and arrived at the above conclusion.To some issues,e.g.regarding Schemes No.2 and No.8,the jury members were not able to come to an unanimous conclusion because of the limitation of time allowed.The jury members would express their opinions in an appendix to this report if they considered necessary.

Jury of China"National Grand Theatre"
Architectural Design Competition
November 17,1998
Museum of Chinese Revolution,Beijing,China.

实 施 方 案
Implement Scheme

法国巴黎机场公司(法国)

ADP Aeroports de Paris(France)

清华大学协助 （中国）

Joint Collaborator: Tsinghua University(China)

法国巴黎机场公司（法国）
ADP Aeroports de Paris(France)
清华大学协助
（中国）
Joint Collaborator:
Tsinghua University
(China)

这是一座全新的建筑。

它完全由曲线组成，宛如湖上仙阁。

这座仙阁是敞开还是关闭着？令人难以定夺。但见灰色的钛合金属板和玻璃制成的多角立面与昼夜的光芒交相辉映。这两种材料的颜色在不同的时间里变幻莫测。玻璃墙如同拉开的幕布，使建筑物内部的剧场、散步小径和展览厅依稀可辨。同时，部分区域在钛板的覆盖保护下又显得更为隐秘。

这一巨型建筑的造型如同珠宝盒、果壳或含苞欲放的花朵。人们透过钛金属的表面可觑见内部深红色的木饰，歌剧院半透明的金壳内的猩红大厅以及大剧院观众的所有活动。音乐厅和剧场的外部为银色立面，从外部可眺望大剧院内部的街道、广场、楼厅、休息室内活动的人群，以及把水中仙阁与地面连接起来的透明走廊。大剧院内部的不断运动体现出建筑物的活力。它是一个既简单又复杂，既明晰又隐秘的肌体。

冬季，溜冰爱好者们将在冰面上留下无数道弧线。大剧院的地点和设计堪称举世无双。它的诞生和存在与特定的地点息息相关，世界上没有任何建筑物能与之相比。如果它出现在任何其它的地方，那简直是件无法想象的事。它表述的是和谐、宁静、简洁这些古老的对立统一的哲理。

我们现在想谈一下天安门广场周围的市区建设问题，并阐述我们对于这一总体研究的方向性思考。

我们可首先指出这一新方案的圆形设计可避免不适当地缩小人民大会堂西门的入口面积，并从紫禁城的入口处隐去。我们还可指出，绒线胡同向地下的延伸可避免其偏离对称的人民大会堂轴线的问题。除此之外，鉴于前一阶段更为自由的抉择和十分新颖的国家大剧院设计，我们有意以此为契机，在保持历史连续性的前提下，继续为这段历史增加一个新章节。面对围绕着天安门广场的完全对称，风格近乎划一的建筑群，如何在紫禁城的对面插入一个延续这一古老传奇的历史的章节呢？我们的意图自然是完成理想的设计，但同时还要考虑中国和世界建筑历史发展的不同背景。

自从19世纪以来，把自然园林引入城市的观念似乎从未象今天那样强烈。减少污染，建造花园、公园和绿地，向市民提供更为舒适和谐的生活环境。在尊重历史的前提下，以大公园内的现代建筑来环抱广场及其周围统一风格的古典建筑，这不正是对巨大的天安门花岗石广场的一种补偿吗？这个公园将成为一散步的乐园和市中心的肺脏。

我们力图避免一些现代剧院所走过的弯路。它们出于功能的目的设计了一些难看的侧门和后门，造成了建筑上的败笔。而国家大剧院将是公园最主要的，也许还是最大的入口。湖面本身便是公园的第一个景点。这里由于邻近长安街而显得较为喧闹，不够悠闲，但可用于其它较为喧闹的活动，如冬季溜冰等。

我们并非想要通过这一公园的设计来彻底解决这一需要更长，更深入思考的问题。但是，该公园至少对人民大会堂和国家大剧院的周围环境，以及将大剧院场地后移70m的决定来说具有重大意义。

It's a new building!

It is a building made of curves that emerges like an island at the center of a lake.

Is it open or is it closed?It is both at once.Its exterior surface is in glass and gray tianium:the glass opens the building like a curtain that is drawn aside so that you can see the theaters,promenades and exhibition spaces inside;the titanium protects and covers,thereby creating more secret shadowy areas.The colors of the titanium and the glass change with the hour of the day and night as light moves over its surface and plays on its countless facets.

You can see that this huge volume is a shell,the enclosing case,the par-

tially opened outer covering of a fruit or flower.You can see that the titanium is faced with an inner skin of dark red wood,and that the bright red interior of the opera house is itself enclosed by a partially transparent gilt covering so that the movements of the audience entering and leaving it are visible from outside.You can see that the outer surface of the concert hall and the theater are silver.And everywhere you can see people moving about,along the inner streets,on the plazas,in the upper lounges and walking just below the surface of the water through the transparent underpass that connect the building to the shore.These continual movements convey the life of a building that has become like a living organism,at once simple and complex,evident and secretive.

In winter,on the frozen surface of the water,iceskaters will trace the pattern of countless other curves all around the building.

There is no other building like it in the world.It is a unique building,born in unque circumstances for a singular place,inconceivable anywhere else.Its design expresses serenity and simplicity of the most ancient harmony between opposites.

We would now like to raise the issue of urban planning in the area around Tien An Men Square and express our thoughts concerning the possible direction that an eventual development study could take.

Why should we raise this subject? We could content ourselves with noting that the round shape of the new project accomplishes the goals of ensuring that the space behind the Great Hall of the People is not unpleasantly reduced and that the view of it from the entrance to the Forbidden City is highly discreet.We could further content ourselves with noting that by laying Rong Xian Hutong Street partially below ground we avoided having an offset road disturb the symmetry of the Gread Hall of the People.

But we feel that it is necessary to go farther.A building as new as the one we are proposing for the Beijing National Grand Theater of China,whose design was born as a result of the greater leeway decided at the end of the preceding phase of the competition,marks the desire to pursue the development of this immense part of the city and to do so not by breaking with the continuity of its ancient history but rather by opening a new chapter in it.What did the building around Tien An Men Square do,in their symmetrical layout and nearly total stylistic homogeneity,if not introduce,in face of the Forbidden City,a new chapter continuing an ancient and prestigious history? Such an aim is just as reasonable today,if in a very different national and international architectural context.

Neversince the 19th century has the desire to bring nature back into cities been so strong. The determination to reduce pollution and to make life in cities more comfortable and harmonious has led to the creation of gardens, parks and open spaces. Would it not be fitting to compensate for the inorganic quality of Tien An Men Square with a zone of buildings incorporated into an enormous park. Such a zone, without betraying or turning its back on the history of the homogenous buildings on Tien An Men Square, could be the vital lumgs of the city centre.

The National Grand theater, which is equally beautiful from every angle (unlike many recent theaters with unseemly side or rear facades due to badly handled functional imperatives) will then look like one of the entrances to this park, probably one of the biggest. The lake itself will be the first component of the park, keeping in mind that the din of traffic from Chang An Avenue nearby makes the site too unpleasant for activities that are not themselves noisy, such as skating in winter. The plan that we have drawn for this park is not by any means intended as a solution to a question that we know demands much lengthy, indepth consideration.But, at least in the area around the Great Hall of the People and the National Grand Theater, we feel that it brings out the full sense and impact of the decision to move the new building's front back by 70 meters.

南侧透视
PERSPECTIVE OF SONTH ELEVATION

鸟瞰图
BIRD'S - EYE VIEW

北侧透视
PERSPECTIVE OF NORTH
ELEVATION

56

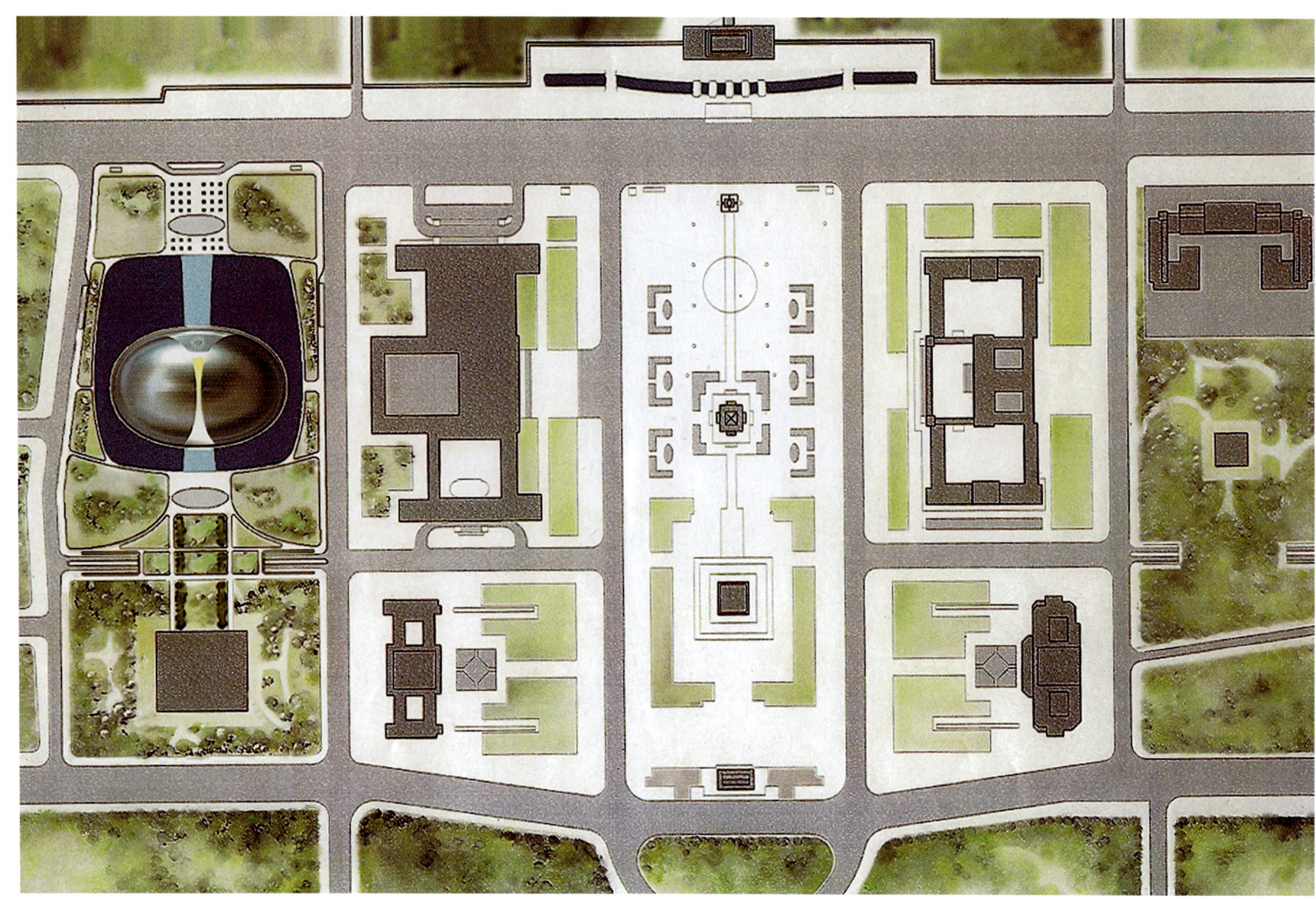

天安门广场总体规划图
GENERAL PLAN OF THE TIAN'AN MEN SQUARE

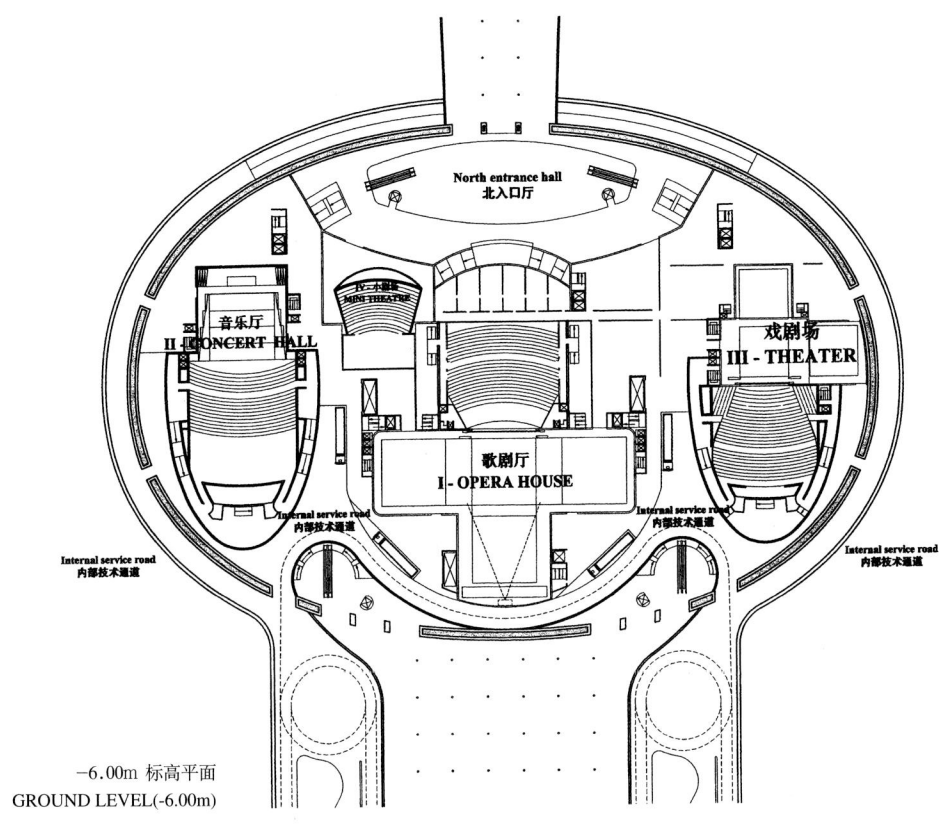

−6.00m 标高平面
GROUND LEVEL(-6.00m)

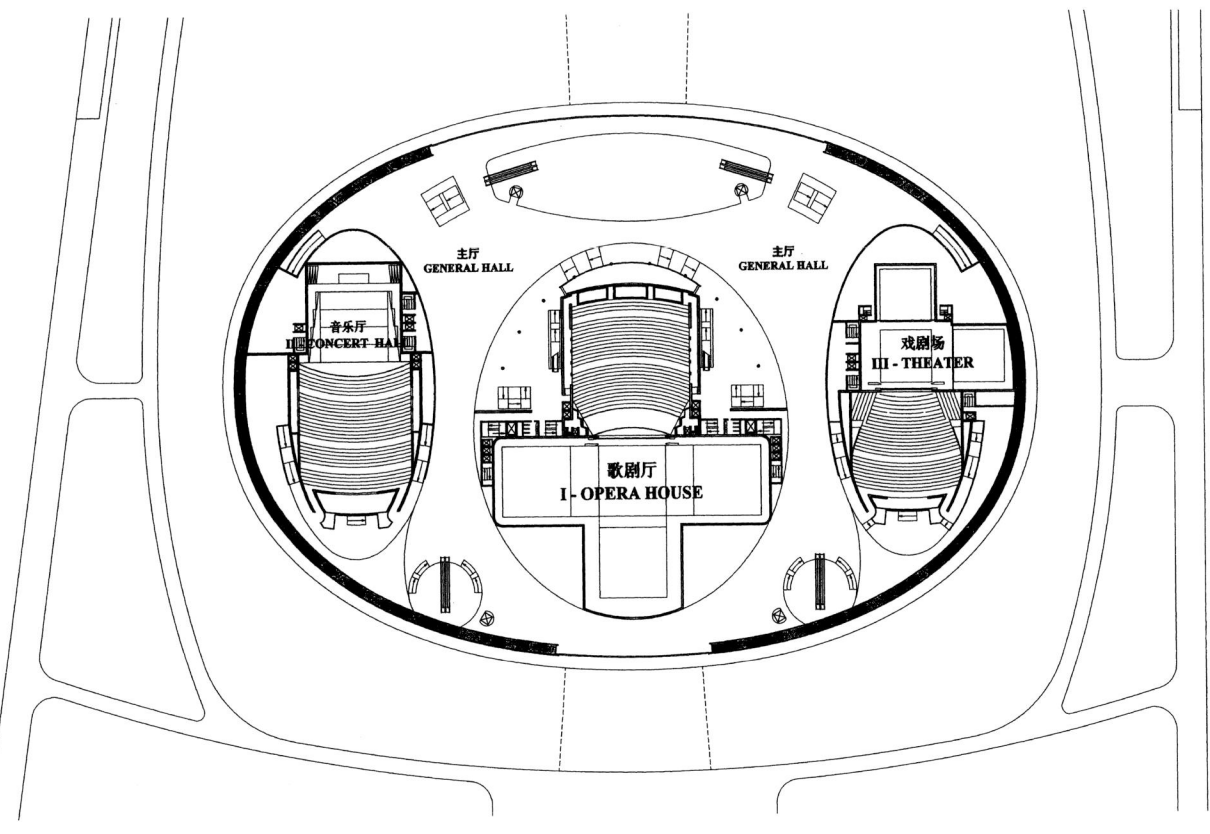

±0.00m 标高平面
GROUND LEVEL(± 0.00m)

歌剧院观众厅
PERSPECTIVE OF THE OPERA HOUSE AUDITORIUM

剖面
SECTION

北大厅内部透视
INTERIOR PERSPECTIVE OF NORTH LOBBY

在湖面下的入口走廊透视
PERSPECTIVE OF THE UNDER LAKE ENTRANCE

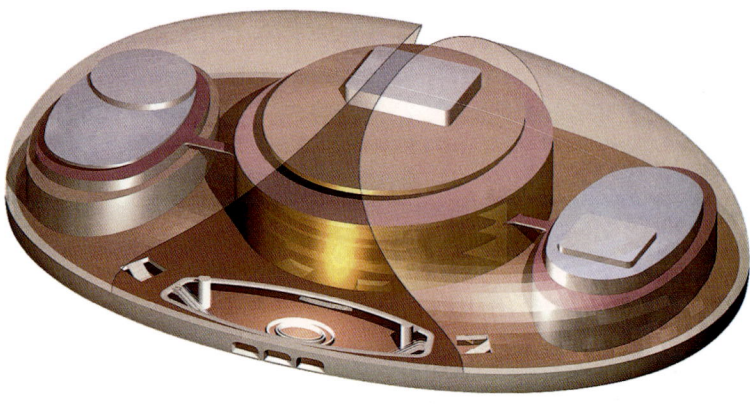

内部空间剖视
AXONOMETRY OF THE INTERIOR VOLUMES

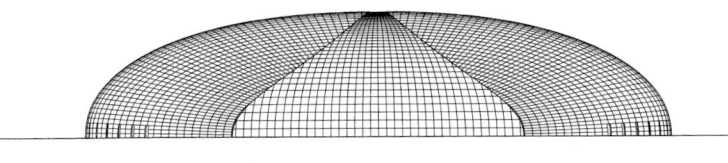

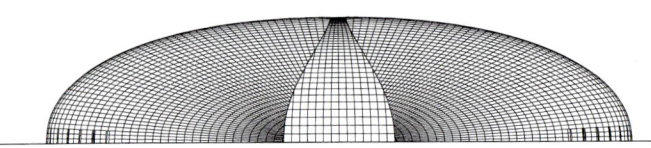

立面
ELEVATION

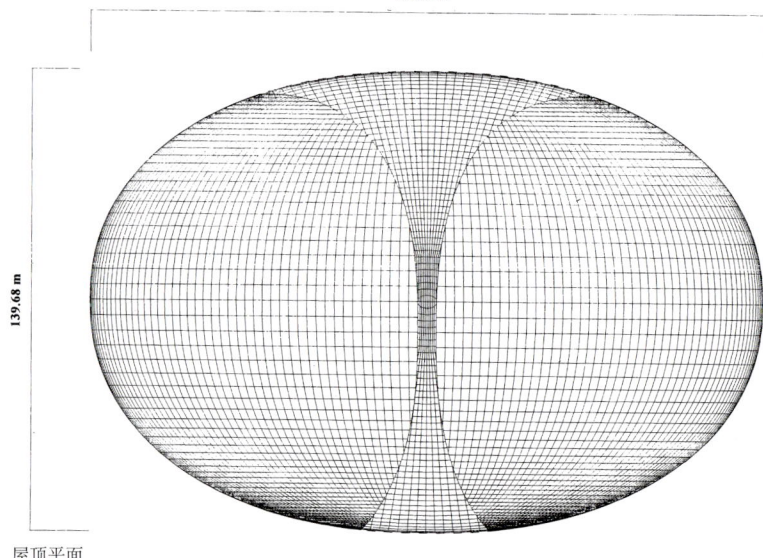

210.00 m

139.68 m

屋顶平面
ROOF PLAN

各观众厅数据
Data for Auditoriums

		歌剧院 Opera House	音乐厅 Concert Hall	戏剧院 Theater	小戏场 Mini Theater
观众厅面积(m²) Floor Area of Auditorium		2400	1600	1200	980
观众厅体积(m³) Volume of Auditorium		19500	21500	9000	11000
座席数 Seating Capacity	总座席数 Seating Capacity(Seats)	2500	2000	1200	520
	其中：池座 Among Them:Auditorium	1630	1450	810	300/500
	楼座 Floor Seat	870	550	390	
	休息厅面积(m²) Lounge Floor Area	4400	4000	3200	1462
舞台尺寸 Stage Dimentions (m²)	主舞台 Main Stage	970	480	570	700
	左右侧台 Left and Right Side Stages	1200	300	640	
	后舞台 Back Stage	680		320	
	台仓 Under Stage Storage	1600	200	570	960
	升降乐池 Elevating Orchestra Pit	120	320	75	
最大视距(m) Max. Visual Distance		40	33	26	
最大俯角(°) Max. Angle of Depression		28	22	30	

国家大剧院总面积
Grand National Theatre Areas Breakdown

场地北部　North Side Site

停车场　Parking lots	20000m²
商业面积 (包括入口、自行车停放处及留给人民大会堂使用的5000m²，如要，此商业面积可改成停车场）。 Commercial Area (including main access, bicycle stalls, and 5000m² for utility of people's Hall; if needed commercial areas can be transformed in Parking Areas).	16000m²

国家大剧院　Grand National Theatre

地下层 (化妆室、贵宾入口、贵宾停车场及贵宾室等等)。 Underground Levels(dressing rooms, VIP park lots, VIP access and lounges, etc).	37000m²
主入口层(入口厅、舞台、摊位、服务通道) Main Access Level (entrance halls, stages, stalls, service road)	29500m²
主厅层　General Hall Level	20000m²
休息厅、展览厅、排演室 Foyers/Exhibition rooms/ Rehearsal rooms	16000m²
总面积　Total floor area	102500m²

上报候选方案（2 项）
Candidate plans for the higher level's approual (two items)

塔瑞·法若建筑师事务所(英国)

Terry Farell & partners (UK)

北京市建筑设计研究院(中国)

Beijing Institute of Architectural Design & Research (China)

清华大学(中国)

Tsinghua University (China)

法国巴黎机场公司协助(法国)

Joint collaborator: ADP Aeroports de Paris(France)

塔瑞·法若建筑师
事务所（英国）
Terry Farrell &
Partners (UK)
北京市建筑设计
研究院（中国）
Beijing Institute of
Architectural Design &
Research (China)

纵深化：线性组织

北京的城市格局强烈地体现了中国注重南北朝向的传统。这种格局使小尺度的庭院胡同服从于由故宫为主构成的延续十几公里的巨大城市尺度的轴线，形成极其明显的几何形。

但是这种格局不是西方古典主义风格的轴线，而是一种纵深化的格局，就是将建筑和院落按南北向排列，并突出建筑的南北立面，这种严整序列的美已经贯穿于我们方案的各个部分。这样，我们的建筑就成为北京城市肌理的一个重要部分，同其血脉一起融合成巨大的肌体。对天安门广场建筑群中的任何一个新部分，其结构和肌理的整合性至关重要，其它不同的几何体都将是外来的，不相关的，无法成为这一整体的一部分。

此外最重要的是，每个小几何体也应是非常完美的部分，这大大地启发和丰富了我们的方案。从整体来看，我们的方案分北部公园、建筑本身和南部公园三大部分，它们由一条完全传统的南北向轴线贯穿始终。沿着轴线漫步，人们可以领略和体验到建筑各个部分的不同魅力。不仅能体验到南北方向的轴线，而且还能体验到中心在向各个方向开放。

基座和屋顶

除了施工和功能方面的优点之外，中国传统建筑受到广泛认同的两个部分是基座和屋顶，它表达了两个世界、两种精神。

基座，理性而严整，代表了地上人间的世俗世界；屋顶，浪漫而抒情，代表了空中仙境的超俗世界。

我们不照搬传统形式，也不抛弃其精髓，我们确信我们找到了一种表达这两个部分的方式，它既忠实于任务书的要求，又使建筑充满了现代感和前瞻性。

理性的基座

我们的方案将开敞的休息大厅放在主入口一侧，剧场后台部分由石材面柱墙形成封闭的围合，虚实结合的长方形几何体是我们前几轮方案的鲜明特征。然而，在上一轮方案中，我们在基座之上屋顶之下增加了一层上人平台（如同天安门城楼），它形成了我们方案中崭新的令人激动的，最具有原创性的部分。观众可登上24m高的基座顶，如同登上了景山山顶，北京城市中心的壮阔景观尽收眼底。人们既可以从建筑内的电梯和自动扶梯登上基顶，也可以从建筑东西两侧的大台阶爬上基座，以体验登高望远的情趣。

巨大的休息厅

中国传统的大型公共建筑和公共空间都有着巨大而丰富的尺度。从一开始我们就设想创造一个世界最大的休息厅，而它又不应有罗马圣保罗教堂、罗马万神庙等西方巨大室内空间那种封闭的感觉。我们经常为室内外连续的空间感所激动，而这正是中国传统建筑的精髓所在。因此，从完全开放的地面广场到玻璃和柱框围成的巨大休息厅，到每个观台厅四周的挑廊和内部休息廊，再进到完全封闭的充满艺术变幻的观众厅中，这就形成了丰富多变的空间序列。

每个表演空间都有一个开敞的外部升起平台作为观众聚集汇合的场所，它们也可以作为额外表演的场地。巨大的休息厅也是一个戏剧化的空间，来往穿梭的人流以及非正式的表演，其本身就充满了戏剧性效果。大休息厅还可以灵活分割，连同各观众厅一起单独使用，形成有单独出入口的空间。这些想法大都在过去的报告中已有陈述，而这些想法在方案的各个阶段都贯穿始终。

LAYERING:THE LINEAR ORGANISATION

There is wonderful pattern to the city of Beijing, based on traditional Chinese recognition of the importance of the North/South orientation. This pattern is omnipresent and makes an extraordinarily insistent geometry that underlies the smallest scale of the courtyard hutong to the great city scale axis of the imperial Palace and chang'an Avenue, which reach out many kilometres across the city.

But this pattern is not primarily axial in the western Beaux-Arts style, but is a layering where the buildings and its rooms are laid in blocks running East/West,giving priority to North and South elevations. The beauty of this filigreed order we have let penetrate to all parts of our planning, and in this way i believe the building is critically part of the very tissue of Beijing, with its arteries and veins integrated into the larger body. This integration of structure and fabric is vital i believe for any new part of the Tiananmen complex, as other different geometries would be foreign, alien and not a part of the whole.

Additionally, and most importantly, the underlying geometry is a very beautiful one, and informs and enriches our scheme. In this way at the large scale are three components, the North Park, the Building itself and the South Park, and in the true tradition a great North/South route or axis connects right from one end to the other. travelling down this directly accessible axis takes one on an unfolding journey through all and every part of the experience of the scheme. It is vitally critical to this experience not just that there are East/West layers but that the centre is open all the way.

TWO PART VERTICAL HEIRARCHY: BASE AND ROOF

Going beyond mere construction and functional advantages,Chinese architecture has always recognised these two building elements, base and roof, as having the ability, and presenting the opportunity to express two worlds:two spirits.

One, the base, is rational and ordered,of the ground, the temporal, the real world of human life. The other, the roof, is the romantic, the lyrical, of the air and the sky.

Without in any way being formally derivative and completely rejecting pastiche, I believe we have achieved a composition expressing these two elements in a way wholly truthful to the programme and with the resultnt building that is contemporary and forward looking.

THE RATIONAL BASE

The design is layered with open foyers to the entry side and solid stone plinth/walls enclosing the rear theatres and backstage. Part solid, part open, the rectilinear geometry is very clear and consistent with all our earlier stage submissions. However, at the last stage we added a level flat table top platform below the roof, which is like the tops of the base of major Chinese buildings(like Tiananmen Gate for example).This platform provides a spectacularly new and exciting and highly orginal element to our scheme. It provides a completely new dimension as the public can rise up to 24M(80 feet high)above the roof tops, to see form an expansive viewing platform in this historic heart of Beijing. It adds a dimension like the hill as (name to be researched) at the north of the Forbidden City, which is so enormously popular with visitors. It is accessible from great staircases on the East and West

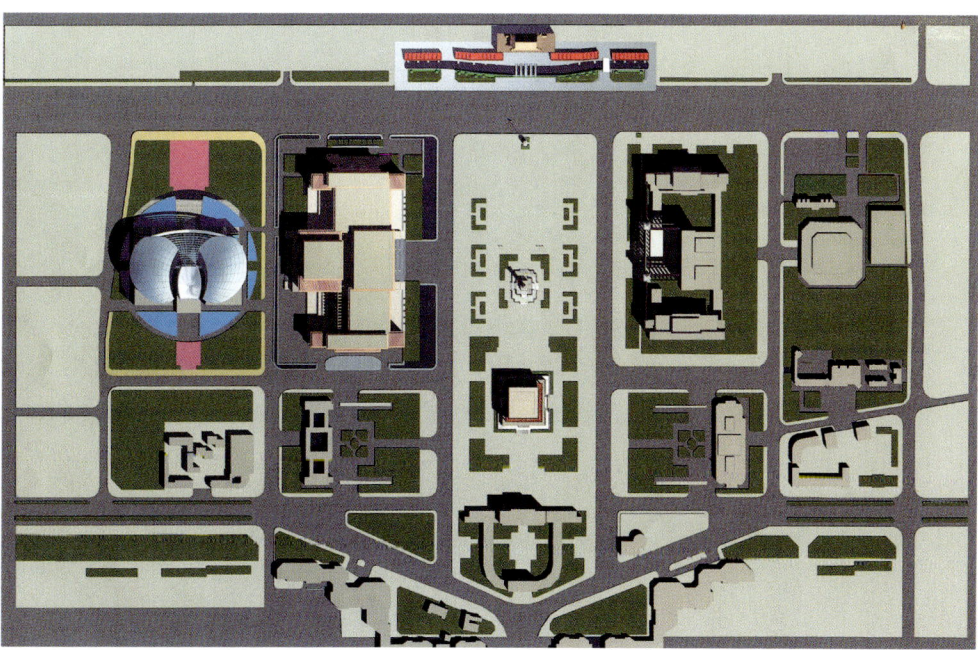

鸟瞰图
BIRD'S-EYE VIEW

sides,which will be adventurous and fun to rise up.(But there are also lifts and escalators within the building for those who prefer easier means of access.)

THE GREAT FOYER

Great Chinese public buildings and public spaces reflect a truly large and generous scale. Form our very first thoughts we wanted to create a foyer that would be one of the great internal spaces of the world. Yet it should not have a sense of being enclosed like great western internal spaces such as St. Peter's Cathedral Nave in Rome, or the Pantheon in Rome. Rather I have always been thrilled by the spatial continuity of outside to inside that is the essence of traditional architecture haere. So there is a changing sequence from the very openness of the external northern square to the glass and columned frame construction grand foyer, to the balconies and internal lobbies of each auditorium,and then to the genuinely enclosed world of fantasy and art, of make believe and magic within the auditoriums themselves.

Each performance space has an open external raised platform to gather and meet and mingle. On these open levels additional performances can be held. The great Foyer is a theatrical space of people arriving and leaving and meeting, with informal performances, but it can also be subdivided logically and efficiently to give flexibility of use, with each venue separately accessed. Much of this thinking is described in our original reports as our intentions have been consistent here through all stages.

天安门广场总平面
GENERAL PLAN OF THE TIAN'AN MEN SQUARE

北侧透视
PERSPECTIVE OF NORTH
ELEVATION

24m 屋顶平台景观
VIEW FROM DECK ROOF (+24.0m)

室内大厅透视
PERSPECTIVE OF INDOOR LOBBY

草图
SKETCH

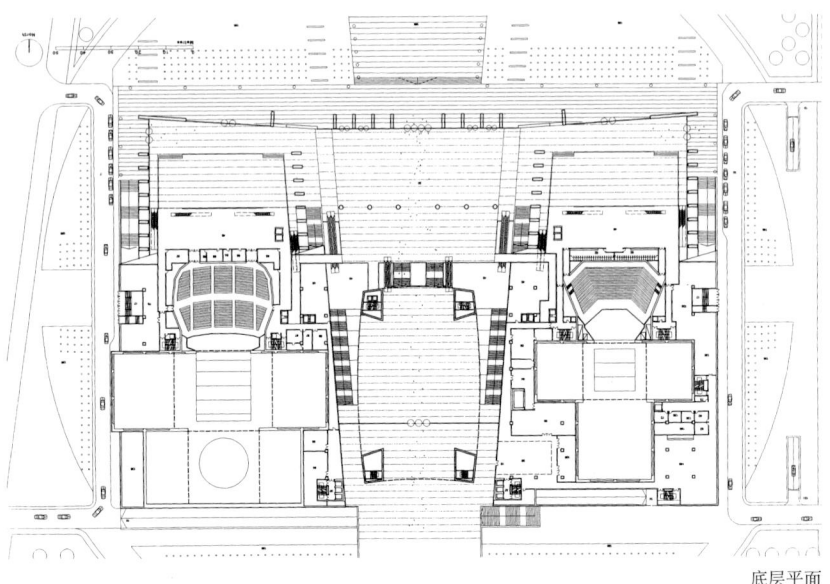

底层平面
GROUND FLOOR PLAN

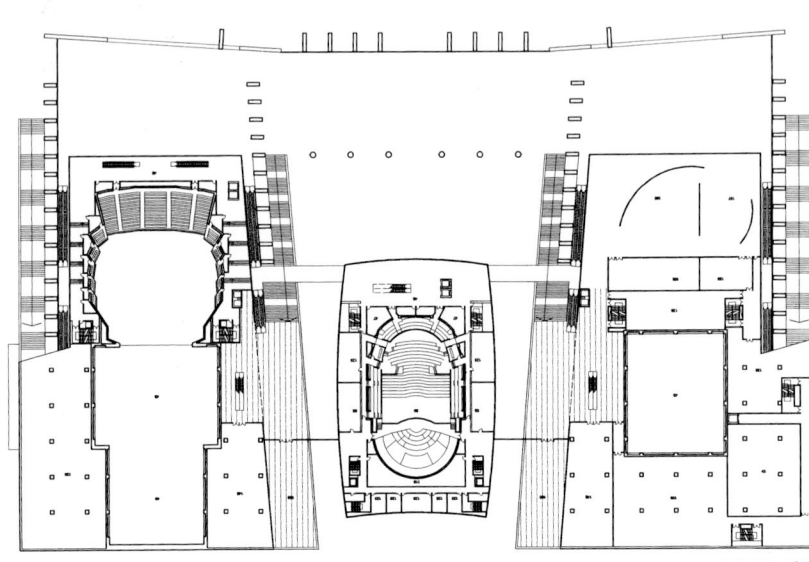

+15.00m平面
LEVEL (+15.0m)

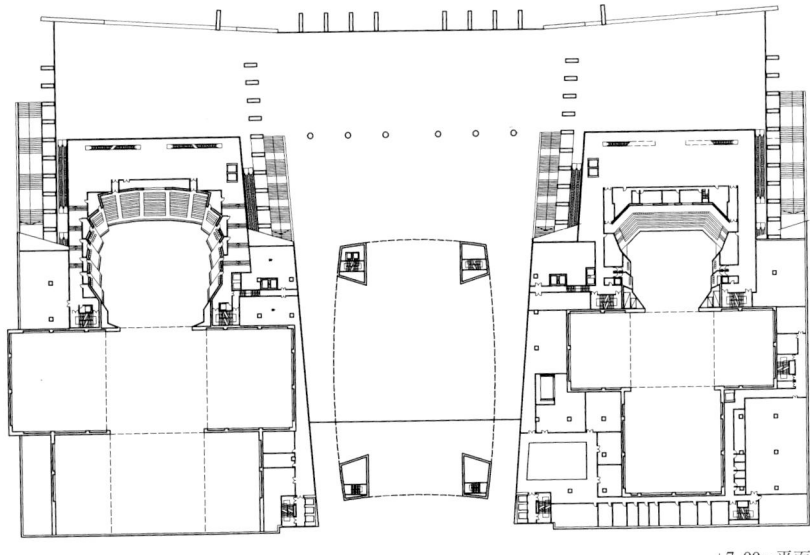

+7.00m平面
LEVEL (+7.0m)

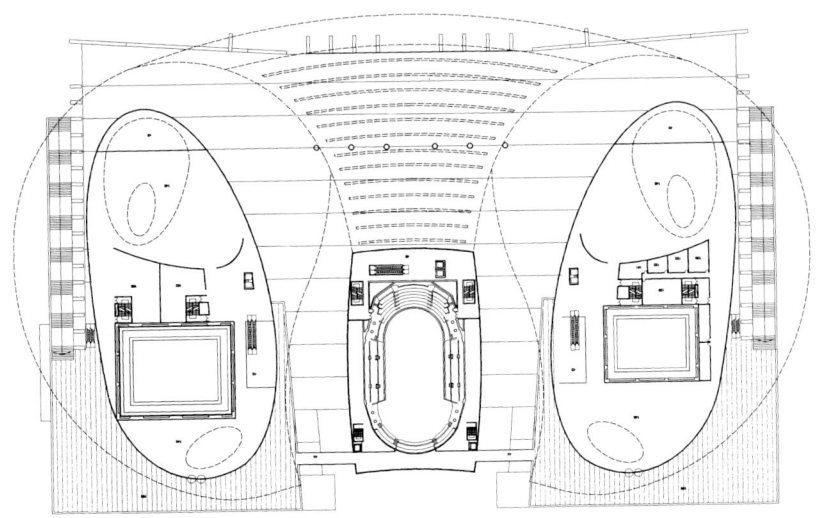

+24.00m平面
LEVEL (+24.0m)

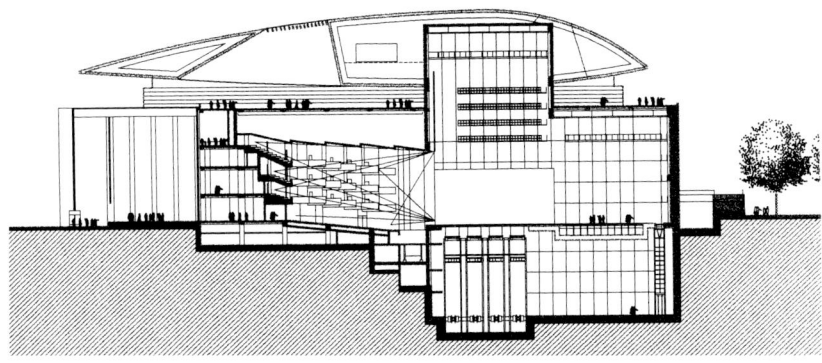

A–A 剖面
SECTION A-A

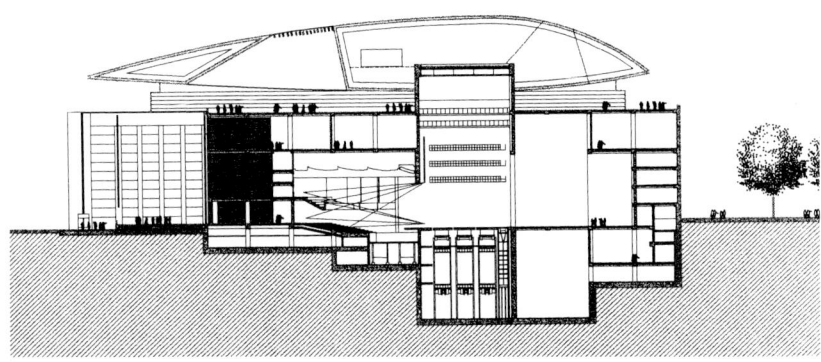

C–C 剖面
SECTION C-C

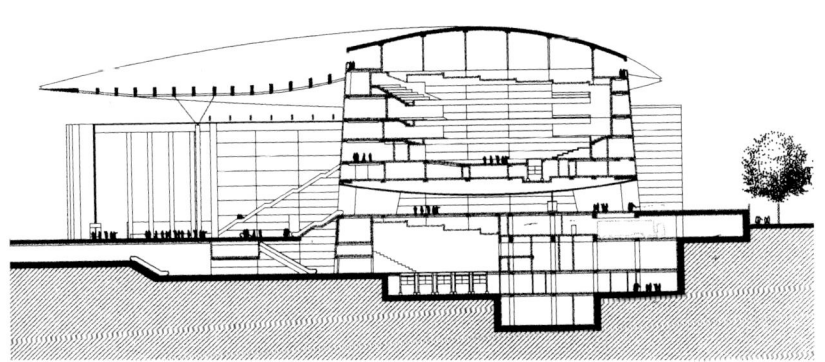

D–D 剖面
SECTION D-D

技术经济指标
Technical Economic Index

基地面积(m²) Site Area	总建筑面积(m²) Total Floor Area	建筑高度(m) Building Height	建筑覆盖率(%) Building Coverage Ratio
37184	地上：70341 地下：33414	30/45(局部)	42.2
容积率 Plot Ratio	绿化覆盖率(%) Green Coverage Ratio	停车数量 Number of Stalls	自行车停车数量 Number of Bicycle Stalls
1.89	14.3	地上：80 地下：100	600

各观众厅数据
Data for Auditoriums

		歌剧院 Opera House	音乐厅 Concert Hall	戏剧院 Theater	小戏场 Mini Theater
	观众厅面积(m²) Floor Area of Auditorium	2170	1053	1027	942.5
	观众厅体积(m³) Volume of Auditorium	19000	22113	9310	9425
座席数 Seating Capacity	总座席数 Seating Capacity(Seats)	2500	2211	1200	500
	其中：池座 Among Them:Auditorium	1200	801	872	500
	楼座 Floor Seat	1300	1410	328	
	休息厅面积(m²) Lounge Floor Area	2100	1152	1008	420
舞台尺寸 (m²) Stage Dimentions	主舞台 Main Stage	36 × 27	26 × 18	27 × 21	30 × 31
	左右侧台 Left and Right Side Stages	22 × 27		17 × 21	
	后舞台 Back Stage	26 × 25		27 × 26	
	台仓 Under Stage Storage	2406	200	1383	1092
	升降乐池 Elevating Orchestra Pit	26 × 6	24 × 16	21 × 11	
	最大视距(m) Max. Visual Distance	44	40	27	15
	最大俯角(°) Max. Angle of Depression	29.2	25.5	22.4	23.0

屋顶平面(第二次修改)
ROOF PLAN
(SCHEM OF THE 2nd MODIFICATION)

北立面(第二次修改)
NORTH ELEVATION
(SCHEM OF THE 2nd
MODIFICATION)

南立面(第二次修改)
SOUTH ELEVATION
(SCHEM OF THE 2nd
MODIFICATION)

东立面(第二次修改)
EAST ELEVATION
(SCHEM OF THE 2nd
MODIFICATION)

清华大学(中国)
Tsinghua University
(China)
**法国巴黎机场公司
(法国)协作**
Joint Collaborator:
ADP Aeroports de
Paris (France)

本方案系根据领导小组的指示,将国家大剧院南移至正对人民大会堂东西向轴线的位置上。设想在宽阔如茵的草坪上、碧波晶莹的瑶池中,浮现出一座玲珑剔透,又凝重华美的艺术殿堂,不但能大幅度改善、提高首都中心区的生态环境质量、造福子孙后代、且在总体布局上会更烘托出天安门广场不凡的宏伟气势,与迈向21世纪我国政治、经济、文化的国际地位更相匹配。

不言而喻,为体现上述精神,国家大剧院的造型既应反映21世纪飞速发展的新、高科技之时代特征,还应反映有5000年历史积淀、博大精深的我中华文化的艺术底蕴。让世人看到:她既是世界的,又是中国的;既是面向未来的,又十分珍惜传统文化中之精髓与气韵,以显现出炎黄子孙的才智生生不息、无穷无尽的演化、融合、延续与创新。

本方案的格局与造型是借鉴了中国传统图案中圆形与方形交替相套的母题而发展出来的。平面布局是:在圆形环廊中套了个十字形平面。而圆形环廊又被套在一个正方形水池中,环廊架空在水面上。方形水池四角设有四根16m高的图腾柱。北侧临长安街、南侧临人大会堂南侧路均有宽阔的大草坪。临南北轴线方向并设置矩形的大水池,水池中有音乐喷泉。水池两侧矗立着历史上最著名的戏剧家与音乐家的雕像。十字形平面内以歌剧院居中,戏剧场在西,音乐厅在东,公共大厅面北,实验小剧场朝南,各占一方。布局主次分明,自然合理,互不干扰而又相互联结;合而不同、分而不隔,各有各的特色。各自均能独立开放,合在一起又能形成有统一秩序感的表演艺术中心。

本方案中的大部分功能使用问题均已有所考虑,在方案深入阶段均能妥善解决,不在此赘述。

According to the requirements of the leadership group of the National Grand Theater, in this design proposal the Grand Theater is set back to the south side right on the east-west axis of the people's Great Hall. The theater is designed as an exquisitely carved, superb and magnificent art palace built with open wide green spaces and azure glittering and translucent pools It will improve ecological environment of capital central district and bring benefit to posterity. In the general layout, the Theater will serve as a foil of vigorous and grand Tian'anmen Square, which matches the international political, economic and cultural position of china towards the next 21st century.

To embody the above spirlt, the appearance of the National Grand Theater should not only reflect the feature of high technology of the 21st century,but also exhibit broad and deep chinese cultural art with 5000years history. The National Grand Theater belongs to china and also belongs to the world, which should cherish the prime of traditional chinese culture and also look on the future.

The general layout of this design proposal adapts the Chinese traditional pattern of 'circle and square' for reference and transfer the pattern with new conception. In the general plan layout, a cross type plan is designed within a circle type corridor, which is located at a square type pool. The circle type corridor is built overhead above the pool. At the four corners of the square type pool, four 16-meter high totem columns are setup. Open wide lawns are designed at the north side along Chang'an Avenue and at the south side along the Great Hall South Road. Along the south-north axis, open square thpe pools are set up with music fountains and the statures of tamous thespians and musicians in history stand tall and upright over the pools. In the cross type plan layout the Opera House is located at the middle, the Theater and the Concert Hall are set up in the west side and the east side of the building. The public entrance hall is set up at the north and the Mini Theater faces to the south. The Opera House, the Theater, the Concert Hall and the Mini Theater are functioning independently, they can be used separately and can still be connected as one unite performance art center for easy management.

北侧透视
PERSPECTIVE OF NORTH ELEVATION

鸟瞰图
BIRD'S-EYE VIEW

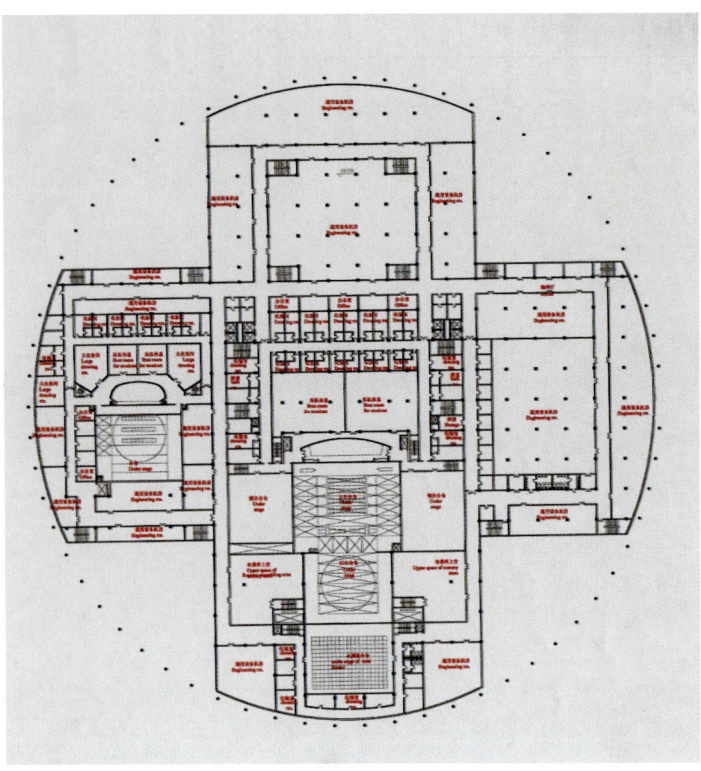

地下一层平面
BASEMENT PLAN

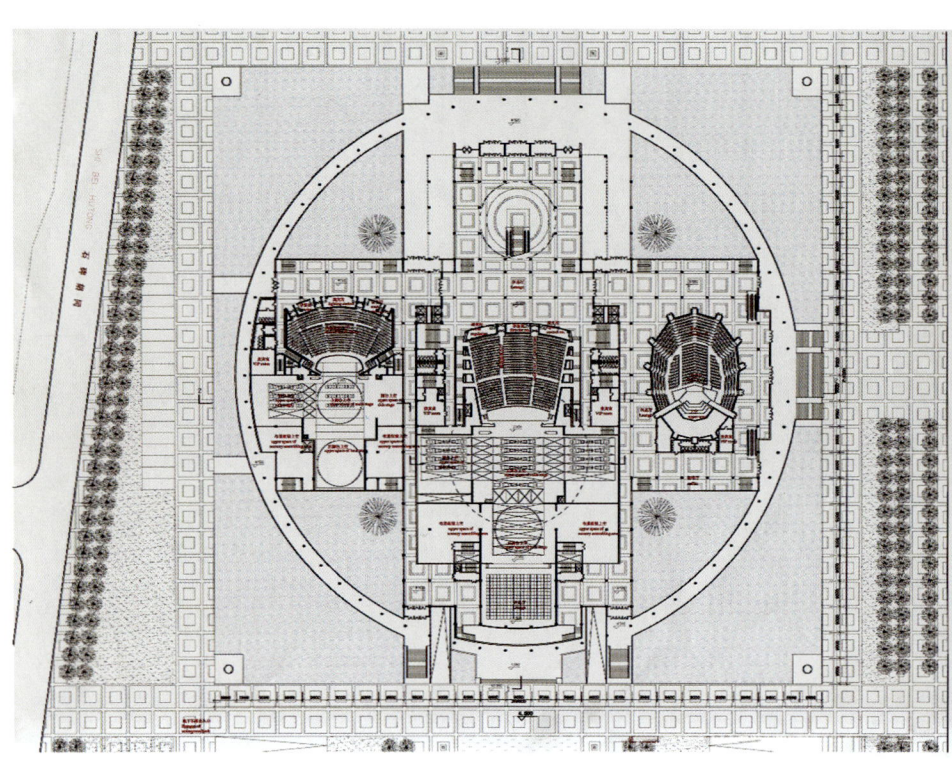

一层平面
GROUND FLOOR PLAN

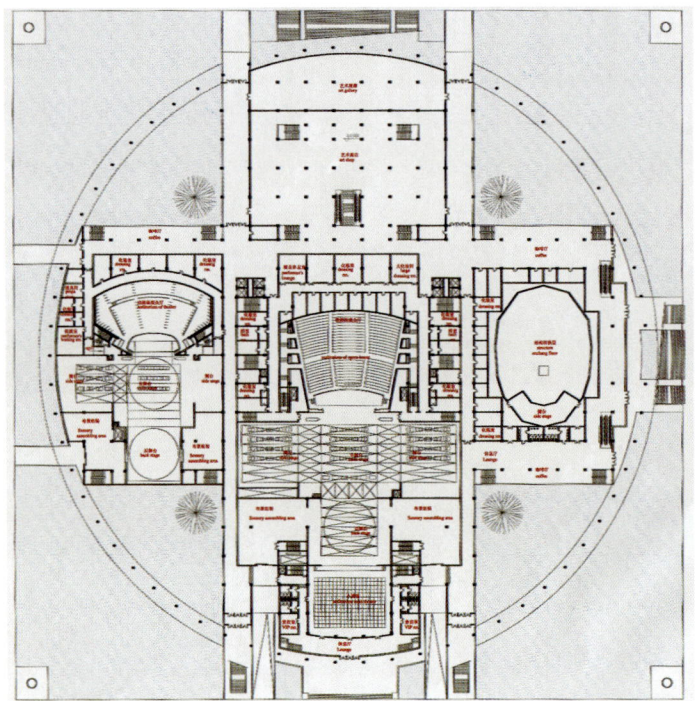

地面层平面
2nd FLOOR PLAN

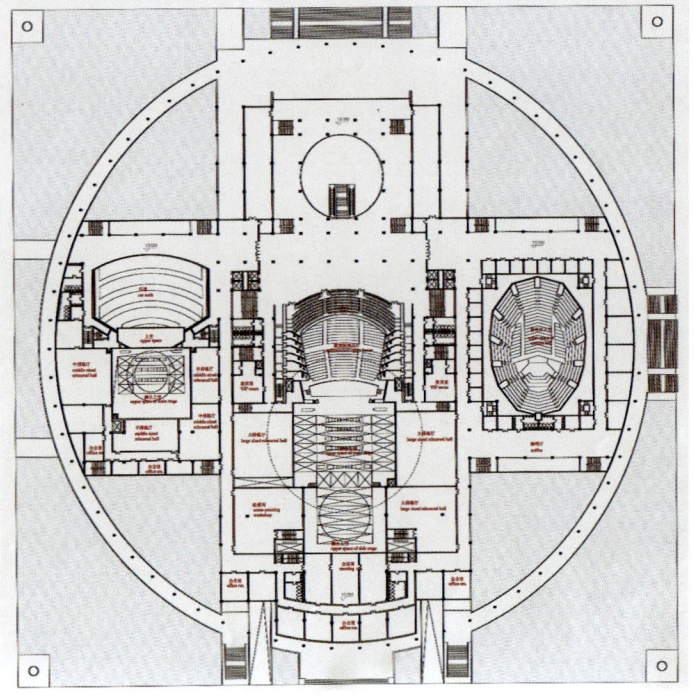

三层平面
3rd FLOOR PLAN

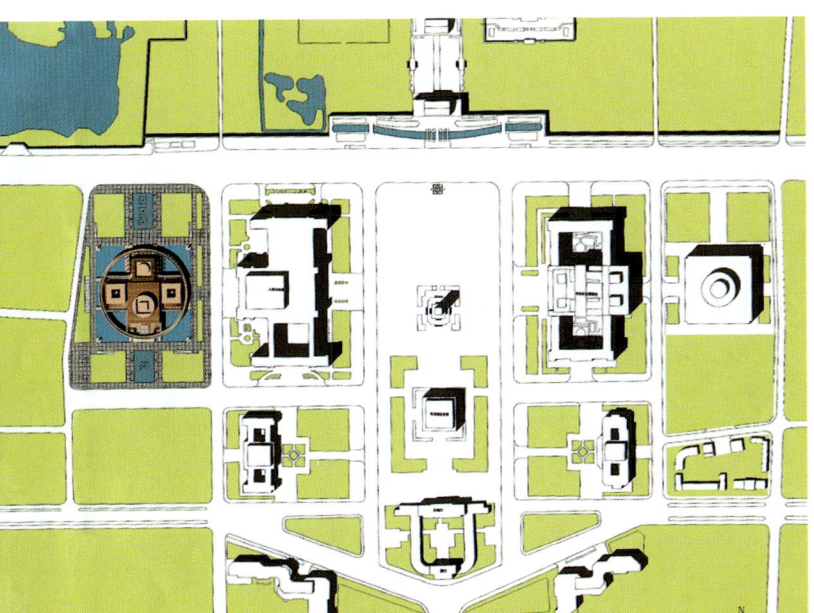

总平面
OVERALL SITE PLAN

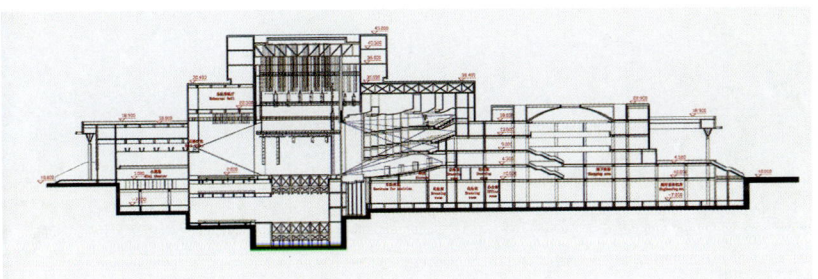

1—1 剖面
SECTION 1-1

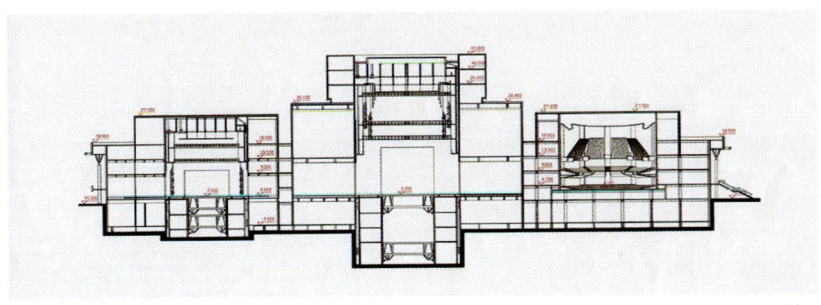

2—2 剖面
SECTION 2-2

技术经济指标
Technical Economic Index

基地面积(m²) Site Area	总建筑面积(m²) Total Floor Area	建筑高度(m) Building Height	建筑覆盖率（%） Building Coverage Ratio
	129899		

容积率 Plot Ratio	绿化覆盖率(%) Green Coverage Ratio	停车数量 Number of Stalls	自行车停车数量 Number of Bicycle Stalls
		地上：20 地下：976	1000

各观众厅数据
Data for Auditoriums

		歌剧院 Opera House	音乐厅 Concert Hall	戏剧院 Theater	小戏场 Mini Theater
	观众厅面积(m²) Floor Area of Auditorium	2000	1200	1070	574
	观众厅体积(m³) Volume of Auditorium	19800	20700	5995	4018
座席数 Seating Capacity	总座席数 Seating Capacity(Seats)	2518	2006	1209	300-500
	其中：池座 Among Them:Auditorium	1385	1082	889	可变成 各种方式
	楼座 Floor Seat	1133	924	320	
	休息厅面积(m²) Lounge Floor Area				
舞台尺寸(m²) Stage Dimentions	主舞台 Main Stage	33 × 27	16 × 12	26 × 21	
	左右侧台 Left and Right Side Stages	左：24 × 20 右：24 × 20		左：18 × 17.7 右：9.1 × 17.7	
	后舞台 Back Stage	27 × 20.8		18 × 11	
	台仓 Under Stage Storage	主：33 × 27 后：25 × 20.8		主：26 × 21	
	升降乐池 Elevating Orchestra Pit	22 × 6 可升降部分		15 × 6 可升降部分	
	最大视距(m) Max. Visual Distance	36	24	24	
	最大俯角(°) Max. Angle of Depression	25	16	27	

第一轮竞赛方案(44项)

方案编号	设计单位中文名称	方案编号	设计单位中文名称
101	法国巴黎机场公司(法国)	306	清华大学建筑设计研究院(中国)
102	法国建筑工作室(法国)	307	东南大学建筑设计研究院(中国)
103	欧博迈亚设计咨询有限公司＋戴尔曼教授建筑设计事务所(德国)	308	东南大学建筑设计研究院(中国)
		401	竹中工务店株式会社(日本)
104	北京市建筑设计研究院(中国)	402	亨克－施利克－陈建筑师事务所(奥地利)
105	北京市建筑设计研究院(中国)	403	清华大学建筑设计研究院(中国)
106	塔瑞·法若建筑师事务所(英国)	404	豪斯保尔建筑师事务所(奥地利)
107	上海现代建筑设计(集团)有限公司(中国)	405	东南大学建筑设计研究院(中国)
108	上海现代建筑设计(集团)有限公司(中国)	406	格里高里国际公司(意大利)
201	矶崎新建筑师事务所(日本)	407	许李严建筑师有限公司(中国香港)
202	卡洛斯·奥特建筑师事务所(加拿大)	408	汉斯豪莱建筑师＋海兹诺曼建筑师(奥地利)
203	王欧阳(香港)有限公司(中国香港)	501	HPST 设计事务所(美国)
204	日本设计株式会社(日本) 建设部建筑设计院(中国)	502	山崎实建筑师事务所(美国)
		503	深圳大学建筑设计研究院(中国)
205	建设部建筑设计院(中国)	504	清华大学安地建筑设计顾问有限公司(中国)
206	清华大学建筑设计研究院(中国)	505	安东内奇设计所(意大利)
207	浙江省建筑设计研究院(中国)	506	北京有色冶金设计研究总院(中国)
208	浙江省建筑设计研究院(中国)	507	HPP 国际建筑设计有限公司(德国)
301	刘荣广伍振民建筑师事务所(香港)有限公司(中国香港)	508	北京三磊建筑设计有限公司(中国) 法国建筑家协会(法国)
		509	尼克拉斯建筑师事务所(希腊)
302	让·奴威尔建筑师事务所(法国)	601	设计集团(意大利)
303	广州市设计院(中国)	603	同济大学建筑设计研究院(中国)
304	刘荣广伍振民建筑师事务所(香港)有限公司(中国香港)	604	GFU 建筑与城市设计(美国)
305	中国建筑科学研究院(中国)		

Schemes of the First Round competition (forty-four items)

number of the plan	desighning organi zation	number of the plan	desighning organi zation
101	ADP Aeroports de Paris (France)	401	Takenaka Corporation (Japan)
102	Architecture · Studio (France)	402	Henke-Schreieck-Chen-Architekten (Austria)
103	Design Group:Obermeyer+Deilmann (Germany)	403	Architectual Design & Research Institute of Tsinghua University (China)
104	Beijing Institute of Architectural Design & Research (China)		
105	Beijing Institute of Architectural Design & Research (China)	404	Wilhelm Holzbauer, Architect (Austria)
106	Terry Farrell & Partners (UK)	405	Architectural Design & Research Institute of Southeast University (China)
107	Shanghai Modern Architectural Design (Group)Co., Ltd. (China)		
108	Shanghai Modern Architectural Design (Group) Co.,Ltd. (China)	406	Gregotti Associati International srl (Italy)
201	Arata Isozaki & Associates (Japan)	407	Rocco Design Limited (China Hong Kong)
202	Carlos Ott & Associates, Architects (Canada)	408	Design Group:Prof. Hans Hollein /Architect Heinz Neumann/ Architect Planungsgemeinschaft (Austria)
203	Wong & Ouyang (h.k.) Ltd. (China Hong Kong)		
204	Nihon Sekkei, Inc.(Japan) Architectural Design Institute of Ministry of Construction (China)	501	Hakomori Pali Stenfors Tang (USA)
		502	Minoru Yamasaki Associates, Inc (USA)
205	Architectural Design Institute of Ministry of Construction (China)	503	Shenzhen University, The College of Architecture & Civil Engineering (China)
206	Architectural Design & Research Institute of Tsinghua University (China)		
		504	Ande Architectural Consultant Ltd. Co. (China)
207	Zhejiang Building Design And Research Institute (China)	505	Antonaci Partners (Italy)
208	Zheijiang Building Design And Research Institute (China)	506	Central Engineering & Research Institute For Non-Ferrous Metallurgical Industries (China)
301	Dennis Lau & NG Chun Man Architects & Engineers (H.K.) Limited (Chian Hong Kong)		
		507	HPP International Planungsgesellschaft mbH (Germany)
302	Architecture Jean Nouve l(France)	508	Sunlight Architects & Engineers (China) Partenaires Architects (France)
303	Guangzhou Design Institute (China)		
304	Dennis Lau & NG Chun Man Architects & Engineers (H.K.)	509	Nicolas Petropoulos, Architect Designer (Greece)
305	China Architecture Science Research Istitute(China) Limited (China Hong Kong)	601	Design Group:Studio Valle Progettazioni Percy Thomas Partnership SGA Design Studio
306	Architectural Design & Research Institute of Tsinghua University (China)	603	The Architectural Design & Research Institute of Tongji University (China)
307	Architectural Design & Research Institute of Southeast University (China)	604	GFU Architect Urban & Interiors (USA)
308	Architectural Design & Research Institute of Southeast University (China)		

观众有的乘汽车或步行，汇聚于大街。他们拾步走上气势雄伟的台阶，它伸入到一敞开向城市、拱券式的大台。台阶上空的探照灯迎照他们，他们进入了表演及梦幻的境界。

大剧院既气势磅礴又华贵古朴。它闪光的穹顶将与天安门广场上的著名历史建筑物交相辉映，尽显城市古典建筑之风采。

我们设计的大剧院共有两个主立面，以使它在市中心的巍峨雄姿在任何角度都敞向北京市民。

休息厅是一个由玻璃材料制成的金色的穹顶，高达20多米。顶上覆盖着开有孔洞的金属网罩，它构成了我们设计的主体。

By car, subway or walking, spectators are gathering on the avenue, they climb the long stairways which enter a huge arch of scenery open towards the city. On the top, in the spotlight, they cross the border towards spectacle and dream.

The result is a majestic theater of monumental proportions, richyet simple, crowned by a brilliant dome. It echoes the prestigious buildings which frame Tian An Men Square, in an extension of its classical urban architecture.

The theater has two main fronts because we wanted it to take up a proud stance in the city and not appear from whatever perspective to be turning its back on the people of Beijing.

The golden dome, a 20 meter high structure in glass with an open-work metal covering, crowns the building and is one of the fundamental elements in its composition.

北侧透视 PERSPECTIVE OF NORTH ELEVATUON

瞭望厅透视 LOBBY PERSPECTIVE

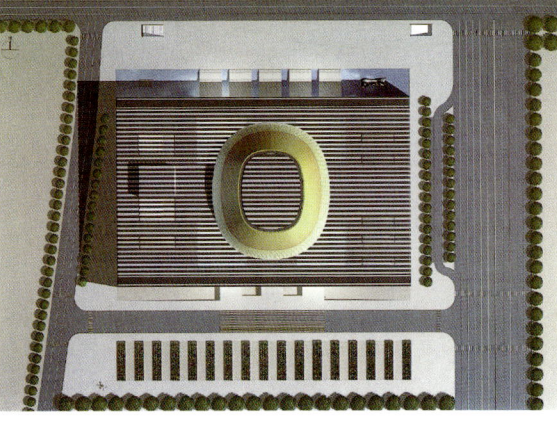

建筑平面及周围环境
OVERALL PLAN & SURROUNDING AREAS

歌剧院观众厅透视
PERSPECTIVE OF THE OPERA HOUSE AUDITORIUM

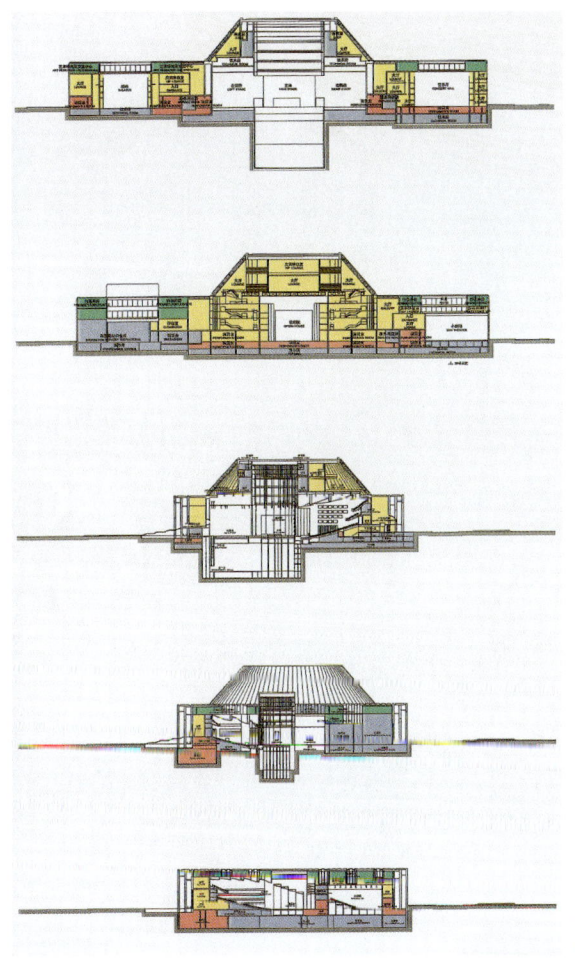

功能分区
FUNCTION ZONE

功能分区
FUNCTION ZONE

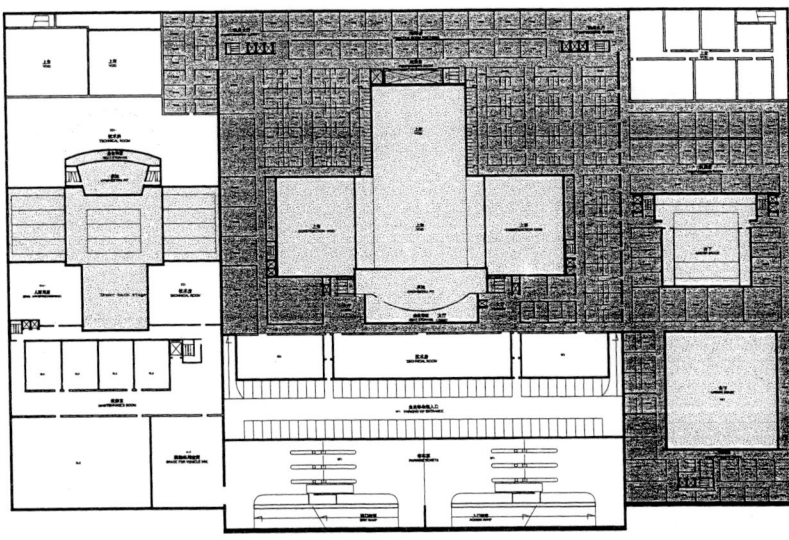

演员化妆层(−4.23m)
PERFORMERS DRESSING-ROOM LEVEL(-4.23m)

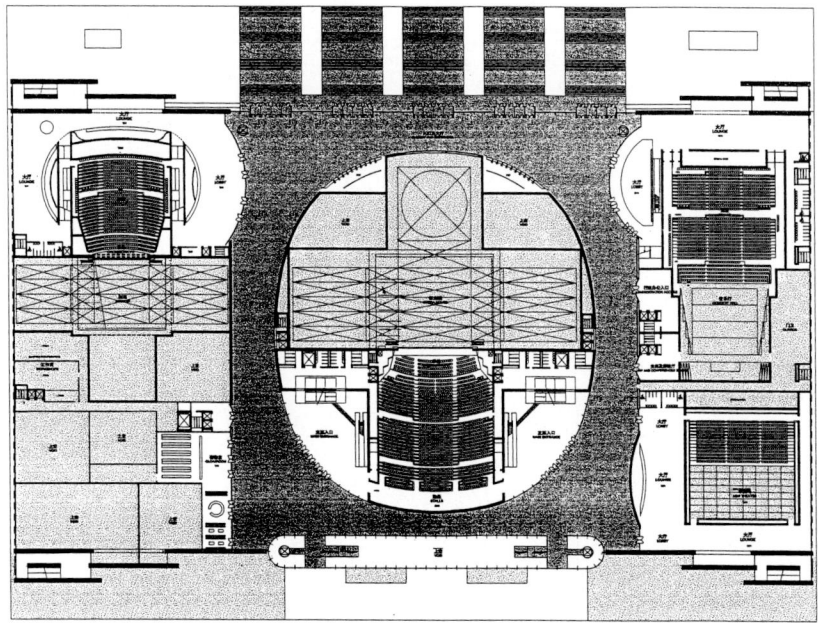

观众入口层(+4.37m)
PUBLIC ENTRANCE LEVEL(+4.37m)

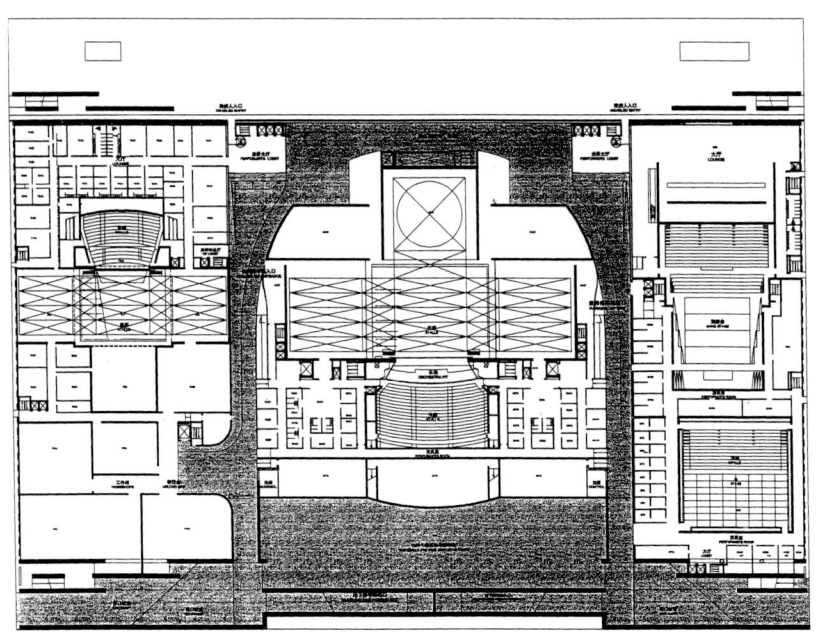

舞台层(−0.63m)
STAGE LEVEL(-0.63m)

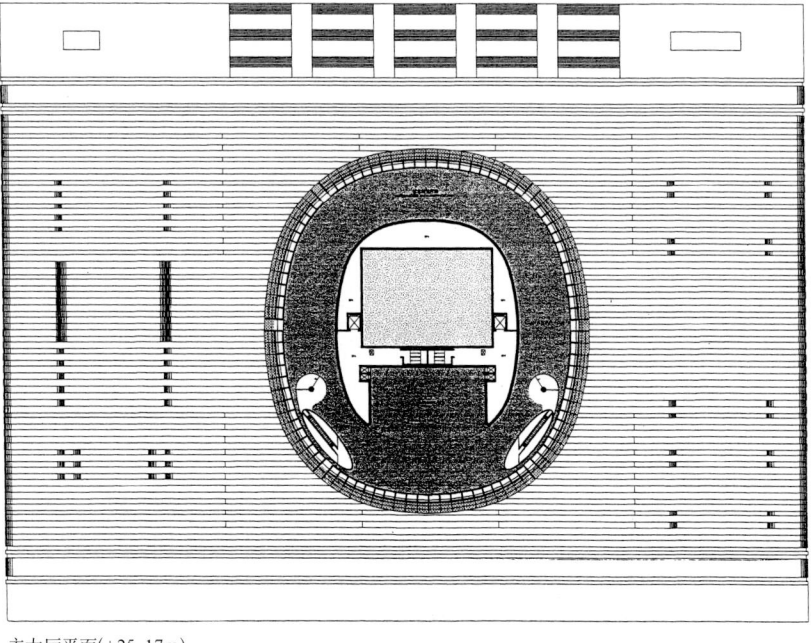

主大厅平面(+25.17m)
MAIN LOBBY LEVEL(+25.17m)

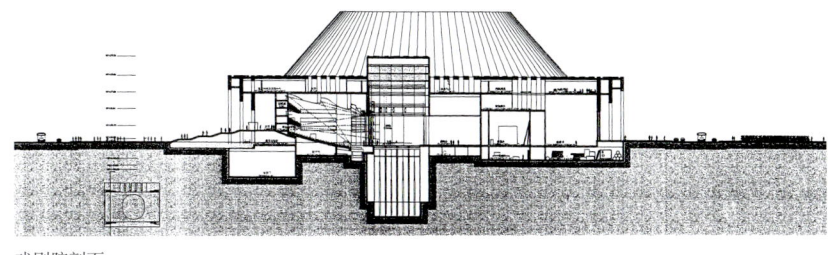

戏剧院剖面
SECTION OF THE THEATER

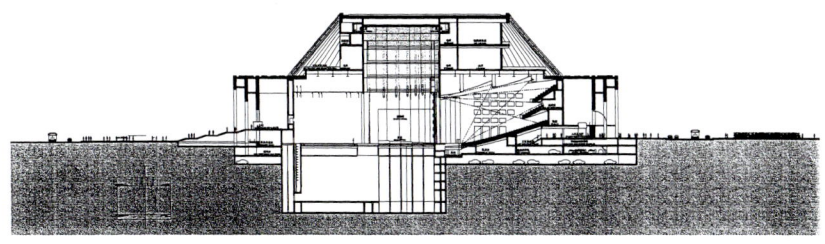

歌剧院剖面
SECTION OF THE OPERA HOUSE

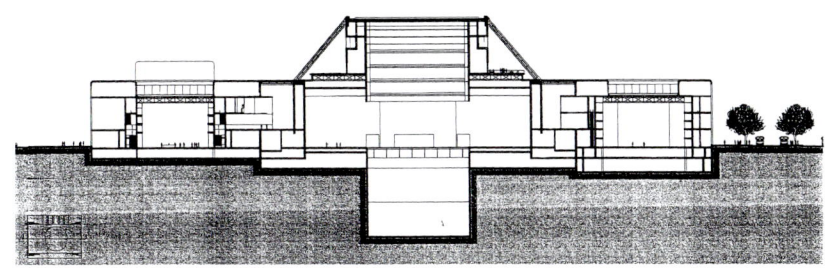

国家大剧院纵剖面
LONGITUDINAL SECTION OF THE NATIONAL GRAND THEATER

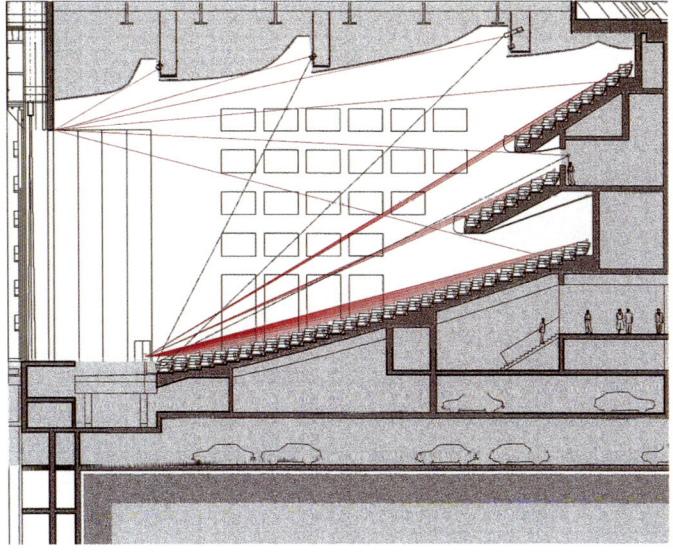

歌剧院视线剖面
VISUAL SECTION OF THE OPERA HOUSE

技术经济指标
Technical Economic Index

基地面积(m²) Site Area	总建筑面积(m²) Total Floor Area	建筑高度(m) Building Height	建筑覆盖率 (%) Building Coverage Ratio
43200	地上：56850 地下：73687	23/29.5/45(局部)	72
容积率 Plot Ratio	绿化覆盖率 (%) Green Coverage Ratio	停车数量 Number of Stalls	自行车停车数量 Number of Bicycle Stalls
3.3		地上：550 地下：20	1200

各观众厅数据
Data for Auditoriums

		歌剧院 Opera House	音乐厅 Concert Hall	戏剧院 Theater	小戏场 Mini Theater
	观众厅面积(m²) Floor Area of Auditorium	2067	1335	842	961
	观众厅体积(m³) Volume of Auditorium	19500	21500	9000	11000
座席数 Seating Capacity	总座席数 Seating Capacity(Seats)	2543	1920	1200	520
	其中：池座 Among Them:Auditorium	1632	1348	840	
	楼座 Floor Seat	910	572	360	
	休息厅面积(m²) Lounge Floor Area	51000	3000	1300	600
舞台尺寸 (m²) Stage Dimentions	主舞台 Main Stage	972	480	67	961
	左右侧台 Left and Right Side Stages	1188	300	640	
	后舞台 Back Stage	676		324	
	台仓 Under Stage Storage	1648	200	576	961
	升降乐池 Elevating Orchestra Pit	116	324	75	
	最大视距(m) Max. Visual Distance	40	38.8	26.40	48
	最大俯角(°) Max. Angle of Depression	30.64	21	33.6	23

　　三座大门，三架拱桥，三道彩虹。

　　这三个建筑实体正如由同一座伸展向天际的石阶承载着的三架拱桥。

　　石阶高台及其各个阶段代表着中国的传统和历史，建筑物便矗立其上。屋顶则如反映蓝天的开敞湖面。

　　如"紫禁城"前的金水桥，三架拱形顶象征性地通向另一个世界，似三条彩虹，连接着和平与未来、传统与现代、东方与西方。连接三座桥的三个大门设立在长安街畔。

　　在这一空间里，通过音乐、戏剧、舞蹈和歌剧，人与宇宙融为一体。

Three doors, three bridges, three rainbows

These three volumes resemble three bridges elevated and held by a single pedestal like stone stairway that rises toward the sky.

This stairway with its various landings expresses the history and tradition of the China on which the whole edifice rests. The roof resembles an open lake mirroring the sky.

Similar to the entrance of the Forbidden City, the three bridges evoke the symbolic passage to another world. Like three rainbows, they associate and link peace and the future, tradition and modernity, and the west and the east. The three doors standing on Chang An Avenue reinforce this trinity.

In this defined space, man synthesizes the universe through the medium of music, drama, dance and opera.

北侧模型照片
NORTH SIDE OF THE MODEL

歌剧院观众厅透视
PERSPECTIVE OF THE OPERA HOUSE AUDITORIUM

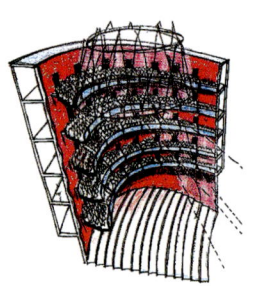

观众厅草图
AUDITORIUM SKETCH

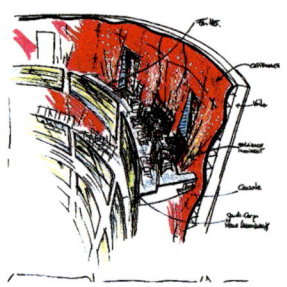

观众厅草图
AUDITORIUM SKETCH

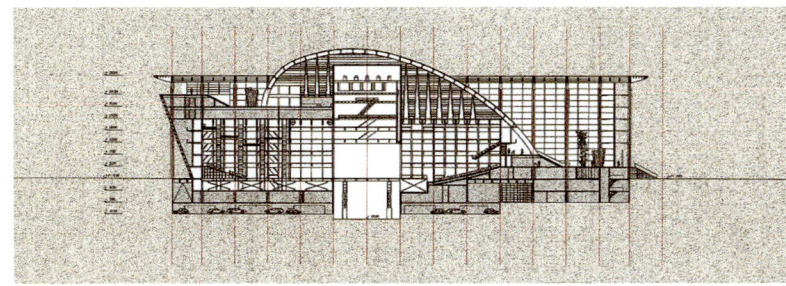

剖面图
SECTION

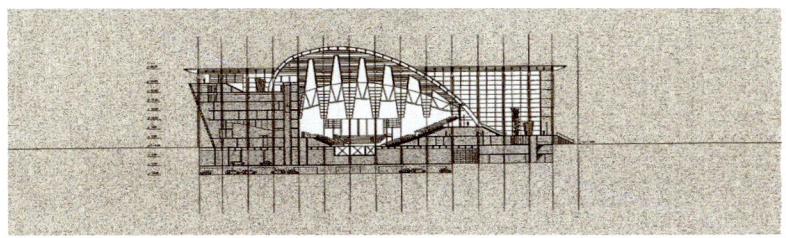

音乐厅剖面
SECTION OF THE CONCERT HALL

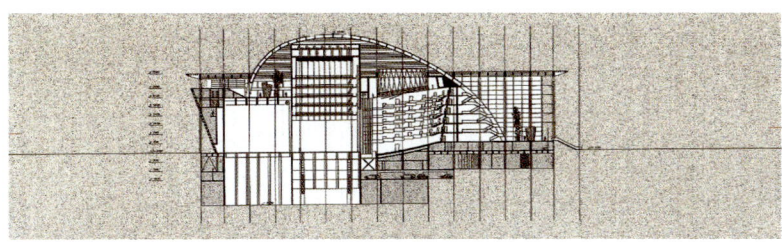

歌剧院剖面
SECTION OF THE OPERA HOUSE

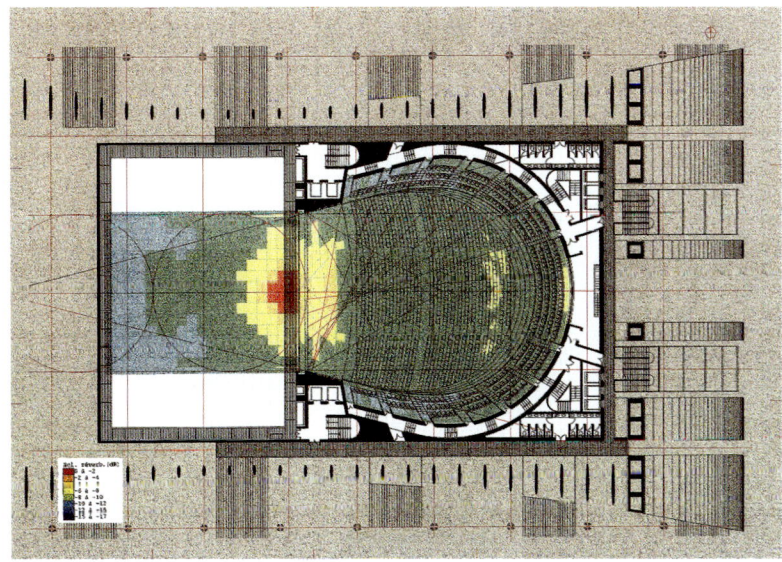

歌剧院声学分析
ACOUSTICAL ANALYSIS OF THE OPERA HOUSE

技术经济指标
Technical Economic Index

基地面积(m²) Site Area	总建筑面积(m²) Total Floor Area	建筑高度(m) Building Height	建筑覆盖率(%) Building Coverage Ratio
38900	地上：78952 地下：52324	30/45(局部)	76
容积率 Plot Ratio	绿化覆盖率(%) Green Coverage Ratio	停车数量 Number of Stalls	自行车停车数量 Number of Bicycle Stalls
3.37	12	地上：50 地下：570	1060

各观众厅数据
Data for Auditoriums

		歌剧院 Opera House	音乐厅 Concert Hall	戏剧院 Theater	小戏场 Mini Theater
	观众厅面积(m²) Floor Area of Auditorium	1840	1857	867	788
	观众厅体积(m³) Volume of Auditorium	19610	22340	12950	11850
座席数 Seating Capacity	总座席数 Seating Capacity(Seats)	2506	2020	1203	456
	其中：池座 Among Them:Auditorium	1064		851	
	楼座 Floor Seat	1442		352	
	休息厅面积(m²) Lounge Floor Area	5005	3037	2950	1595
舞台尺寸 Stage Dimentions (m²)	主舞台 Main Stage	955	400	522	
	左右侧台 Left and Right Side Stages	2 × 542	2 × 278	2 × 300	
	后舞台 Back Stage	637		353	
	台仓 Under Stage Storage	1710	330	522	1153
	升降乐池 Elevating Orchestra Pit	144		195	
	最大视距(m) Max. Visual Distance	33	27	29	25
	最大俯角(°) Max. Angle of Depression	45	20	26	可变

欧博迈亚设计咨询
有限公司＋
戴尔曼教授建筑
设计事务所(德国)
Design Group:
Obermeyer+Deilmann
(Germany)

本方案设计的国家大
剧院，看上去像一个前面
带有"柱廊"极有吸引力
的"艺术殿堂"。从连成一
体的宏大建筑造型可以让
人一目了然地看出其不同
的内涵。北入口区前按建
筑艺术要求设置的露天设
施只是将大剧院同街道交
通分隔开来，而不是完全
阻断。

In its unified overall architectural form-which, how-
ever, also makes comprehensible the differing functions to
be found there the National
Grand Theater as proposed in
this design represents an inviting Temple of the Muses'
with its Urban Loggia portico.
The architecturally designed
open areas fronting the north
entrance will ensure that road
traffic is kept at a distance
without, however, presenting
any hindrance.

北侧透视图
PERSPECTIVE OF NORTH ELEVATION

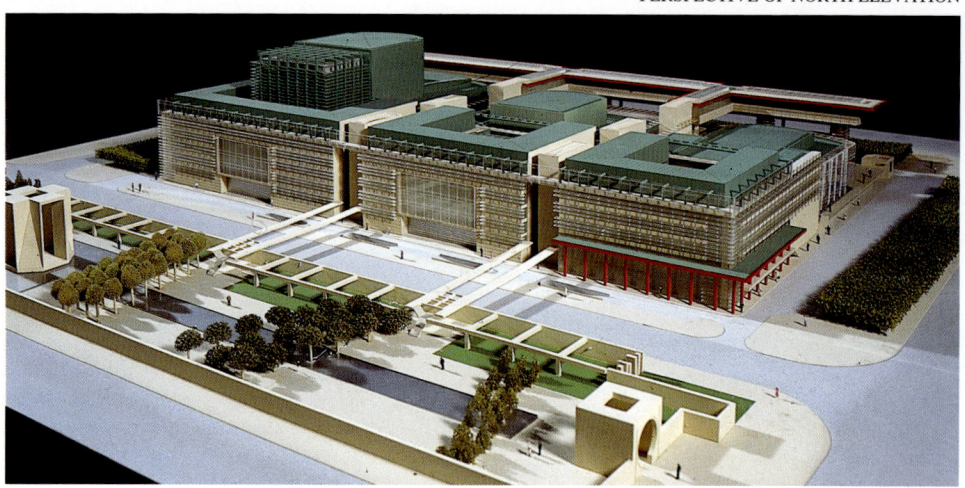

南模透视
PERSPECTIVE OF SOUTH ELEVATION

歌剧院观众厅透视
PERSPECTIVE OF THE OPERA HOUSE AUDITORIUM

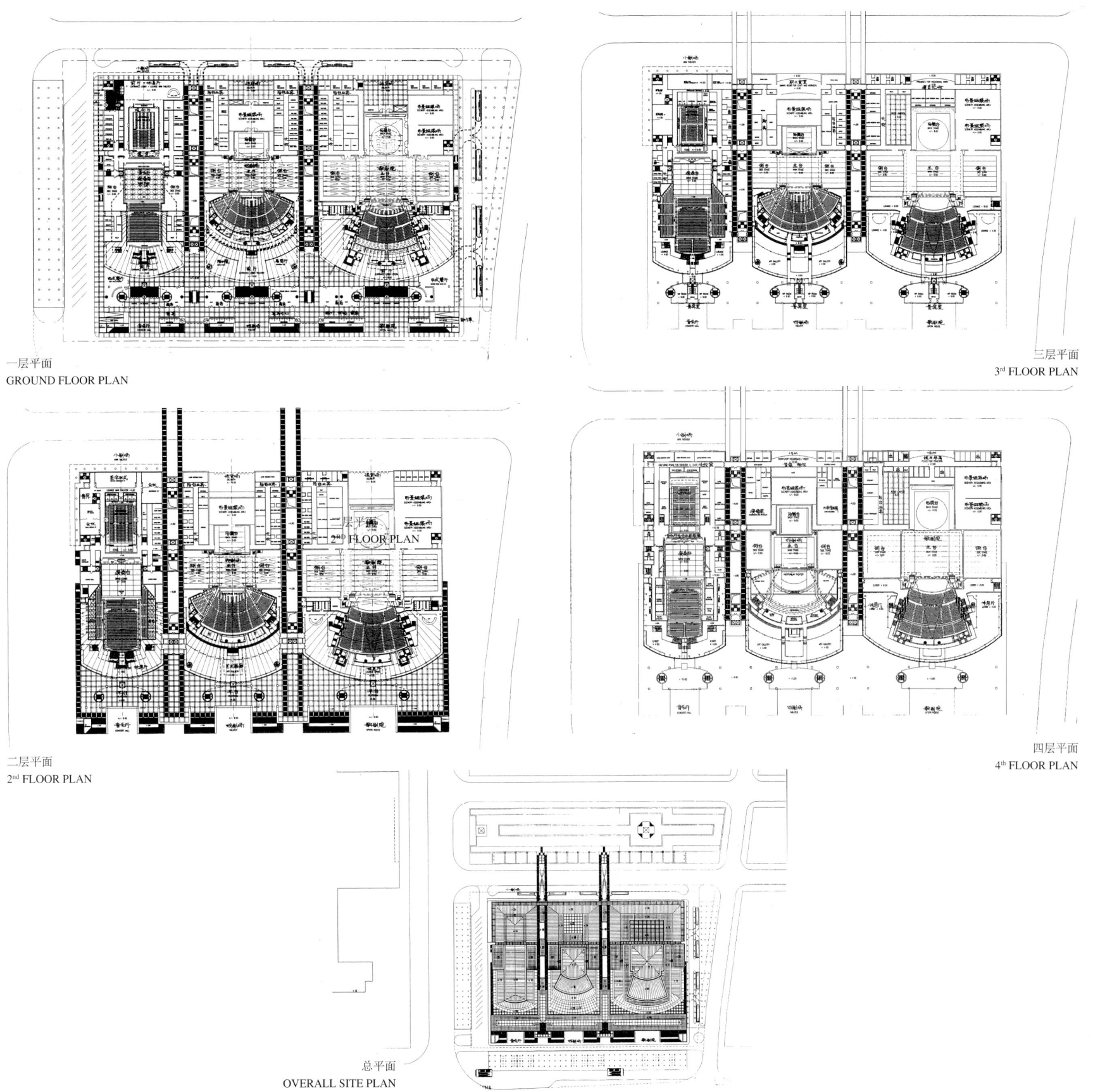

一层平面
GROUND FLOOR PLAN

三层平面
3rd FLOOR PLAN

二层平面
2nd FLOOR PLAN

四层平面
4th FLOOR PLAN

总平面
OVERALL SITE PLAN

87

大厅透视
AUDITORIUM PERSPECTIVE

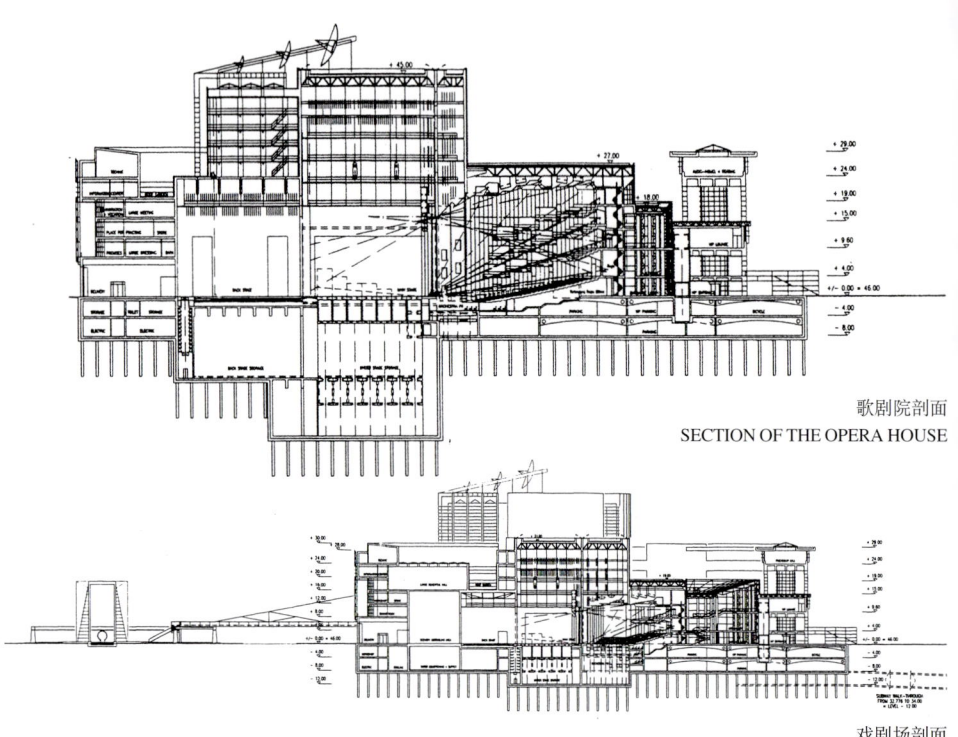

歌剧院剖面
SECTION OF THE OPERA HOUSE

戏剧场剖面
SECTION OF THE THEATER

小剧场和音乐厅剖面
SECTION OF THE MINI-THEATER & THE CONCERT HALL

戏剧场台唇变化
VARIATIONS OF THE THEATER PROSCENIUM

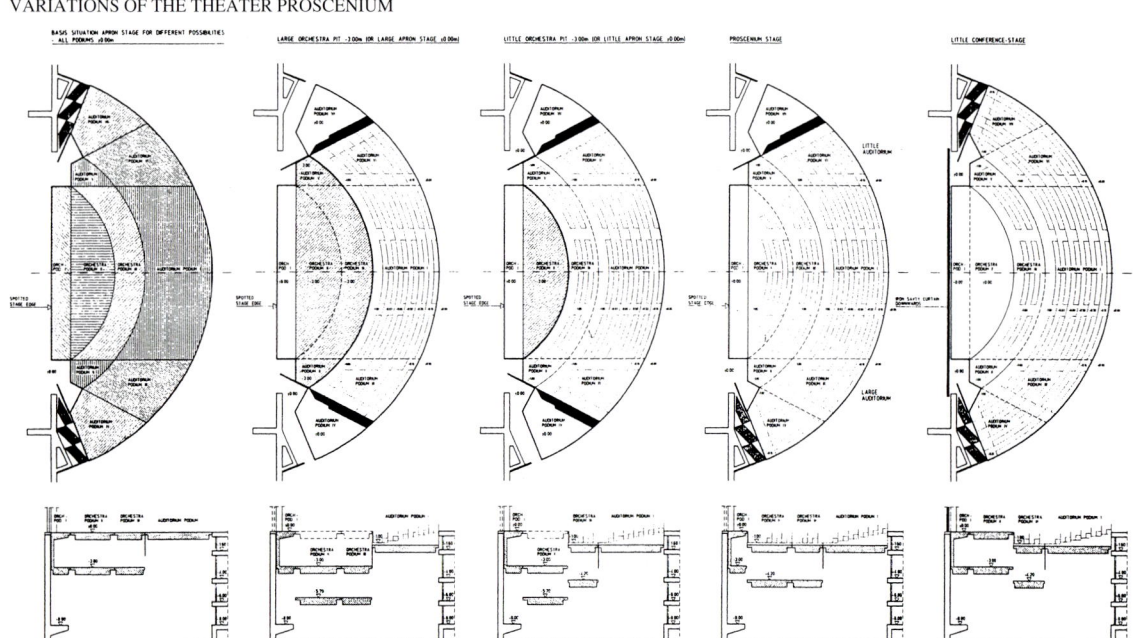

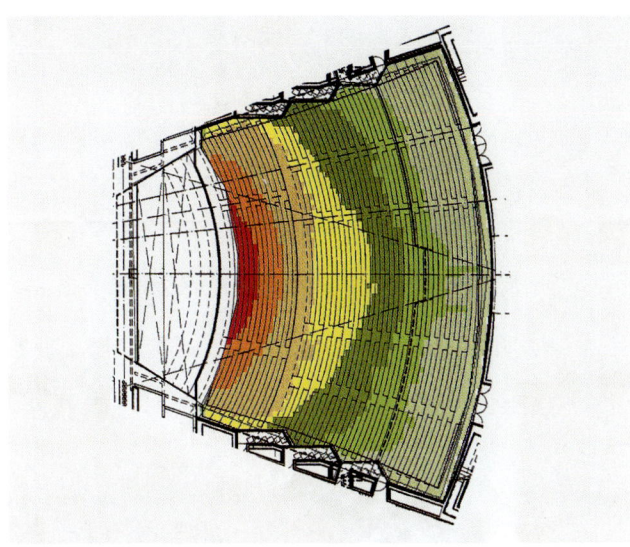

歌剧院观众席声频分布
AUDIO ARRANGEMENT OF THE OPERA HOUSE AUDITORIUM

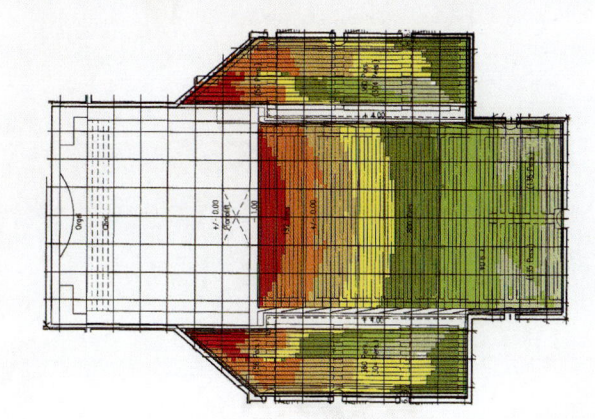

音乐厅观众席声频分布
AUDIO ARRANGEMENT OF THE CONCERT HALL AUDITORIUM

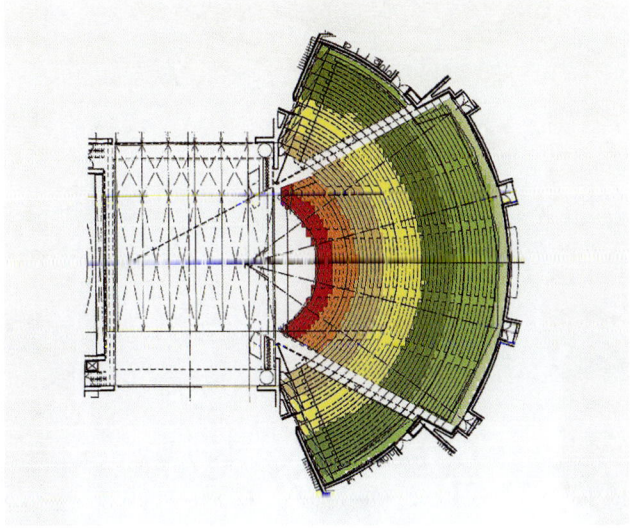

戏剧场观众席声频分布
AUDIO ARRANGEMENT OF THE THEATER AUDITORIUM

技术经济指标
Technical Economic Index

基地面积（m²） Site Area	总建筑面积（m²） Total Floor Area	建筑高度（m） Building Height	建筑覆盖率（%） Building Coverage Ratio
38916	地上：83607800 地下：43892500	30/ 歌剧院舞台 塔楼 45	72.8

容积率 Plot Ratio	绿化覆盖率（%） Green Coverage Ratio	停车数量 Number of Stalls	自行车停车数量 Number of Bicycle Stalls
2.1	6	地上：74 地下：578	1200

各观众厅数据
Data for Auditoriums

		歌剧院 Opera House	音乐厅 Concert Hall	戏剧院 Theater	小戏场 Mini Theater
	观众厅面积（m²） Floor Area of Auditorium	1866	1428	920	1783
	观众厅体积（m³） Volume of Auditorium	25022	25162	15776	4837
座席数 Seating Capacity	总座席数 Seating Capacity(Seats)	2518	1972	1210	500
	其中：池座 Among Them:Auditorium	1258	990	968	440
	楼座 Floor Seat	1260	982	242	60
	休息厅面积（m²） Lounge Floor Area	3018	2452	1329	621
舞台尺寸（m²） Stage Dimentions	主舞台 Main Stage	972	437	540	100*
	左右侧台 Left and Right Side Stages	1175	472	627	
	后舞台 Back Stage	650		330	
	台仓 Under Stage Storage	954	195	490	225
	升降乐池 Elevating Orchestra Pit	159		200	
	最大视距（m） Max. Visual Distance	37	35.5	27.0	25
	最大俯角（°） Max. Angle of Depression	29.5	17	23.5	11.5

北京市建筑设计
研究院（中国）
Beijing Institute of
Architectural Design &
Research (China)

本方案布局考虑的原则是：

1.建筑体形应是较完整、较对称的，方能与天安门广场及其周围的公共建筑取得协调。

2.大剧院的主要入口应朝向最主要的大街——长安街，因为这是观众来剧院最主要的交通路线，同时也是城市景观最重要的立面。与此同时也要兼顾南侧集中绿地的景观需要，设相应的入口，避免南侧成为背立面的感觉。

3.建筑形式及其风格要与天安门广场及其周围的建筑协调，但又要体现出跨世纪的时代精神，使其成为首都具有标志性的文化建筑。

This design has taken the following principles into consideration.

1. The building should be designed in a comparatively intact and symmetrical shape to coordinate with the Square and surrounding buildings.

2. The main entrance of the National Grand Theater should be on the north side facing the most important avenue, Chang An Avenue, which will be the main access road for the visitors. The north elevation of the Theater will be one of the most attracting spots of the city scenery. Meanwhile, the south elevation should also be treated properly by providing south entrance facing the green area on the south side to avoid a sense of coming to the back of the Theater.

3. The building form and style should be integrated with those of Tian An Men Square and surrounding buildings. At the same time the Theater should be designed to express the spirit of the times at the threshold of the 21st century and make it a symbolic cultural building of the capital.

北侧透视
PERSPECTIVE OF NORTH ELEVATION

南侧透视
PERSPECTIVE OF SOUTH ELEVATION

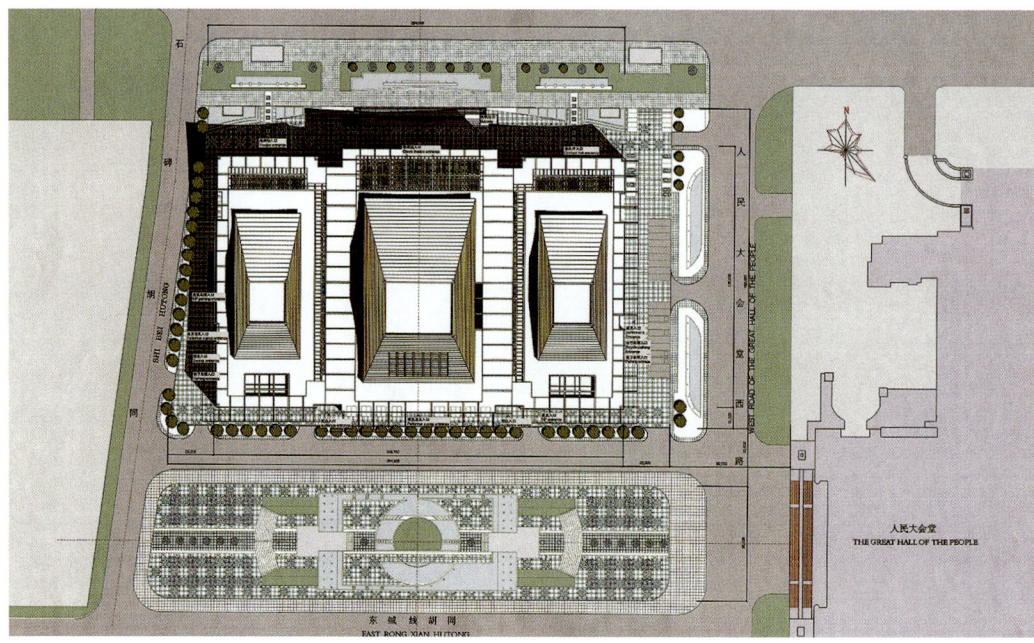

总平面
OVERALL SITE PLAN

歌剧院观众厅透视
PERSPECTIVE OF THE OPERA HOUSE
AUDITORIUM

休息大厅透视
FOYER PERSPECTIVE

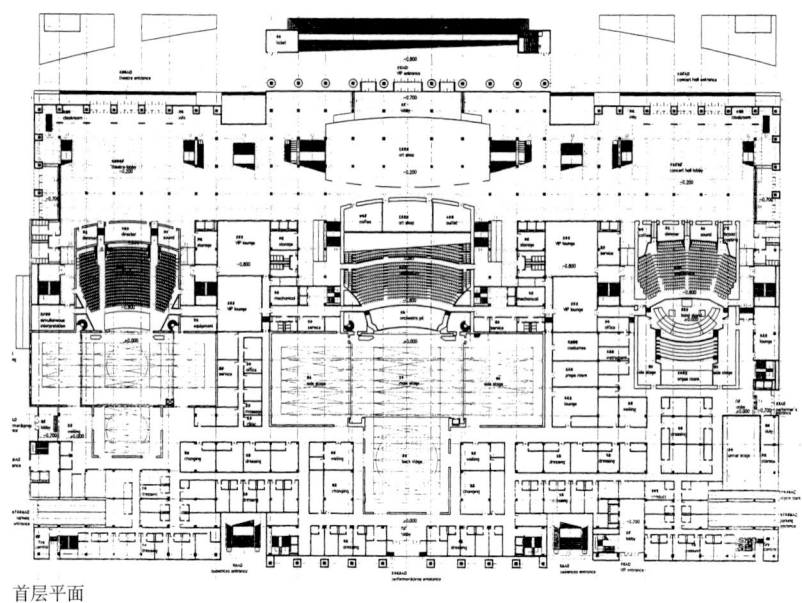

首层平面
GROUND FLOOR PLAN

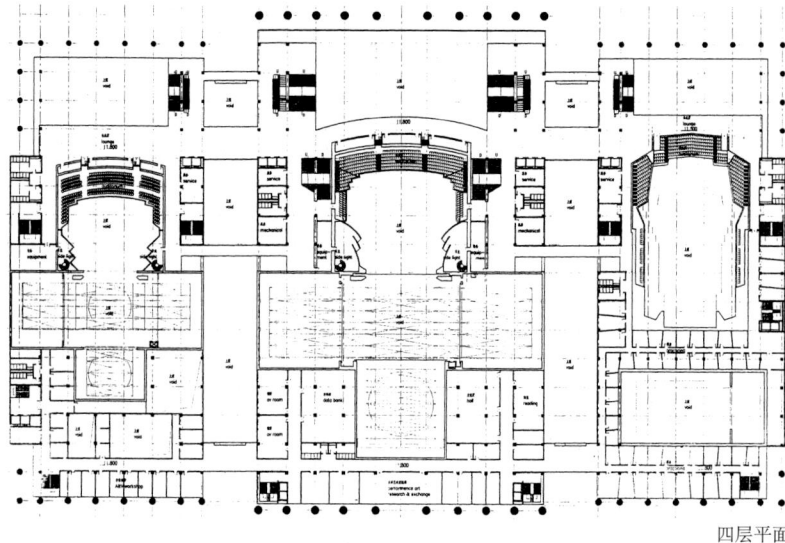

四层平面
4th FLOOR PLAN

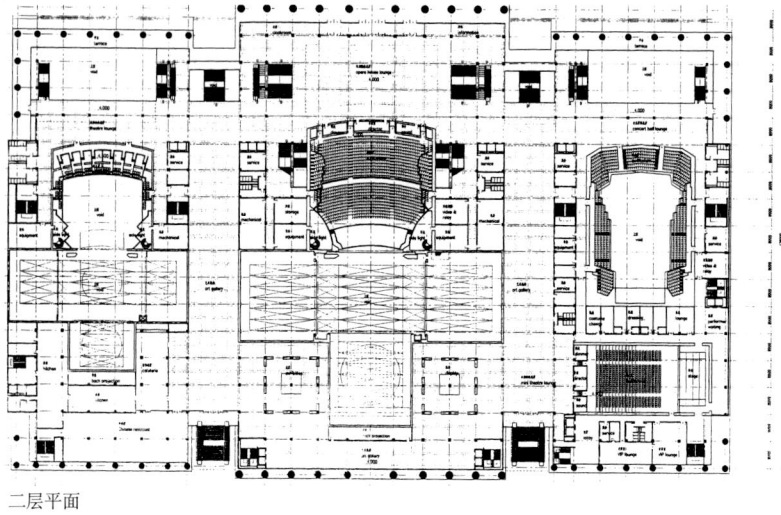

二层平面
2nd FLOOR PLAN

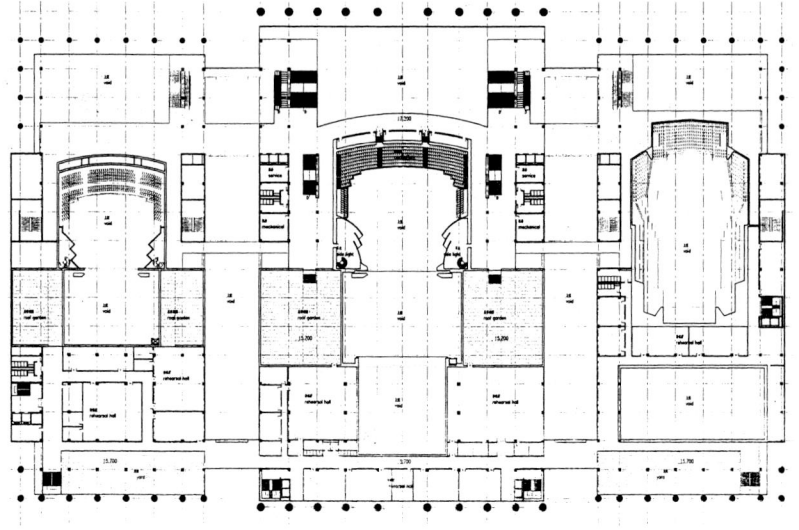

五层平面
5th FLOOR PLAN

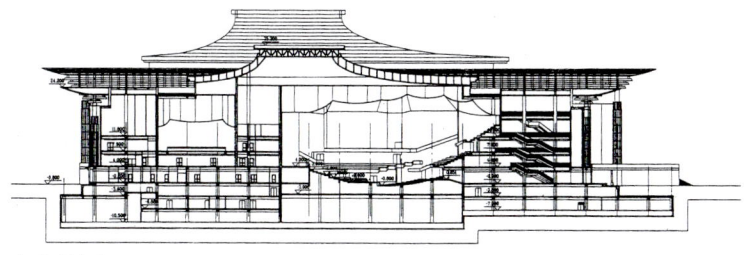

音乐厅剖面
SECTION OF THE CONCERT HALL

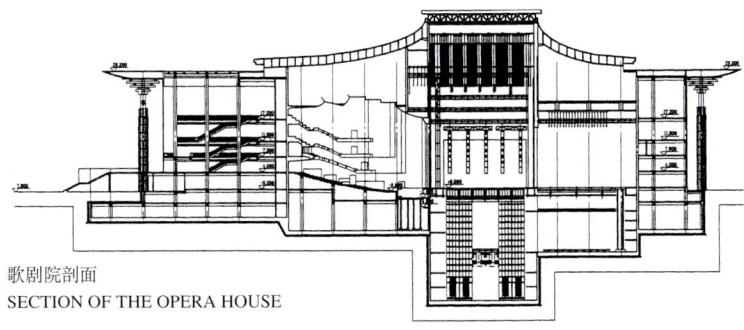

歌剧院剖面
SECTION OF THE OPERA HOUSE

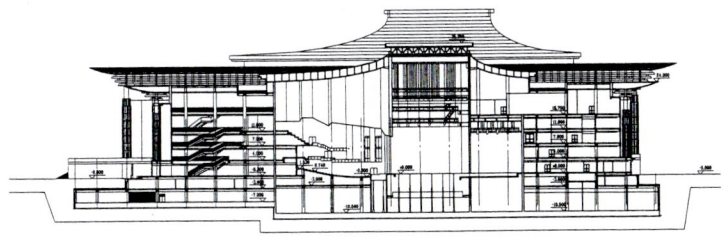

戏剧院剖面
SECTION OF THE THEATER

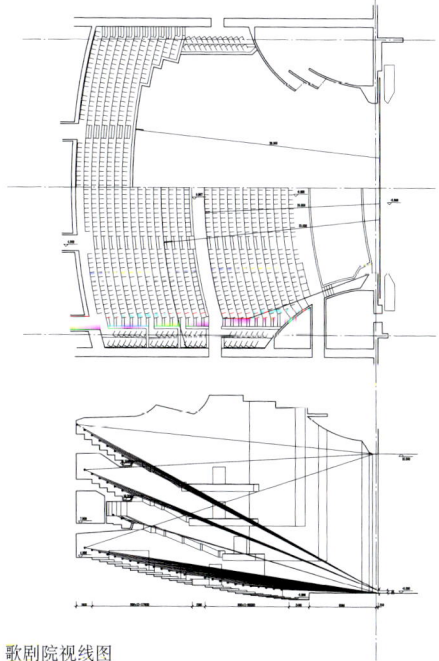

歌剧院视线图
VISUAL ANALYSIS DRAWING OF THE OPERA HOUSE

技术经济指标
Technical Economic Index

基地面积(m²) Site Area	总建筑面积(m²) Total Floor Area	建筑高度(m) Building Height	建筑覆盖率（%） Building Coverage Ratio
38900	地上：77620 地下：57000	30	79.5

容积率 Plot Ratio	绿化覆盖率(%) Green Coverage Ratio	停车数量 Number of Stalls	自行车停车数量 Number of Bicycle Stalls
2	42(用地内包括东、南绿地)	地上：64 地下：567	500

各观众厅数据
Data for Auditoriums

		歌剧院 Opera House	音乐厅 Concert Hall	戏剧院 Theater	小戏场 Mini Theater
观众厅面积(m²) Floor Area of Auditorium		2175	1602	718	549
观众厅体积(m³) Volume of Auditorium		20900	20800	9478	4282
座席数 Seating Capacity	总座席数 Seating Capacity(Seats)	2515	2032	1364	592
	其中：池座 Among Them:Auditorium	1372	640	784	480
	楼座 Floor Seat	1143	1392	580	112
	休息厅面积(m²) Lounge Floor Area	4780	3150	3000	1500
舞台尺寸(m²) Stage Dimentions	主舞台 Main Stage	34 × 27	24 × 18	27 × 22	12 × 9.1
	左右侧台 Left and Right Side Stages	23 × 27		左22 × 12.5 右22 × 14.6	4 × 9.1
	后舞台 Back Stage	25 × 25		19.5 × 14.5	
	台仓 Under Stage Storage	34 × 27 × 28		27 × 22 × 12.6	21.5 × 9.1 × 4.2
	升降乐池 Elevating Orchestra Pit	6.5		16.4 × 5.6	
最大视距(m) Max. Visual Distance		35.45	34	29.2	22
最大俯角(°) Max. Angle of Depression		29.11	24.75	24	9

北京市建筑设计研究院（中国）
Beijing Institute of Architectural Design & Research (China)

本方案布局考虑的原则是：

1. 建筑体形应是较完整、较对称的，方能与天安门广场及其周围的公共建筑取得协调。

2. 大剧院的主要入口应朝向最主要的大街——长安街，因为这是观众来剧院最主要的交通路线，同时也是城市景观最重要的立面。与此同时也要兼顾南侧集中绿地的景观需要，设相应的入口，避免南侧成为背立面的感觉。

3. 建筑形式及其风格要与天安门广场及其周围的建筑协调，但又要体现出跨世纪的时代精神，使其成为首都具有标志性的文化建筑。

This design has taken the following principles into consideration.

1. The building should be designed in a comparatively intact and symmetrical shape to coordinate with the Square and surrounding buildings.

2. The main entrance of the National Grand Theater should be on the north side facing the most important avenue, Chang An Avenue, which will be the main access road for the visitors. The north elevation of the Theater will be one of the most attracting spots of the city scenery. Meanwhile, the south elevation should also be treated properly by providing south entrance facing the green area on the south side to avoid a sense of coming to the back of the Theater.

3. The building form and style should be integrated with those of Tian An Men Square and surrounding buildings. At the same time the Theater should be designed to express the spirit of the times at the threshold of the 21st century and make it a symbolic cultural building of the capital.

北侧透视
PERSPECTIVE OF NORTH ELEVATION

南侧透视
PERSPECTIVE OF SOUTH ELEVATION

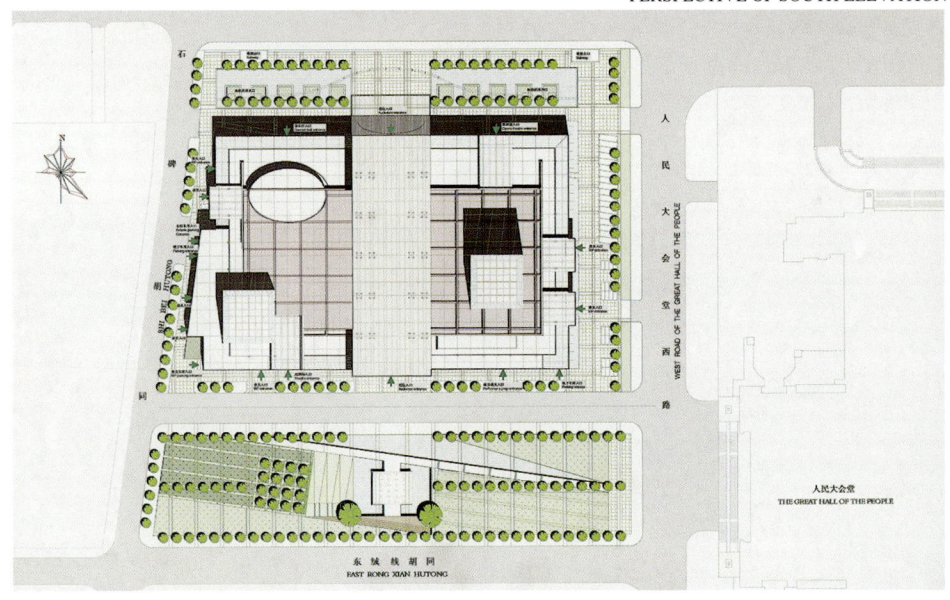

总平面
OVERALL SITE PLAN

观众厅透视
AUDITORIUM PERSPECTIVE

休息大厅透视
FOYER PERSPECTIVE

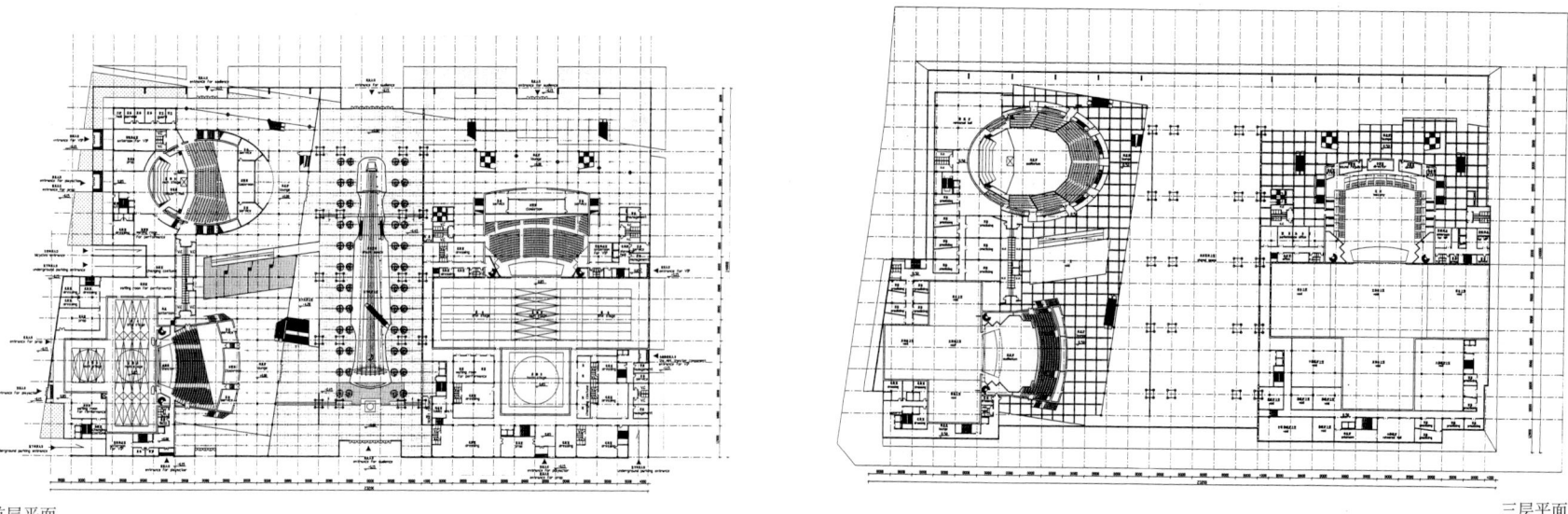

首层平面
GROUND FLOOR PLAN

三层平面
3rd FLOOR PLAN

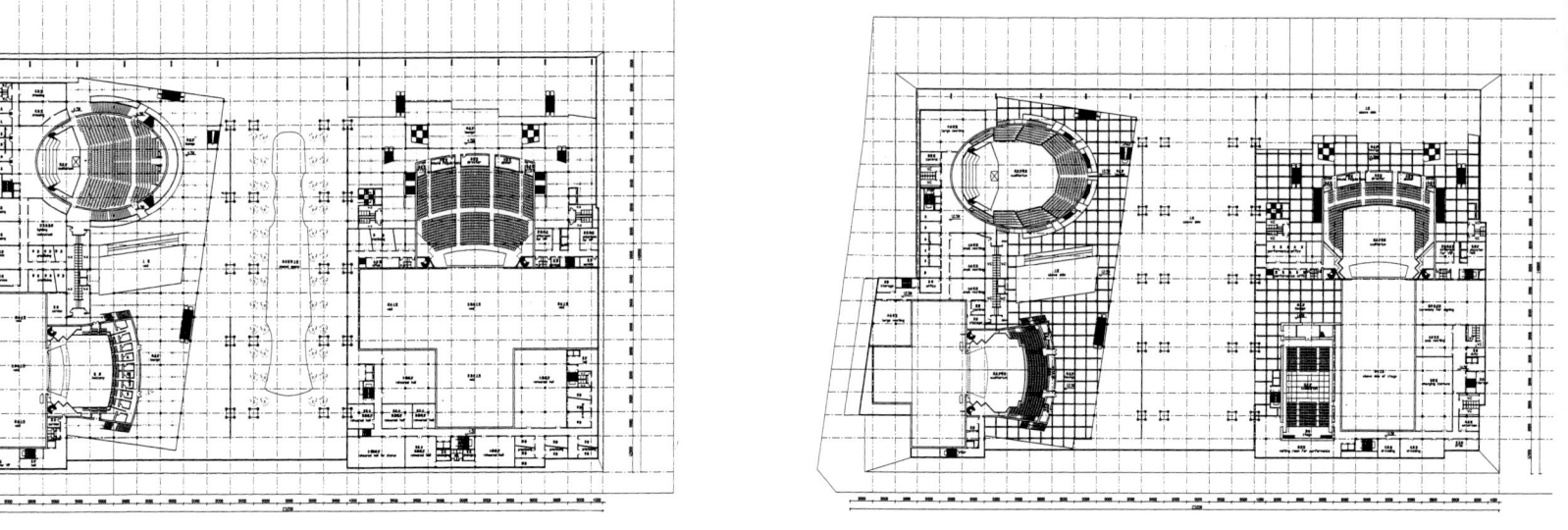

二层平面
2nd FLOOR PLAN

四层平面
4th FLOOR PLAN

Ⅱ－Ⅱ 剖面
SECTION Ⅱ－Ⅱ

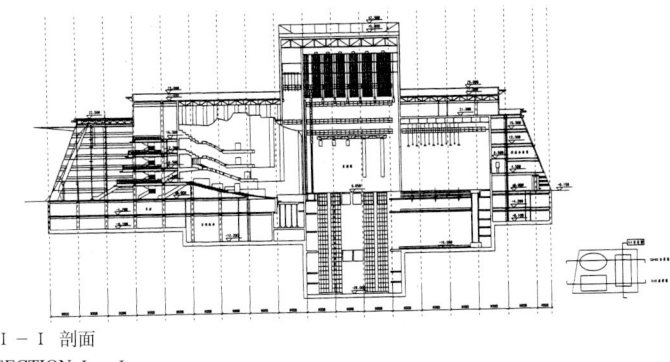

Ⅰ－Ⅰ 剖面
SECTION Ⅰ－Ⅰ

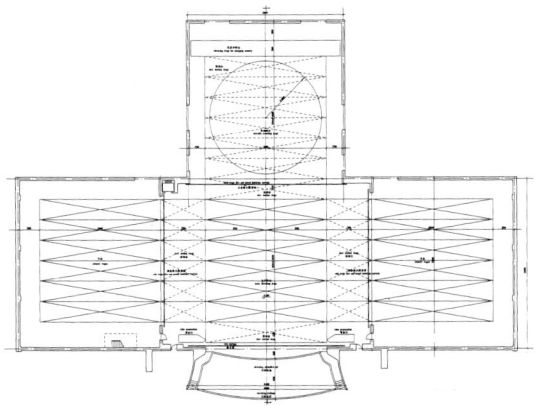

歌剧院舞台平面
STAGE LAYOUT OF THE OPERA HOUSE

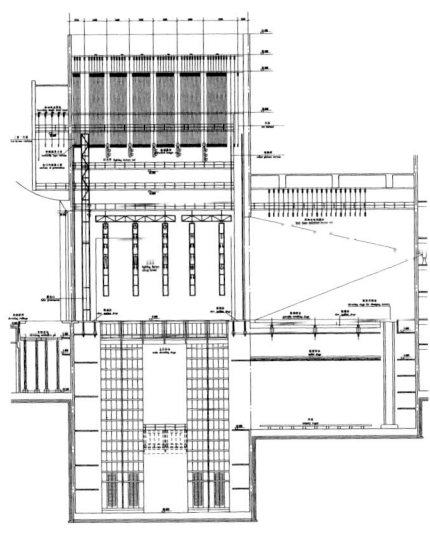

歌剧院舞台剖面
SECTION OF THE OPERA HOUSE

技术经济指标
Technical Economic Index

基地面积 (m²) Site Area	总建筑面积 (m²) Total Floor Area	建筑高度 (m) Building Height	建筑覆盖率 (%) Building Coverage Ratio
38900	地上：70270 地下：61600	29	76.63
容积率 Plot Ratio	绿化覆盖率 (%) Green Coverage Ratio	停车数量 Number of Stalls	自行车停车数量 Number of Bicycle Stalls
1.8	45.74	地上：60 地下：567	1000

各观众厅数据
Data For Auditoria

		歌剧院 Opera House	音乐厅 Concert Hall	戏剧院 Theater	小戏场 Mini Theater
	观众厅面积 (m²) Floor Area of Auditorium	2175	1242	730	591
	观众厅体积 (m³) Volume of Auditorium	20900	19860	9485	4435
座席数 Seating Capacity	总座席数 Seating Capacity(Seats)	2515	2027	1386	476
	其中：池座 Among Them:Auditorium	1372	902	794	396
	楼座 Floor Seat	1143	1125	592	80
	休息厅面积 (m²) Lounge Floor Area	9120	4980	4760	1080
舞台尺寸 (m²) Stage Dimentions	主舞台 Main Stage	34 × 27	23.8 × 16.9	27 × 22	15 × 8
	左右侧台 Left and Right Side Stages	23 × 27		22 × 14.6	
	后舞台 Back Stage	25 × 25		19.5 × 14.5	
	台仓 Under Stage Storage	34 × 27 × 28		27 × 22 × 12.6	34.6 × 17.6 × 4
	升降乐池 Elevating Orchestra Pit	6.5		16.4 × 5.6	
	最大视距 (m) Max. Visual Distance	35.45	34.2	29.2	21.5
	最大俯角 (°) Max. Angle of Depression	29.11	19	25	13～23

106
塔瑞·法若
建筑师事务所
（英国）
Terry Farrell &
Partners(UK)

剧院面向西长安街的三个主要部分联合起来作为一个主要入口及国家大剧院大堂，只要踏进剧院，入口的大窗便如"落幕"把外界分割起来，城市生活旋即转为舞台的幻想世界。

剧院大堂是"人生舞台"，经过的人们穿梭于大堂内的配套设施，包括楼梯、升降机、展览厅、餐厅及茶座等地方，由他们上演一幕幕的聚散离合。

Three sectors facing West Chang An Avenue are linked to form a great public entrance and foyer space where public life of the city is moderated to the fantasy world of the Theater. The great public entrance and foyer itself form a stage to the city outside. The great entrance window is itself a proscenium arch to the world of theater inside.

The entrance foyer is the stage of life. It is a place where the drama is acted out by people on the move, congregating and dispersing. The great hall is animated by stairs, escalators, lifts, exhibition spaces, restaurants and bars.

夜景透视　NIGHT VIEW

北侧　NORTH ELEVATION

大厅透视　LOBBY PERSPECTIVE

南侧　SOUTH ELEVATION

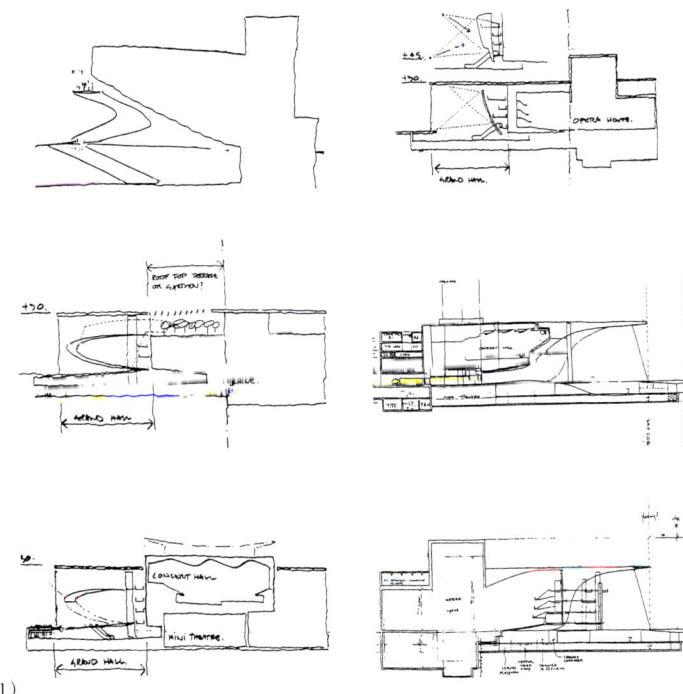

观众厅透视
AUDITORIUM PERSPECTIVE

发展纲要（1）
DESIGN DEVELOPMENT SKETCHES(1)

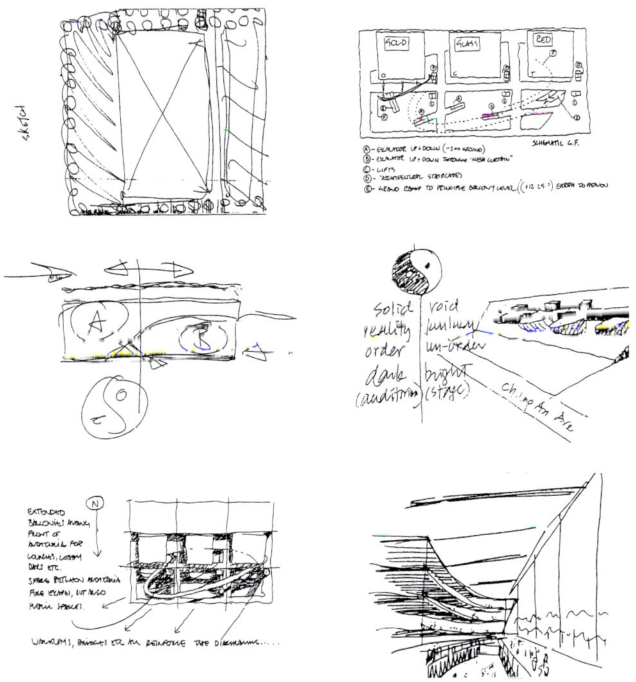

发展纲要（2）
DESIGN DEVELOPMENT SKETCHES(2)

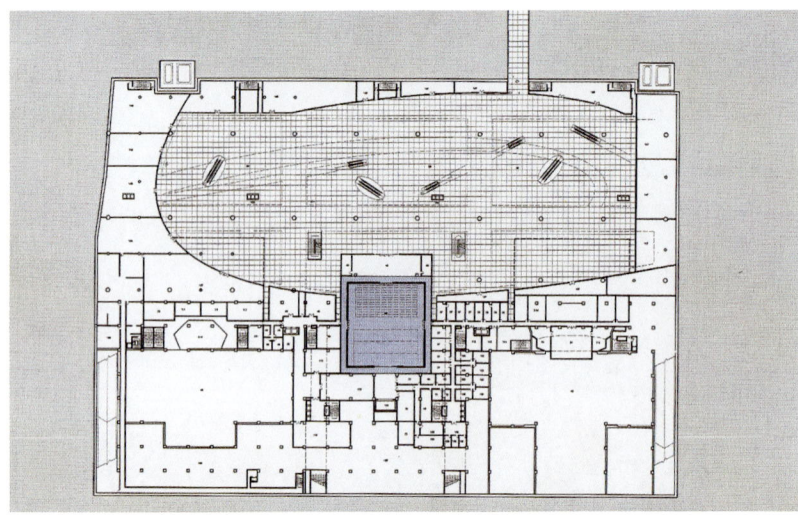

下底层平面　LOWER GROUND FLOOR PLAN

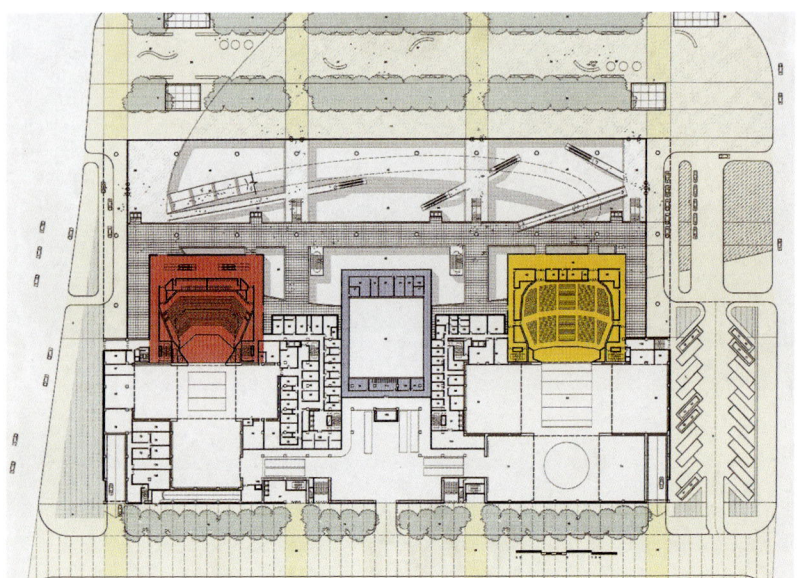

底层平面　GROUND FLOOR PLAN

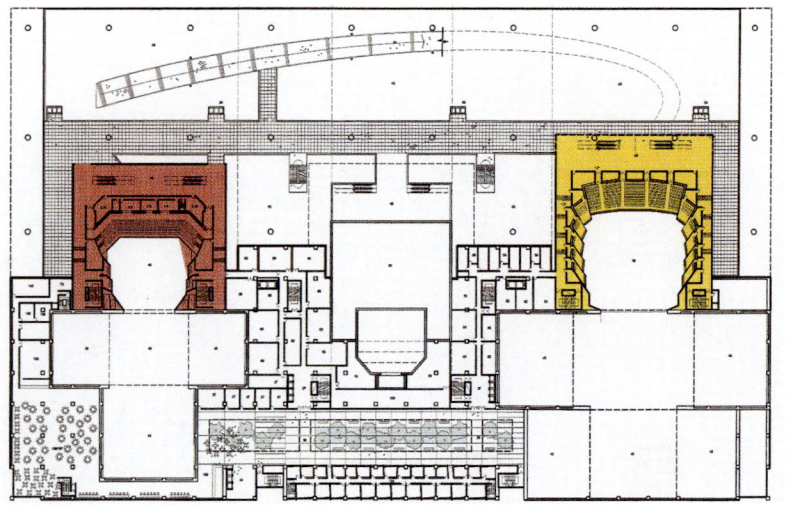

一层平面　FIRST FLOOR PLAN

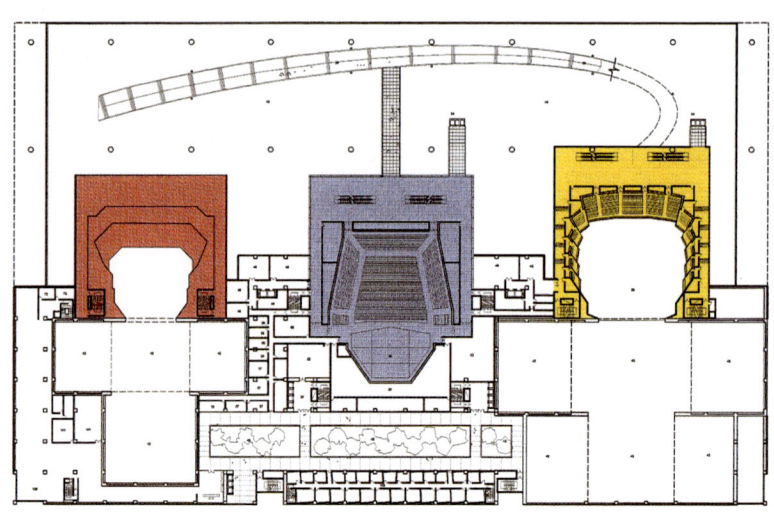

二层平面　2nd FLOOR PLAN

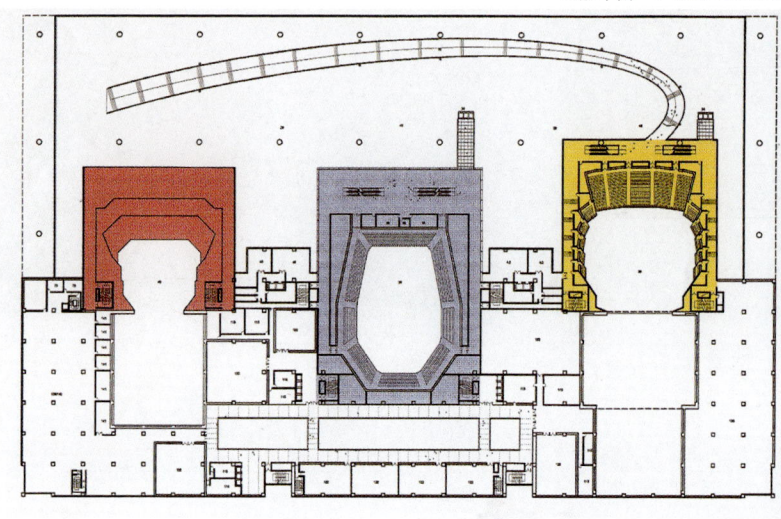

三层平面　3rd FLOOR PLAN

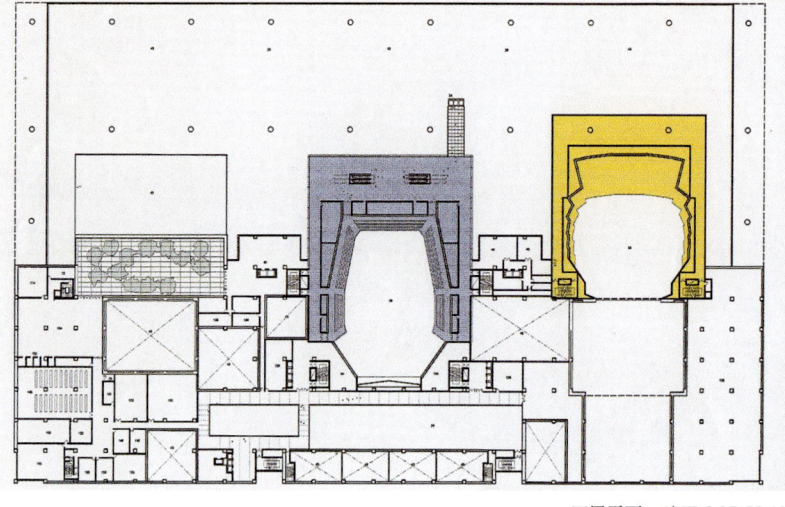

四层平面　4th FLOOR PLAN

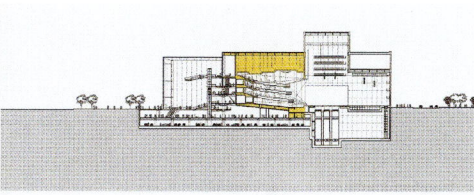

A—A 剖面　SECTION A-A

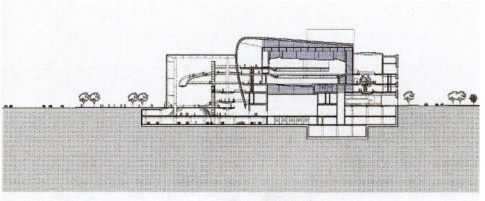

B—B 剖面　SECTION B-B

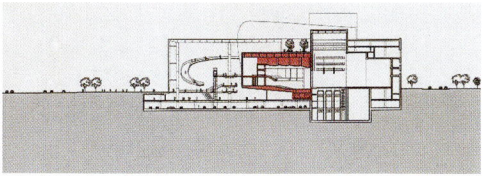

C—C 剖面　SECTION C-C

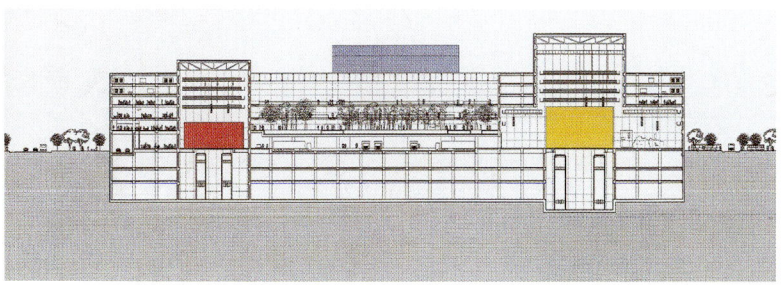

D—D 剖面　SECTION D-D

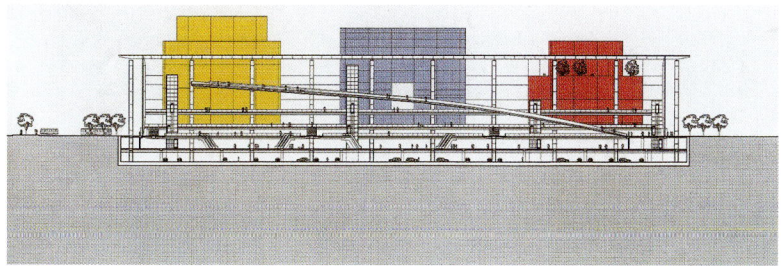

E—E 剖面　SECTION E-E

技术经济指标
Technical Economic Index

基地面积(m²) Site Area	总建筑面积(m²) Total Floor Area	建筑高度(m) Building Height	建筑覆盖率 (%) Building Coverage Ratio
38900	地上：73704 地下：68420	30/45(局部)	82

容积率 Plot Ratio	绿化覆盖率 (%) Green Coverage Ratio	停车数量 Number of Stalls	自行车停车数量 Number of Bicycle Stalls
3.6	10	地上： 地下：500	1000

各观众厅数据
Data For Auditoria

			歌剧院 Opera House	音乐厅 Concert Hall	戏剧院 Theater	小戏场 Mini Theater
	观众厅面积(m²) Floor Area of Auditorium		2170	1950	1027	930
	观众厅体积(m³) Volume of Auditorium		19000	20565	9310	10800
座席数 Seating Capacity	总座席数 Seating Capacity(Seats)		2500	2000	1200	500
	其中：池座 Among Them:Auditorium		1200	1115	872	500
	楼 座 Floor Seat		1300	885	328	
	休息厅面积(m²) Lounge Floor Area		2100	1680	10008	420
舞台尺寸 (m²) Stage Dimentions	主舞台 Main Stage		36 × 27	26 × 18	27 × 21	30 × 31
	左右侧台 Left and Right Side Stages		22 × 27		17 × 21	
	后舞台 Back Stage		26 × 25		27 × 26	
	台 仓 Under Stage Storage		2160	200	580	960
	升降乐池 Elevating Orchestra Pit		26 × 6	24 × 16	21 × 11	
	最大视距(m) Max. Visual Distance		44	28	27	15
	最大俯角（°） Max. Angle of Depression		29.2	25.5	22.4	23.0

107

上海现代建筑设计
（集团）有限公司
（中国）
Shanghai Modern
Architectural Design
(Group)Co., Ltd.(China)

以对称性构图和大尺
度手法体现场所性。

以传统大屋顶的曲线
体现民族性。

以现代的空间、材料
和结构形式体现时代性。

以空间的开放性和舒
适性体现新建筑的人民
性。

Embodying location of the
new building with symmetrical design and large scale ar-
tifice.

Embodying nationality of
the new building with traditional large roof curve.

Embodying contemporary
with modern style of space,
material and structure.

Embodying popularity of
the new building with open
and comfortable space.

北侧透视　PERSPECTIVE OF NORTH ELEVATION

南侧透视　PERSPECTIVE OF SOUTH ELEVATION

大厅透视
LOBBY PERSPECTIVE

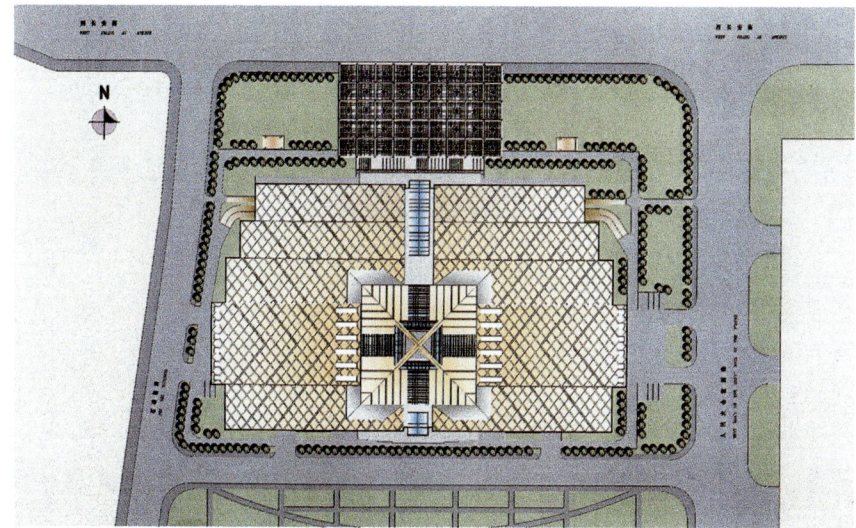

总平面
OVERALL SITE PLAN

观众厅透视
AUDITORIUM PERSPECTIVE

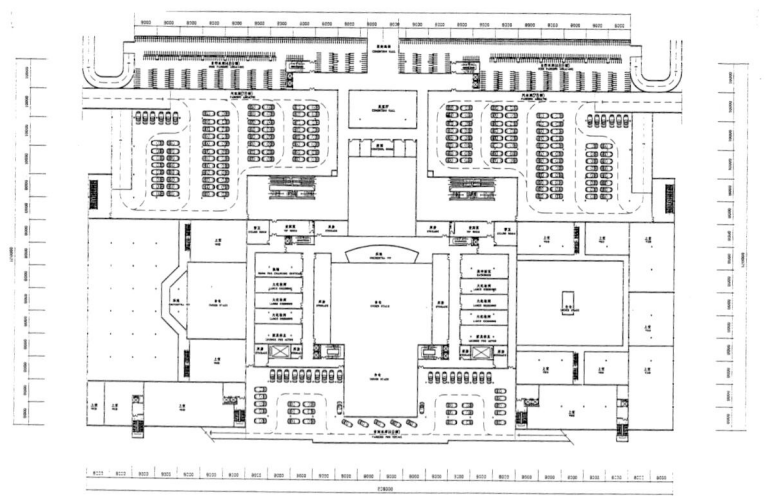

地下一层平面
FIRST BASEMENT PLAN

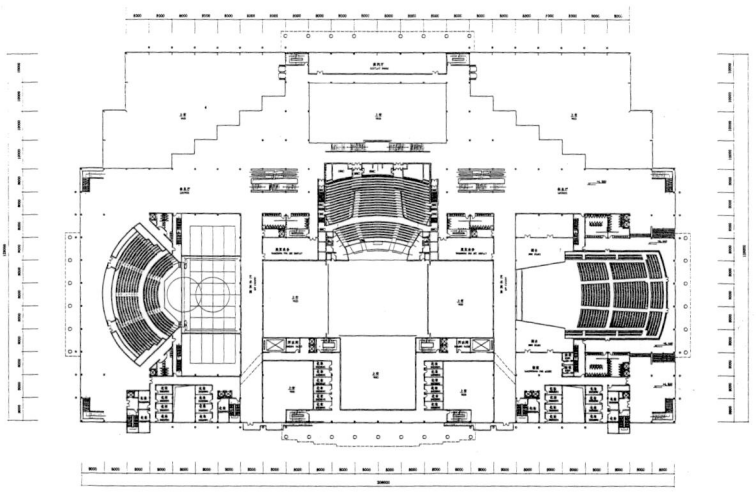

二层平面
2nd FLOOR PLAN

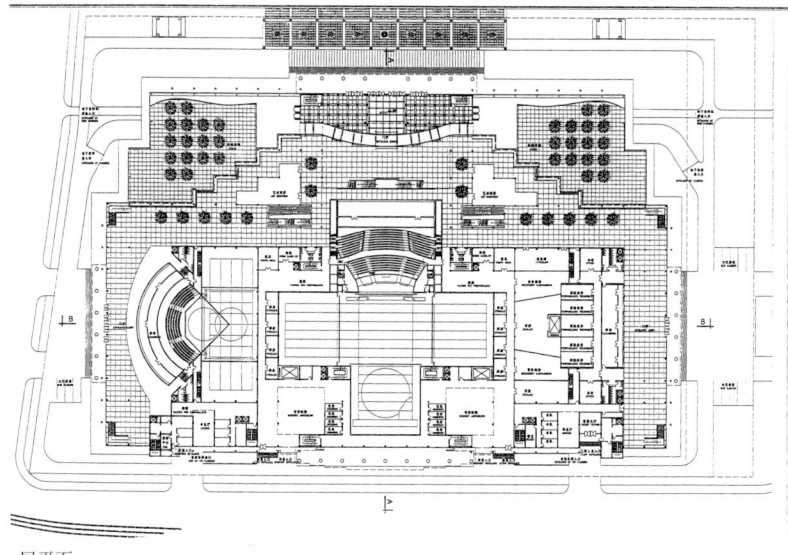

一层平面
GROUND FLOOR PLAN

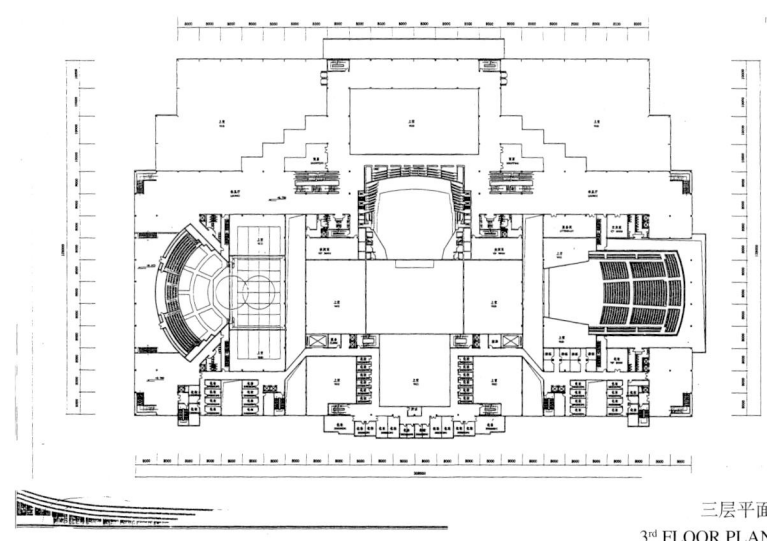

三层平面
3rd FLOOR PLAN

B–B 剖面
SECTION B-B

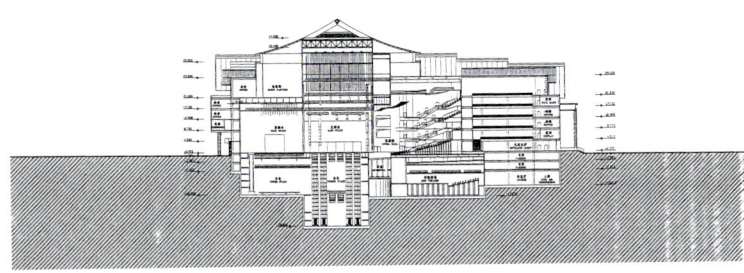

A—A 剖面
SECTION A-A

建筑构思
DESIGN CREATION

建筑构思
DESIGN CREATION

建筑构思
DESIGN CREATION

技术经济指标
Technical Economic Index

基地面积（m²） Site Area	总建筑面积（m²） Total Floor Area	建筑高度（m） Building Height	建筑覆盖率（%） Building Coverage Ratio
38900	地上：84224 地下：45228	45	69.94

容积率 Plot Ratio	绿化覆盖率（%） Green Coverage Ratio	停车数量 Number of Stalls	自行车停车数量 Number of Bicycle Stalls
3.33	10.35	地上：32 地下：471	978

各观众厅数据
Data For Auditoria

		歌剧院 Opera House	音乐厅 Concert Hall	戏剧院 Theater	小戏场 Mini Theater
	观众厅面积（m²） Floor Area of Auditorium	2431	1592	1154	778
	观众厅体积（m³） Volume of Auditorium	20916	23703	7750	5880
座席席数 Seating Capacity	总座席数 Seating Capacity(Seats)	2520	2004	1250	489
	其中：池座 Among Them:Auditorium	1208	1256	870	370
	楼 座 Floor Seat	1312	748	380	119
	休息厅面积（m²） Lounge Floor Area				
舞台尺寸（m²） Stage Dimentions	主舞台 Main Stage	27 × 36	24 × 18	15 × 37	20 × 6
	左右侧台 Left and Right Side Stages	22 × 27	12 × 18	15 × 21	
	后舞台 Back Stage	25 × 26			
	台 仓 Under Stage Storage	27 × 36		15 × 37	16 × 28
	升降乐池 Elevating Orchestra Pit	24 × 7	4 × 8	14 × 6	
	最大视距（m） Max. Visual Distance	34	35	18.5	13.6
	最大俯角（°） Max. Angle of Depression	31	23	30	20

上海现代建筑设计
（集团）有限公司
（中国）
Shanghai Modern
Architectural Design
(Group)Co., Ltd.(China)

本方案的建筑形象只是提供了两个蓬勃向上的拱顶和一个贯穿东西的晶莹通透的块体。两者构成了一个完整的整体，简洁、明快、一目了然，并不具象一个事物，它是雄鹰的双翅？含苞欲放的花朵？翻飞的蝴蝶？远航的白帆？舞动的水袖？都是，都不是。正因为它不具象某一事物，才留给人们想象的余地。

建筑物中部的玻璃大厅的斜屋面与圆弧侧墙交切正好产生了一个类似大屋顶的形式，这个大屋顶仅是轮廓如宫殿的外形，却由玻璃构成，隐喻着"艺术的殿堂"这个主题。

The image of this scheme consists of two vigorous arches and one glittering atrium, which forms an entirety---simple, sprightly and clear at a glance. It is not like just one thing---is it the wing of hawk?Butterfly in flying?Junk on sailing?Long sleeve in waving? All yes, and all no.Just because it is not only like one thing, it gives people space to imagine.

The pitched roof of the glass hall in the central building meets the arc side wall formed a roof exactly similar to the traditional Chinese curved roof style, which made by glass is only apparently similar to the palace---metaphor of the theme"Art Palace".

北側透视
PERSPECTIVE OF NORTH ELEVATION

观众厅透视 AUDITORIUM PERSPECTIVE

总平面
OVERALL SITE PLAN

休息厅
LOBBY PERSPECTIVE

夜景
NIGHT VIEW

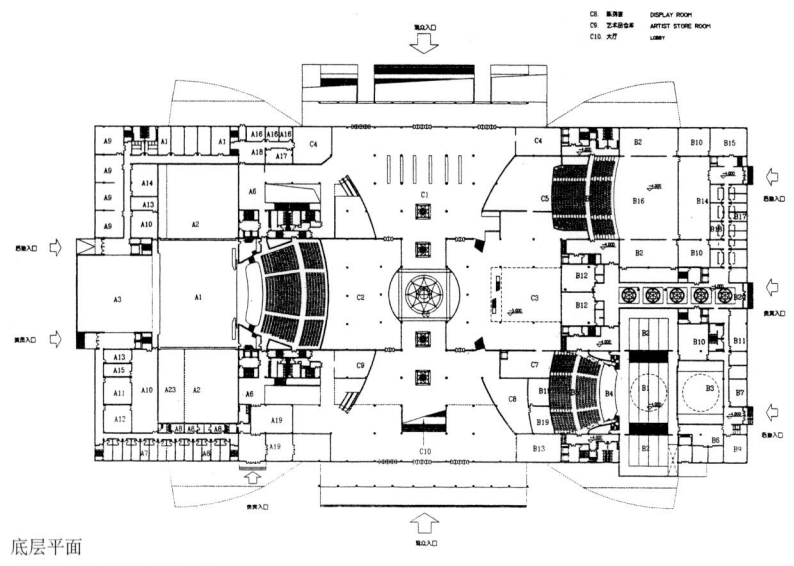

C8. 陈列室　　　DISPLAY ROOM
C9. 艺术品仓库　ARTIST STORE ROOM
C10. 大厅　　　　LOBBY

底层平面
FIRST BASEMENT PLAN

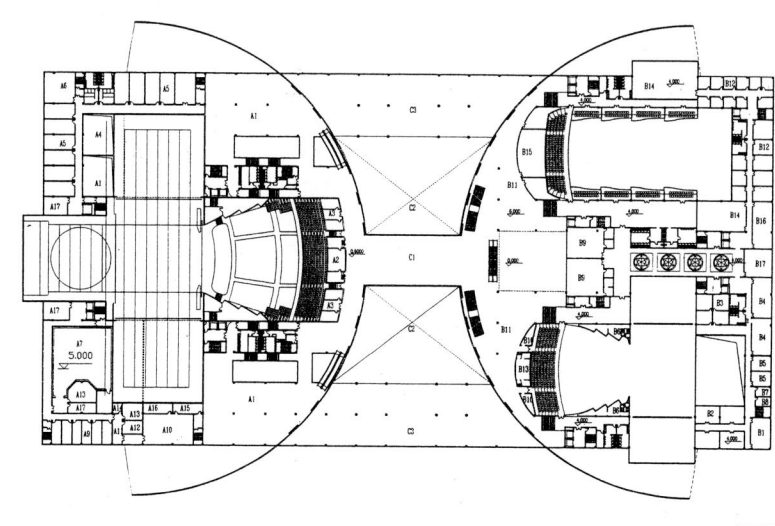

二层平面
2nd FLOOR PLAN

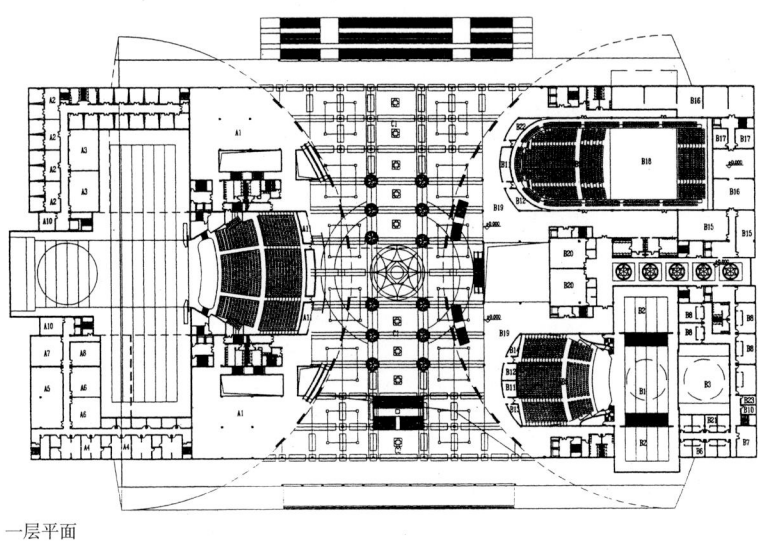

一层平面
GROUND FLOOR PLAN

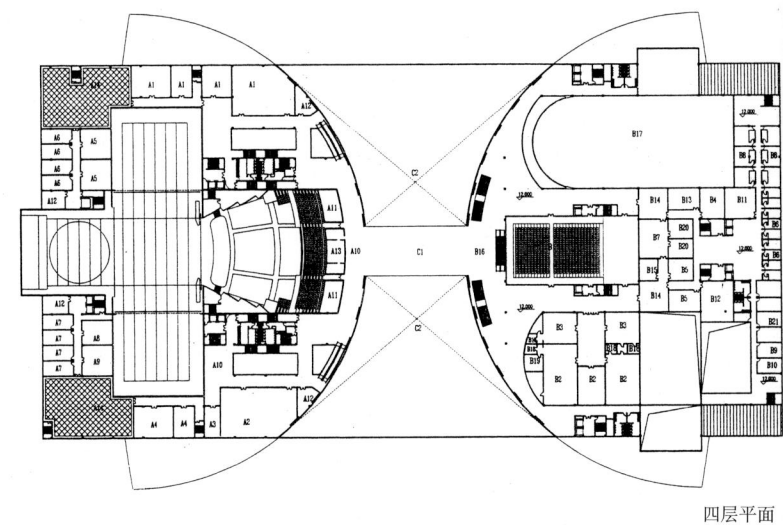

四层平面
4th FLOOR PLAN

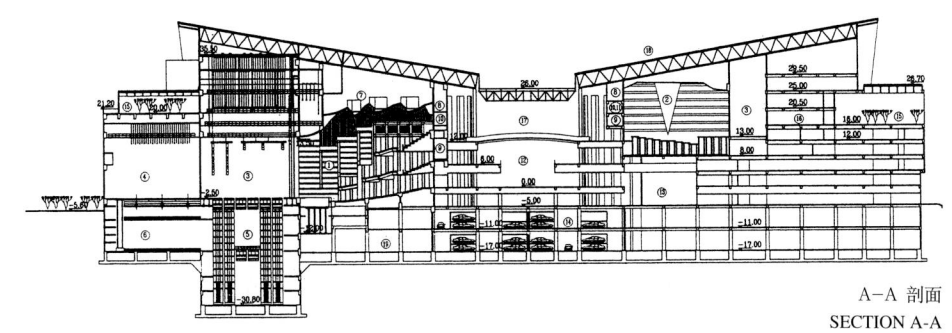

A—A 剖面
SECTION A-A

B

C-C 剖面 SECTION C-C

D-D 剖面 SECTION D-D

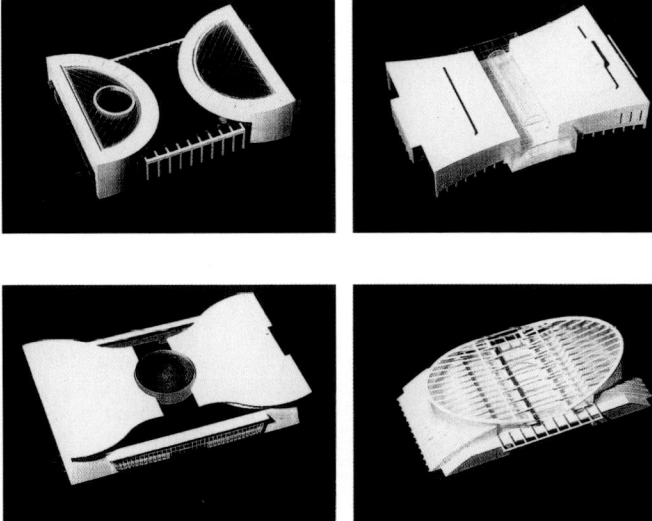

分析图 ANALYSIS DRAWING

技术经济指标
Technical Economic Index

基地面积(m²) Site Area	总建筑面积(m²) Total Floor Area	建筑高度(m) Building Height	建筑覆盖率 (%) Building Coverage Ratio
38900	地上：74050 地下：47420	31.6/46.6(局部)	62.4
容积率 Plot Ratio	绿化覆盖率 (%) Green Coverage Ratio	停车数量 Number of Stalls	自行车停车数量 Number of Bicycle Stalls
3.12	15.8	地上：107 地下：520	1150

各观众厅数据
Data For Auditoria

		歌剧院 Opera House	音乐厅 Concert Hall	戏剧院 Theater	小戏场 Mini Theater
	观众厅面积(m²) Floor Area of Auditorium	1960	1671	946	560
	观众厅体积(m³) Volume of Auditorium	21200	20368	7618	6720
座席数 Seating Capacity	总座席数 Seating Capacity(Seats)	2494	2030	1215	598
	其中：池座 Among Them:Auditorium	1386	1540	859	598
	楼座 Floor Seat	1180	490	356	
	休息厅面积(m²) Lounge Floor Area	3642	2356	1527	852
舞台尺寸 Stage Dimentions (m²)	主舞台 Main Stage	36 × 27	27 × 19.5	27 × 19.5	20 × 10
	左右侧台 Left and Right Side Stages	27 × 27	919.5	13.5 × 19.5	
	后舞台 Back Stage	27 × 24		21 × 15	
	台仓 Under Stage Storage	36 × 27	27 × 19.5	27 × 19.5	20 × 38
	升降乐池 Elevating Orchestra Pit	4.5 × 20		9.75 × 19.5	
	最大视距(m) Max. Visual Distance	36	29	28	26
	最大俯角（°) Max. Angle of Depression	26	22.5	21.2	

我们的中国国家大剧院设计方案，是在继承了天安门广场的历史和空间的连续性的前提下，作为新中国的国家象征，拥有全新的造型。这里，我们吸取了天安门广场的台座、列柱、屋顶的三个空间特征要素，运用计算机技术将其变形为现代形象。

Our proposal for the National Grand Theater extends the historical and spatial continuity of Tian An Men Square. On this basis, a completely new architectural form is created to symbolize to contemporary China. Three major architectural elements which characterize the visual impression of Tiam An Men Square are extracted: plinth, colonnade and roof. By virtue of the advancing computer technology, the three elements are transformed into contemporary figures.

北侧模型透视　VIEW FROM NORTH SIDE OF THE MODEL

入口大厅 LOBBY

立面比较　ELEVATIONS AT THE SAME SCALE

观众厅透视 AUDITORIUM PERSPECTIVE

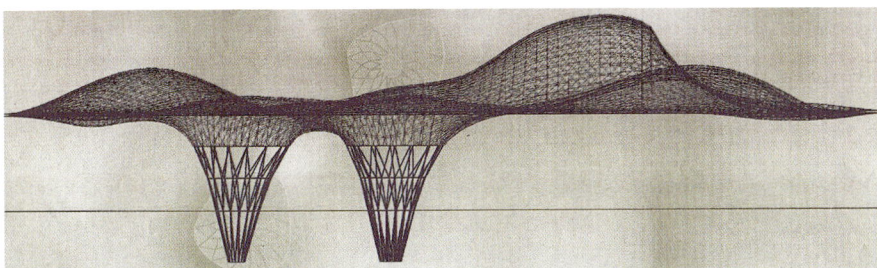

屋顶侧面 SILHOUETTE OF THE ROOF

休息厅透视 LOBBY PERSPECTIVE

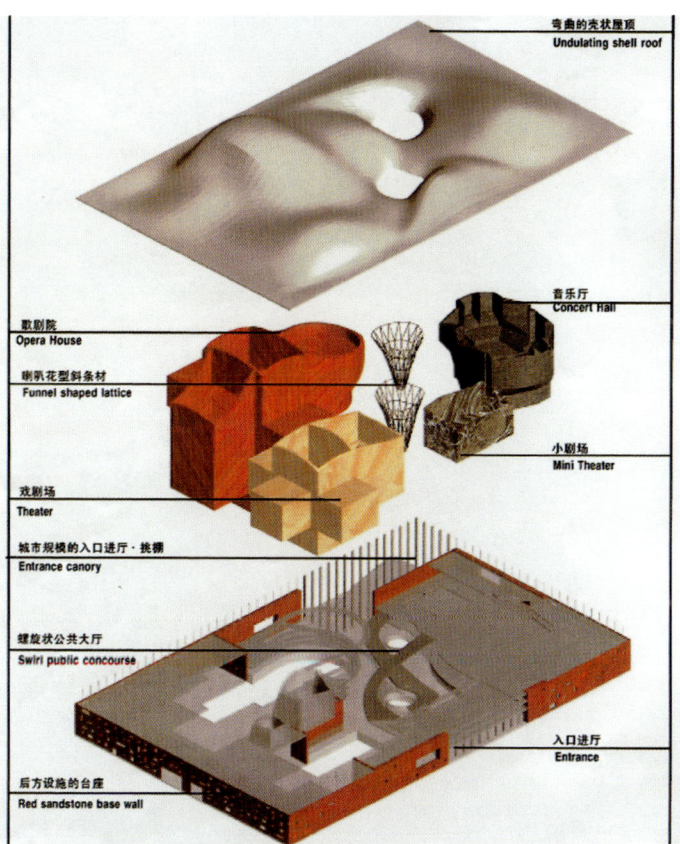

弯曲的壳状屋顶
Undulating shell roof

音乐厅
Concert Hall

歌剧院
Opera House

喇叭花型斜条材
Funnel shaped lattice

小剧场
Mini Theater

戏剧场
Theater

城市规模的入口进厅·挑棚
Entrance canopy

螺旋状公共大厅
Swirl public concourse

入口进厅
Entrance

后方设施的台座
Red sandstone base wall

整体构成 GENERAL COMPOSITION

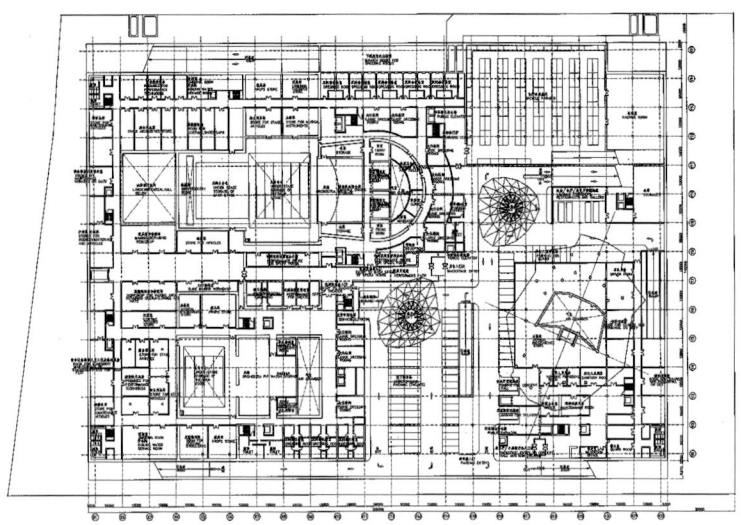

地下一层平面(-4.5m) FIRST BASEMENT LEVEL(-4.5m)

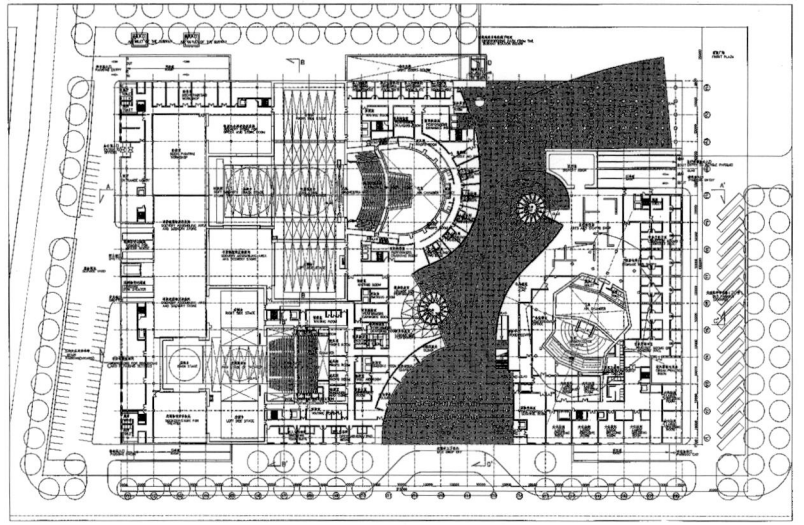

一层平面(±0.00m) GROUND LEVEL(±0.00m)

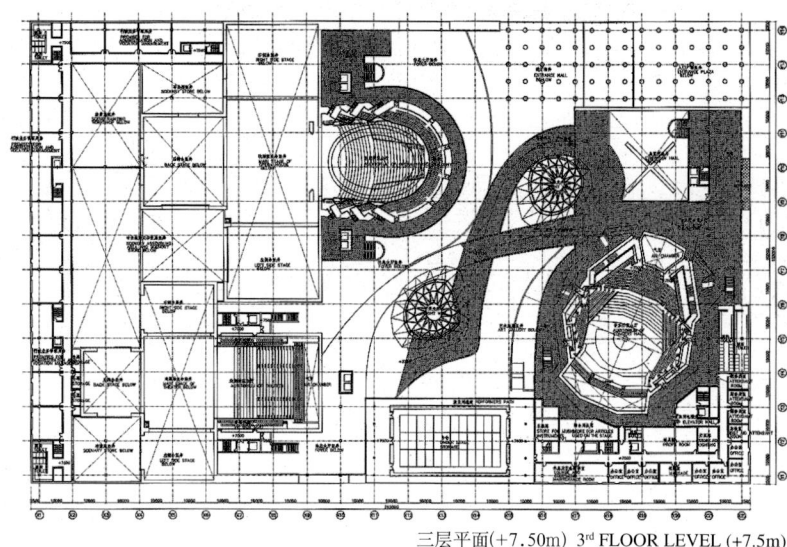

三层平面(+7.50m) 3rd FLOOR LEVEL (+7.5m)

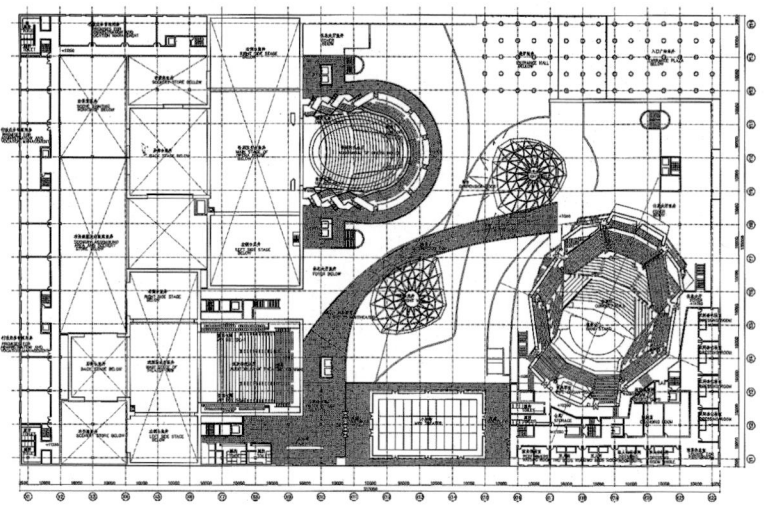

四层平面(11.25m) FOURTH FLOOR LEVEL (+11.25m)

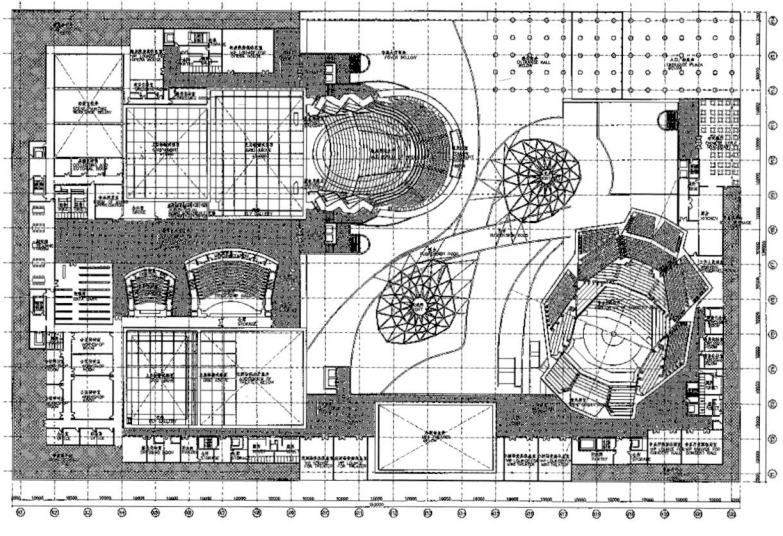

五层平面(+15.00m) FIFTH FLOOR LEVEL (+15.00m)

二层平面(3.75m) 2nd FLOOR LEVEL (+3.75m)

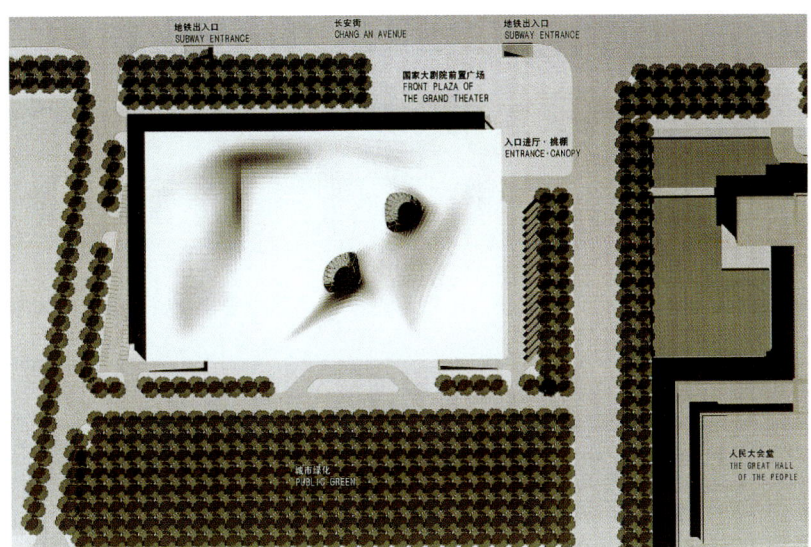

总平面
OVERALL SITE PLAN

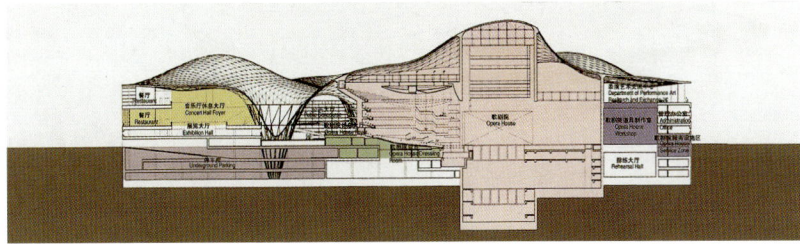

歌剧院剖面
SECTION OF THE OPERA HOUSE

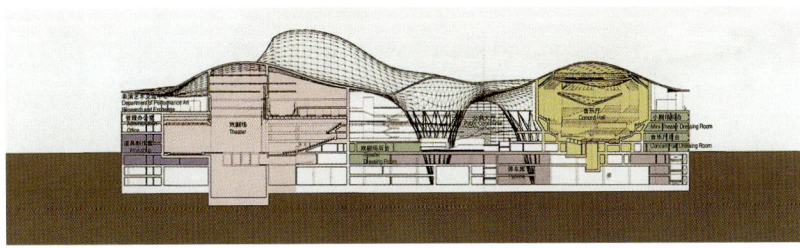

戏剧院、音乐厅剖面
SECTIONS OF THE THEATER & THE CONCERT HALL

技术经济指标
Technical Economic Index

基地面积（m²） Site Area	总建筑面积（m²） Total Floor Area	建筑高度（m） Building Height	建筑覆盖率（%） Building Coverage Ratio
38916	地上：70960 地下：58860	45	74.58

容积率 Plot Ratio	绿化覆盖率（%） Green Coverage Ratio	停车数量 Number of Stalls	自行车停车数量 Number of Bicycle Stalls
3.33	19.4	地上：72 地下：496	1000

各观众厅数据
Data For Auditoria

			歌剧院 Opera House	音乐厅 Concert Hall	戏剧院 Theater	小戏场 Mini Theater
	观众厅面积（m²） Floor Area of Auditorium		1973	1930	1102	6782
	观众厅体积（m³） Volume of Auditorium		20300	32800	8440	6782
座席数 Seating Capacity	总座席数 Seating Capacity(Seats)		2520	2050	1252	550
	其中：池座 Among Them:Auditorium		782	2050	804	500
	楼座 Floor Seat		1738		448	50
	休息厅面积（m²） Lounge Floor Area		5208	3861	2593	1543
舞台尺寸（m²） Stage Dimentions	主舞台 Main Stage		36 × 27	24 × 18	27 × 21	30 × 15
	左右侧台 Left and Right Side Stages		22 × 27	12.5 × 11 8.5 × 13	15 × 21	
	后舞台 Back Stage		26 × 25		18 × 18	
	台仓 Under Stage Storage		36 × 27	24 × 12.5	27 × 21	30 × 15
	升降乐池 Elevating Orchestra Pit		23 × 6.5		21 × 10.5	
	最大视距（m） Max. Visual Distance		32.5/44	35.0	25/28	
	最大俯角（°） Max. Angle of Depression		27	22	21	

中国特色的建筑：

采用曲线屋面是受到了中国古代传统建筑艺术的启发，特别是受到了"紫禁城"（故宫）中的宫殿屋顶的艺术启发而产生。建筑周边的壕沟也是受到"紫禁城"四周护城河启发而设计。

一座真正的歌剧院：

国家大剧院必须具有一个与众不同的性格外形，并使其能成为国家和城市首都的特征和标志。

在北京：

国家大剧院将建在首都，因此她必须能显示出北京这个首都城市的特殊地位。

天安门广场：

国家大剧院的建筑必须重点突出以天安门为城市中轴线的布局。创作出每每相互对称的，设有前置广场的建筑群。

对称性：

天安门广场是以正南、正北为中轴线的正交型布局。

未来的国家大剧院将保持这一城市的秩序。

外形：

故宫是四周围着高墙的庭院式建筑布局。

社会主义革命和建设时期的建筑是新古典主义正交式的设计布局。新的国家大剧院将采用一种自由式的，又必须是对称型延续历史的，以表达人民共和国未来的综合式布局。

建筑材料：

新的国家大剧院应该采用与附近西长安街南面一些建筑物相类似的建筑材料——新一代的石材。以表达人民共和国的连续性。

空间场地：

新的国家大剧院是一个有着很大面积活动场地的建筑群，以易于达到新的国家大剧院与天安门广场这个历史性的场所及其周围的建筑相匹配。

A Chinese Building

The curved roofs are inspired by the traditional roofs of China, particoularly those of the Forbidden city. The perimeter moat is also inspired by the moat of the Forbidden City.

An Opera House

The National Grand Theater must have a distinct form in order to become a symbol of the city and the country.

In Beijing

Beijing is the capital city. The National Grand Opera must reflect the monumental scale of Beijing.

In Tian An Men Square

The building must reflect the axial layout of Tian An Men, creating a symmetrical building with a large frontal public space.

Symmetry

Tian An Men Square is an

北侧透视
PERSPECTIVE OF NORTH ELEVATION

北侧模型
NORTH SIDE OF THE MODEL

歌剧院观众厅透视
PERSPECTIVE OF THE OPERA HOUSE AUDITORIUM

公共大厅透视
PUBLIC HALL PERSPECTIVE

orthogonal composition with a strong north-south axis.

The future building must preserve this urban order.

Form

The imperial buildings are organized around enclosed courtyards.

The socialist buildings are neoclassical orthogonal compositions.

The future buildings should be free but symmetrical forms, suggesting the future of the people's republic.

Materials

New buildings should be clad in the same limestone as used for the recent buildings, which are on all south of west Chang An Avenue, and represent thecontinuity of the people's republic.

Space

Similar to the historic buildings on Tian An Men Square, the new building should be an accessible building with a large public space.

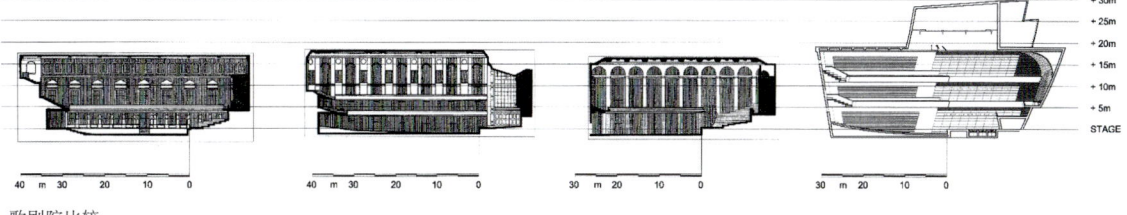

歌剧院比较
COMPARISON OF OPERA HOUSES

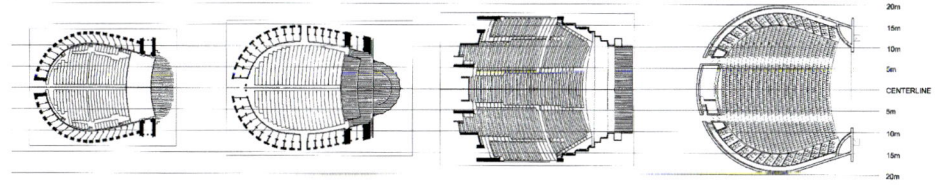

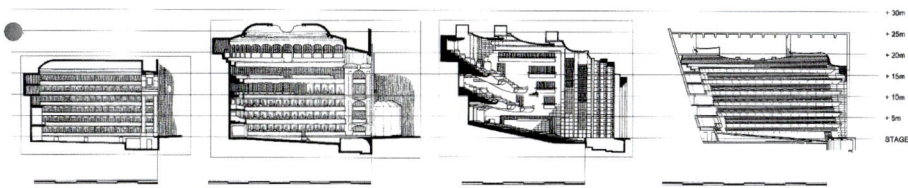

音乐厅比较
COMPARISON OF CONCERT HALLS

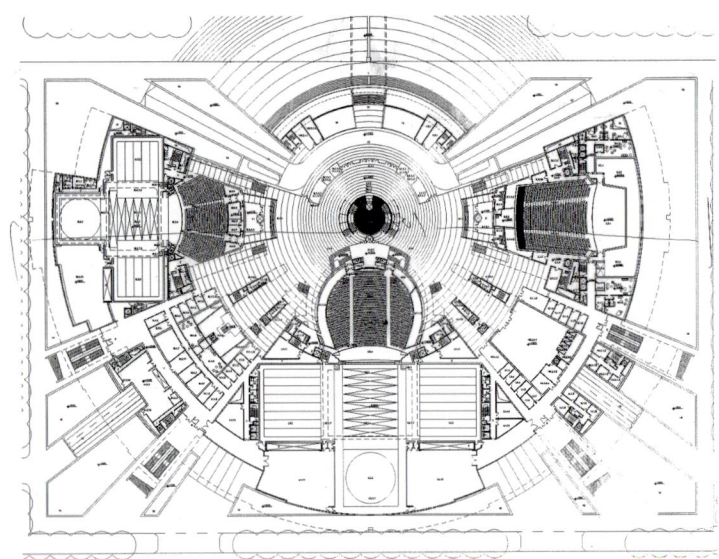

首层平面（± 0.00m）GROUND FLOOR LEVEL(± 0.00m)

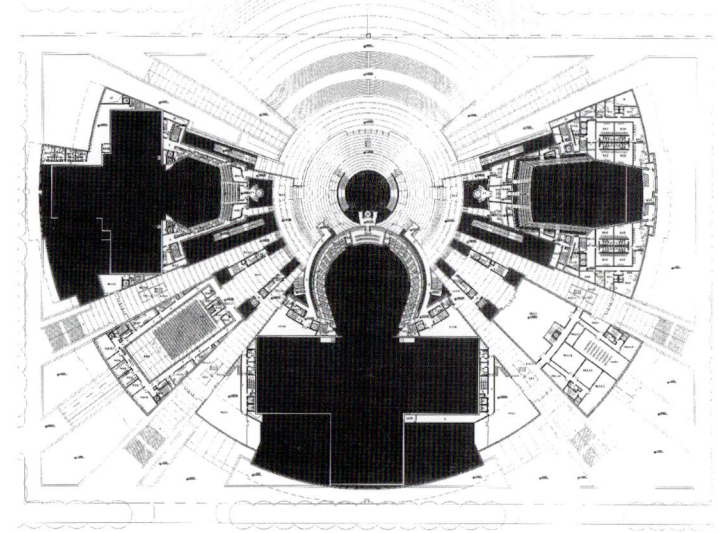

三层平面(+5.10m)　3rd FLOOR LEVEL(+5.1m)

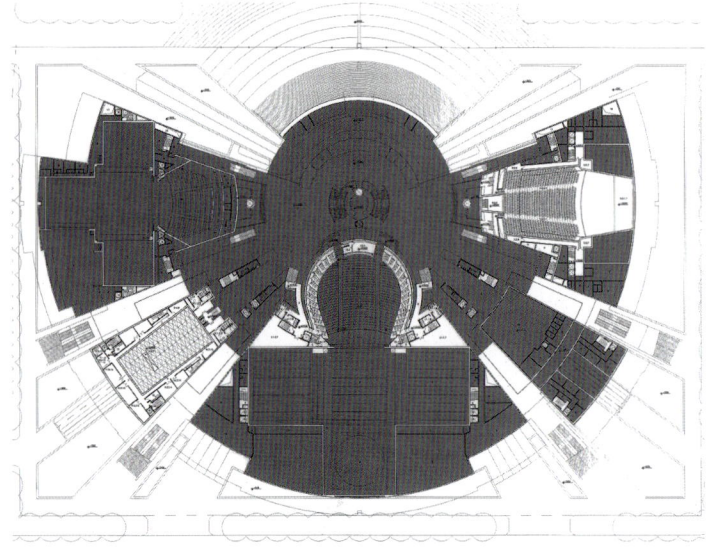

二层平面(+2.50m)　2nd FLOOR LEVEL(+2.50m)

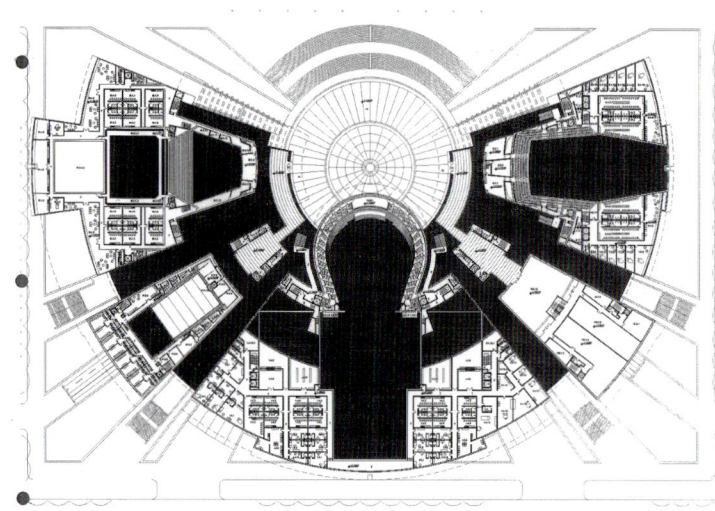

四层平面(+12.90m)　4th FLOOR LEVEL(+12.9m)

总平面图　OVERALL SITE PLAN

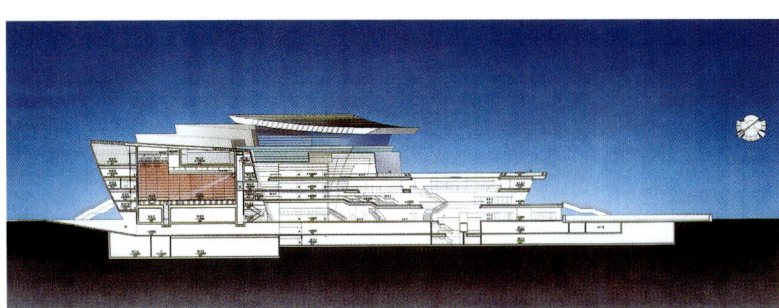

剖面
SECTIONS

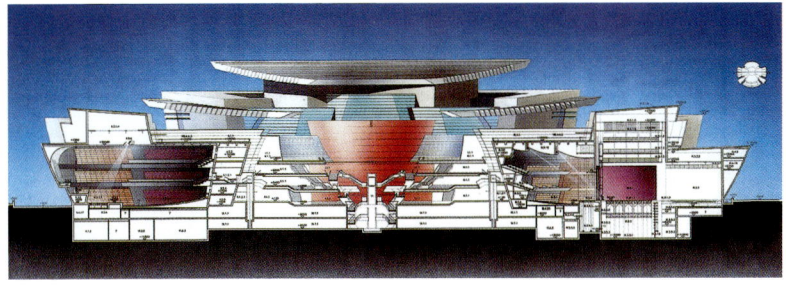

剖面
SECTIONS

剖面
SECTIONS

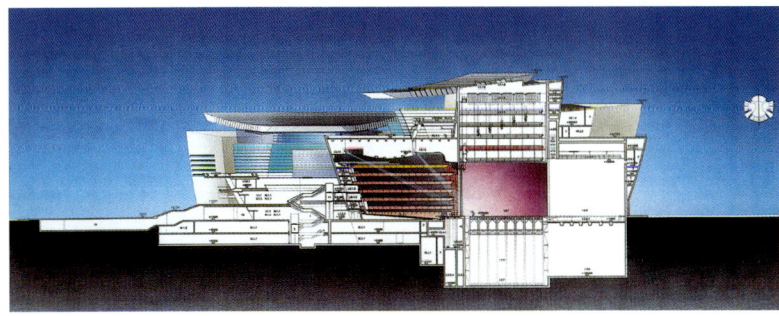

剖面
SECTIONS

技术经济指标
Technical Economic Index

基地面积(m²) Site Area	总建筑面积(m²) Total Floor Area	建筑高度(m) Building Height	建筑覆盖率 (%) Building Coverage Ratio
38915	129415	30/45(局部)	0.46
容积率 Plot Ratio	绿化覆盖率 (%) Green Coverage Ratio	停车数量 Number of Stalls	自行车停车数量 Number of Bicycle Stalls
3.32	19.5	地上：70 地下：533	1000(地下)

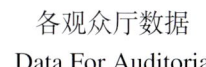

各观众厅数据
Data For Auditoria

		歌剧院 Opera House	音乐厅 Concert Hall	戏剧院 Theater	小戏场 Mini Theater
	观众厅面积(m²) Floor Area of Auditorium	2794	1710	863	482/283
	观众厅体积(m³) Volume of Auditorium	22250	20500	9384	2800/2143
座席数 Seating Capacity	总座席数 Seating Capacity(Seats)	2514	1880	1178	495/383
	其中：池座 Among Them:Auditorium	2794	1710	863	482/283
	楼 座 Floor Seat	1319	1052	583	241/186
	休息厅面积(m²) Lounge Floor Area	5931	4746	2848	2628
舞台尺寸 Stage Dimentions (m²)	主舞台 Main Stage	700	438	566	165/111
	左右侧台 Left and Right Side Stages	730	438	548	
	后舞台 Back Stage	652		324	00/45
	台 仓 Under Stage Storage	710	239	566	300/300
	升降乐池 Elevating Orchestra Pit	91(可调) 60(固定)	131(可调) 110(固定)	81(可调) 30(固定)	132/55
	最大视距(m) Max. Visual Distance	39	27.8	28.8	21.40/17.10
	最人俯角（°） Max. Angle of Depression	22	27	23	9/9

王欧阳(香港)有限
公司(中国香港)
Wong & Ouyang(HK)
Ltd.(China Hong Kong)

我们尝试揉合中国建筑体形上的飞瀑般流动的曲线屋顶和飘逸的重叠的飞檐，传统水墨画的神似而非绝对的形似，用抽象提炼的手法加以变形，结合现代高科技、新材料所提供的设计灵活性和造型的自由度，赋予大剧院以现代感。

The major architectural characteristic, however, is an abstraction of the traditional roof line that, in the same manner as the Chinese tradition of ink painting and poems, aims at the spirit and not the form. Integrated with a high technology approach, the elegant and fluid profile symbolizes the vital duality of tradition and its breakthrough.

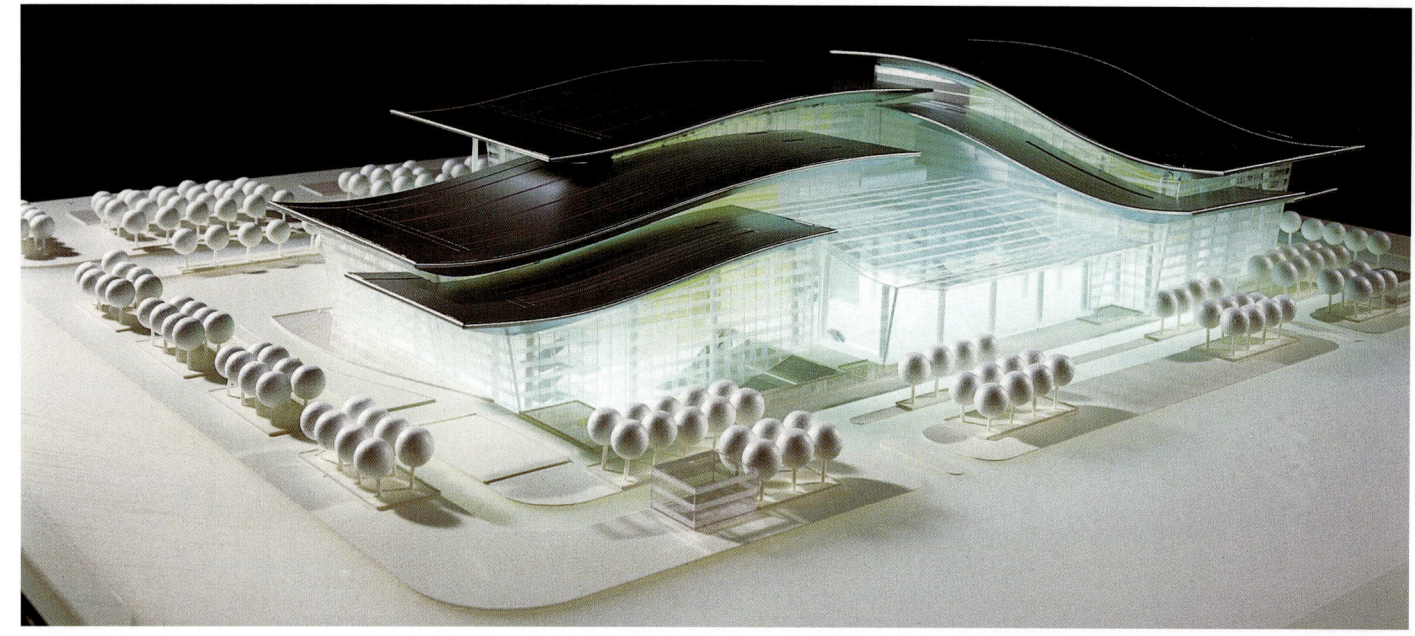

北侧模型 VIEW FROM NORTH SIDE OF THE MODEL

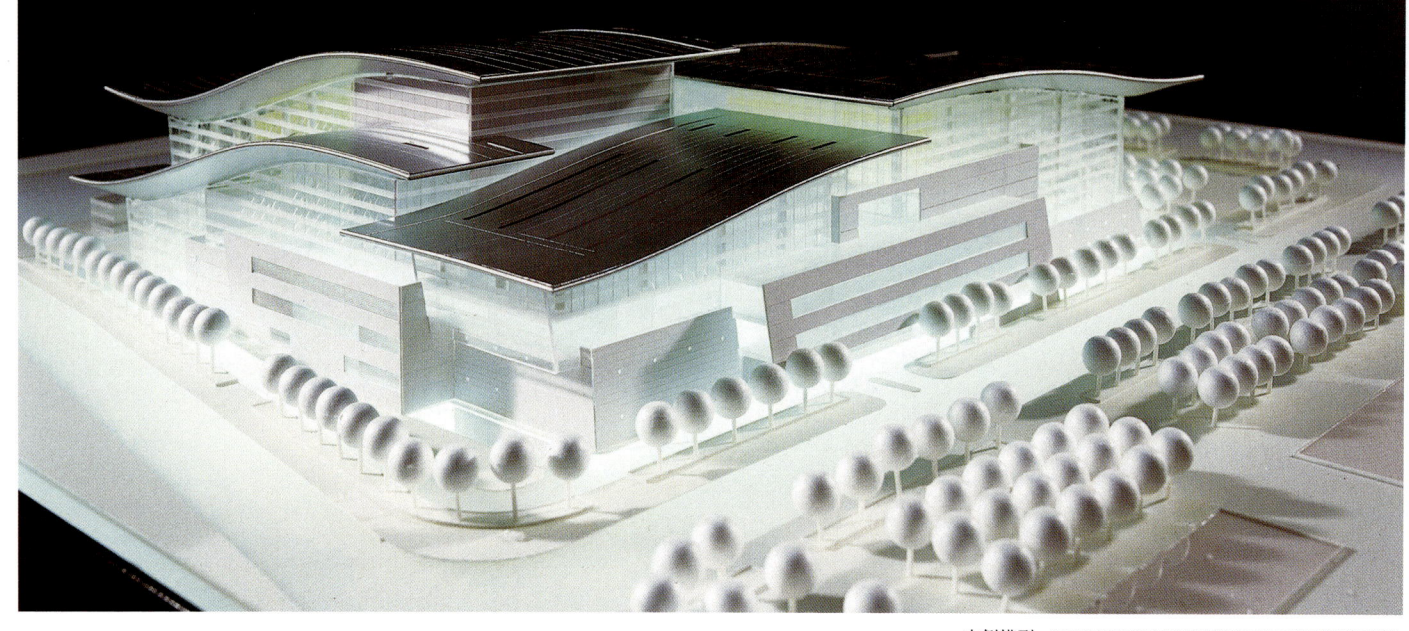

南侧模型 VIEW FROM SOUTH SIDE OF THE MODEL

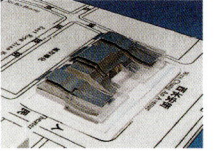
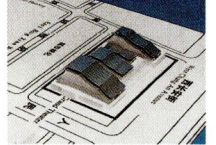
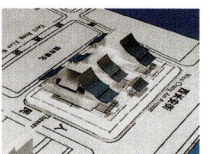
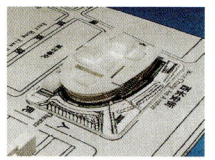
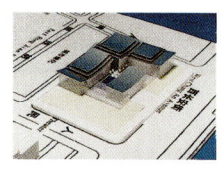
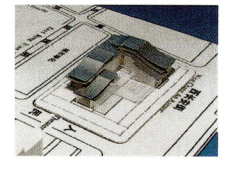
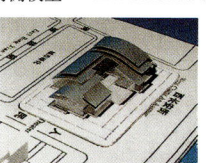
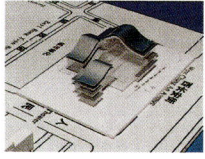
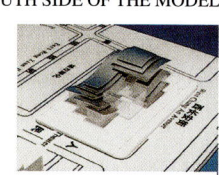
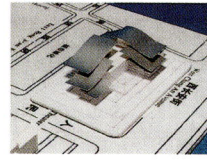

第一阶段 FIRST PHRASE

第三阶段 THIRD PHRASE

第二阶段 SECOND PHRASE

第四阶段 FOURTH PHRASE

公共大厅透视
PUBLIC HALL PERSPECTIVE

歌剧院观众厅透视
PERSPECTIVE OF THE OPERA HOUSE AUDITORIUM

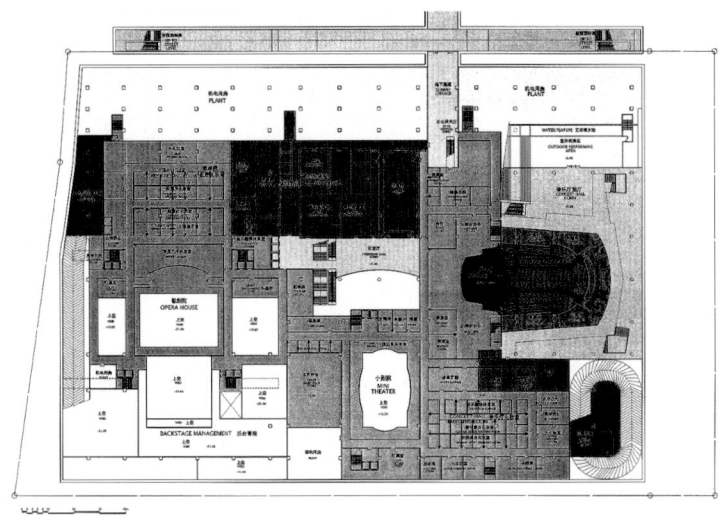

地下一层平面 FIRST BASEMENT PLAN

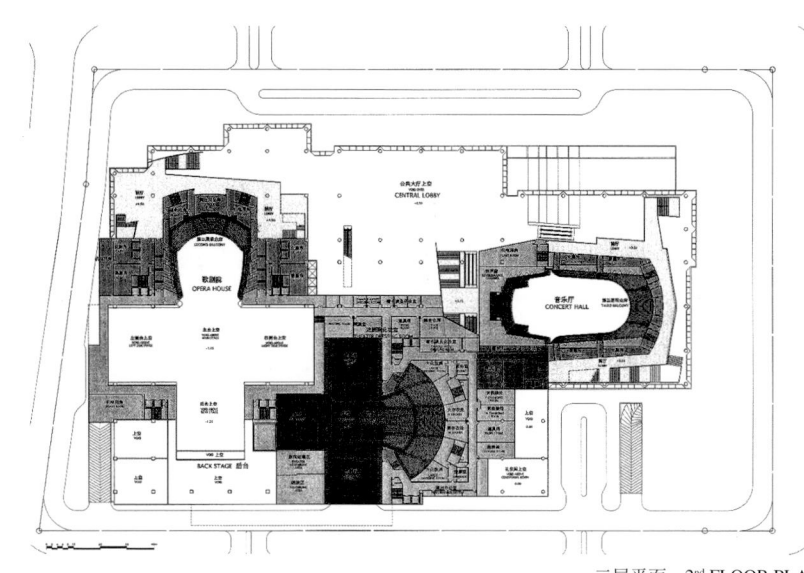

二层平面 2nd FLOOR PLAN

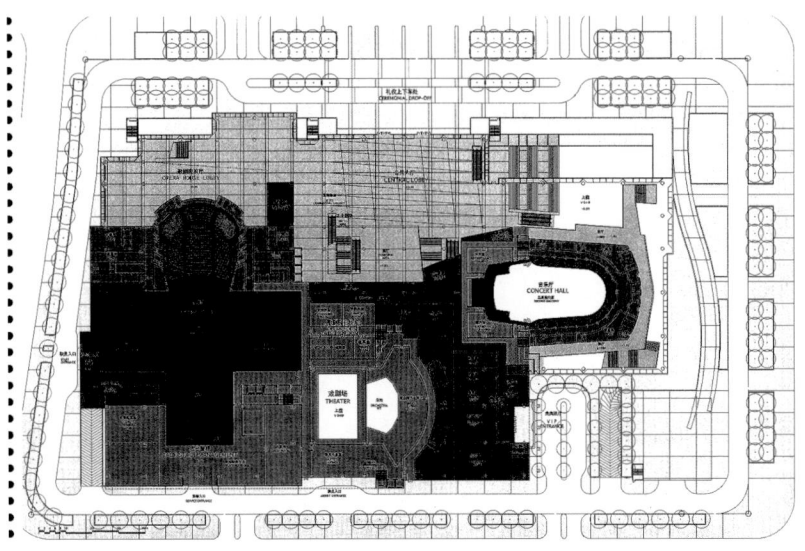

首层平面 GROUND FLOOR PLAN

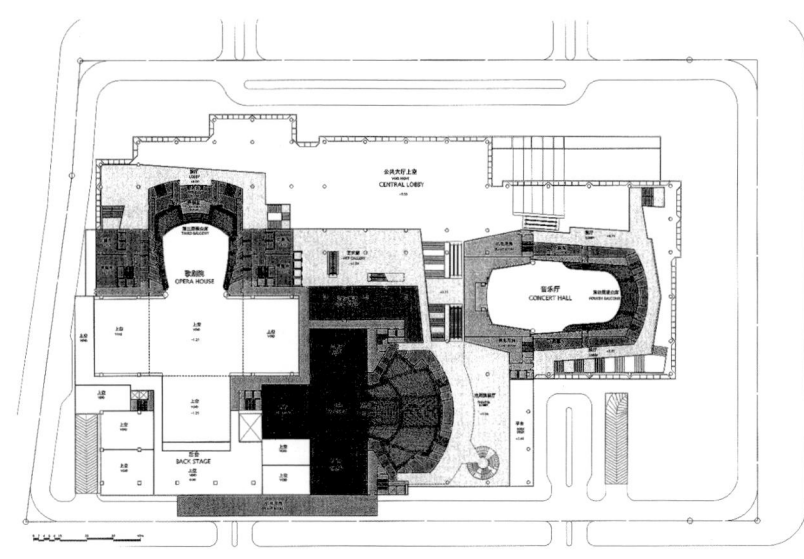

三层平面 3rd FLOOR PLAN

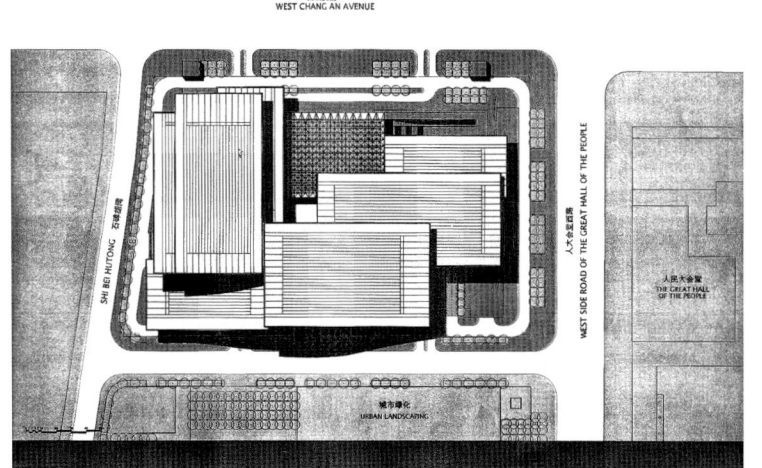

总平面 OVERALL SITE PLAN

剖面1—1
SECTION 1-1

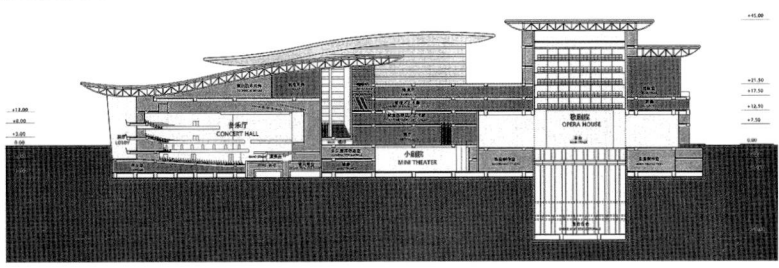

剖面2—2
SECTION 2-2

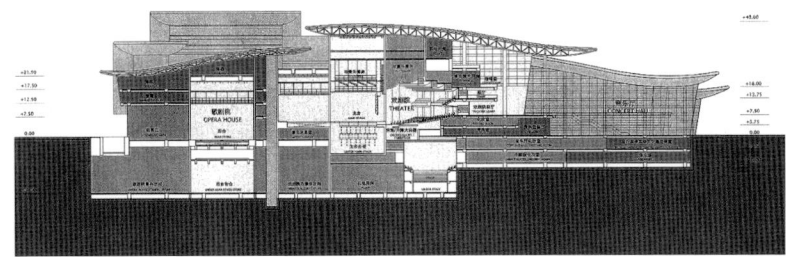

剖面3—3
SECTION 3-3

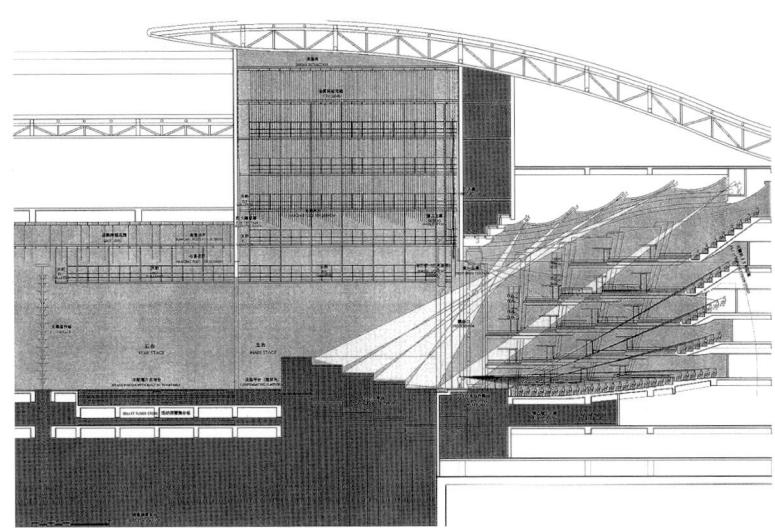

歌剧院舞台视线
STAGE SIGHT LINES OF THE OPERA HOUSE

技术经济指标
Technical Economic Index

基地面积(m²) Site Area	总建筑面积(m²) Total Floor Area	建筑高度(m) Building Height	建筑覆盖率(%) Building Coverage Ratio
38900	地上：68052 地下：51728	30/45(局部)	78
容积率 Plot Ratio	绿化覆盖率(%) Green Coverage Ratio	停车数量 Number of Stalls	自行车停车数量 Number of Bicycle Stalls
3	41	地上：80 地下：500	1000

各观众厅数据
Data For Auditoria

		歌剧院 Opera House	音乐厅 Concert Hall	戏剧院 Theater	小戏场 Mini Theater
	观众厅面积(m²) Floor Area of Auditorium	2175	1600	960	520
	观众厅体积(m³) Volume of Auditorium	21500	21200	16800	5663
座席数 Seating Capacity	总座席数 Seating Capacity(Seats)	2502	2006	1200	498
	其中：池座 Among Them:Auditorium	1075	1066	938	320
	楼座 Floor Seat	1427	940	262(378)	178
	休息厅面积(m²) Lounge Floor Area	3600	2800	1600	600
舞台尺寸(m²) Stage Dimentions	主舞台 Main Stage	916	432	526	
	左右侧台 Left and Right Side Stages	1242	432	632	
	后舞台 Back Stage	650		356	
	台仓 Under Stage Storage	1566	290	526	
	升降乐池 Elevating Orchestra Pit	160		205	
	最大视距(m) Max. Visual Distance	38	39	26.5	21.5
	最大俯角(°) Max. Angle of Depression	35	23	31	22

204

日本设计株式
会社（日本）
Nihon Sekkei,Inc.(Japan)

建设部建筑
设计院（中国）
Architectural Design
Institute of Ministry of
Construction (China)

建筑造型继承中国传统的建筑技术与设计。同时，充分利用现代最先进科技，以创造出全球独一无二的既现代化又风格独特的城市景观。

外墙面设计为在长城、天安门等中国最著名的传统建筑上所采用的，略为倾斜，雄伟而庄重的大墙壁。

空间设计为无柱的大空间，可通过最先进的高透明性的大面积的吊挂式玻璃幕墙获得丰富的自然采光。

在几个主要入口和观众休息厅上部设置了多处反弧面的金属饰面屋顶，其造型新颖、独特，和大面积的具传统象征意义的红墙壁形成强烈对比，使立面更加生动而富于变化。

A design statement created upon the foundations of traditional Chinese design and the latest modern design technologies in vicinnical century.

As can be seen in the Tian An Men Rostrum and the Great wall, the walls will be tilted slightly and the stone masonry of the large exterior wall will have the same depth, prestige and grandeur.

The design will feature a large space with high ceilings and no columns. Large, highly transparent glass walls and a massive glass roof will infuse the space with natural sunlight.

At the upper part of several main entrances and foyers the roof is decorated by many of the reflex arc metals. The novel and unique shape is strong comparison to the large area of traditional symbolized red wall; which made the elevation more active and varied.

北侧透视　PERSPECTIVE OF NORTH ELEVATION

南侧透视　PERSPECTIVE OF SOUTH ELEVATION

立面　SECTIONS

歌剧院透视
PERSPECTIVE OF THE OPERA HOUSE

休息大厅透视
FOYER PERSPECTIVE

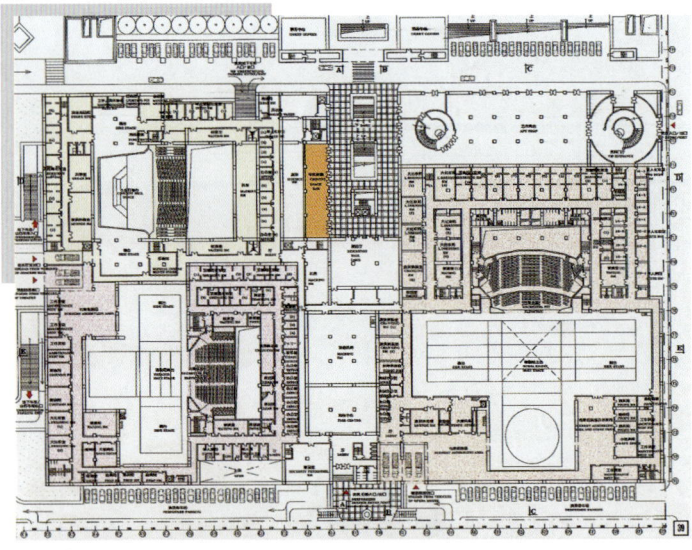

一层平面　GROUND FLOOR PLAN

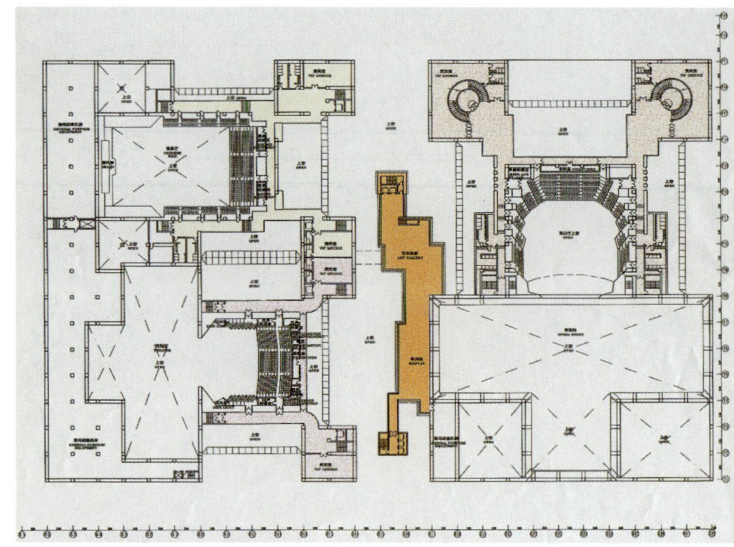

三层平面　3rd FLOOR PLAN

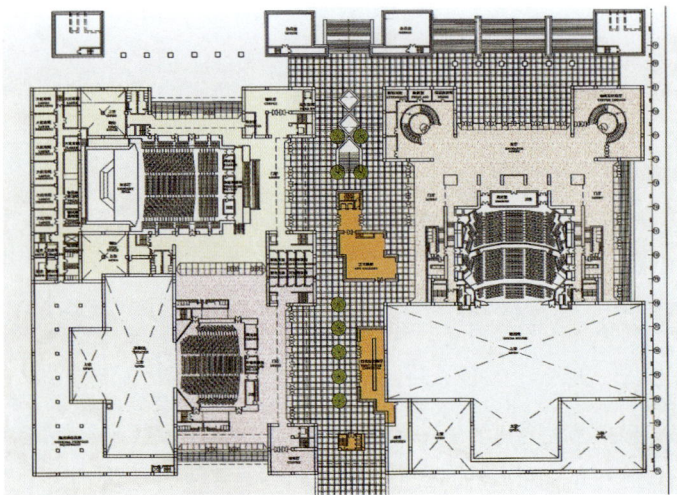

二层平面　2nd FLOOR PLAN

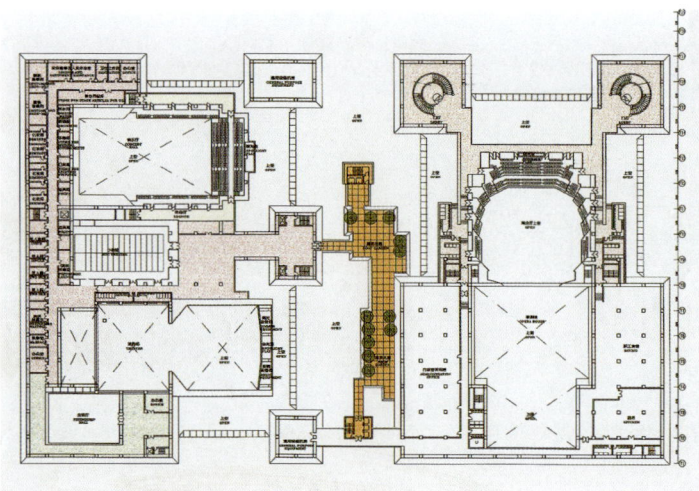

四层平面　4th FLOOR PLAN

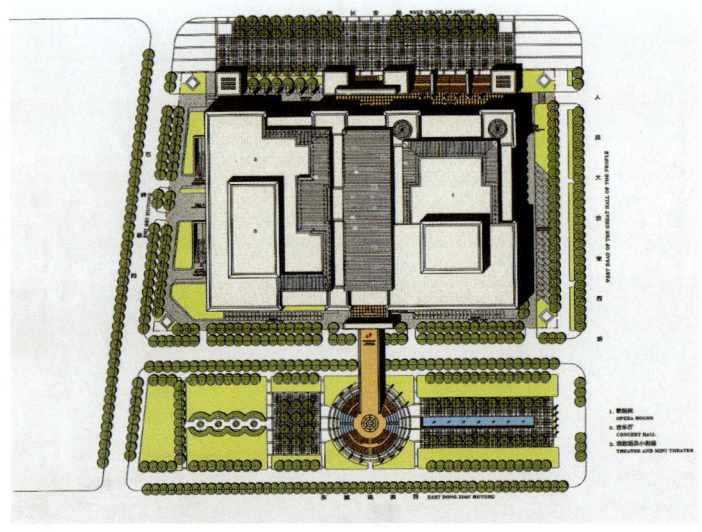

总平面　OVERALL SITE PLAN

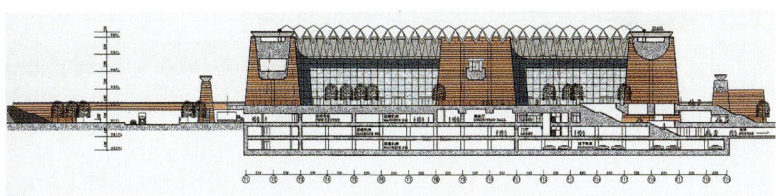

A—A 剖面　SECTION A-A

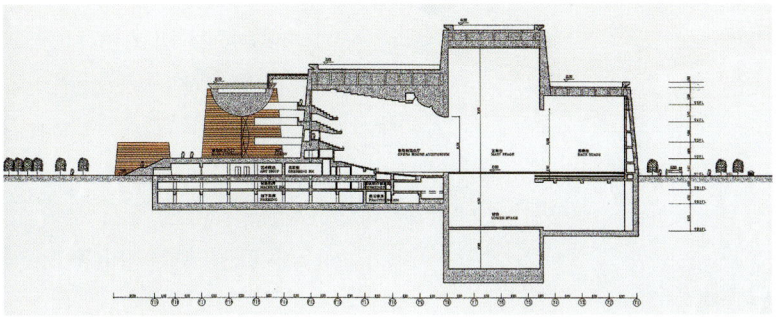

C—C 剖面　SECTION C-C

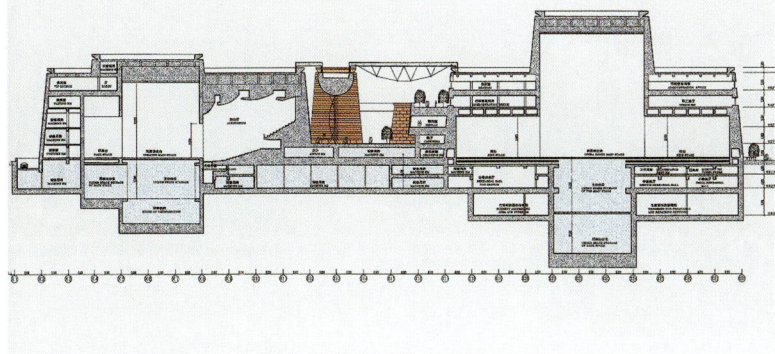

D—D 剖面　SECTION D-D

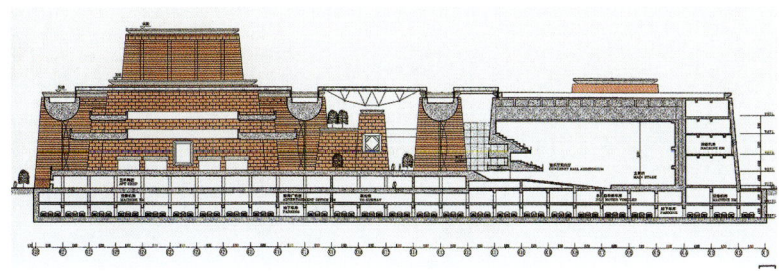

E—E 剖面　SECTION E-E

技术经济指标
Technical Economic Index

基地面积(m²) Site Area	总建筑面积(m²) Total Floor Area	建筑高度(m) Building Height	建筑覆盖率（%） Building Coverage Ratio
36700	地上：74820 地下：56180	45	79.5

容积率 Plot Ratio	绿化覆盖率(%) Green Coverage Ratio	停车数量 Number of Stalls	自行车停车数量 Number of Bicycle Stalls
3.6	12	地上：111 地下：501	1000

各观众厅数据
Data For Auditoria

		歌剧院 Opera House	音乐厅 Concert Hall	戏剧院 Theater	小戏场 Mini Theater
	观众厅面积(m²) Floor Area of Auditorium	2384	1608	962	1998
	观众厅体积(m³) Volume of Auditorium	18810	15230	9020	6080
座席数 Seating Capacity	总座席数 Seating Capacity(Seats)	2572	2006	1202	450~522
	其中：池座 Among Them:Auditorium	1246	1146	830	352~424
	楼座 Floor Seat	1326	860	372	98
	休息厅面积(m²) Lounge Floor Area	3720	2846	1880	1652
舞台尺寸(m²) Stage Dimentions	主舞台 Main Stage	972	450	526	78
	左右侧台 Left and Right Side Stages	1242	531	736	
	后舞台 Back Stage	725		345	
	台仓 Under Stage Storage	1344	207	660	
	升降乐池 Elevating Orchestra Pit	168		96	
	最大视距(m) Max. Visual Distance	43	36	28	25
	最人俯角（°） Max. Angle of Depression	26	22	20	21

205

建设部建筑
设计院（中国）
Architectural Design
Institute of Ministry of
Construction (China)

国家——"古琴"的意象表达对中国源远流长的民族艺术的象征。

首都——主体建筑形象表达对作为中国古典都市与建筑杰作的北京的敬意与融合。

天安门广场——背景建筑形象表达对人民大会堂、革命历史博物馆、毛主席纪念堂等天安门广场建筑群的延续。

"琴与琴盒"——主体建筑与背景建筑的插合关系，形成本方案的基本意念与特征。两把琴分别是歌剧院与音乐厅、戏剧场与小剧场的组合体；它们是产生美好艺术的地方，是艺术与艺术家的象征；琴盒是方正的柱廊，呼应着邻近的人民大会堂，是人民的象征；琴与琴盒的关系表达了艺术产生于人民，艺术家从人民中走出又必将回归人民中间的道理。

"琴弦"——琴弦从建筑屋顶的肋状结构延伸到广场与公共大厅之中，变成广场上的线状桥和地面的图案，是人们经常活动的地方。他们在"琴弦"上驻足，犹如在演奏美妙的乐曲，表达出人民参与艺术，人民是艺术的创造者的含义。

戏剧意象——沿长安街经过大剧院，建筑形象在人们面前的渐变展开，犹如经历了一场

内院透视　PERSPECTIVE VIEW FROM THE INSIDE

北侧模型
VIEW FROM NORTH SIDE OF THE MODEL

休息厅透视　FOYER PERSPECTIVE

设计理念
DESIGN CONCEPT

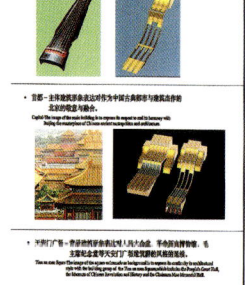

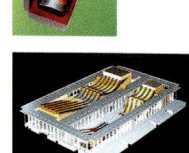

由大幕缓缓拉开直到谢幕的戏剧。

State---The image of "Ancient Qin"symbolizes the longstanding and traditional Chinese national arts.

Capital---The image of the main building is to express its respect to and its harmony with Beijing---the masterpiece of Chinese ancient metropolitan and architecture.

Tian An Men Square---The image of the square colonnade as background is to express its continuity in architectural style with the building group of the Tian An Men square, which includes the People's Great Hall, the Museum of Chinese Revolution and History and the Chairman Mao Memorial Hall.

"Qin"and its case---The insertion of the main building and the background building forms the basic concept and character of this design. Two Qins are the combinations of the opera house with the concert hall, and the theater with the mini-theater respectively. They are the places where excellent arts are created, and they are the symbols of arts and artists."Case"is the square colonnade which in accordance with the People's Great Hall nearby. It is the symbol of people.The relationship between Qin and its case is to express that arts are created from people and the artists who come from people must go back to people.

南侧透视
PERSPECTIVE OF SOUTH ELEVATION

String of "Qin"---Strings flow down all the way from the rib structure of the building roof to the public hall and the square, then change into the narrow bridges and the floor paving. These are the places where people wander around. They could stand on the "strings" as if playing beautiful music. This expresses the meaning that people take part in arts and people are the creator of arts.

Opera Sensation---Passing by the National Grand Theater from the Chang An Avenue, the gradual emerging of the building is just like enjoying an opera from curtain open to curtain close.

歌剧院观众厅
PERSPECTIVE OF THE OPERA HOUSE AUDITORIUM

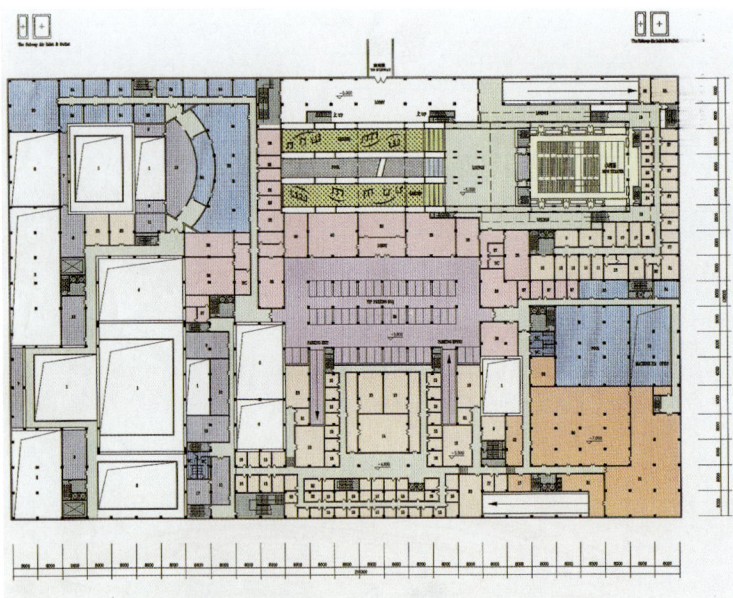

地下一层平面　FIRST BASEMENT PLAN

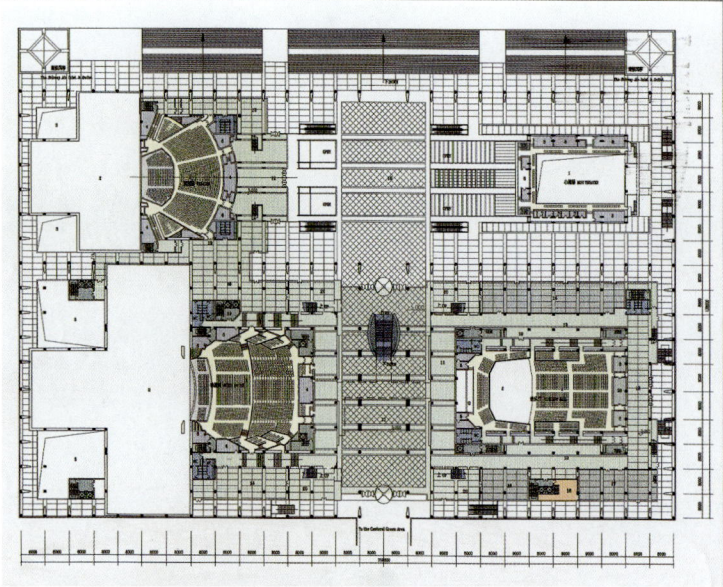

二层平面　2nd FLOOR PLAN

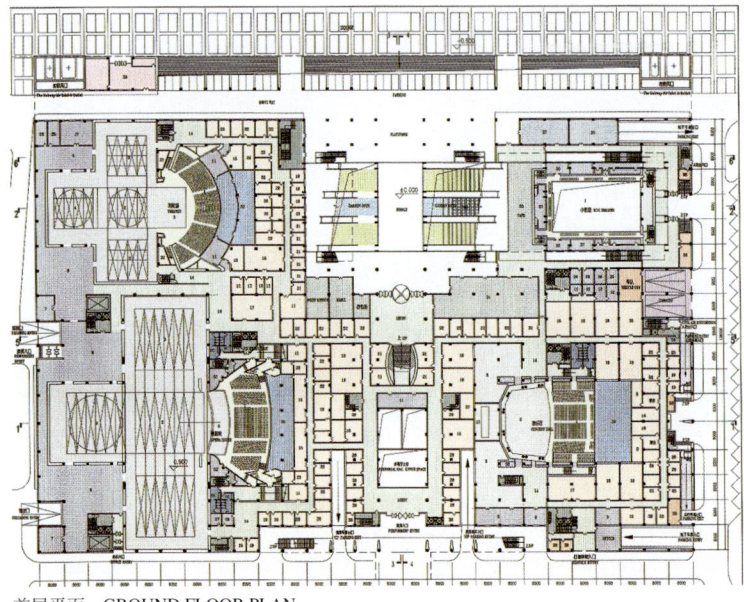

首层平面　GROUND FLOOR PLAN

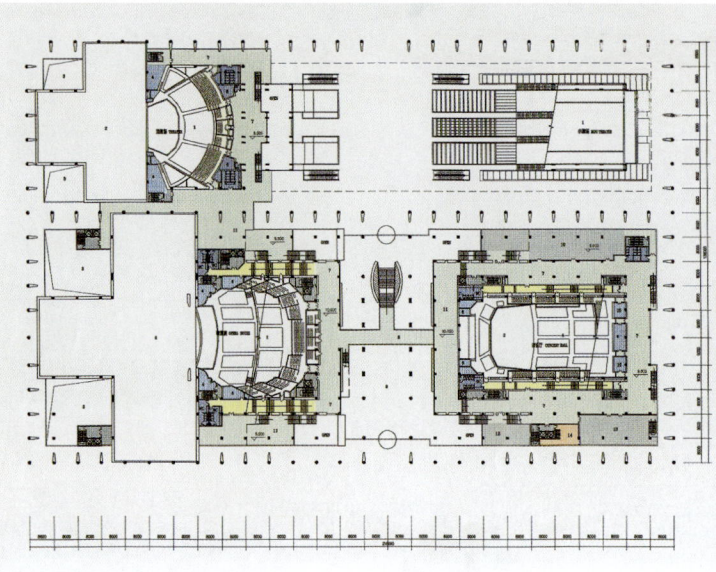

三层平面　3rd FLOOR PLAN

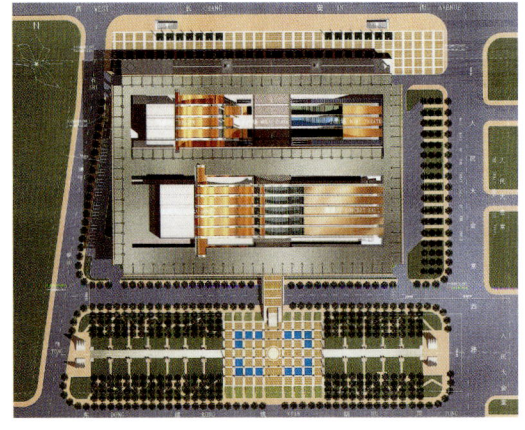

总平面　OVERALL SITE PLAN

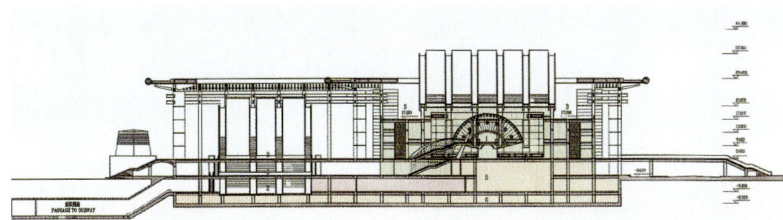

1-1 剖面　SECTION 1-1

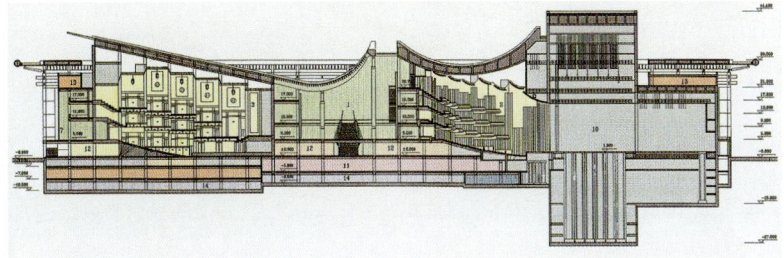

2-2 剖面　SECTION 2-2

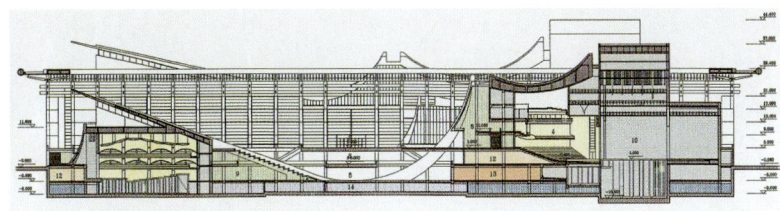

4-4 剖面　SECTION 4-4

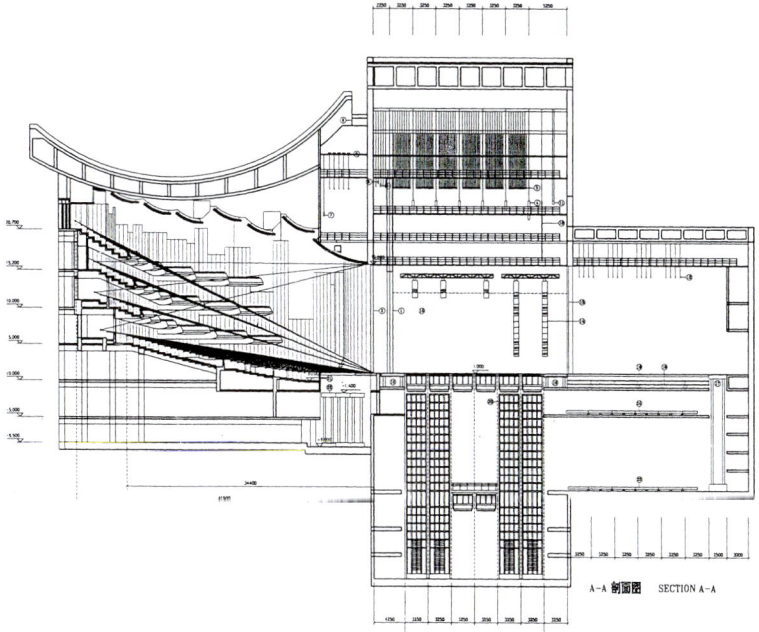

歌剧院视线分析及舞台布置
VISUAL ANALYSIS & STAGE LAYOUT OF THE OPERA HOUSE

技术经济指标
Technical Economic Index

基地面积(m²) Site Area	总建筑面积(m²) Total Floor Area	建筑高度(m) Building Height	建筑覆盖率 (%) Building Coverage Ratio
38900	128928(合计)	30/45(局部)	77

容积率 Plot Ratio	绿化覆盖率 (%) Green Coverage Ratio	停车数量 Number of Stalls	自行车停车数量 Number of Bicycle Stalls
2.51 (不含地下二、三层)	7	地上：140 地下：497	1045

各观众厅数据
Data For Auditoria

		歌剧院 Opera House	音乐厅 Concert Hall	戏剧院 Theater	小戏场 Mini Theater
	观众厅面积(m²) Floor Area of Auditorium	2200	1980	940	870
	观众厅体积(m³) Volume of Auditorium	1220	1070	790	630
座席席数 Seating Capacity	总座席数 Seating Capacity(Seats)	2514	1992	1206	474~628
	其中：池座 Among Them:Auditorium	1292	926	1018	314~628
	楼　座 Floor Seat	1100	1066	188	180
	休息厅面积(m²) Lounge Floor Area	4680	4250	2240	960
舞台尺寸(m²) Stage Dimentions	主舞台 Main Stage	34 × 27 × 39	22 × 16 × 21	27 × 21 × 29	64~192
	左右侧台 Left and Right Side Stages	23 × 27 × 12	550	15 × 21 × 12 +10 × 21 × 12	
	后舞台 Back Stage	26 × 25 × 18		23 × 15 × 16	50
	台　仓 Under Stage Storage	34 × 27 × 16	12 × 17 × 6	27 × 21 × 12	28 × 16 × 4
	升降乐池 Elevating Orchestra Pit	20 × 10		14 × 16	
	最大视距(m) Max. Visual Distance	41.5	31	28	
	最大俯角 (°) Max. Angle of Depression	25	25	20	

206
清华大学建筑
设计研究院
（中国）
Architectural Design
& Research Institute
of Tsinghua
University(China)

建筑总体布局说明：

1．体形设计简洁完整而又丰富了城市景观。

2．建筑内部南北贯通，融入环境。

3．"一竖两横"的布局。

4．合理的交通组织。

建筑风格和体形轮廓：

1．朴实、高雅的文化气息。

2．传统的东方韵味。

3．艺术的浪漫气质。

4．现代化的时代精神。

General Layout

1.Physical form:being concise and complete; to enrich the city landscape.

2.Inner space: being penetrated from north to south, connected with urban surroundings.

3.Layout:"one vertical and two horizontals".

4.Access: being organized rationally.

Architectural styles and configurable outlines

1.simple, noble and graceful culture flavor

2.traditional eastern charm

3.romantic artistic style

4.modern spirit of the times

北侧透视　PERSPECTIVE OF NORTH ELEVATION

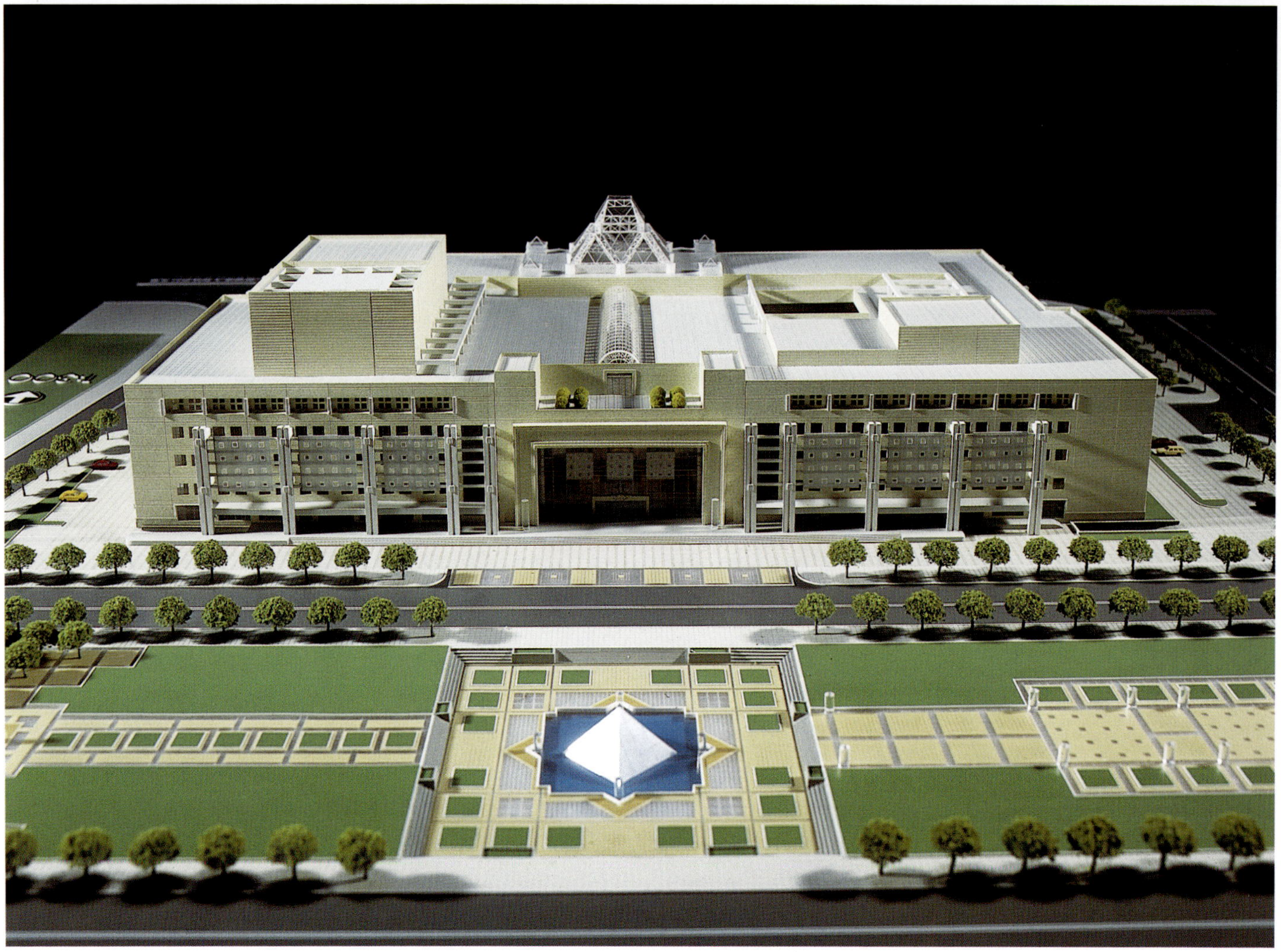

南侧模型照片　VIEW FROM SOUTH SIDE OF THE MODEL

歌剧院观众厅
PERSPECTIVE OF THE OPERA HOUSE AUDITORIUM

音乐厅观众厅透视
PERSPECTIVE OF THE OPERA HOUSE AUDITORIUM

休息大厅透视
FOYER PERSPECTIVE

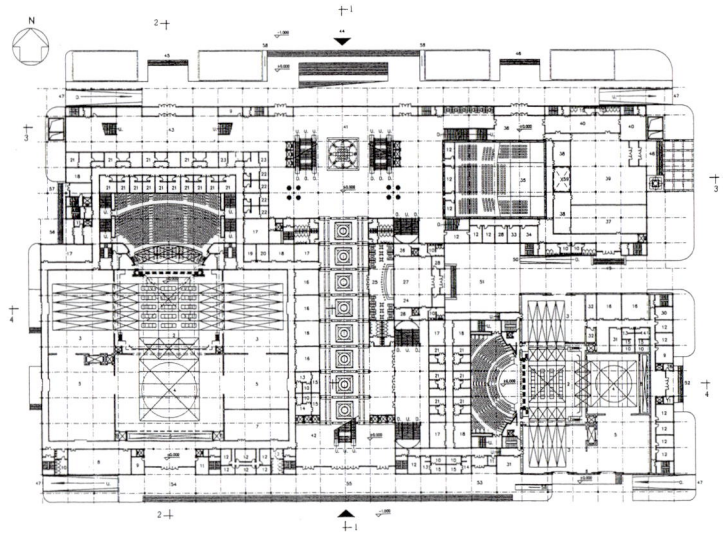

一层平面　GROUND FLOOR PLAN

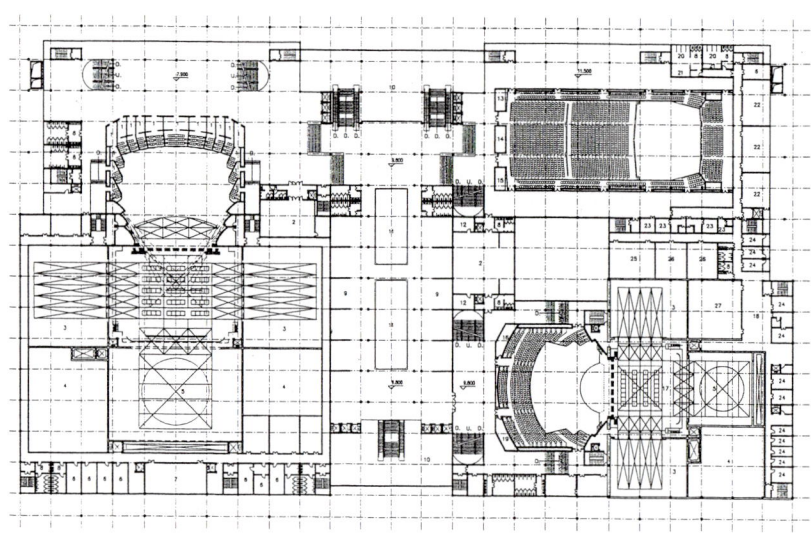

三层平面　3rd FLOOR PLAN

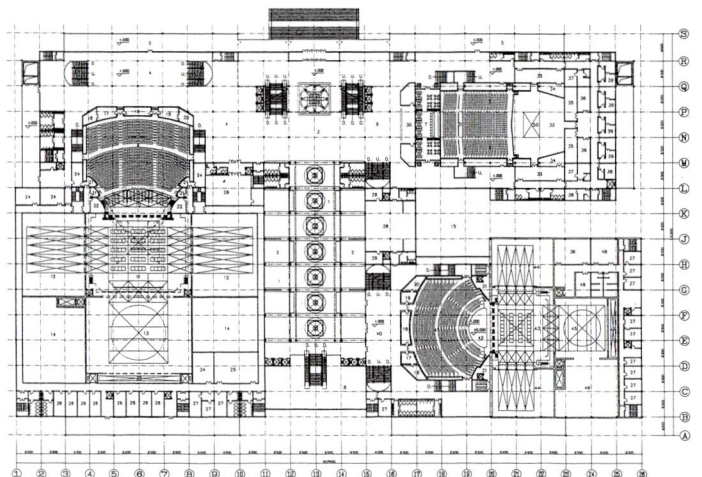

二层平面　2nd FLOOR PLAN

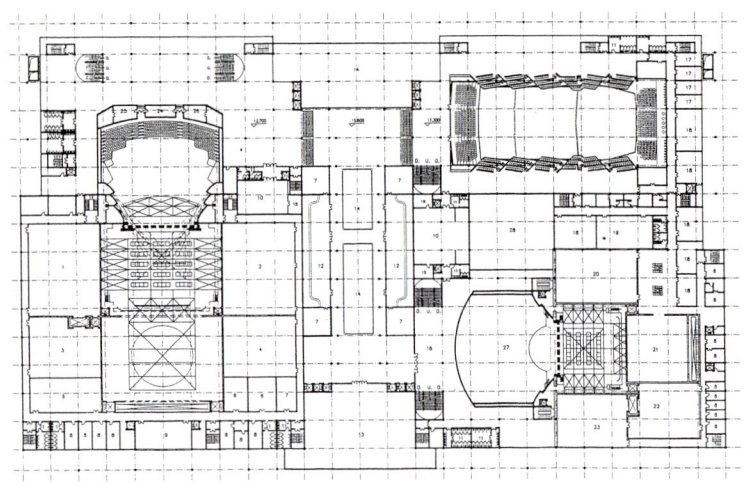

四层平面　4th FLOOR PLAN

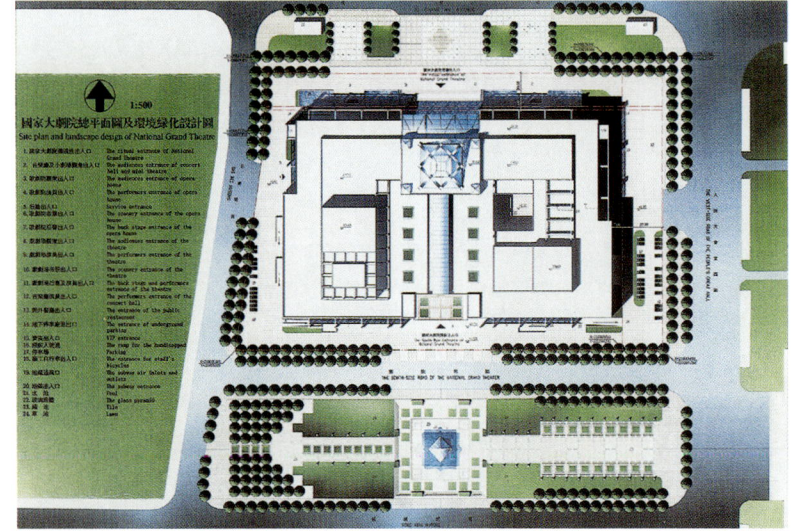

总平面　OVERALL SITE PLAN

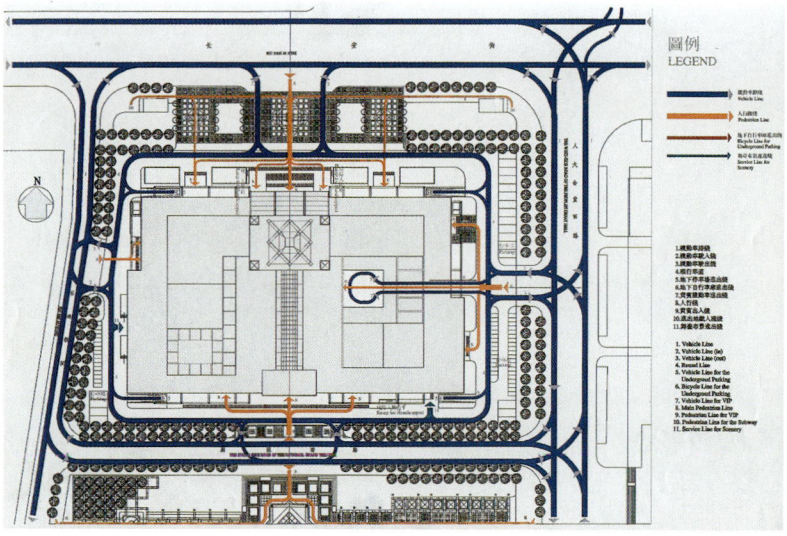

交通图　TRAFFIC PLAN

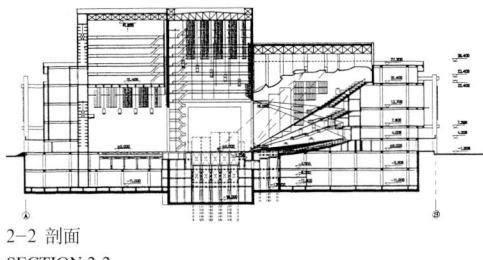

2-2 剖面
SECTION 2-2

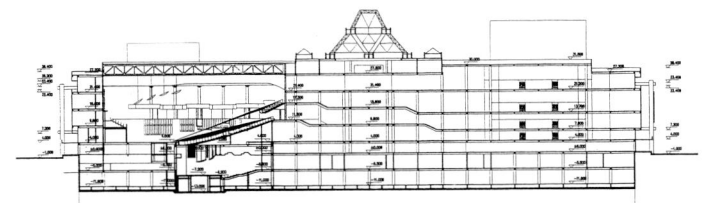

3-3 剖面
SECTION 3-3

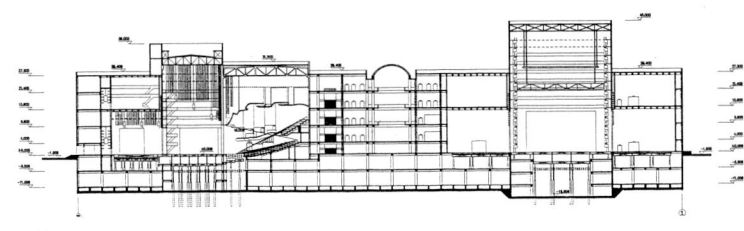

4-4 剖面
SECTION 4-4

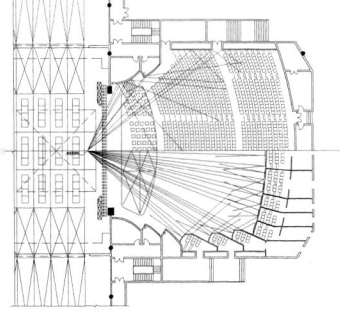

歌剧院声学分析平面
ACOUSTICAL ANALYSIS PLAN OF THE OPERA HOUSE

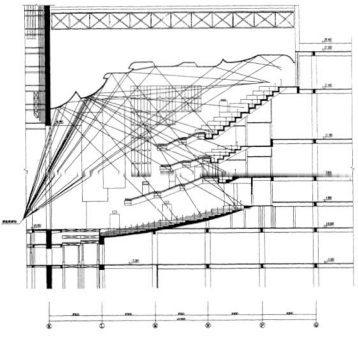

声学分析剖面
ACOUSTICAL ANALYSIS SECTION

技术经济指标
Technical Economic Index

基地面积(m²) Site Area	总建筑面积(m²) Total Floor Area	建筑高度(m) Building Height	建筑覆盖率(%) Building Coverage Ratio
3.89	地上：80000 地下：48000	28.4/45(局部)	60.8

容积率 Plot Ratio	绿化覆盖率(%) Green Coverage Ratio	停车数量 Number of Stalls	自行车停车数量 Number of Bicycle Stalls
3.34	7.1	地上：71 地下：541	1000

各观众厅数据
Data For Auditoria

		歌剧院 Opera House	音乐厅 Concert Hall	戏剧院 Theater	小戏场 Mini Theater
	观众厅面积(m²) Floor Area of Auditorium	2792	1750	1103	828
	观众厅体积(m³) Volume of Auditorium	22773	23722	9366	7452
座席数 Seating Capacity	总座席数 Seating Capacity(Seats)	2516	2040	1398	366~596
	其中：池座 Among Them:Auditorium	1322	1056	884	
	楼座 Floor Seat	1194	984	514	
	休息厅面积(m²) Lounge Floor Area	5600	3800	1700	1400
舞台尺寸 Stage Dimentions (m²)	主舞台 Main Stage	36×27	21.9×18	27×21	24.9×16.6
	左右侧台 Left and Right Side Stages	21.5×27	5.0×16.6	16×21	
	后舞台 Back Stage	36×25		20×18	
	台仓 Under Stage Storage	36×27(主) 36×25(后)	8.3×5.8	27×21(主) 20×18(后)	24.9×16.6
	升降乐池 Elevating Orchestra Pit	21×6		12×6	
	最大视距(m) Max. Visual Distance	37.0	36.9	24.5	
	最大俯角(°) Max. Angle of Depression	30.5	20	27	

207
浙江省建筑设计
研究院(中国)
Zhejiang Building
Design and Research
Institute (China)

非对称式格局，歌剧院布置在西侧，靠石碑胡同，以免由于体量过大与东侧人大会堂造成拥塞感，为了考虑环境文脉的因素，最终还是采用了对称式格局，以歌剧院为中心统领整个建筑群。

A symmetrical the Opera House will be built in the west of the Grand Theater. Close to Shibe Hutong so as to avoid the sense of jamming and clogging caused by the huge volume of the Great Hall of People on the east In view of the traditional factors of the surroundings, a symmetrical pattern is adopted, with the Opera House as the core to command the whole complex.

北侧模型
VIEW FROM NORTH SIDE OF THE MODEL

南侧模型
VIEW FROM SOUTH SIDE OF THE MODEL

歌剧院观众厅透视
PERSPECTIVE OF THE OPERA HOUSE AUDITORIUM

休息大厅透视
FOYER PERSPECTIVE

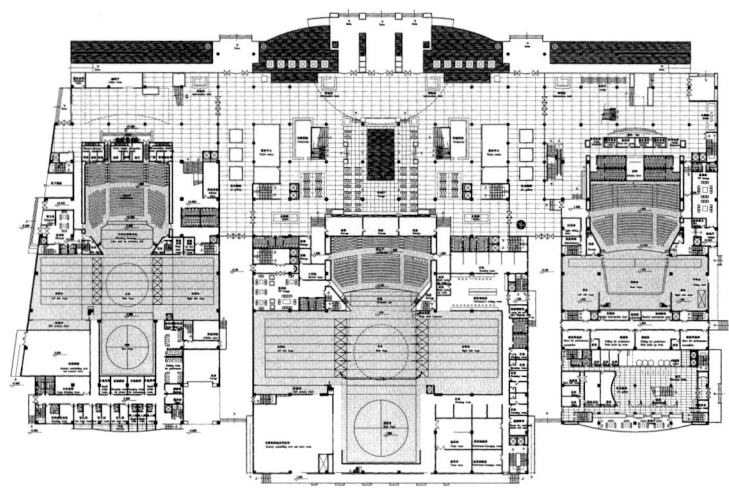

一层平面　GROUND FLOOR PLAN

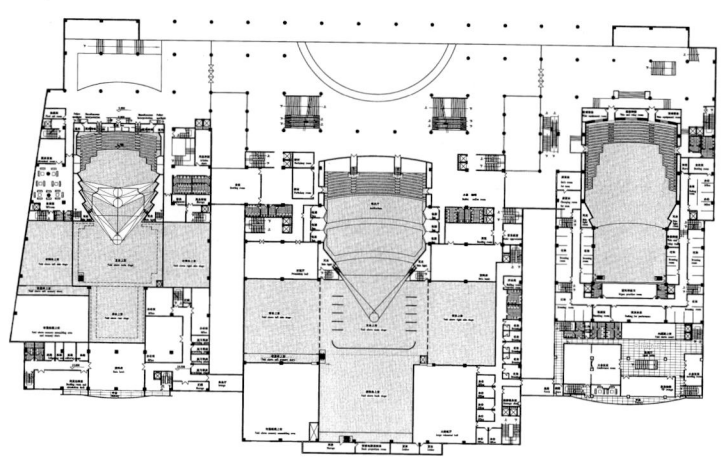

三层平面　3rd FLOOR PLAN

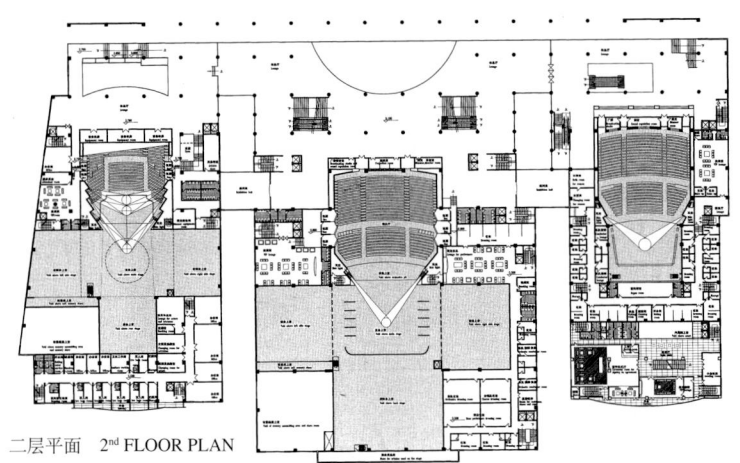

二层平面　2nd FLOOR PLAN

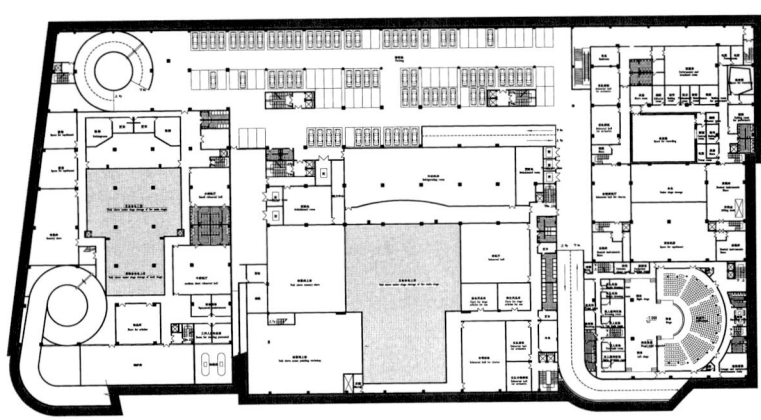

地下二层平面　2nd BASEMENT PLAN

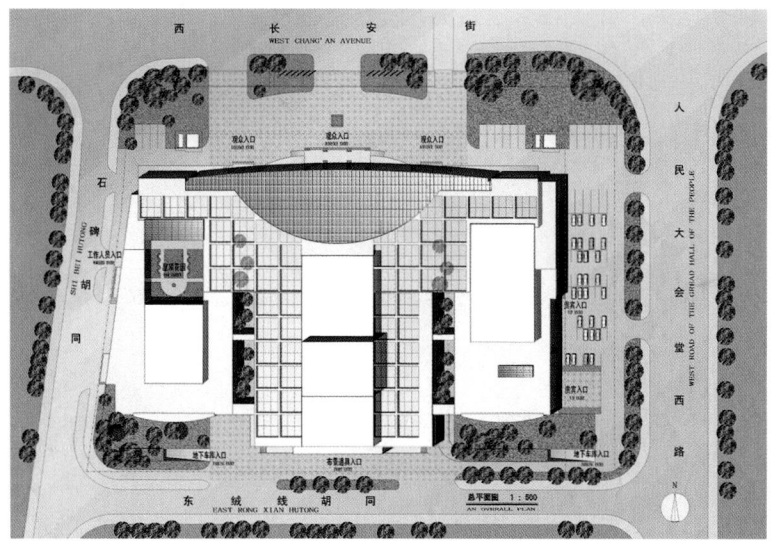

总平面　OVERALL SITE PLAN

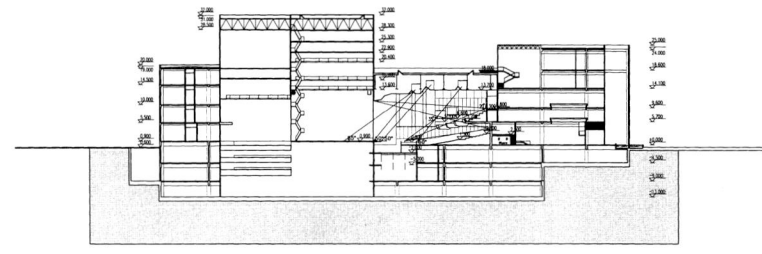

1-1 剖面
SECTION 1-1

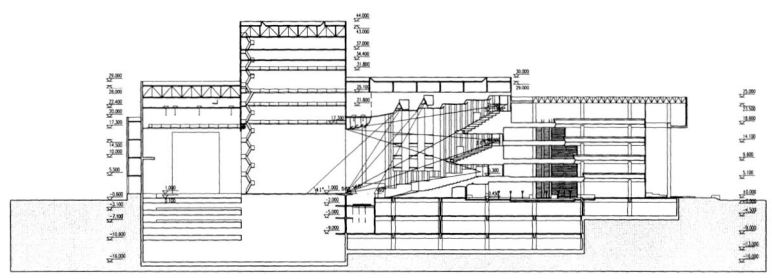

2-2 剖面
SECTION 2-2

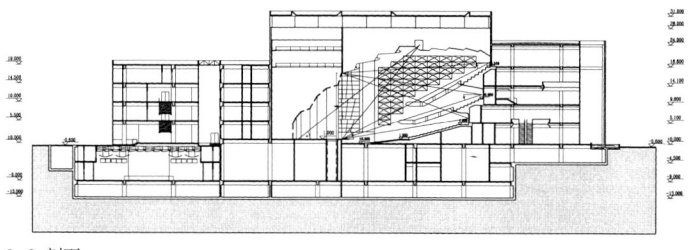

3-3 剖面
SECTION 3-3

技术经济指标
Technical Economic Index

基地面积(m²) Site Area	总建筑面积(m²) Total Floor Area	建筑高度(m) Building Height	建筑覆盖率 (%) Building Coverage Ratio
38900	地上：72059 地下：61400	25/44(局部)	65.6
容积率 Plot Ratio	绿化覆盖率 (%) Green Coverage Ratio	停车数量 Number of Stalls	自行车停车数量 Number of Bicycle Stalls
1.85	11.9	地上：63 地下：498	1236

各观众厅数据
Data For Auditoria

		歌剧院 Opera House	音乐厅 Concert Hall	戏剧院 Theater	小戏场 Mini Theater
	观众厅面积(m²) Floor Area of Auditorium	1476	1323	860	450
	观众厅体积(m³) Volume of Auditorium	16460	21200	7400	3000~3360
座席数 Seating Capacity	总座席数 Seating Capacity(Seats)	2509	2009	1208	383
	其中：池座 Among Them:Auditorium	636	710	628	341
	楼 座 Floor Seat	1873	1299	580	42
	休息厅面积(m²) Lounge Floor Area	2523	2216	2028	370
舞台尺寸 Stage Dimentions (m²)	主舞台 Main Stage	36 × 27	24.75 × 18	27 × 21	
	左右侧台 Left and Right Side Stages	27 × 22	164	15.5 × 21	
	后舞台 Back Stage	26 × 25	7.5 × 12	19 × 19	
	台 仓 Under Stage Storage	1899	270	302	215
	升降乐池 Elevating Orchestra Pit	6 × 20	3 × 18	6.2 × 16	
	最大视距(m) Max. Visual Distance	40	37	34	
	最大俯角（°） Max. Angle of Depression	32	28	29	34

浙江省建筑设计
研究院(中国)
Zhejiang
Building Design
and Research Institute
(China)

本设计方案构思总的
努力有三:其一,建筑从
局部到整体的功能合理和
有机性;其二,国家大剧
院在天安门广场特殊条件
下与四周建筑和环境的协
调性与有机性;第三,国
家大剧院在建筑文脉延续
和建筑风格创新方面,努
力争取做到时代性与继承
中、外优秀建筑文化传统
的统一。

The general efforts of the
design conceptions include
three aspects. Firstly, the rational function and the organic
feature of the architecture
from the part to the whole.
Secondly, the harmonious relationship and the organic feature with environmental surroundings under the special
situation at the Tian An Men
Square. Thirdly, to strive for
the unification of cultural context with design creation to fit
the days as well as inheriting
and bringing forth the out
standing chinese and foreign
culture traditions.

北侧透视　PERSPECTIVE OF NORTH ELEVATION

休息大厅透视
FOYER PERSPECTIVE

南侧透视
PERSPECTIVE OF SOUTH ELEVATION

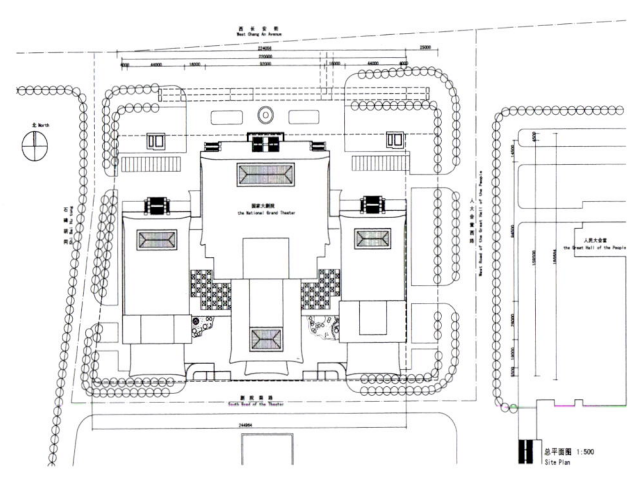

总平面
OVERALL SITE PLAN

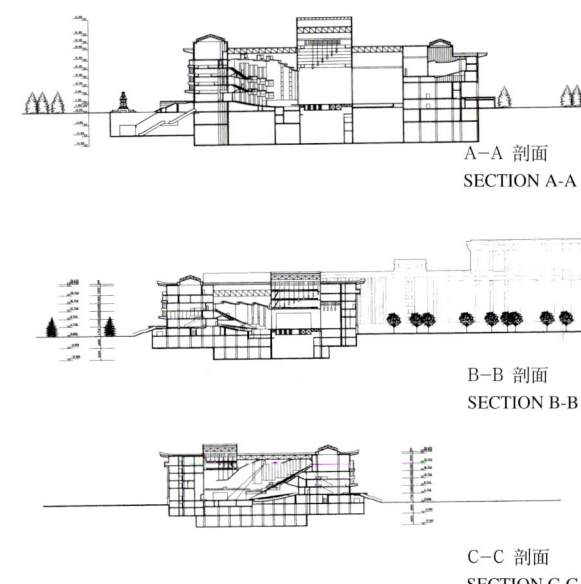

A-A 剖面
SECTION A-A

B-B 剖面
SECTION B-B

C-C 剖面
SECTION C-C

一层平面　GROUND FLOOR PLAN

二层平面　2nd FLOOR PLAN

四层平面　4th FLOOR PLAN

技术经济指标
Technical Economic Index

基地面积(m²) Site Area	总建筑面积(m²) Total Floor Area	建筑高度(m) Building Height	建筑覆盖率(%) Building Coverage Ratio
38900	地上：94938 地下：35069	30/45(局部)	71.5
容积率 Plot Ratio	绿化覆盖率(%) Green Coverage Ratio	停车数量 Number of Stalls	自行车停车数量 Number of Bicycle Stalls
3.34	23.4	地上：152 地下：460	1500

各观众厅数据
Data For Auditoria

		歌剧院 Opera House	音乐厅 Concert Hall	戏剧院 Theater	小戏场 Mini Theater
	观众厅面积(m²) Floor Area of Auditorium	1935	1620	1071	450
	观众厅体积(m³) Volume of Auditorium	20607	18090	8779	2806
座席数 Seating Capacity	总座席数 Seating Capacity(Seats)	2516	2003	1244	378
	其中：池座 Among Them:Auditorium	1158	1004	800	378
	楼座 Floor Seat	1358	999	444	
	休息厅面积(m²) Lounge Floor Area	3247	960	1458	
舞台尺寸 Stage Dimentions (m²)	主舞台 Main Stage	36 × 27	24 × 18	27 × 21	12 × 25
	左右侧台 Left and Right Side Stages	22 × 27	14.4 × 18	27 × 16.4	12 × 12.4
	后舞台 Back Stage	25 × 25		16.4 × 20	
	台仓 Under Stage Storage	36 × 27	24 × 8.4	27 × 21	
	升降乐池 Elevating Orchestra Pit	192.74		171.2	
	最大视距(m) Max. Visual Distance	33.5	34.34/41.2	21.5	19.5
	最大俯角(°) Max. Angle of Depression	29	28	27	10

刘荣广、伍振民
建筑师事务所
（香港）有限公司
（中国香港）
Dennis Lau & NG Chun
Man Architects & Engi-
neers (H.K.) Limited
(China Hong Kong)

整项设计为不同功能的空间组合，采纳"分散集团式"的布局，以求在功能组织上既能展现各建筑空间独立个体但亦可联成一气。

设计造型与功能分布以一东西走向的主干建筑体作发展中心，此主干建筑体位处各剧院、音乐厅、戏剧场等台唇位置，清楚界定主从空间，公共或服务用房的分区，继而按设计任务书的各部门内容延展建筑群。

The building develops from a concise design agenda which addresses the clarity of the circulation sequences, the manipulation of the transition between different scales and the use of an architectural vocabulary derived from the elegance of its construction techniques.

The bar building at the proscenium of the opera house, concert hall and theater forms the street wall of the sunken court which connects the east west ends of the development in the form of a visual corridor. This urban scale passage to the Great Hall of the People celebrates the ceremonial importance of the neighboring environment.

北侧模型照片
VIEW FROM NORTH SIDE OF THE MODEL

户外展廊
OUTSIDE GALLERY

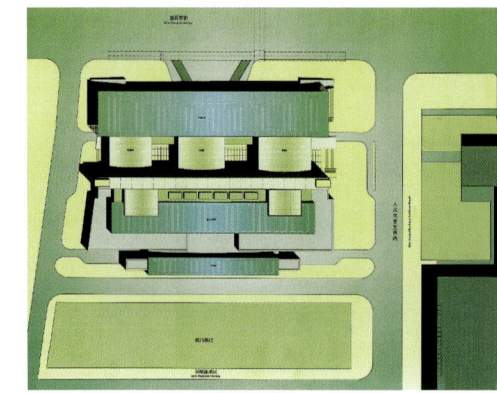

总平面
OVERALL SITE PLAN

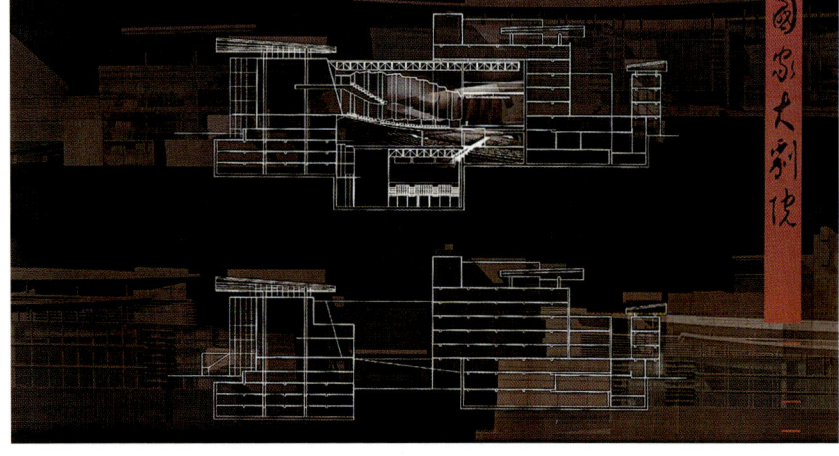

纵剖面
LONGITUDINAL SECTION

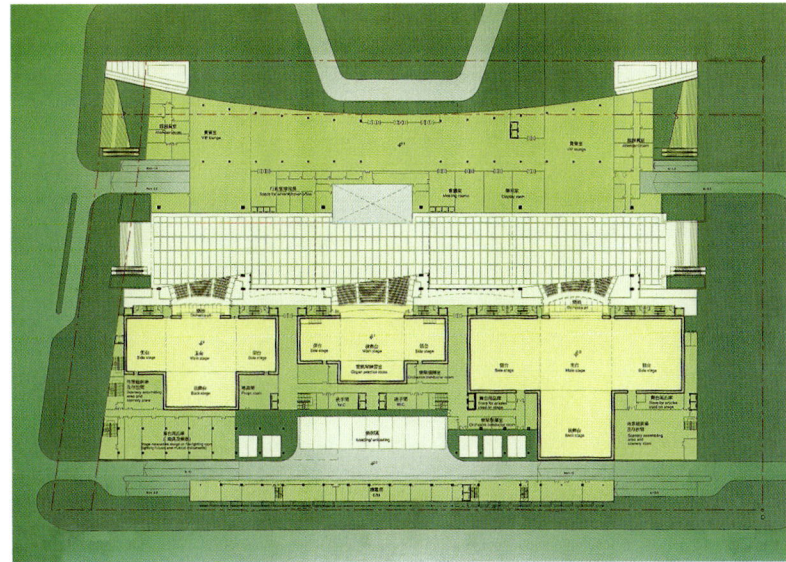

地面层　GROUND FLOOR PLAN

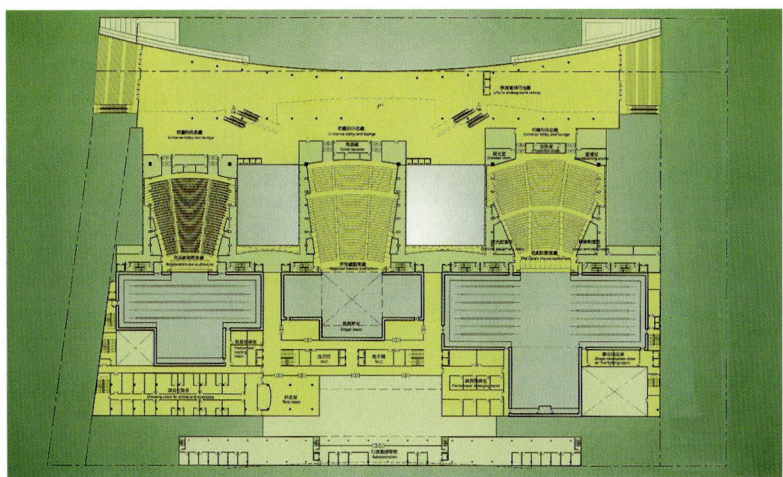

一层平面　2nd FLOOR PLAN

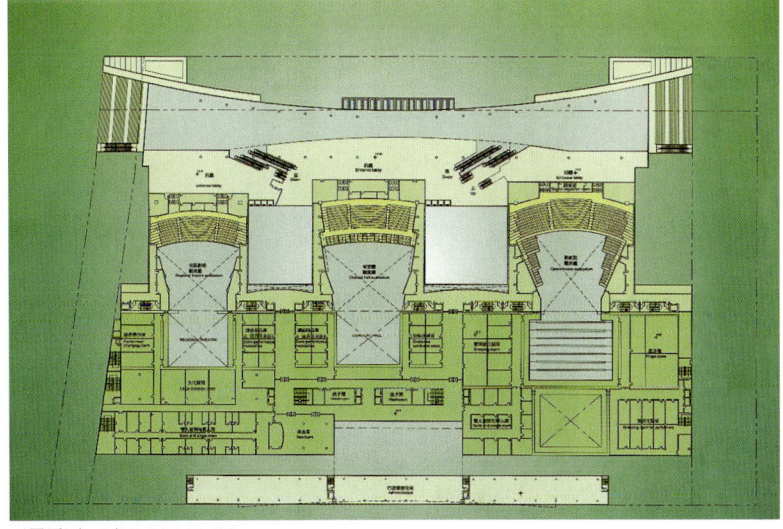

三层平面　3rd FLOOR PLAN

技术经济指标
Technical Economic Index

基地面积（m²） Site Area	总建筑面积（m²） Total Floor Area	建筑高度（m） Building Height	建筑覆盖率（%） Building Coverage Ratio
38900	地上：75450 地下：49150	30/45（局部）	74.5
容积率 Plot Ratio	绿化覆盖率（%） Green Coverage Ratio	停车数量 Number of Stalls	自行车停车数量 Number of Bicycle Stalls
3.2	40.9	地上： 地下：522	1003

各观众厅数据
Data For Auditoria

		歌剧院 Opera House	音乐厅 Concert Hall	戏剧院 Theater	小戏场 Mini Theater
	观众厅面积（m²） Floor Area of Auditorium	2581	1947	1525	1071
	观众厅体积（m³） Volume of Auditorium	67203	38946	30496	21428
座席数 Seating Capacity	总座席数 Seating Capacity(Seats)	2505	1975	1248	40 seats/stall 14 stalls
	其中：池座 Among Them:Auditorium	1419	1513	995	
	楼座 Floor Seat	1086	462	253	
	休息厅面积（m²） Lounge Floor Area	3000	2600	2000	
舞台尺寸（m²）Stage Dimentions	主舞台 Main Stage	36 × 27	24 × 18	27 × 21	200s.m
	左右侧台 Left and Right Side Stages	22 × 27 × 2	216 × 2	21 × 15 × 2	
	后舞台 Back Stage	26 × 25	24 × 4.5	27 × 13	
	台仓 Under Stage Storage	1622	24 × 8	567	
	升降乐池 Elevating Orchestra Pit	160		26 × 8	
	最大视距（m） Max. Visual Distance	49	49	41	
	最大俯角（°） Max. Angle of Depression	24	18	17	

141

让·奴威尔
建筑师事务所
（法国）
Architectures Jean Nouvel(France)

中国国家大剧院应以现代性来表达中国的过去和现在的价值。建筑物总是某一文化时刻的表现，它长久地象征着一个时代。

中国建筑的独特之处在于屋顶。我们无法想象象征中国，象征北京的一个建筑没有屋顶，从而落入世界上庸俗低级的平行六面体的俗套中。

中国建筑的特色还有颜色，红色的墙壁，屋顶和装潢有时也是红色，或者绿色和蓝色。

中国的象征与神话亦包括五行：水、气、木、金、火。中国国家大剧院据此将主要入口设于南面，因水自北向南流。

中国建筑往往在屋顶或入口处装饰有图形、人物和动物的雕刻。

中国国家大剧院要巧妙利用这种抽象／具象的关系，将三大象征纳入其中，每一个象征一个厅：俄尔修斯作为歌剧厅的象征，贝多芬象征古典音乐，第三个厅装饰上京剧脸谱。

中国国家大剧院也象征着知识——中国人民带给全人类的贡献。主要的参照为星相术、罗盘和机械钟。大帘子是活动的，它们随着太阳的转动而移动，以保护休息室的大玻璃窗。

北面，三面金色、红黑色、红色的大门令人联想到中国伟大建筑物的雄伟庄严的大门。各种色彩和绘画象征框饰和钉饰。

The chinese National Grand Theater must express with modernity the values of China yester-

南侧模型　VIEW FROM SOUTH SIDE OF THE MODEL

南侧透视　PERSPECTIVE OF SOUTH ELEVATION

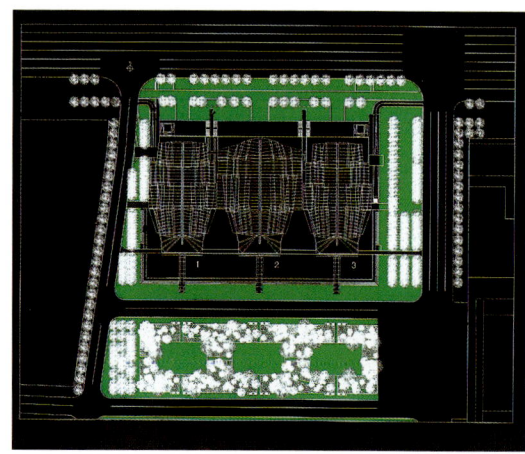

总平面　OVERALL SITE PLAN

day and today. Architecture is always the petrifaction of a moment of culture, a period that it will symbolize for a long time.

The identity of Chinese architecture is symbolized by the roof. We can hardly imagine an architecture to symbolize China (the National Theater)and Beijing without a specific roof, and not one reflecting the anonymity of the trivial international order.

The identity of Chinese architecture is marked by color, or rather by colors:the reds of walls and roofs and painted decor, greens and blues.

Other Chinese symbols and myths refer to the five elements: water, air, wood, steel and fire. The Chinese National Grand Theater is also attentive to favorable directions, with the main entrance to the south and water flowing from north to south.

Chinese architecture also often makes use of figurative paintings, and of sculpted characters or animals aligned on roofs or entrances.

The Chinese National Theater uses to ad vance this relationship between abstraction and metaphor by integrating the three large icons symbolizing each of the three auditoriums: Orpheus for opera, Beethoven for classical music, and a madeup mask for the Beijing Opera, in the third auditorium.

The Chinese National Theater also symbolizes the wisdom that Chinese have brought to humankind. The main reference is linked to astrology to the compass, and to the mechanical clock. The large blinds are mobile and follow the course of the sun, shielding the glass of the foyers.

On the north front the three great doors in gold, red-black and red evoke the monumental doors of Chinese monuments. A play of colors and of painting represents linings and nails.

北侧模型
VIEW FROM NORTH SIDE OF THE MODEL

南侧夜景透视
NIGHT VIEW FROM SOUTH SIDE

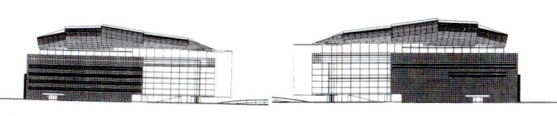

西立面
PERSPECTIVE OF WEST ELEVATION

东立面
PERSPECTIVE OF EAST ELEVATION

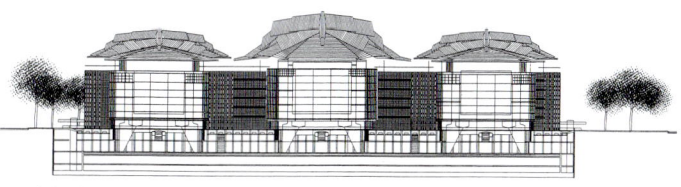

南立面
PERSPECTIVE OF SOUTH ELEVATION

143

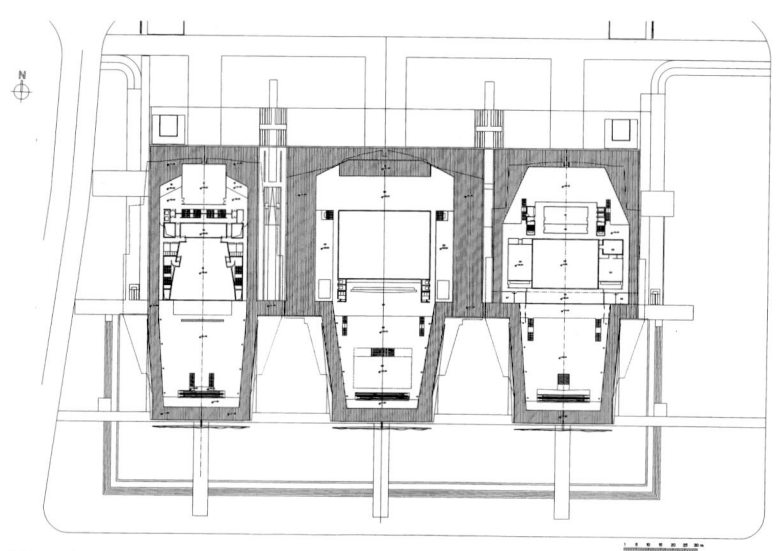

底层平面　GROUND FLOOR PLAN

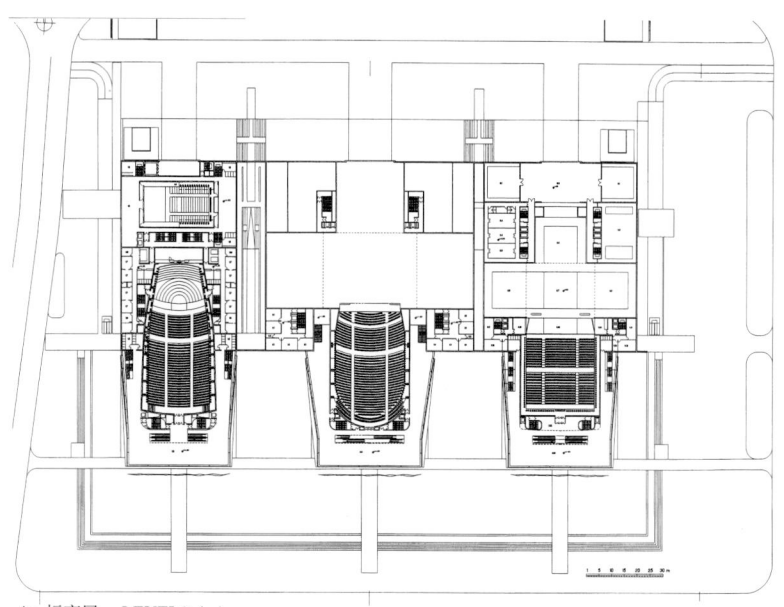

4m 标高层　LEVEL(+4m)

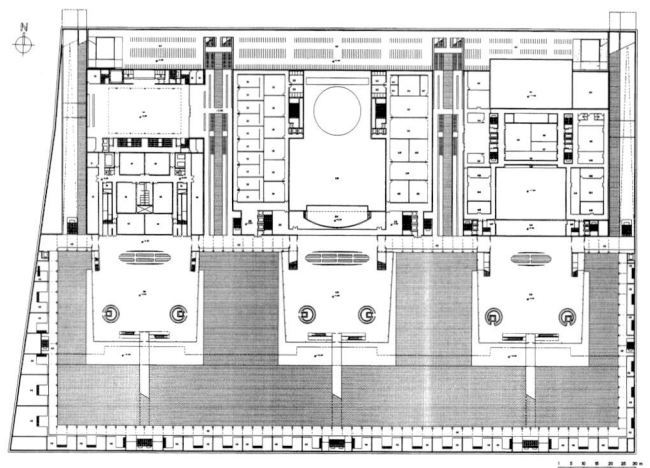

地下一层平面　FIRST BASEMENT PLAN

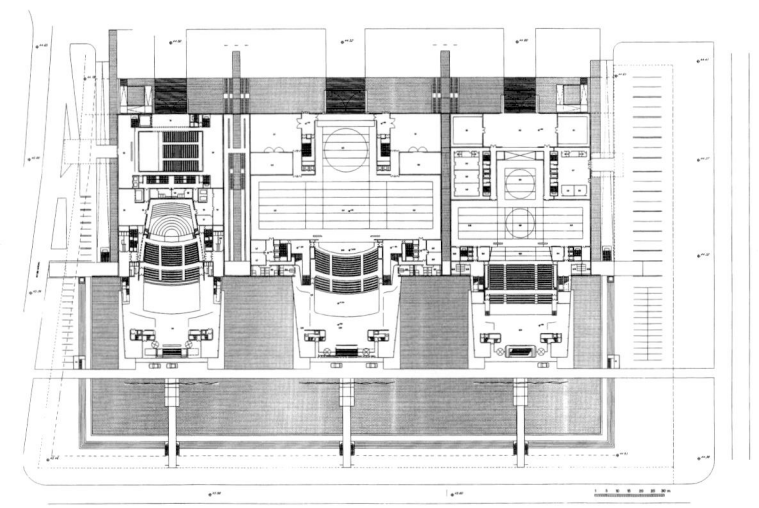

33m 标高层　LEVEL(+33m)

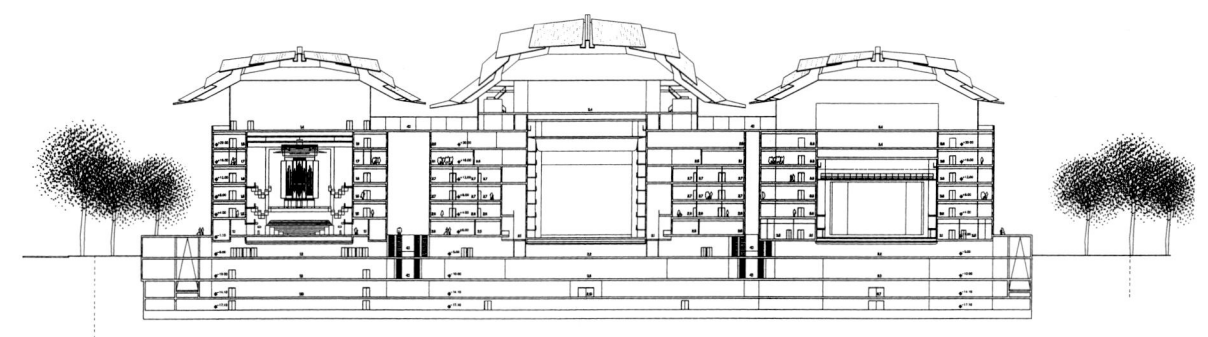

观众厅横剖面　CROSS SECTION OF THE AUDITORIUM

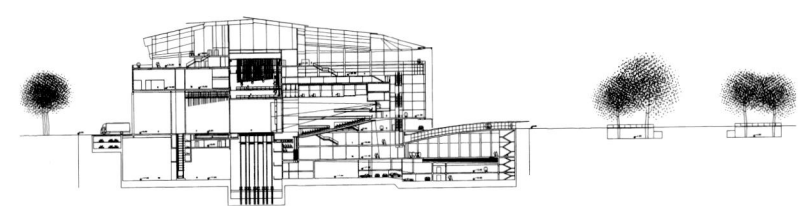

戏剧院剖面
SECTION OF THE THEATER

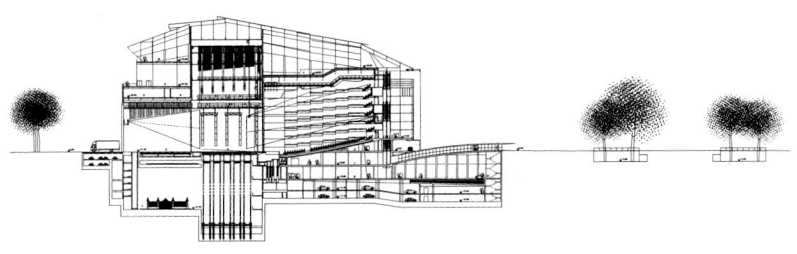

歌剧院剖面
SECTION OF THE OPERA HOUSE

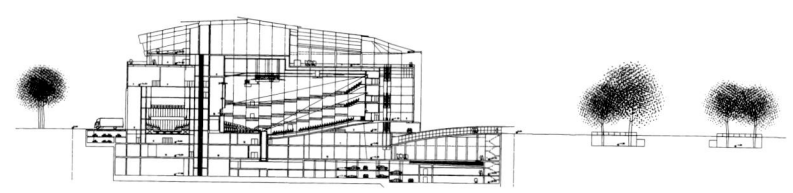

音乐厅剖面
SECTION OF THE CONCERT HALL

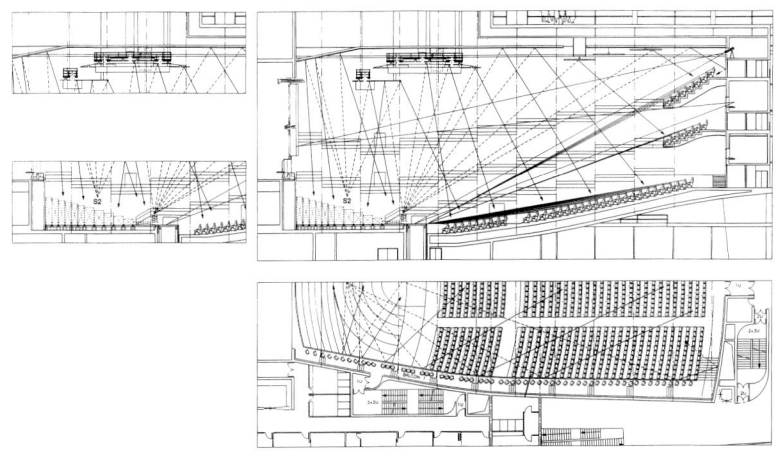

音乐厅声学与视角分析
ACOUSTICAL & VISUAL ANALYSIS OF THE CONCERT HALL

技术经济指标
Technical Economic Index

基地面积(m²) Site Area	总建筑面积(m²) Total Floor Area	建筑高度(m) Building Height	建筑覆盖率 (%) Building Coverage Ratio
43342	地上：69815 地下：71711	51.5	77.44
容积率 Plot Ratio	绿化覆盖率 (%) Green Coverage Ratio	停车数量 Number of Stalls	自行车停车数量 Number of Bicycle Stalls
3.62	41.9	地上：59 地下：540	1396

各观众厅数据
Data For Auditoria

		歌剧院 Opera House	音乐厅 Concert Hall	戏剧院 Theater	小戏场 Mini Theater
	观众厅面积(m²) Floor Area of Auditorium	2701	2897,43	1205	557.82
	观众厅体积(m³) Volume of Auditorium	23987.60	25525	11224	4629.9
座席数 Seating Capacity	总座席数 Seating Capacity(Seats)	2622	2045	1106	485
	其中：池座 Among Them:Auditorium	1482	1319	1008	442
	楼 座 Floor Seat	1140	726	98	43
	休息厅面积(m²) Lounge Floor Area	6814	5200.84	1686	1802.05
舞台尺寸 Stage Dimentions (m²)	主舞台 Main Stage	36 × 27 × 36	20 × 14.5 × 20	27 × 19.4 × 22	16.2 × 27 × 12
	左右侧台 Left and Right Side Stages	22 × 27 × 12.5	15.5 × 13 × 3	16.25 ×19.5×15	
	后舞台 Back Stage	26 × 25 × 18.7		20.1 ×60.5 ×12.8	16.2 × 4.9 × 15
	台 仓 Under Stage Storage	26 × 25 × 27	15.5 × 19 × 3.5	20×60.5×20.5	
	升降乐池 Elevating Orchestra Pit	37 × 6		27 × 5.7	
	最大视距(m) Max. Visual Distance	46	40	30	18.40
	最大俯角(°) Max. Angle of Depression	32	27.9	17.5	15.1

303
广州市设计院
（中国）
Guangzhou
Design Institute
(China)

珠联璧合的整体构思
为了协调天下第一广
场——天安门广场的巨大
空间尺度，为了协调人民
大会堂300多米长的连续
檐口及巨大体积，国家大
剧院设计方案首先排除了
将歌剧院、音乐厅、戏剧
场孤立呆板地排列在一起
的做法，而是通过一定的
中性空间将三大主厅组合
在一起，框定在统一的屋
架下，形成有相当空间尺
度的巨大单体，从而使得
大剧院能与天安门广场建
筑在尺度上既主次分明又
协调统一。

In order to be in harmony
with the spacious size of Tian
An Men Square as well as the
300-meter long continuous
eaves of the People's Great
Hall, this proposal gives no
consideration from the beginning on an isolated and awk-
ward arrangement of the opera house, concert hall and
theater. By way of certain
neutral space, this proposal
combines the three main halls
under one united roof to form
a single unit with more sizable
space, which achieves the pri-
mary and secondary as well as
harmony among the Theater
and its surrounding buildings
at Tian An Men Square.

北侧透视
PERSPECTIVE OF NORTH ELEVATION

南侧透视
PERSPECTIVE OF SOUTH ELEVATION

歌剧院观众厅透视
PERSPECTIVE OF THE OPERA HOUSE AUDITORIUM

公共大厅透视
PUBLIC HALL PERSPECTIVE

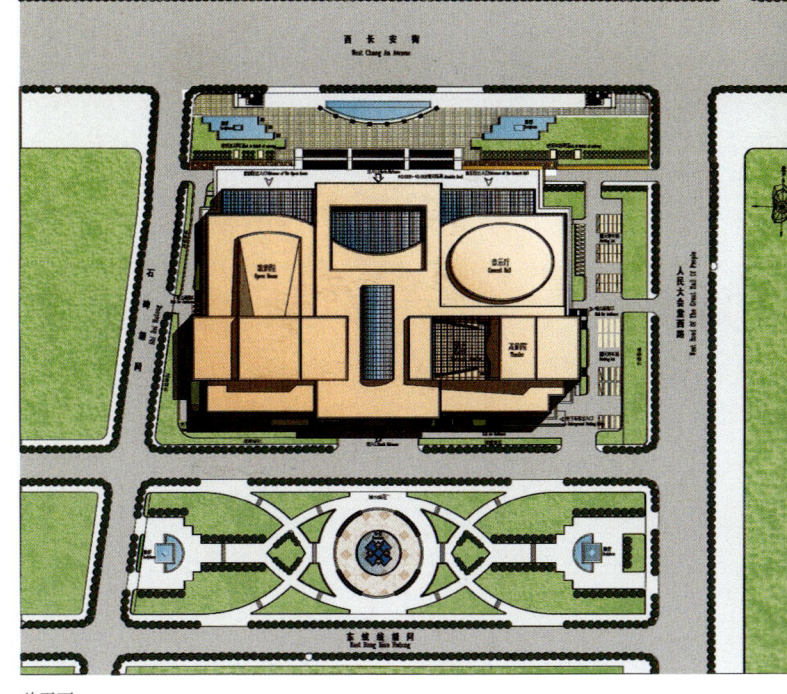

总平面
OVERALL SITE PLAN

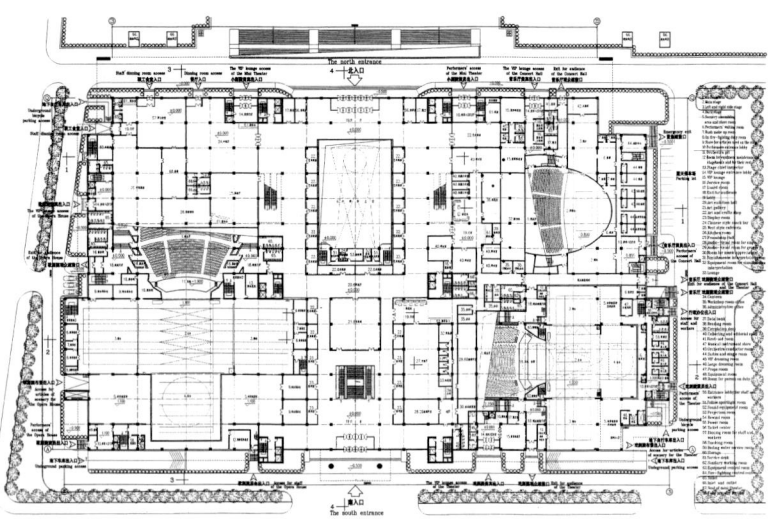

±0.00m平面
GROUND LEVEL(±0.00m)

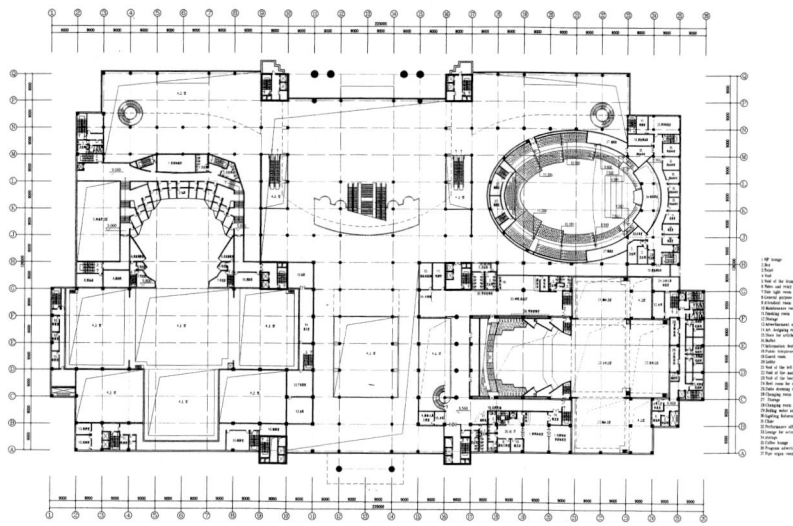

+9.00m平面
LEVEL(+9.00m)

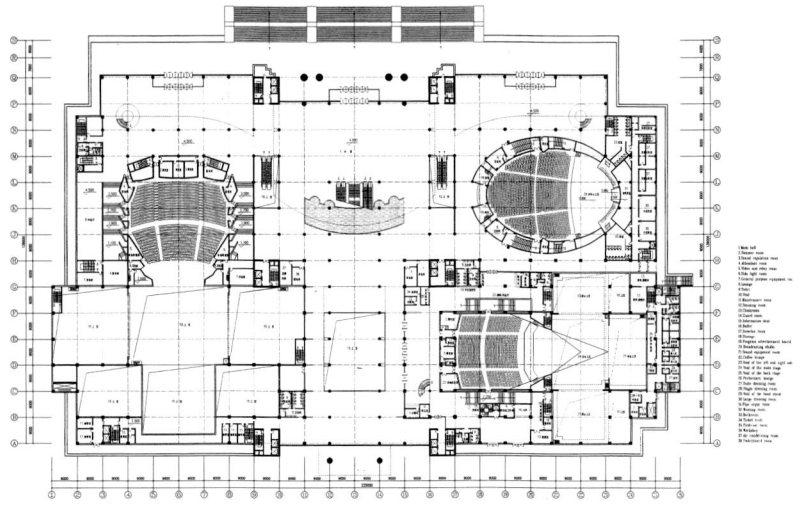

+4.50m平面
LEVEL(+4.50m)

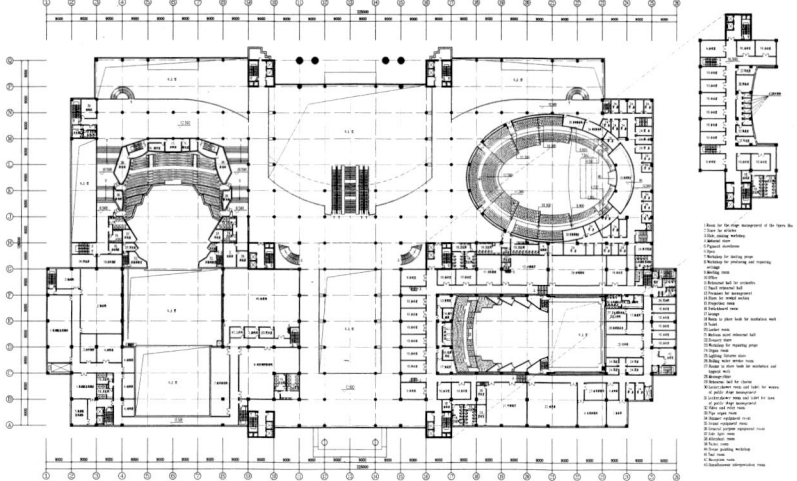

+12.50m平面
LEVEL(+12.50m)

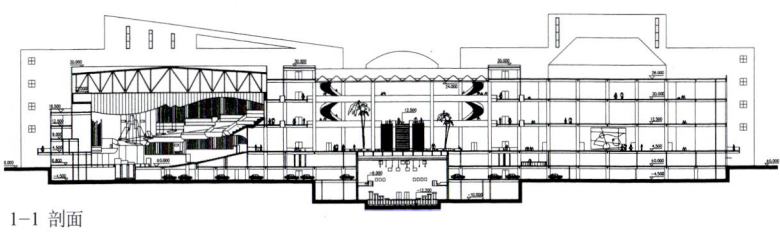

1-1 剖面
SECTION 1-1

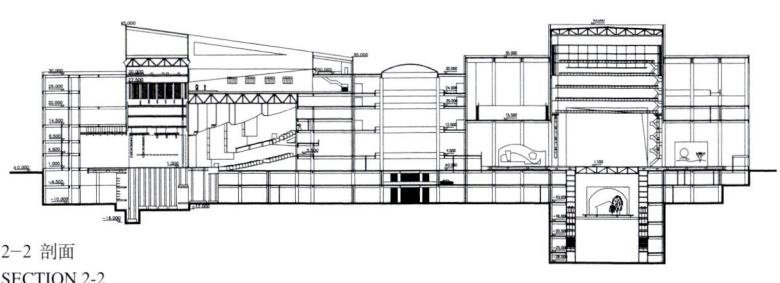

2-2 剖面
SECTION 2-2

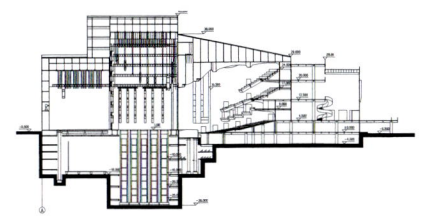

3-3 剖面
SECTION 3-3

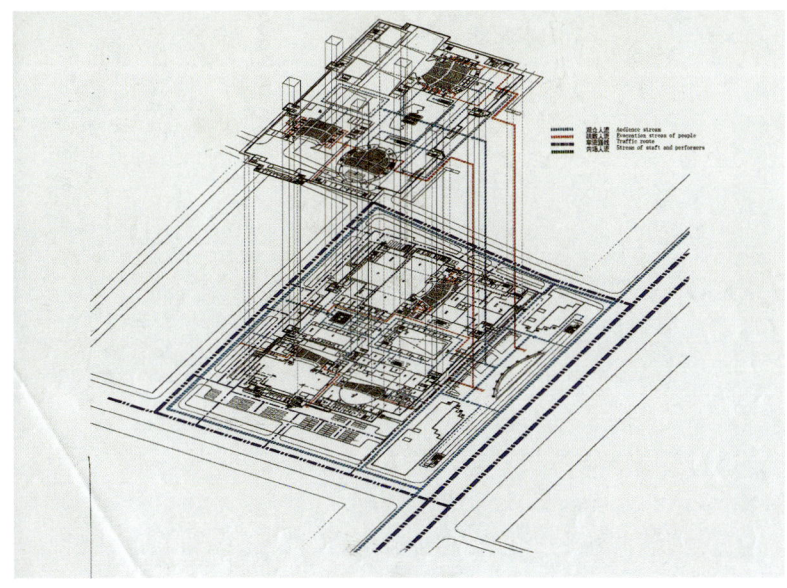

交通分析
TRAFFIC ANALYSIS PLAN

技术经济指标
Technical Economic Index

基地面积(m²) Site Area	总建筑面积(m²) Total Floor Area	建筑高度(m) Building Height	建筑覆盖率 (%) Building Coverage Ratio
38900	地上：90400 地下：39500	30/45(局部)	55.6

容积率 Plot Ratio	绿化覆盖率 (%) Green Coverage Ratio	停车数量 Number of Stalls	自行车停车数量 Number of Bicycle Stalls
2.32	11	地上：72 地下：499	1016

各观众厅数据
Data For Auditoria

		歌剧院 Opera House	音乐厅 Concert Hall	戏剧院 Theater	小戏场 Mini Theater
观众厅面积(m²) Floor Area of Auditorium		2050	1742	951	1053
观众厅体积(m³) Volume of Auditorium		24125	23000	7722	13689
座席数 Seating Capacity	总座席数 Seating Capacity(Seats)	2520	2026	1200	482
	其中：池座 Among Them:Auditorium	1312	1223	872	322
	楼座 Floor Seat	1208	803	328	160
	休息厅面积(m²) Lounge Floor Area	3802	2498	1934	756
舞台尺寸(m²) Stage Dimentions	主舞台 Main Stage	36 × 27	27 × 18	27 × 20	
	左右侧台 Left and Right Side Stages	22 × 27	11 × 21 14 × 18	14 × 20 10 × 20	
	后舞台 Back Stage	26 × 25		27 × 14	
	台仓 Under Stage Storage	36 × 27 26 × 25	18 × 16	27 × 20	30 × 21
	升降乐池 Elevating Orchestra Pit	21 × 45		183	
最大视距(m) Max. Visual Distance		42	36	34	30
最大俯角（°） Max. Angle of Depression		27	22	18	

304

刘荣广、伍振民
建筑师事务所
（香港）有限公司
（中国香港）
Dennis Lau & NG Chun
Man Architects & Engi-
neers (H.K.) Limited
(China Hong Kong)

初步设计意念，是在网格状的北京城市规划中，以圆形为首，发展成复杂而有规模的设计。圆形之内渐生中空，成为所有戏剧性及人性表演的场地，犹如京剧脸谱，于无尽百变的色彩及花纹，在舞台上演释人生百态。

Design concept:

The initial gesture consists of a circle within the uniformly girded urban structure of Beijing to enclose the comprehensive and complex program requirements. The area enclosed within the circular envelope becomes a voided center---a void which allows for all possible theatrical and human experiences to take place, not unlike the painted faces of Beijing Opera, allowing for an infinite human expressions to take place on stage.

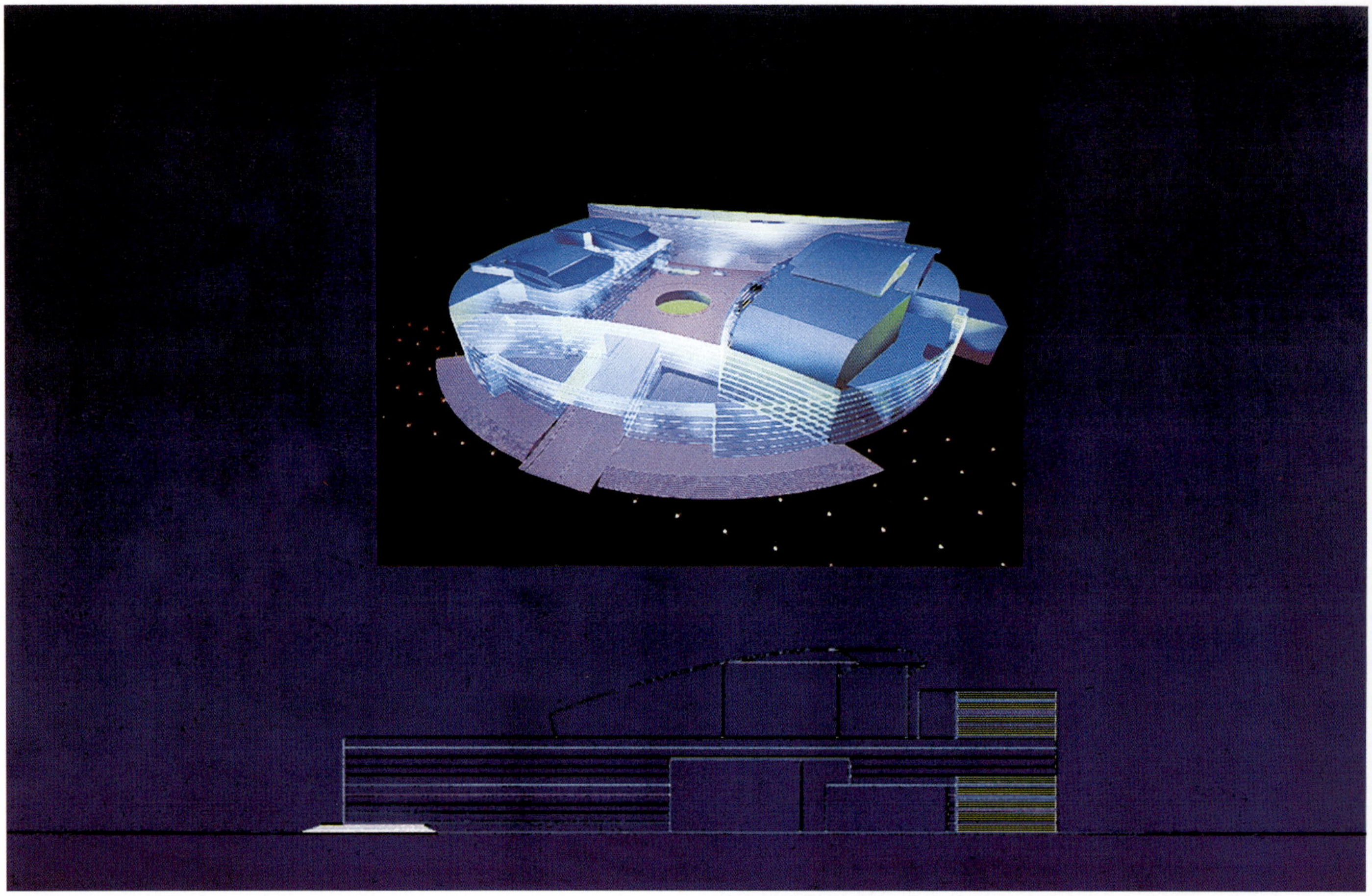

北侧透视　PERSPECTIVE OF NORTH ELEVATION

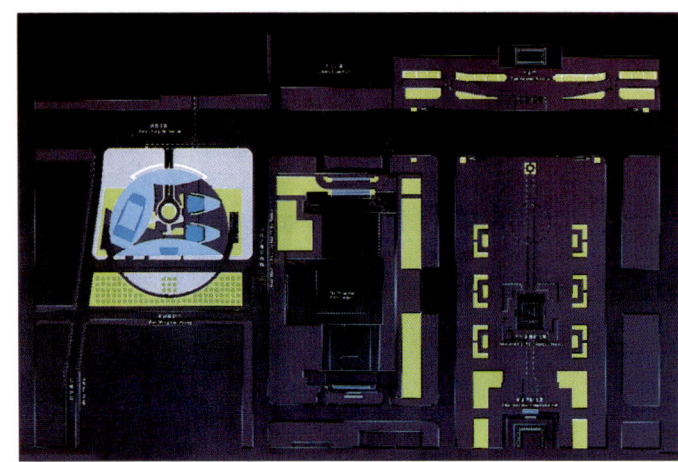

地域图　SURROUNDINGS MAP

歌剧院观众厅透视　PERSPECTIVE OF THE OPERA HOUSE AUDITORIUM

150

底层平面
GROUND
FLOOR PLAN

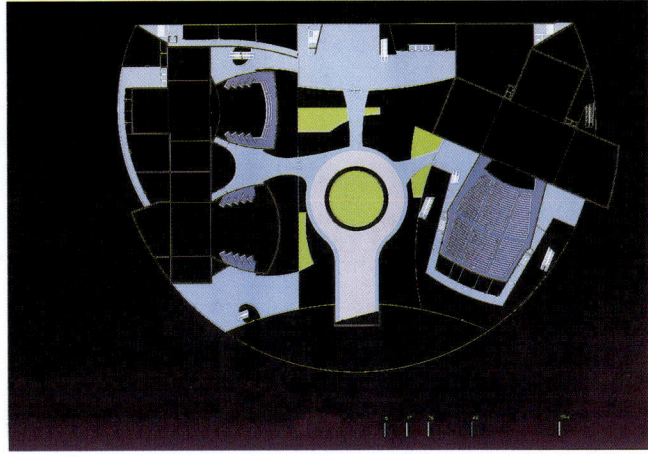

一层平面
1st FLOOR PLAN

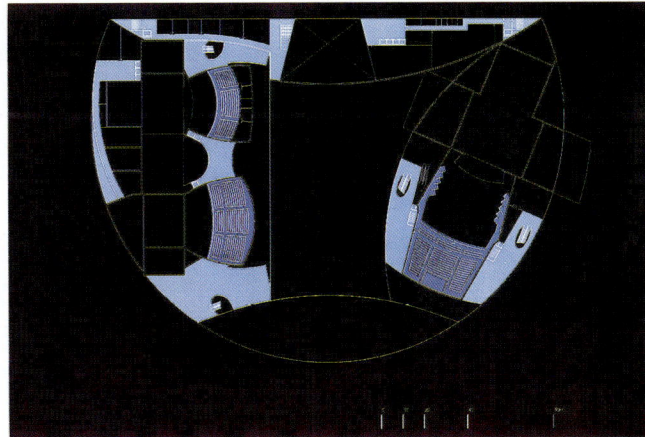

二层平面
2nd FLOOR PLAN

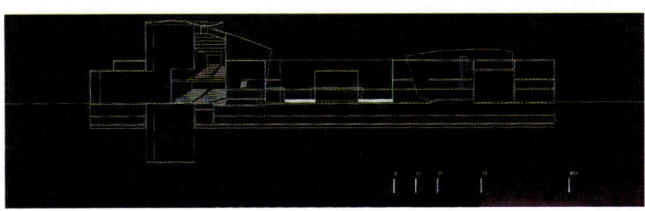

歌剧院剖面
SECTION OF
THE OPERA HOUSE

技术经济指标
Technical Economic Index

基地面积（m²） Site Area	总建筑面积（m²） Total Floor Area	建筑高度（m） Building Height	建筑覆盖率（%） Building Coverage Ratio
38900	地上：56765 地下：68212	30/40	78.29
容积率 Plot Ratio	绿化覆盖率（%） Green Coverage Ratio	停车数量 Number of Stalls	自行车停车数量 Number of Bicycle Stalls
21	21.71	地上：50 地下：500	1006

各观众厅数据
Data For Auditoria

		歌剧院 Opera House	音乐厅 Concert Hall	戏剧院 Theater	小戏场 Mini Theater
	观众厅面积（m²） Floor Area of Auditorium	2495	1679	1287	1960
	观众厅体积（m³） Volume of Auditorium	42100	17500	13772	31360
座席数 Seating Capacity	总座席数 Seating Capacity(Seats)	2520	2030	1253	300~500
	其中：池座 Among Them:Auditorium	1802	1153	696	N/A
	楼座 Floor Seat	718	877	557	N/A
	休息厅面积（m²） Lounge Floor Area	3241	3234	2068	1108
舞台尺寸（m²） Stage Dimentions	主舞台 Main Stage	36 × 27	24 × 18	27 × 19.5	N/A
	左右侧台 Left and Right Side Stages	594+691	216+216	315+315	N/A
	后舞台 Back Stage	26 × 25	120	20 × 16.5	N/A
	台仓 Under Stage Storage	1622	218	526	N/A
	升降乐池 Elevating Orchestra Pit	167	N/A	87	N/A
	最大视距（m） Max. Visual Distance	58.9	33.5	32.3	N/A
	最大俯角（°） Max. Angle of Depression	14.97	26.60	30.10	N/A

中国建筑科学研究院（中国）
China Architecture Science Research Institute(China)

一座纯粹方方正正的建筑，

一点京城中的明朗洁净空间，

一个洗炼的方正构成体，

一片清纯的长安街景画面。

红色柱廊回响着"紫禁城"的脚步声，

水晶空间流动着中南海的清泉，

屋顶雕塑体牵动了北海的白塔，

红墙辉映着庄严雄伟的天安门。

环形地灯映射出宇宙的苍穹，

八卦通道打开了未来世界的大门，

大红灯笼喜气洋洋，

城市走廊鼓乐齐鸣。

长江、黄河歌唱了一千个世纪，

华夏大地歌舞升平一万年。

祖祖辈辈的心声，世世代代的梦想——

中国人心中的国家大剧院。

……

A pure, square and upright building,

A clean and bright space in Beijing,

A cubic composition, simple and succinct,

A picture of Chang An Street, fresh and innocent.

In the red colonnade the footsteps in the Forbidden City can be heard,

In the crystal space the Zhong Nan Hai stream is flowing by,

The sculptural roof, echoing with the White Tower in the park of Bei Hai,

The red walls, reflecting the Tian An Men Gate, solemn and great.

Ground lamps in circle glittering like stars in universal sky,

The passages in the Eight Diagrams leading to the future ahead;

Big red lanterns bringing happiness, hanging so high,

In the corridor of the city comes the loud music of drums and pipes.

Chanting for centuries, Changjiang and Huanghe Rivers,

Blessing our motherland, with peace and prosperity forever,

It is a dream passing down from generations,

In bottom of our heart is the National Grand Theater in Beijing.

……

北立面透视
PERSPECTIVE OF NORTH ELEVATION

城市走廊
CITY CORRIDOR

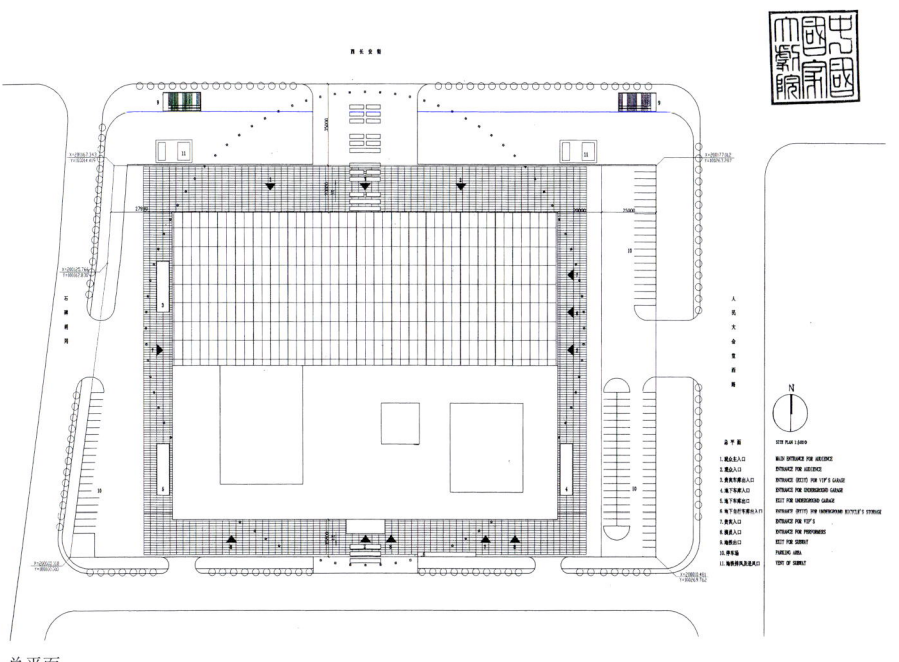

歌剧院观众厅透视
PERSPECTIVE OF THE OPERA HOUSE AUDITORIUM

总平面
OVERALL SITE PLAN

夜景透视
NIGHT VIEW

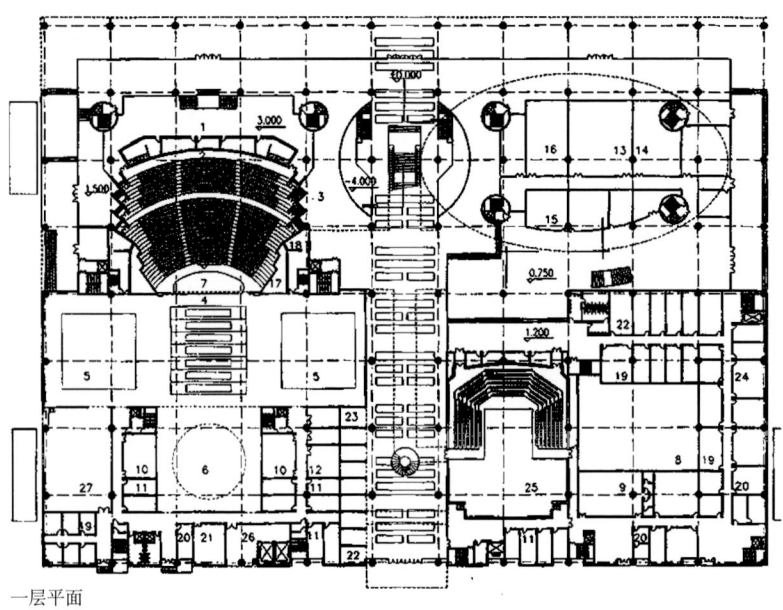

一层平面
GROUND FLOOR PLAN

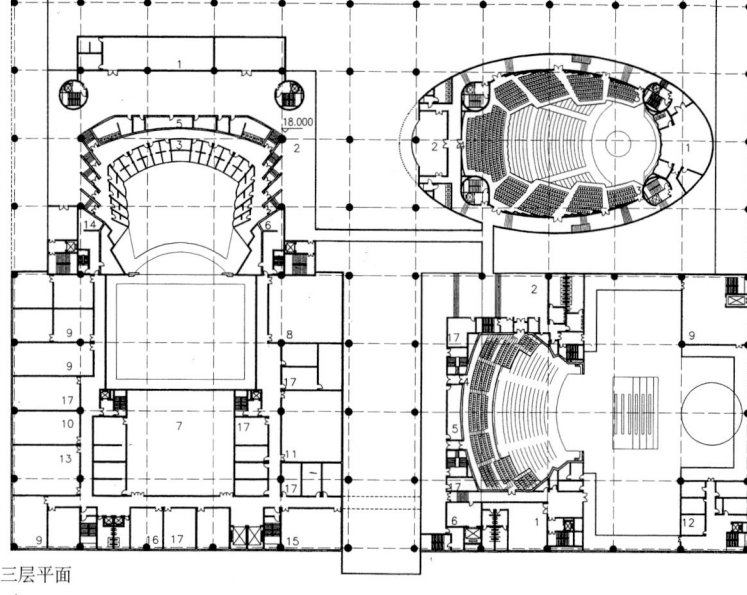

三层平面
3rd FLOOR PLAN

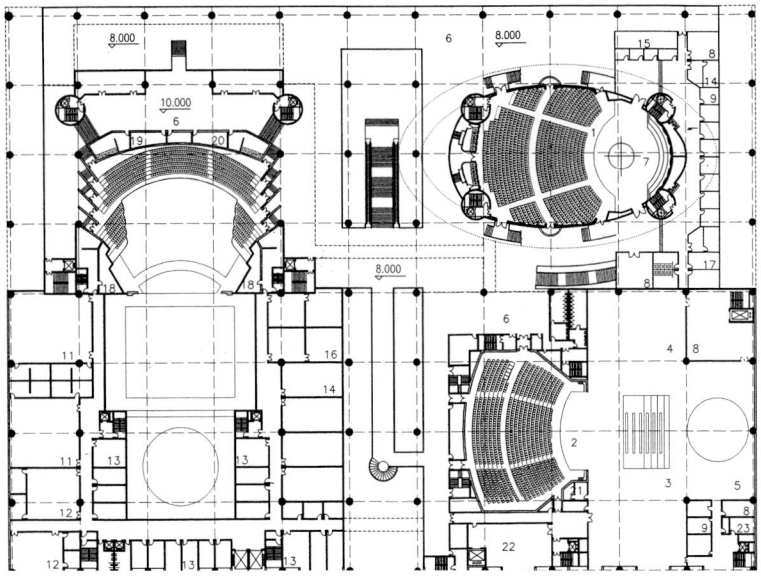

二层平面
2nd FLOOR PLAN

四层平面
4th FLOOR PLAN

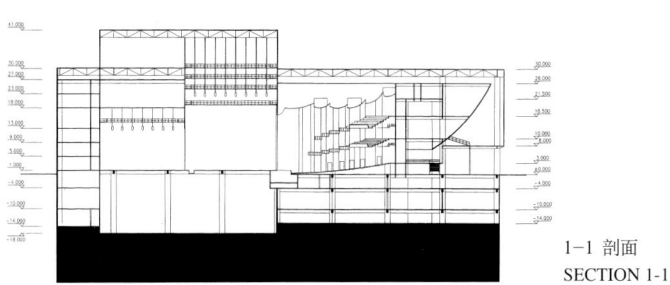

1-1 剖面
SECTION 1-1

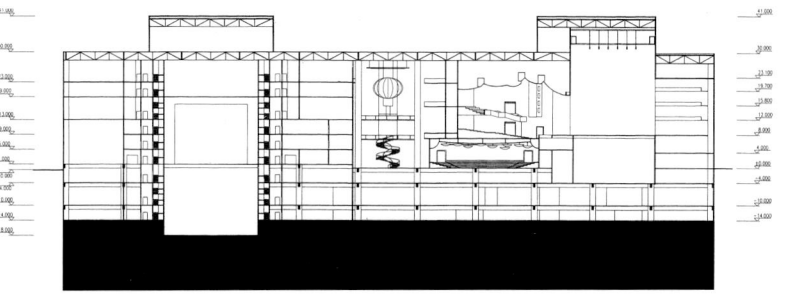

3-3 剖面
SECTION 3-3

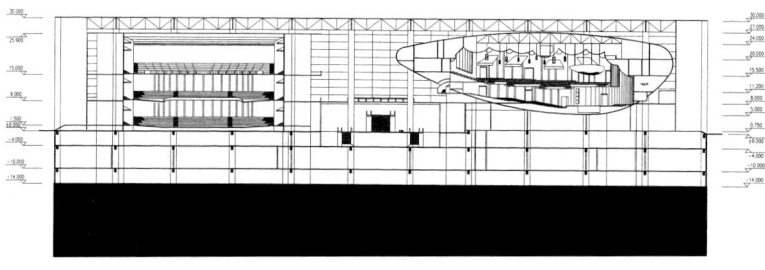

4-4 剖面
SECTION 4-4

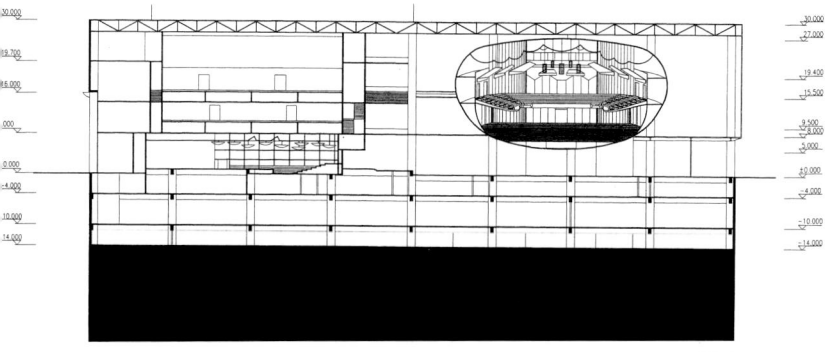

5-5 剖面
SECTION 5-5

技术经济指标
Technical Economic Index

基地面积（m²） Site Area	总建筑面积（m²） Total Floor Area	建筑高度（m） Building Height	建筑覆盖率（%） Building Coverage Ratio
38900	地上：66023 地下：65441	30/41（局部）	63.4

容积率 Plot Ratio	绿化覆盖率（%） Green Coverage Ratio	停车数量 Number of Stalls	自行车停车数量 Number of Bicycle Stalls
3.38	2	地上：101 地下：576	1020

各观众厅数据
Data For Auditoria

		歌剧院 Opera House	音乐厅 Concert Hall	戏剧院 Theater	小戏场 Mini Theater
	观众厅面积（m²） Floor Area of Auditorium	2161	1621	1075	590
	观众厅体积（m³） Volume of Auditorium	18500	22900	7600	1998
座席数 Seating Capacity	总座席数 Seating Capacity(Seats)	2540	1998	1342	472
	其中：池座 Among Them:Auditorium	1513	1123	896	472
	楼座 Floor Seat	1037	875	446	
	休息厅面积（m²） Lounge Floor Area	2890	2340	1130	660
舞台尺寸（m²） Stage Dimentions	主舞台 Main Stage	36 × 27	25 × 18	27 × 19.5	25 × 12
	左右侧台 Left and Right Side Stages	22 × 27	15 × 12	23 × 16	10 × 9
	后舞台 Back Stage	26 × 25		32 × 16	
	台仓 Under Stage Storage	36 × 27	10 × 10	30 × 20	25 × 12
	升降乐池 Elevating Orchestra Pit	18 × 6.5			
	最大视距（m） Max. Visual Distance	33	34	32	17
	最大俯角（°） Max. Angle of Depression	24	32	31	

清华大学建筑设计研究院
（中国）
Architectural Design
& Research Institute
of Tsinghua
University(China)

功能布局——合理·便捷·明晰·秩序·整体
●实轴——东西脊轴，由此有机生长出建筑的各个部分，同时成为造型上各部分联系的桥梁和纽带。
●虚轴——南北贯通，空间上建筑各部分联系的桥梁和纽带。
●多样的公共空间——开放性。市民共同参与演出的场所，人流聚散的舞台。
●造型设计——原创·明晰·秩序·整体·开放·融合·承继·象征·戏剧

Functional Layout —
·Rational·Convenient·
Clear·Orderly·Wholeness

●Solid Axis — Vertebra from east to west, which is the bridge and link of each components in shapes. Each part of the Grand Theater generates from it organically.

●Void Axis — Penetrated from north to south which is the bridge and link of each part in space.

●Rich and varied public space — The place where people perform together; the stage where people gather and scatter.

●Exterior Design — Public·Original·Transparent·Orderly·Ensemble·Symbolic·Contextual·Harmonious·Dramatic

北侧透视　PERSPECTIVE OF NORTH ELEVATION

南侧透视　PERSPECTIVE OF SOUTH ELEVATION

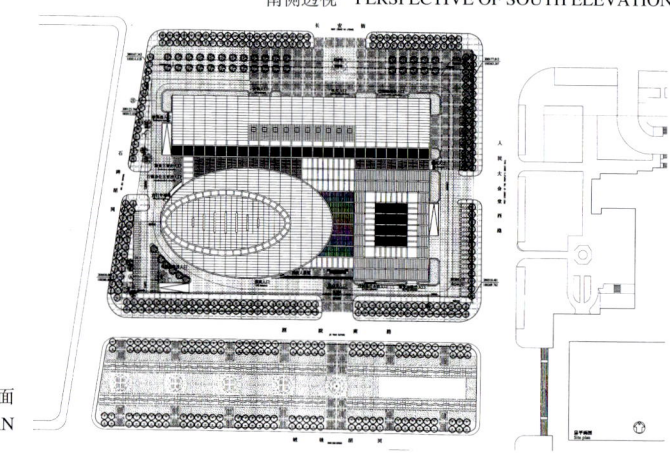

总平面
OVERALL SITE PLAN

大厅室内透视
LOBBY PERSPECTIVE

观众厅室内透视
AUDITORIUM PERSPECTIVE

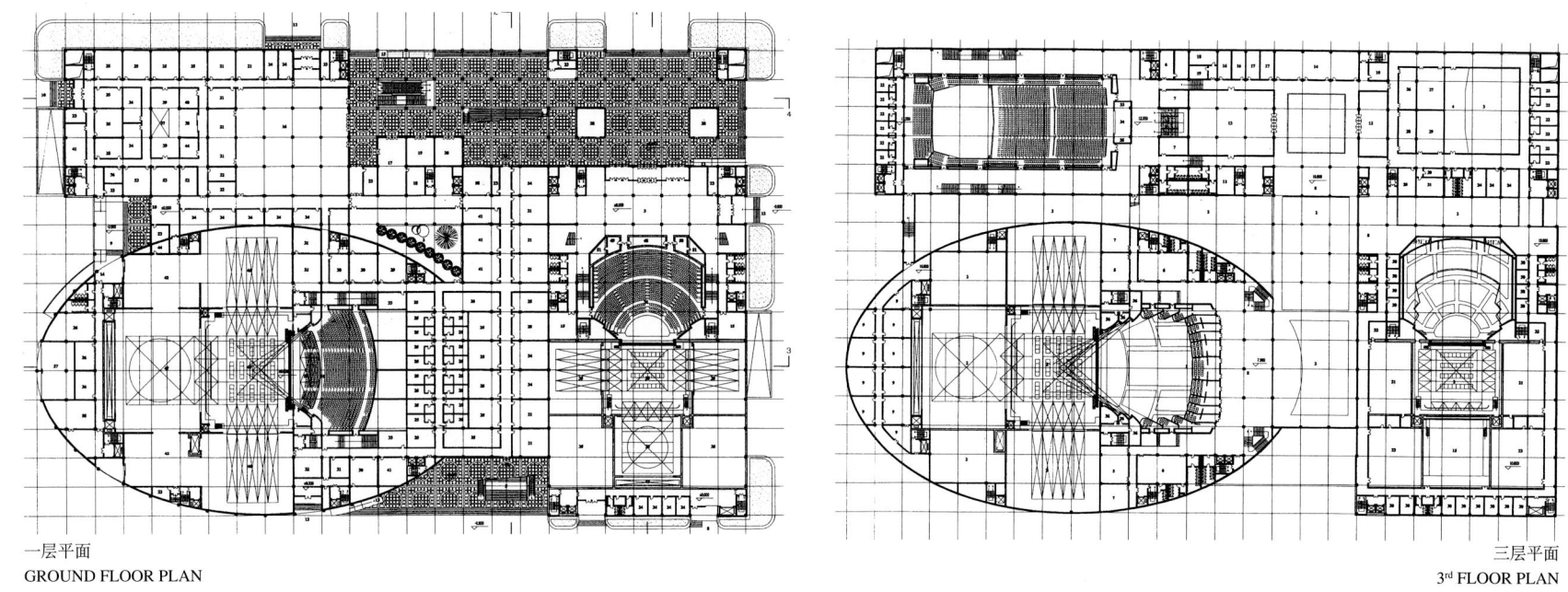

一层平面
GROUND FLOOR PLAN

三层平面
3rd FLOOR PLAN

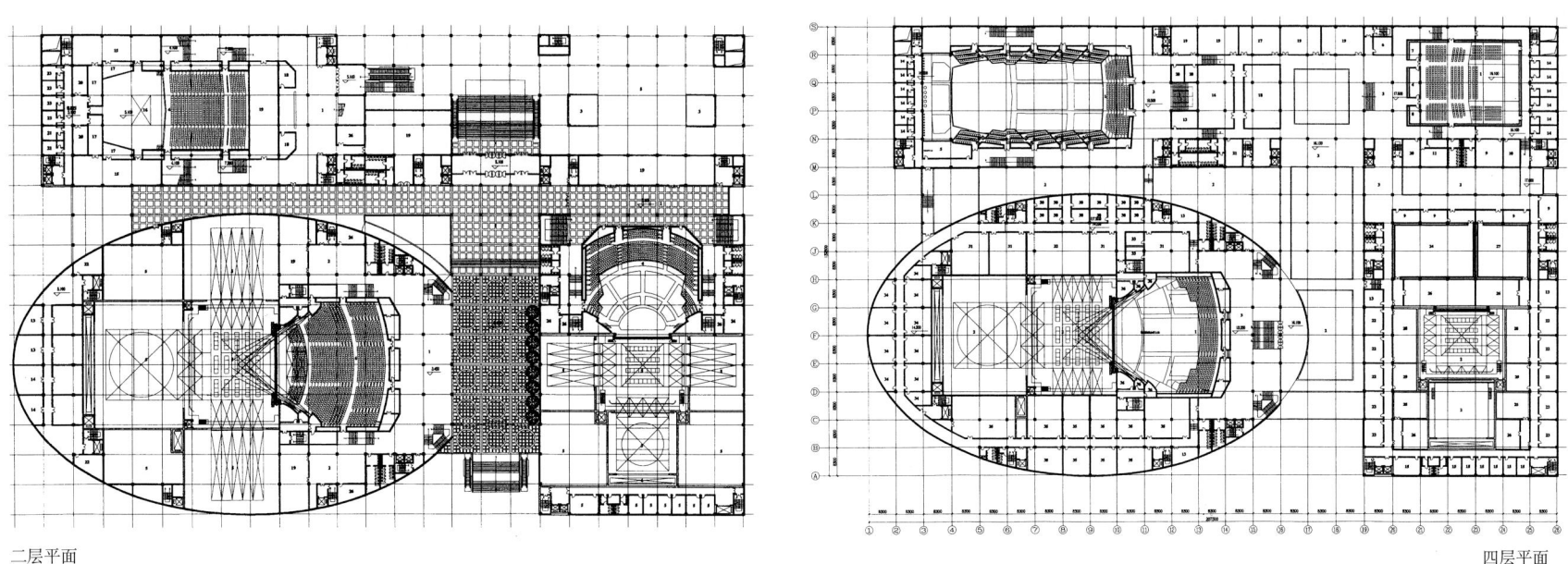

二层平面
2nd FLOOR PLAN

四层平面
4th FLOOR PLAN

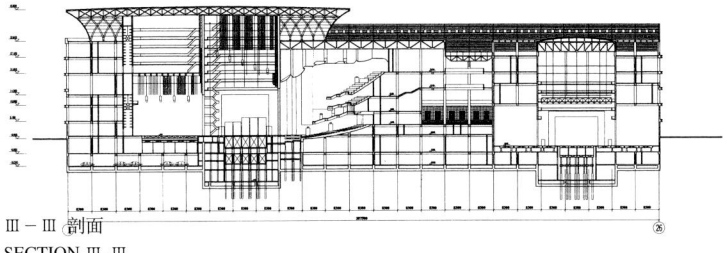

Ⅰ－Ⅰ 剖面
SECTION Ⅰ-Ⅰ

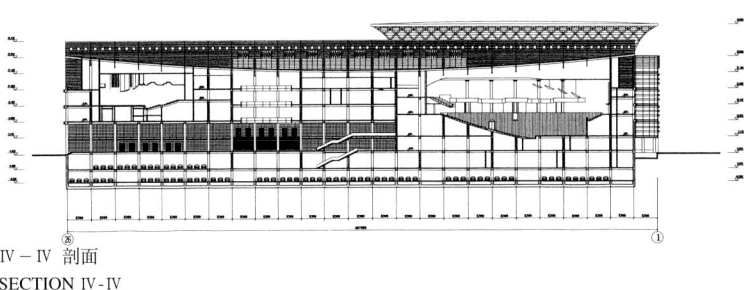

Ⅲ－Ⅲ 剖面
SECTION Ⅲ-Ⅲ

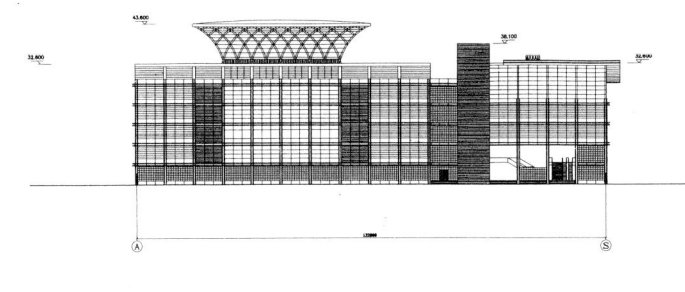

Ⅳ－Ⅳ 剖面
SECTION Ⅳ-Ⅳ

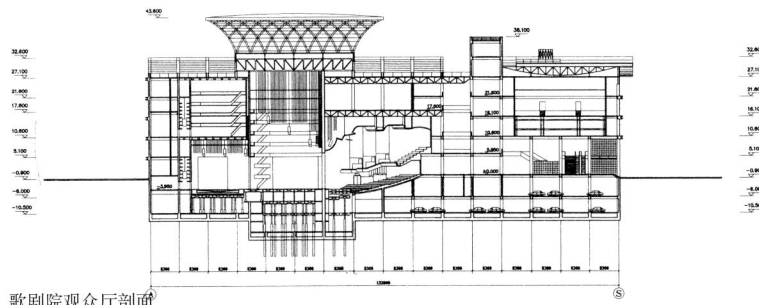

歌剧院观众厅剖面
SECTION OF THE OPERA HOUSE AUDITORIUM

技术经济指标
Technical Economic Index

基地面积（m²） Site Area	总建筑面积（m²） Total Floor Area	建筑高度（m） Building Height	建筑覆盖率（%） Building Coverage Ratio
38900	地上：82500 地下：46000	30/44.6(局部)	63

容积率 Plot Ratio	绿化覆盖率（%） Green Coverage Ratio	停车数量 Number of Stalls	自行车停车数量 Number of Bicycle Stalls
3.30	8.5	地上：71 地下：502	1000

各观众厅数据
Data For Auditoria

		歌剧院 Opera House	音乐厅 Concert Hall	戏剧院 Theater	小戏场 Mini Theater
	观众厅面积(m²) Floor Area of Auditorium	2792	1750	1103	828
	观众厅体积(m³) Volume of Auditorium	22773	23722	9366	7452
座席数 Seating Capacity	总座席数 Seating Capacity(Seats)	2516	2040	1398	366~596
	其中：池座 Among Them:Auditorium	1322	1056	884	
	楼座 Floor Seat	1194	984	514	
	休息厅面积(m²) Lounge Floor Area	5600	3800	1700	1400
舞台尺寸 Stage Dimentions (m²)	主舞台 Main Stage	36×27	21.9×18	27×21	24.9×16.6
	左右侧台 Left and Right Side Stages	21.5×27	10×24.9	16×21	
	后舞台 Back Stage	36×25		20×18	
	台仓 Under Stage Storage	36×27(主) 36×25(后)	8.3×5.8	27×21(主) 20×18(后)	24.9×16.6
	升降乐池 Elevating Orchestra Pit	21×6		12×6	
最大视距（m） Max. Visual Distance		37.0	36.9	24.5	
最大俯角（°） Max. Angle of Depression		30.5	20	27	

307
东南大学建筑设计研究院（中国）
Architectural Design & Research Institute of Southeast University (China)

本方案沿长安街采用了对称的布局。北部礼仪入口的设计是北立面的重点。左右两侧以大开间的钢结构及大面积玻璃帷幕外墙使墙面开间的高宽比例上体现传统中国建筑的精神。新技术的应用与对传统花饰纹样的借鉴，使现代化的材料与技术与中国文化有机结合。传统中国建筑常以"重檐"来表现建筑的重要性。本方案深远的屋顶挑檐，斜角玻璃帷幕墙肩部处理、基本开间的比例，使中国传统建筑的基本要素在当代得到新的体现。

Despite the sophisticated interior function, through delicate arrangement, a symmetrical facade will be raised up along the Chang An Street. The design of the north facade flanked by long-span steel structure and crystal curtain wall, so as to reflect the spirit of the Chinese traditional architecture. The application of advanced technologies and use of classical ornaments make the organic unity of modern materials with Chinese national culture. Multiple cornice was always used for presenting the importance of an building in ancient China. Long-projected cornice, oblique wall top, the proportion of bays, all represent the essence of the traditional Chinese architectural style in current time.

北侧模型照片
VIEW FROM NORTH SIDE OF THE MODEL

南侧透视
PERSPECTIVE OF SOUTH ELEVATION

歌剧院观众厅透视
PERSPECTIVE OF THE OPERA HOUSE AUDITORIUM

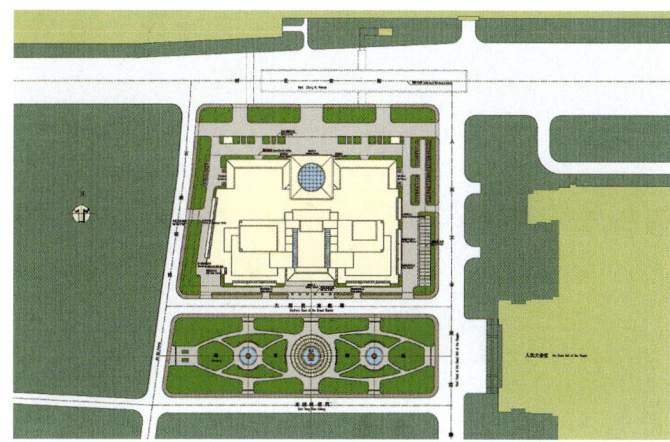

总平面
OVERALL SITE PLAN

公共大厅透视
PUBLIC HALL PERSPECTIVE

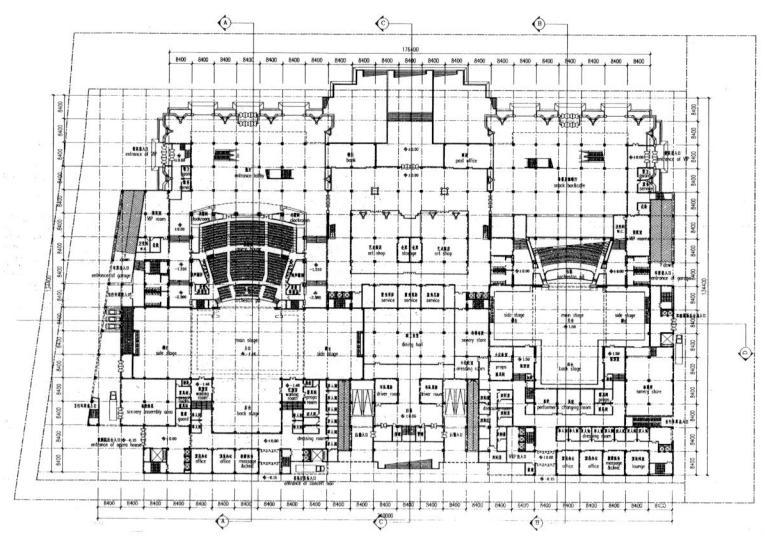

一层平面
GROUND FLOOR PLAN

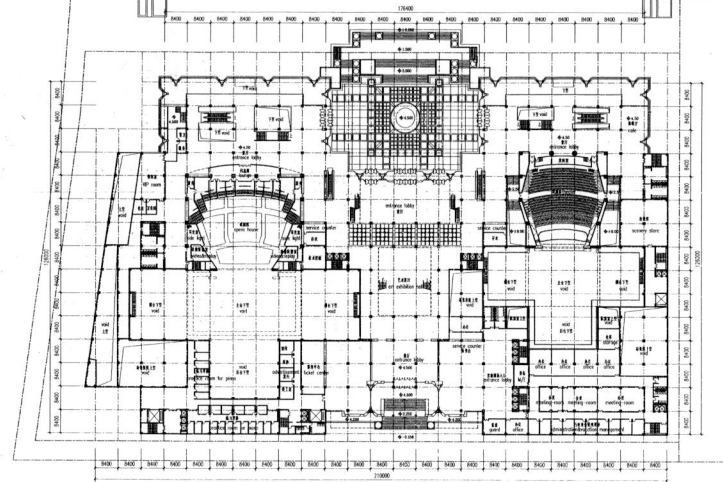

二层平面
2nd FLOOR PLAN

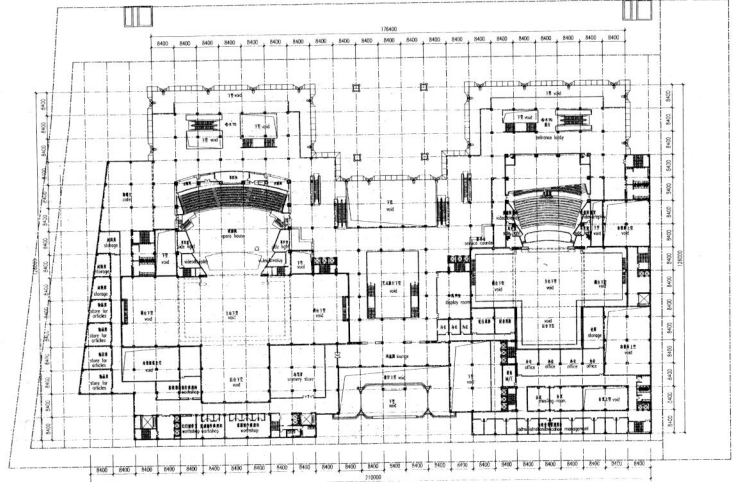

三层平面
3rd FLOOR PLAN

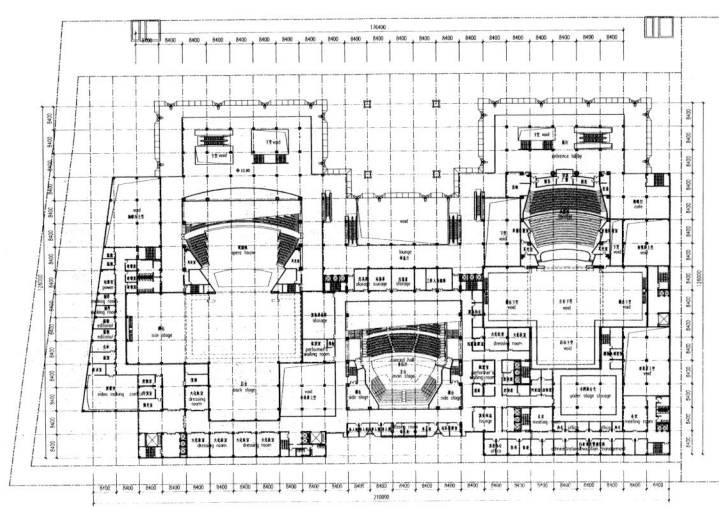

四层平面
4th FLOOR PLAN

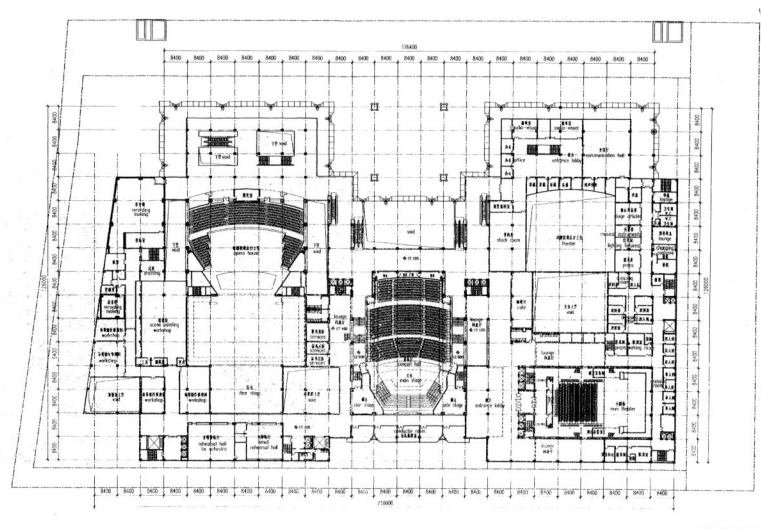

五层平面
5th FLOOR PLAN

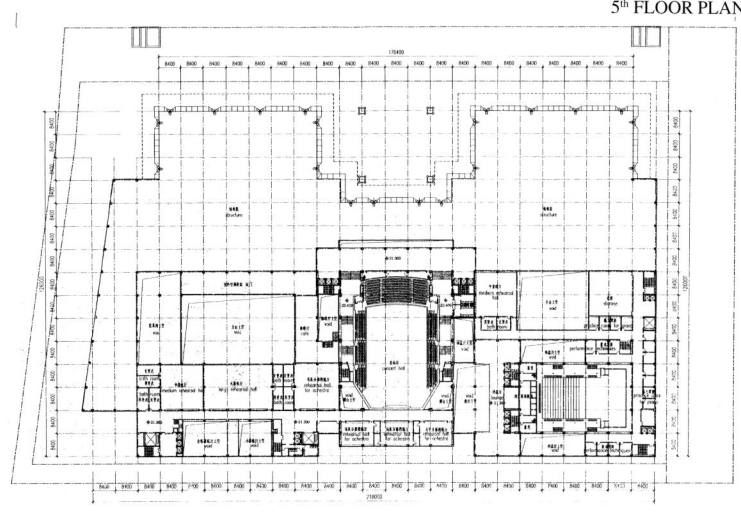

六层平面
6th FLOOR PLAN

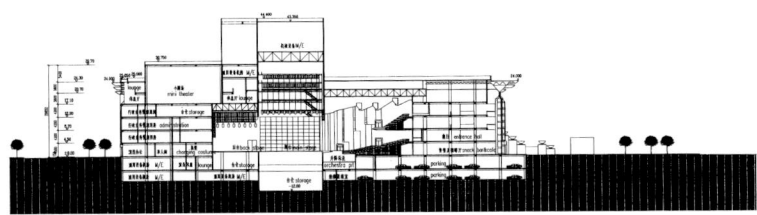

A–A 剖面
SECTION A-A

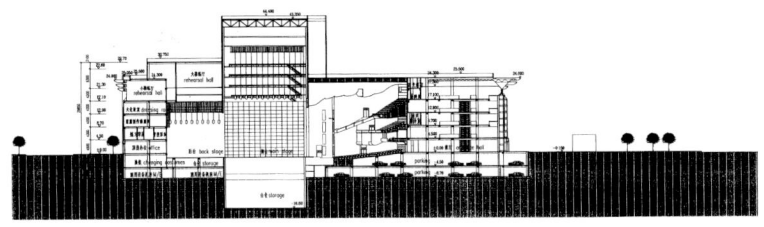

B–B 剖面
SECTION B-B

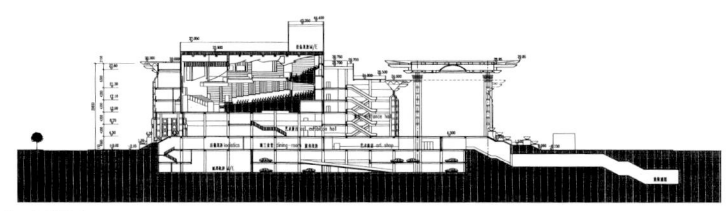

C–C 剖面
SECTION C-C

技术经济指标
Technical Economic Index

基地面积（m²） Site Area	总建筑面积（m²） Total Floor Area	建筑高度（m） Building Height	建筑覆盖率（%） Building Coverage Ratio
38916	地上：83480 地下：45760	43.50	61.9
容积率 Plot Ratio	绿化覆盖率（%） Green Coverage Ratio	停车数量 Number of Stalls	自行车停车数量 Number of Bicycle Stalls
3.32	12.25	地上：102 地下：506	980

各观众厅数据
Data For Auditoria

		歌剧院 Opera House	音乐厅 Concert Hall	戏剧院 Theater	小戏场 Mini Theater
观众厅面积（m²） Floor Area of Auditorium		1023.84	1131.31	718.00	476.8
观众厅体积（m³） Volume of Auditorium		17978.0	15732.6	7681.8	3814.4
座席数 Seating Capacity	总座席数 Seating Capacity(Seats)	2525	2017	1302	487
	其中：池座 Among Them:Auditorium	1187	1223	899	356
	楼座 Floor Seat	40(包厢) 1198(楼座)	794	403	131
	休息厅面积（m²） Lounge Floor Area	3870	2540	2060	1520
舞台尺寸（m²） Stage Dimentions	主舞台 Main Stage	41.2×24.4×36.0	演奏台 最深18.6 最宽28.0	27.0×21.6×25.0	25.2×16.8×10.8
	左右侧台 Left and Right Side Stages	24.4×21.0×10.5(左) 24.4×21.0×14.7(右)		15.9×21.6×15.3	
	后舞台 Back Stage	24.4×19.2×18.0		25.2×12.0×12.8	
	台仓 Under Stage Storage	见舞台机械图			
	升降乐池 Elevating Orchestra Pit				
最大视距（m） Max. Visual Distance		31.7	32.6	22.1	18.4
最大俯角（°） Max. Angle of Depression		29	20.6	30	19

东南大学建筑设计研究院（中国）
Architectural Design & Research Institute of Southeast University (China)

天安门广场南北主轴线是建立环境秩序的首要控制线，广场周界相关建筑都应当受其制约。

基于对该地域环境的上述基本理解，我们认为国家大剧院建筑主体应当具有明显的方向性，既要成为长安街上的重要建筑。同时必须对天安门广场作出恰当的策应。由此形成了主体不对称的基本布局，在用地的东北角留出一块广场作为国家大剧院的主入口广场。面向长安大街和南侧城市绿化广场则相应设置北、南二个重要入口。

The North-South major axis of Tian An Men Square is the prime control line to which the surrounding order introduced, and those constructions around it should be under its control. According to the basic understanding of the local surroundings, we think the direction of the main part of the National Grand Theater must be pure. It ought to be an important construction and also echo properly to Tian An Men Square. So we form an asymmetrical basic distribution of its main body, leaving a square to the region's north-east corner as the main entrance square to the National Grand Theater. The north and south entrances are respectively situated facing to Chang An Street and southern city public green ways.

北侧模型透视
VIEW FROM NORTH SIDE OF THE MODEL

观众厅透视
AUDITORIUM PERSPECTIVE

北侧入口透视
PERSPECTIVE OF NORTH ENTRANCE

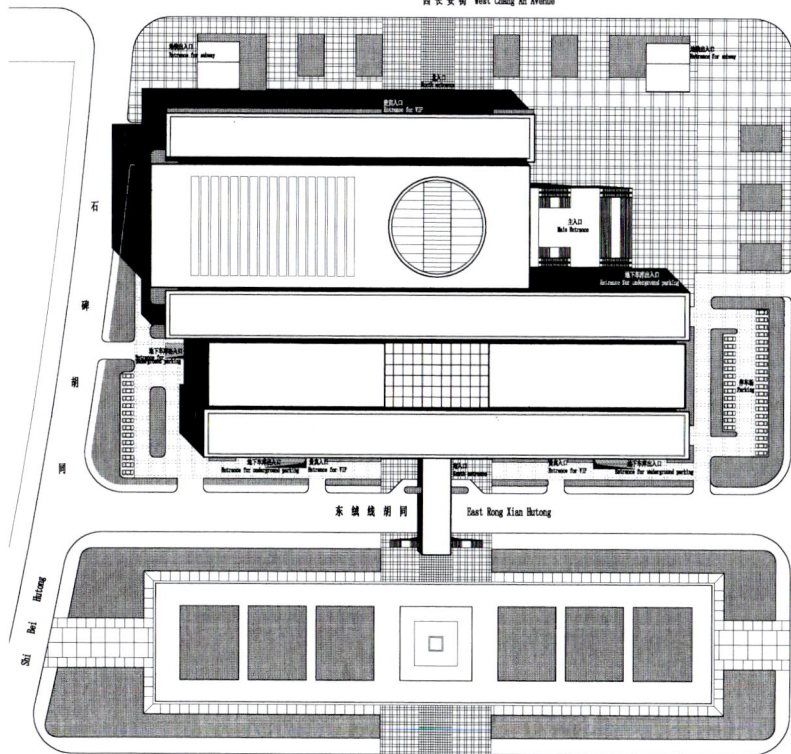

总平面
OVERALL SITE PLAN

设计概念"北京印象"
DESIGN CONCEPT "BEIJING IMPRESSION"

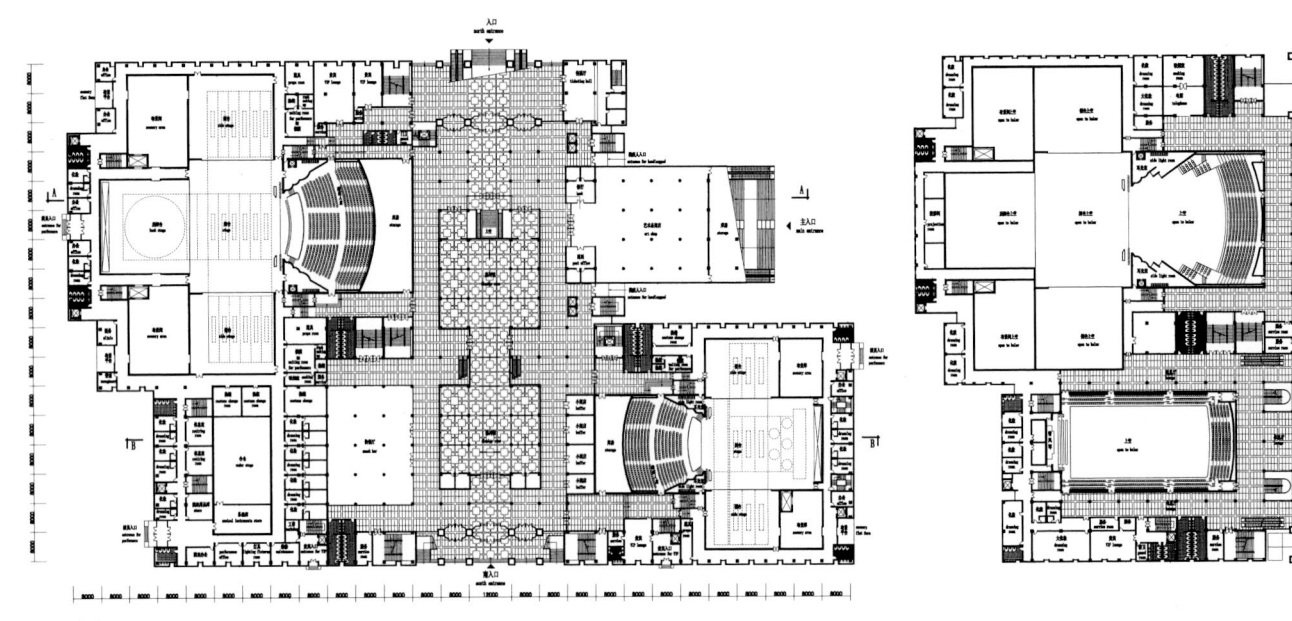

一层平面
GROUND FLOOR PLAN

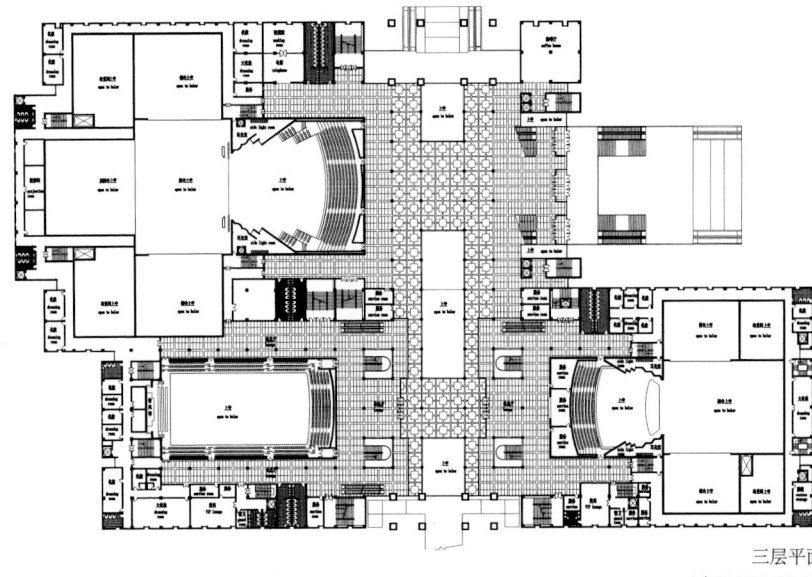

三层平面
3rd FLOOR PLAN

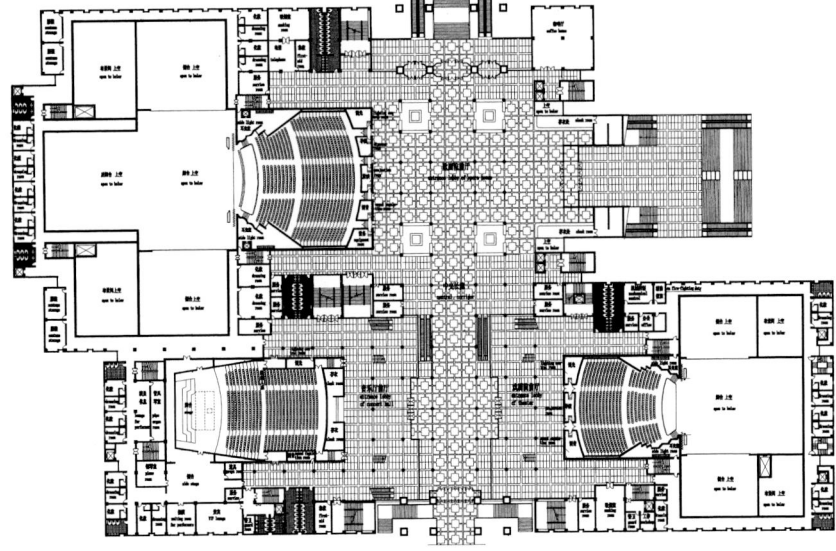

二层平面
2nd FLOOR PLAN

四层平面
4th FLOOR PLAN

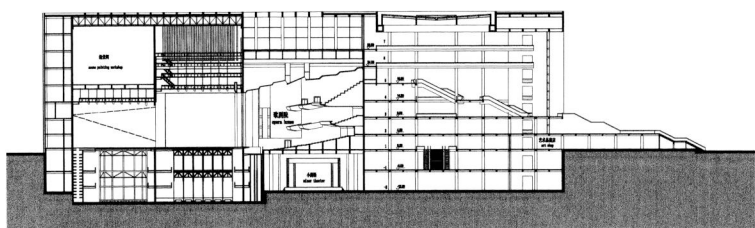

A—A 剖面
SECTION A-A

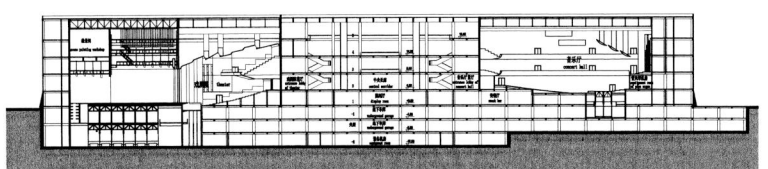

B—B 剖面
SECTION B-B

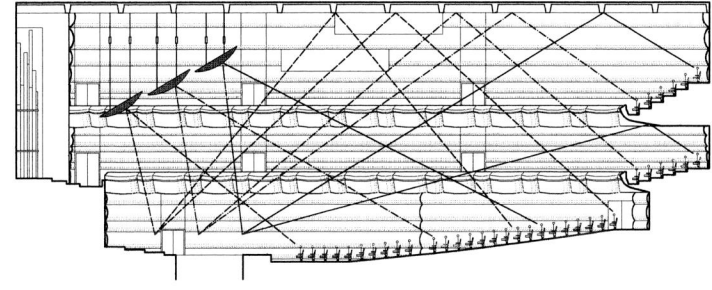

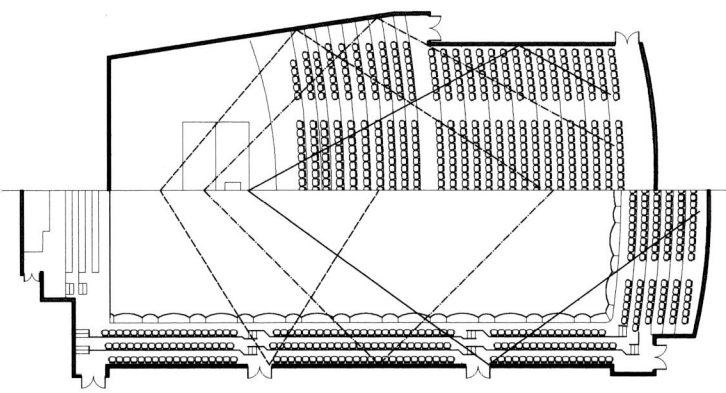

音乐厅平面剖面声线分析
PLAN AND SECTION ACOUSTICAL ANALYSIS OF THE CONCERT HALL

技术经济指标
Technical Economic Index

基地面积（m²） Site Area	总建筑面积（m²） Total Floor Area	建筑高度（m） Building Height	建筑覆盖率 (%) Building Coverage Ratio
38916	地上：82650 地下：48560	40.10	60.18
容积率 Plot Ratio	绿化覆盖率 (%) Green Coverage Ratio	停车数量 Number of Stalls	自行车停车数量 Number of Bicycle Stalls
2.12	15.20	地上：60 地下：510	1050

各观众厅数据
Data For Auditoria

		歌剧院 Opera House	音乐厅 Concert Hall	戏剧院 Theater	小戏场 Mini Theater
	观众厅面积（m²） Floor Area of Auditorium	1932	1456	957	685
	观众厅体积（m³） Volume of Auditorium	16042	19600	6977	2265
座席数 Seating Capacity	总座席数 Seating Capacity(Seats)	2464	1954	1232	409
	其中：池座 Among Them:Auditorium	1194	858	786	409
	楼座 Floor Seat	1270	1096	446	
	休息厅面积（m²） Lounge Floor Area	5100	2900	1740	1694
舞台尺寸（m²） Stage Dimentions	主舞台 Main Stage	27 × 36	18 × 26	21 × 25.4	12 × 24
	左右侧台 Left and Right Side Stages	24.5 × 23	16 × 12	21 × 16	12 × 8
	后舞台 Back Stage	25 × 25.8		23 × 25.4	
	台仓 Under Stage Storage	27 × 36	16 × 24	21 × 25.4	
	升降乐池 Elevating Orchestra Pit	8 × 22		10.5 × 18	
	最大视距（m） Max. Visual Distance	35.91	33.20	29.60	24
	最大俯角（°） Max. Angle of Depression	29.50	22.30	26.02	9.37

**竹中工务店
株式会社(日本)
Takenaka Corporation
(Japan)**

我们把中国国家大剧院命名为举世无双的"艺术五合院",作为一座"剧场庭园"问世。本设计沿用中国自古以来的"围墙式"城市建筑特点。首先重视与以天安门为中心的城市布局之间的直观协调性。围墙内风格各异的剧场环绕四周,形成一个巨大的中厅。其上部由大屋顶覆盖,创造出举世无双的剧场庭园。

We propose an unparalleled theatrical garden, named"Quintet Court Theater",for the National Grand Theater, China. This proposal is based on the basic concept of an "enclosing configuration", a traditional feature of Chinese walled cities. It is intended to harmonize visually with the neat rows of buildings in Tian An Men Square. The enclosure has forum surrounded by four theaters, which expose their individual shapes of distinct functions and are covered by a huge roof to create an unparalleled theatrical garden.

北侧模型透视
VIEW FROM NORTH SIDE OF THE MODEL

南侧模型透视
VIEW FROM SOUTH SIDE OF THE MODEL

歌剧院透视
PERSPECTIVE OF THE OPERA HOUSE

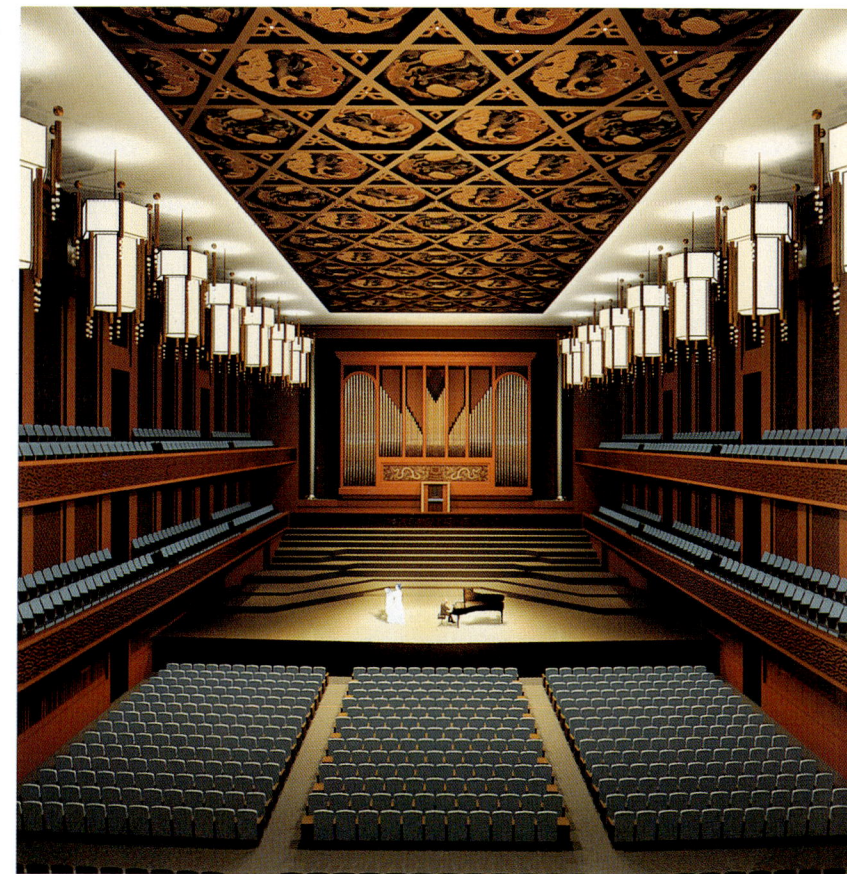

音乐厅透视
PERSPECTIVE OF THE CONCERT HALL

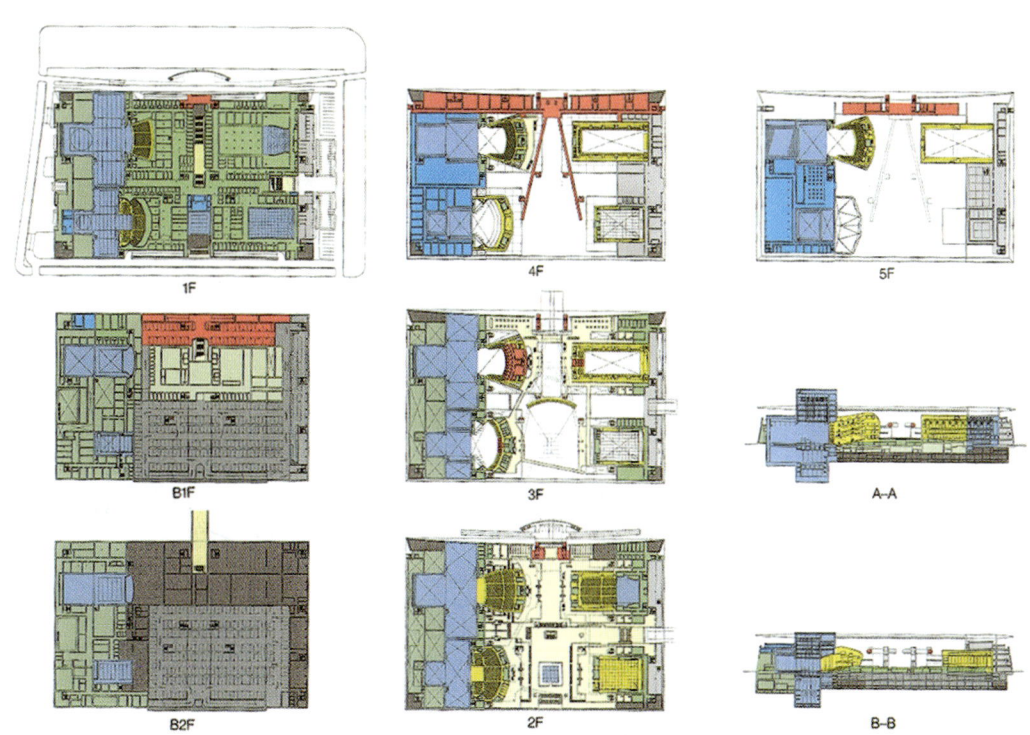

1F

4F

5F

B1F

3F

A-A

B2F

2F

B-B

区域图(功能区)
SECTIONAL PLAN(FUNCTION AREA)

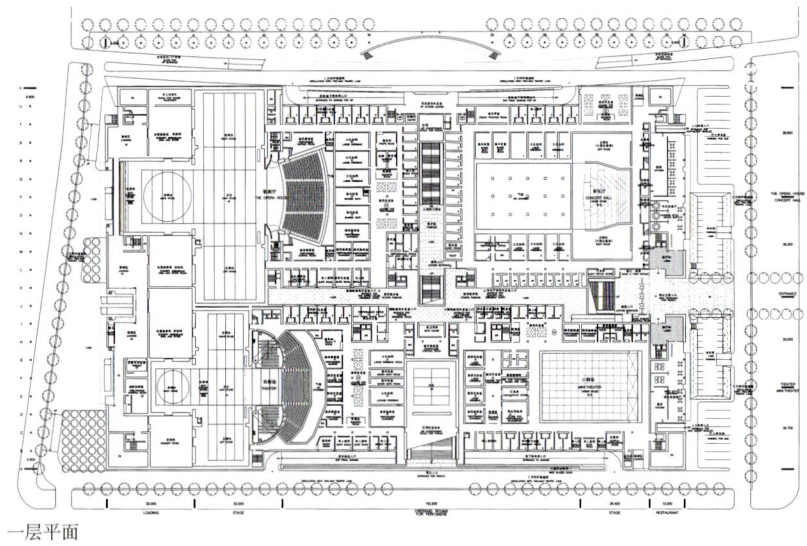

一层平面
GROUND FLOOR PLAN

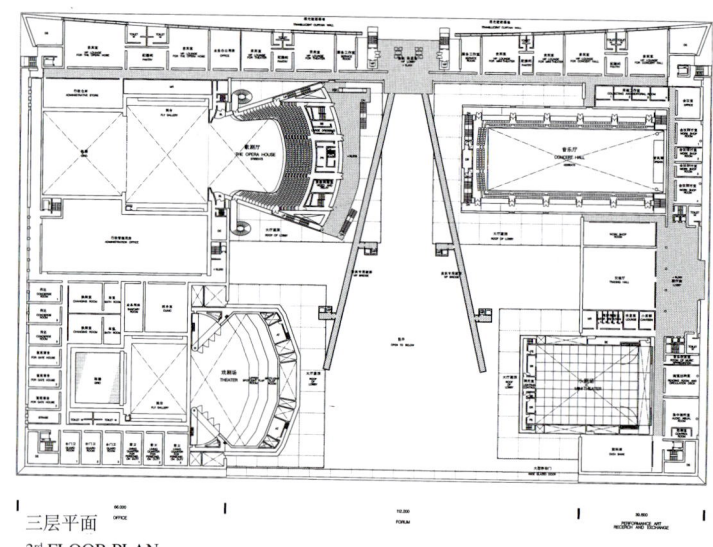

三层平面
3rd FLOOR PLAN

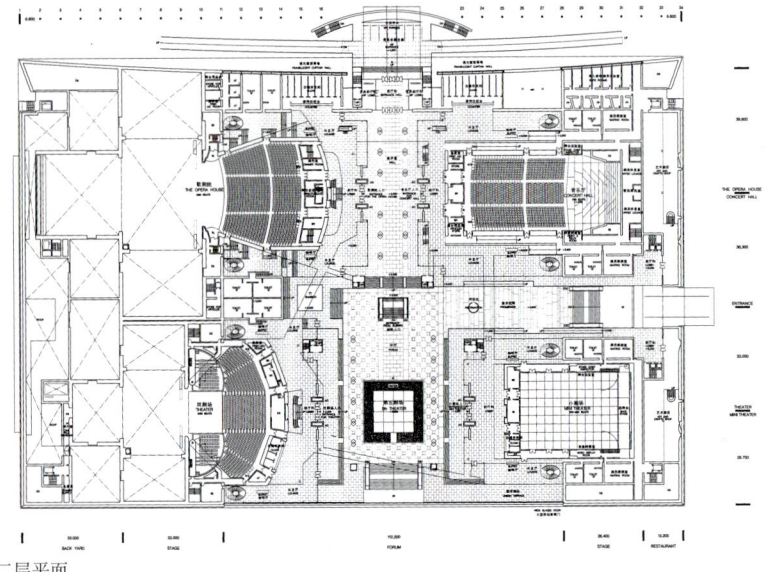

二层平面
2nd FLOOR PLAN

四层平面
4th FLOOR PLAN

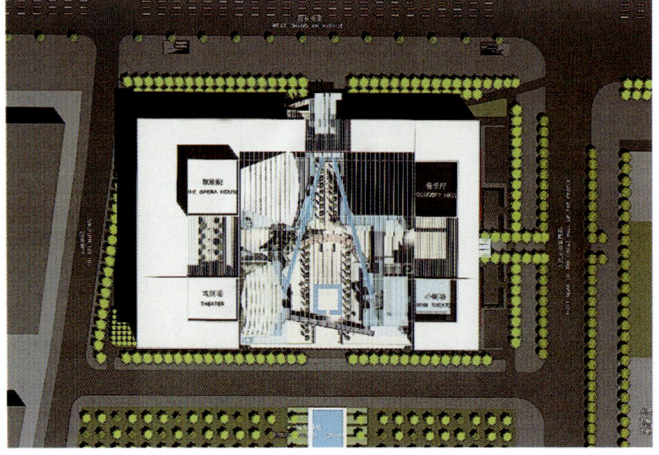

总平面
OVERALL SITE PLAN

170

A-A 剖面
SECTION A-A

B-B 剖面
SECTION B-B

C-C 剖面
SECTION C-C

D-D 剖面
SECTION D-D

技术经济指标
Technical Economic Index

基地面积(m²) Site Area	总建筑面积(m²) Total Floor Area	建筑高度(m) Building Height	建筑覆盖率 (%) Building Coverage Ratio
30187	地上：79167 地下：51706	30/45(局部)	77.60

容积率 Plot Ratio	绿化覆盖率 (%) Green Coverage Ratio	停车数量 Number of Stalls	自行车停车数量 Number of Bicycle Stalls
3.36	18.44	地上：73 地下：506	1000

各观众厅数据
Data For Auditoria

		歌剧院 Opera House	音乐厅 Concert Hall	戏剧院 Theater	小戏场 Mini Theater
观众厅面积(m²) Floor Area of Auditorium		2084	1719	1201	763
观众厅体积(m³) Volume of Auditorium		17600	18300	8600	5500~7100
座席数 Seating Capacity	总座席数 Seating Capacity(Seats)	2605	2006	1235	300~580
	其中：池座 Among Them:Auditorium	1285	1166	1063	300~580
	楼 座 Floor Seat	1320	840		
	休息厅面积(m²) Lounge Floor Area	1438	1098	674	416
舞台尺寸 (m²) Stage Dimentions	主舞台 Main Stage	972	432	567	787
	左右侧台 Left and Right Side Stages	567	247	315	
	后舞台 Back Stage	650		331	94
	台 仓 Under Stage Storage	702		409	
	升降乐池 Elevating Orchestra Pit	95		47~91	
最大视距(m) Max. Visual Distance		38.5	35.5	27	
最大俯角（°） Max. Angle of Depression		33	16	22	

亨克－施利克－陈
建筑师事务所
（奥地利）
Henke-Schreiech-Chen-
Architect(Austria)

本方案试图将各自独立的传统的歌剧院、音乐厅及戏剧院统一在一个内容丰富的大的建筑形体"屋顶"(130m × 220m)下，而形成新的建筑空间。

处于中心的新的活动空间与传统的室内空间相对应。

这块"空的空间"作为公共的、多功能的演出空间，成为内外的联系部分，将北京街市的生活引入内部，成为从"开放"至"封闭"的戏剧形式的过渡。

Through the arrangement of the traditional, independent construction bodies, i.e. opera house, concert hall and theater within an all embracing architectonic form "hard disc"(130 ×220m) a force field of new dimensions with regard to its contents emerges.

The newly gained action spaces in the core of the overall architecture oppose dialectically in dimension and availability the traditional interiors.

As freely accessible, multi-functional performance area this "empty space"(quotation Peter Brook) is predestined to serve as a link between outside and inside-to attract the urban street life of Beijing in all its facets as a sensitive transition from the "open" to the "closed" theater ritual.

北侧模型透视
VIEW FROM NORTH SIDE OF THE MODEL

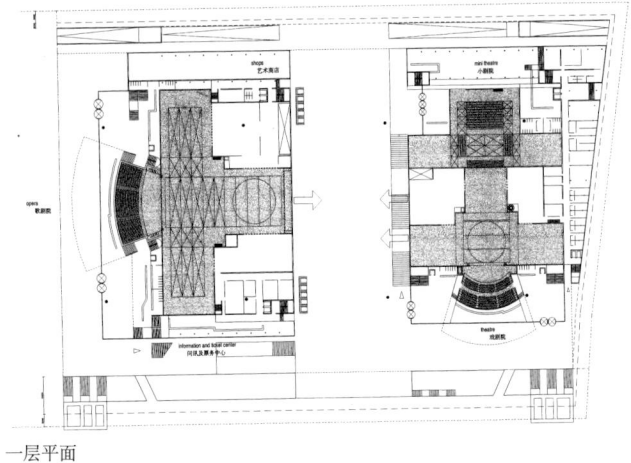

一层平面
GROUND FLOOR PLAN

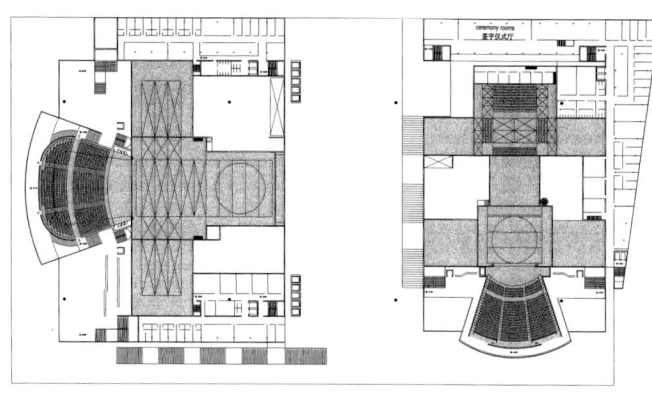

二层平面
2ⁿᵈ FLOOR PLAN

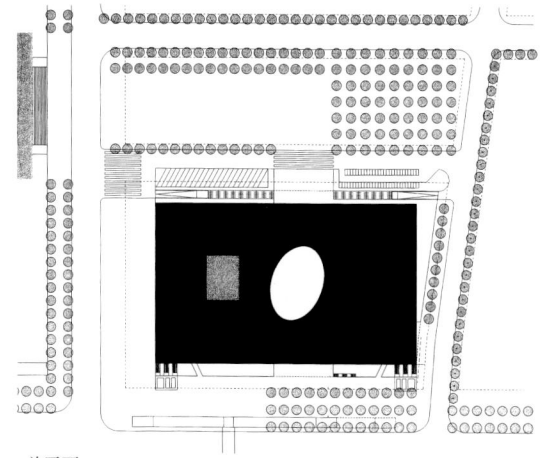

总平面
OVERALL SITE PLAN

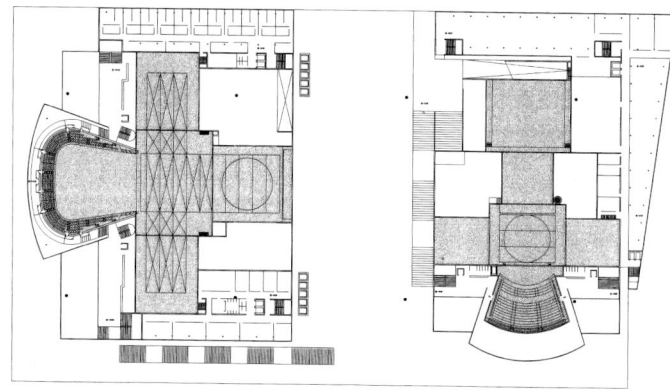

三层平面
3ʳᵈ FLOOR PLAN

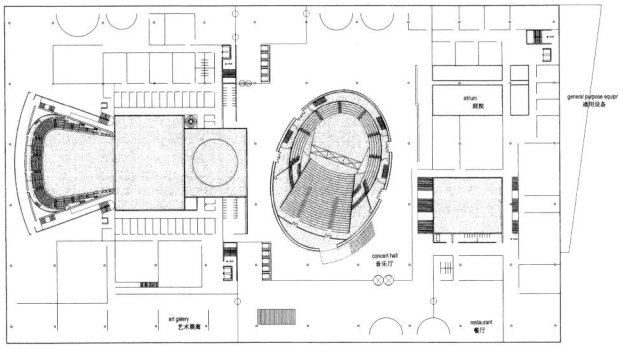

歌剧院观众厅透视
PERSPECTIVE OF THE OPERA HOUSE AUDITORIUM

技术经济指标
Technical Economic Index

基地面积(m²) Site Area	总建筑面积(m²) Total Floor Area	建筑高度(m) Building Height	建筑覆盖率(%) Building Coverage Ratio
38916	地上:684829 地下:146556	30/45(局部)	75.13

容积率 Plot Ratio	绿化覆盖率(%) Green Coverage Ratio	停车数量 Number of Stalls	自行车停车数量 Number of Bicycle Stalls
		地上:77 地下:660	1000

各观众厅数据
Data For Auditoria

		歌剧院 Opera House	音乐厅 Concert Hall	戏剧院 Theater	小戏场 Mini Theater
观众厅面积(m²) Floor Area of Auditorium		800	536	520	270
观众厅体积(m³) Volume of Auditorium		101500	20500	9100	12025
座席数 Seating Capacity	总座席数 Seating Capacity(Seats)	2536	2003	500	500
	其中:池座 Among Them:Auditorium	1064	1523	843	
	楼座 Floor Seat	1472	480	500	360
	休息厅面积(m²) Lounge Floor Area	5364	3984	2367	1375
舞台尺寸(m²) Stage Dimentions	主舞台 Main Stage	972	348	567	140
	左右侧台 Left and Right Side Stages	1174	470	620	460
	后舞台 Back Stage	650			320
	台仓 Under Stage Storage	1622	62	567	140
	升降乐池 Elevating Orchestra Pit	160	60		Variabel
最大视距(m) Max. Visual Distance		33	26	25	13
最大俯角(°) Max. Angle of Depression		33	31	18	Variabel

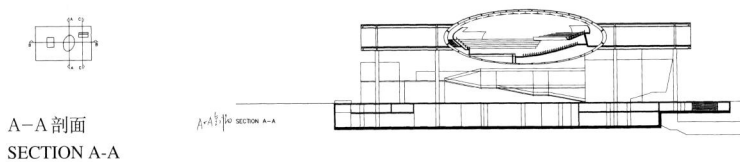

五层平面
5ᵗʰ FLOOR PLAN

A-A 剖面
SECTION A-A

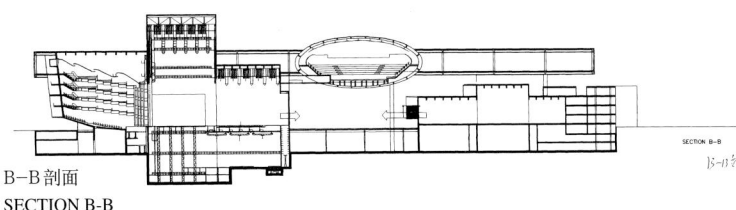

B-B 剖面
SECTION B-B

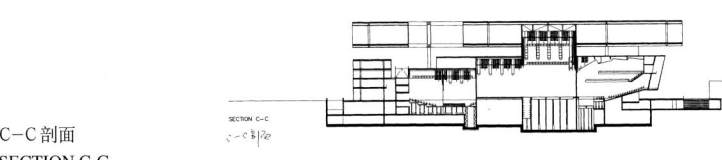

C-C 剖面
SECTION C-C

清华大学建筑设计研究院（中国）

Architectural Design & Research Institute of Tsinghua University(China)

天安门广场的艺术殿堂，中国建筑走向21世纪的标志。构思主题是要在充分弘扬城市整体美的前提下，塑造一座与周围建筑相匹配、有气势的艺术殿堂，既能显现剧场建筑特有的艺术魅力和文化内涵，又能反映中国建筑运用高科技手段迈向21世纪的新形象。

The goal is to achieve a monumental art palace on Tian'anmen Square, a Chinese architectural symbol for the 21st century. The main themes of this design are to merge with the larger environment, to preserve a majestic urban presence, and to coordinate with neighboring buildings. The China National Grand Theater must express Chinese artistic and cultural heritage and, at the same time, it should look forward into the future.

北侧夜景透视　NIGHT VIEW OF NORTH SIDE

南侧透视　PERSPECTIVE OF SOUTH ELEVATION

北侧模型　VIEW FROM NORTH SIDE OF THE MODEL

公共大厅透视
PUBLIC HALL PERSPECTIVE

歌剧院观众厅透视
PERSPECTIVE OF THE OPERA
HOUSEAUDITORIUM

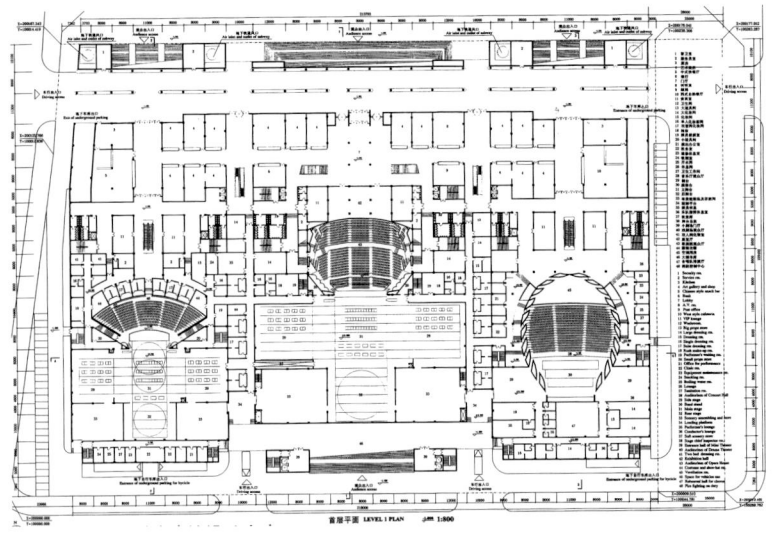

首层平面
GROUND FLOOR PLAN

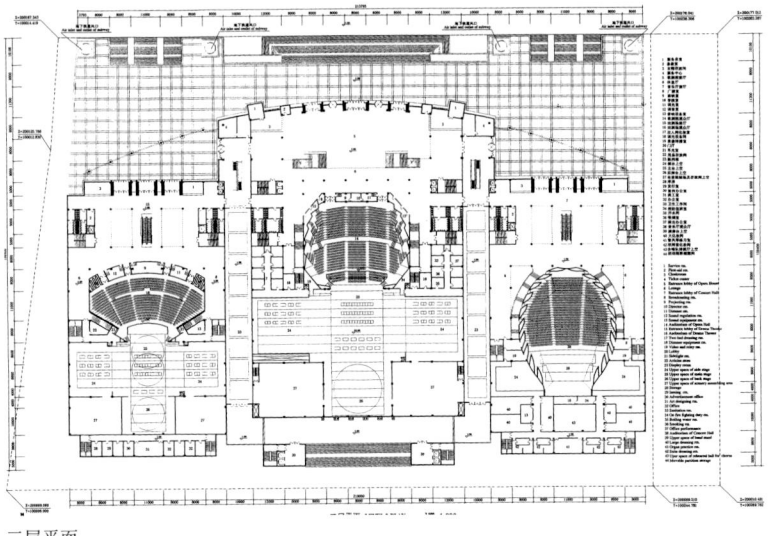

二层平面
2nd FLOOR PLAN

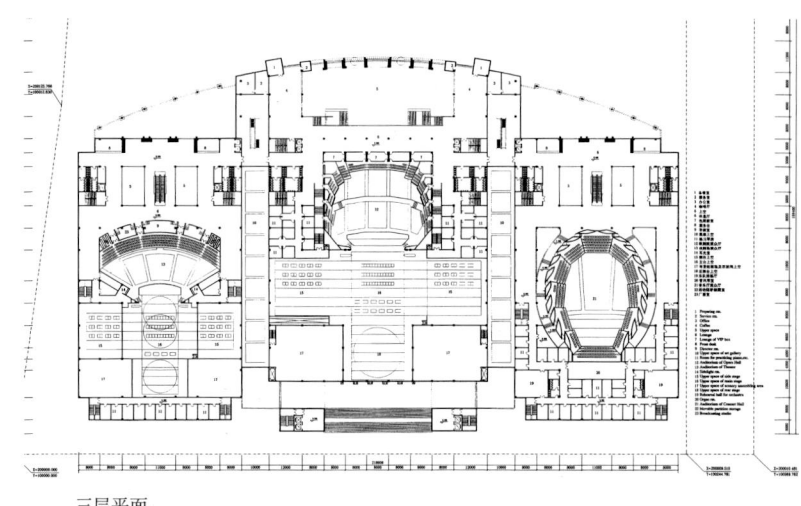

三层平面
3rd FLOOR PLAN

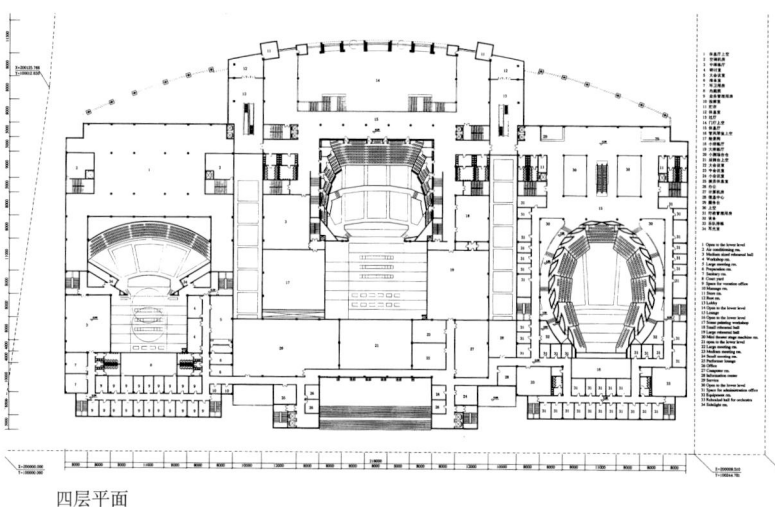

四层平面
4th FLOOR PLAN

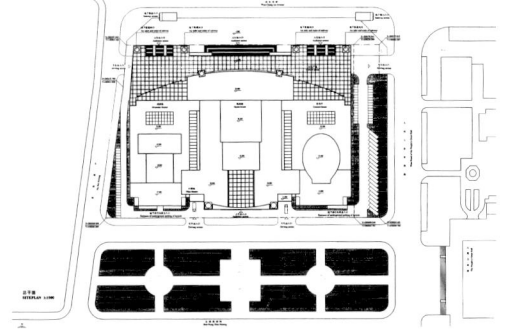

总平面
OVERALL SITE PLAN

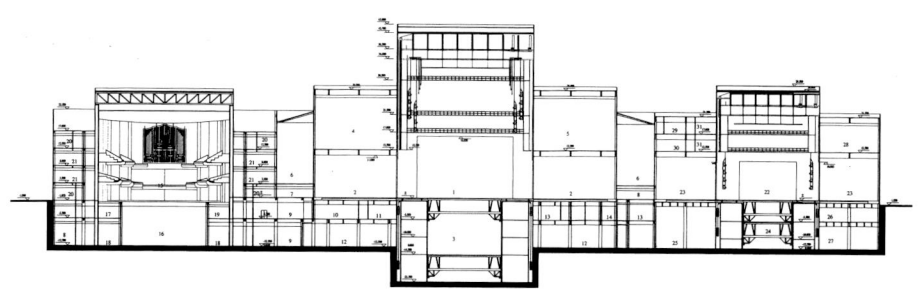

4-4 剖面
SECTION 4-4

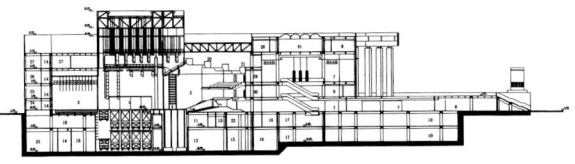

1-1 剖面
SECTION 1-1

2-2 剖面
SECTION 2-2

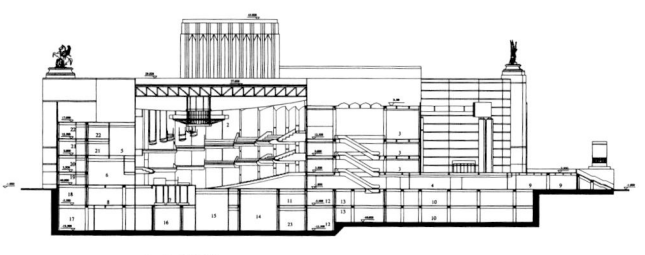

3-3 剖面
SECTION 3-3

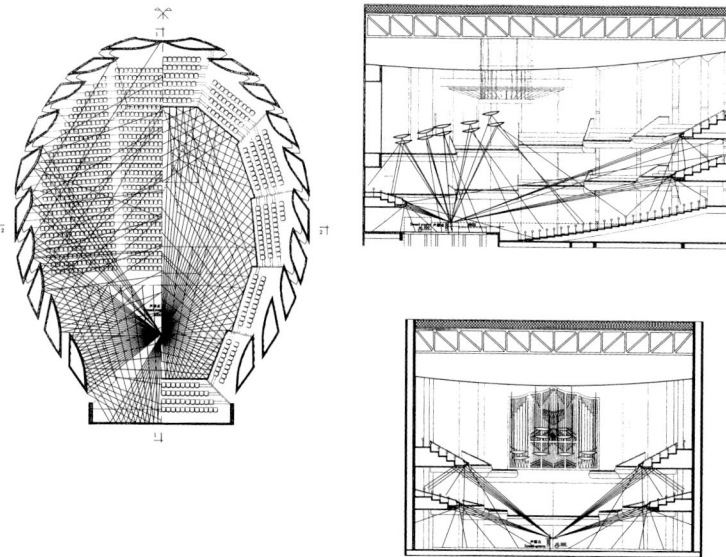

音乐厅声线分析
ACOUSTICAL ANALYSIS DRAWING OF THE CONCERT HALL

技术经济指标
Technical Economic Index

基地面积(m²) Site Area	总建筑面积(m²) Total Floor Area	建筑高度(m) Building Height	建筑覆盖率 (%) Building Coverage Ratio
38900	地上：805586 地下：50465	29/46(局部)	84.4
容积率 Plot Ratio	绿化覆盖率 (%) Green Coverage Ratio	停车数量 Number of Stalls	自行车停车数量 Number of Bicycle Stalls
2.07	4.5	地上：81 地下：508	1012

各观众厅数据
Data For Auditoria

		歌剧院 Opera House	音乐厅 Concert Hall	戏剧院 Theater	小戏场 Mini Theater
观众厅面积(m²) Floor Area of Auditorium		2253.5	2264.4	1045	见详图29小剧场布局图 See detail in drawing 29,Layout of Mini theater
观众厅体积(m³) Volume of Auditorium		21641.76	22415.47	7188	
座席数 Seating Capacity	总座席数 Seating Capacity(Seats)	2742	2185	1249	
	其中：池座 Among Them:Auditorium	1254	1232	892	
	楼 座 Floor Seat	1488	953	357	
	休息厅面积(m²) Lounge Floor Area	5919	3558	2450	
舞台尺寸(m²) Stage Dimentions	主舞台 Main Stage	36 × 27	25 × 15	27 × 19.6	
	左右侧台 Left and Right Side Stages	21 × 20.7	17 × 13	16 × 20	
	后舞台 Back Stage	25.7 × 20		20 × 14	
	台 仓 Under Stage Storage	19.5 × 19.5	19 × 9.6	13 × 13	
	升降乐池 Elevating Orchestra Pit	22 × 5		15 × 7	
最大视距(m) Max. Visual Distance		41.1	27.1	17.6	
最大俯角（°） Max. Angle of Depression		30	29	27	

豪斯保尔建筑师事务所(奥地利)

Wilhelm Holzbauer,
Architect(Austria)

在我们的设计方案中，建筑的各部分集于一个一气呵成的大屋顶之下。之所以这样做，一则为了与毗邻的雄伟壮观的人民大会堂相协调，同时为了与金碧辉煌的故宫屋顶相呼应。

在我们的设计方案中，建筑物三大部分的统一大屋顶呈流线形，至舞台塔楼自然隆起，飘逸洒脱。这是我们对屋顶这一永恒主题交出的新试卷。

建筑主体分为三大块，沿中轴线左右对称。这同样符合中国建筑中借对称追求和谐华贵的基调。

The authors of this project propose a structure that combines all building components under one large roof to give this building a character that is at once monumental in scale to assert itself in relation to the neighboring building mass of the Great Hall of the People, as well as being sensitive to the roof scopes of the Former Imperial palace.

In this project, the flowing lines of three parts of the roof, rising up from the base are an intention to create a new interpretation of this eternal theme of "the roof".

The division of the overall building complex into three parts, arranged symmetrically along a central axis are also an interpretation of the pursuit prevalent in Chinese architecture to achieve regularity and dignity through symmetric.

北侧模型透视
VIEW FROM NORTH SIDE OF THE MODEL

南侧模型透视
VIEW FROM SOUTH SIDE OF THE MODEL

休息厅
FOYER

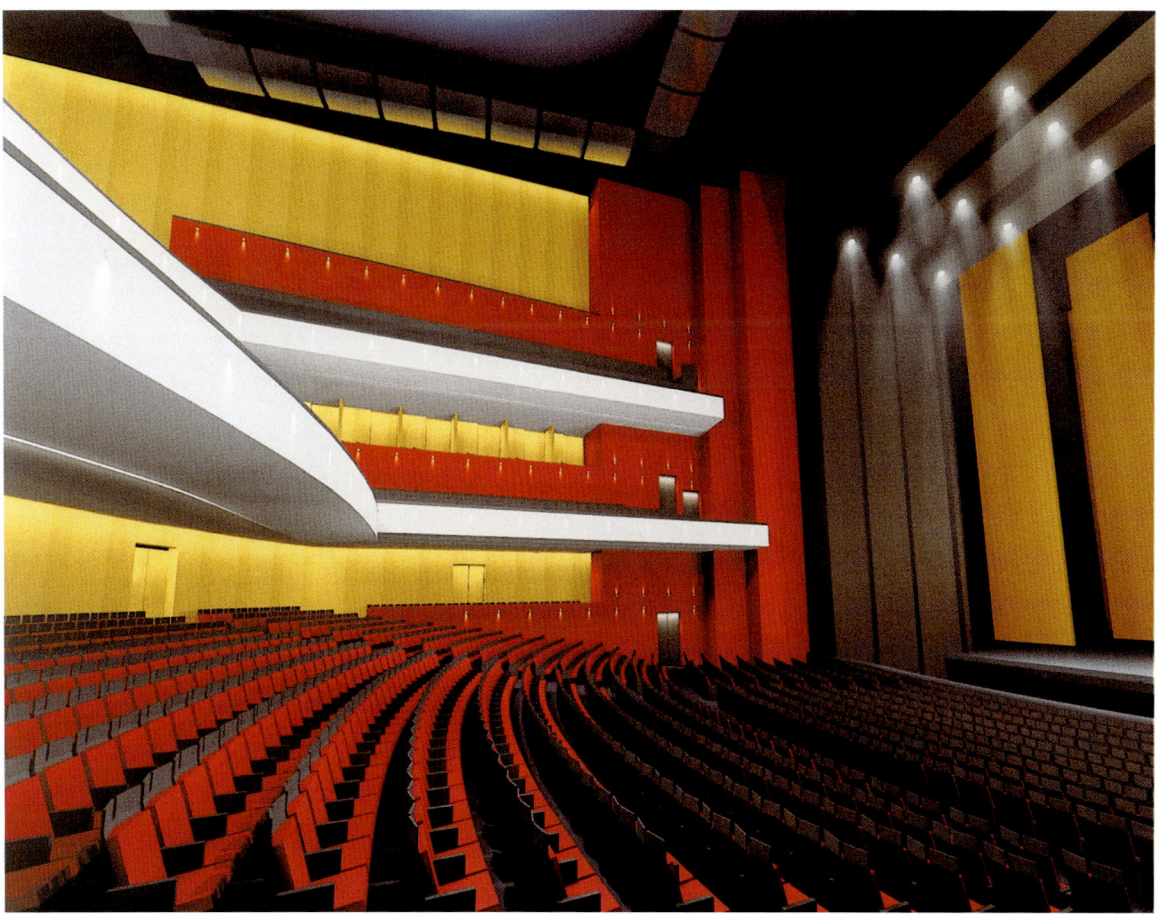

观众厅
AUDITORIUM

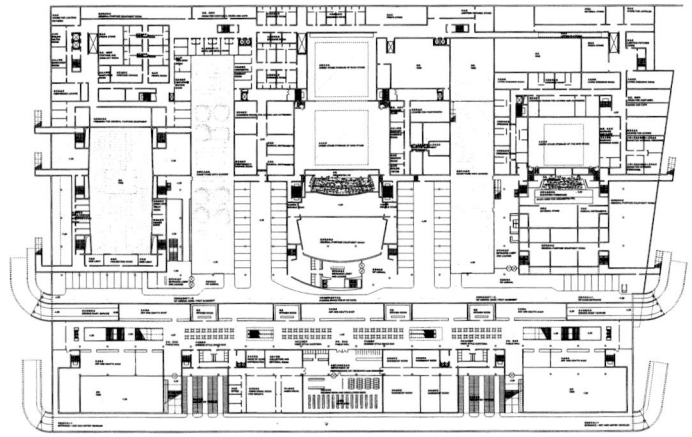

地下一层平面
FIRST BASEMENT PLAN

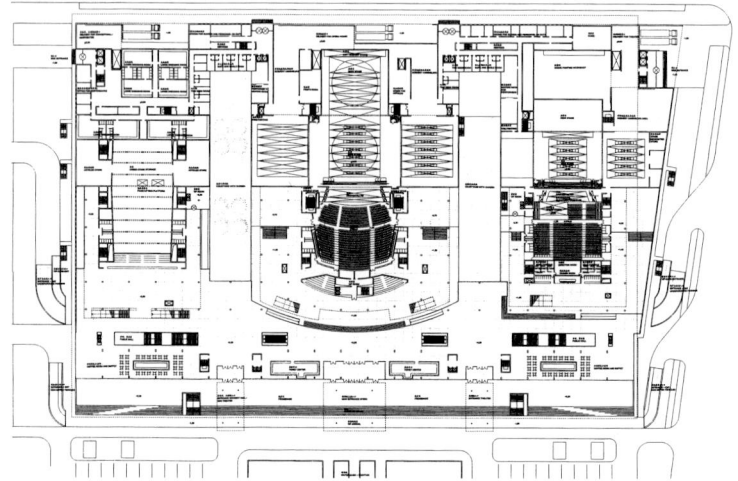

一层平面
GROUND FLOOR PLAN

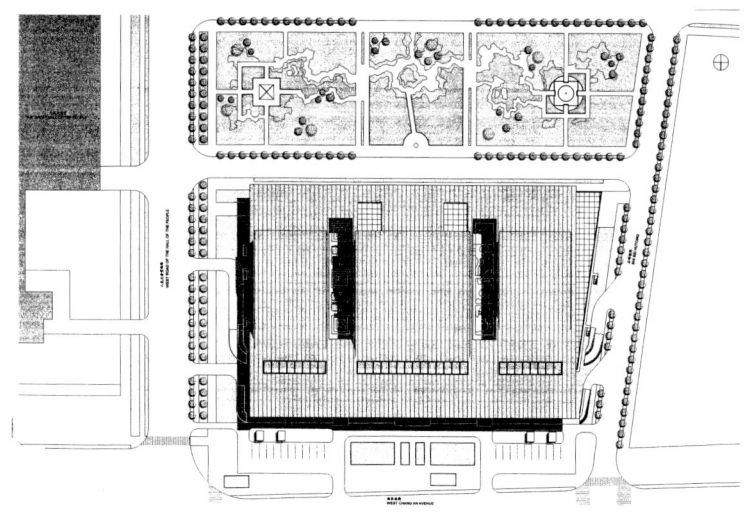

总平面
OVERALL SITE PLAN

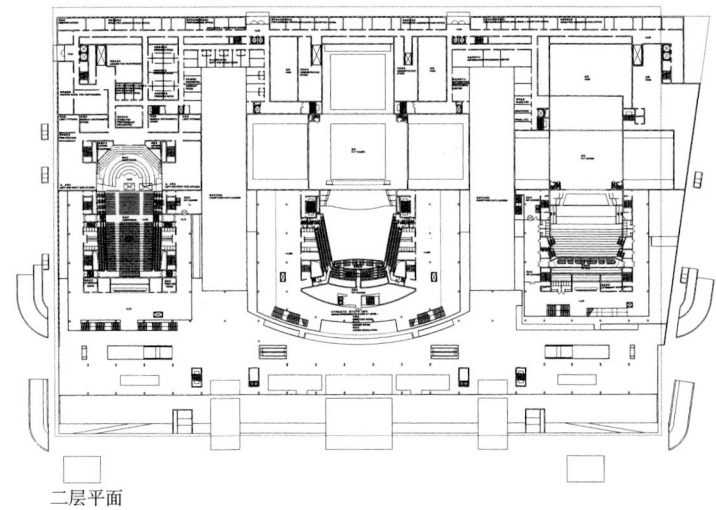

二层平面
2nd FLOOR PLAN

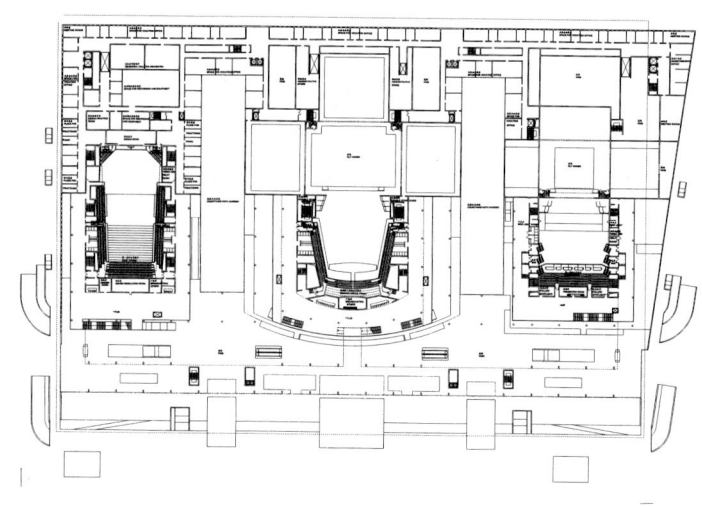

三层平面
3rd FLOOR PLAN

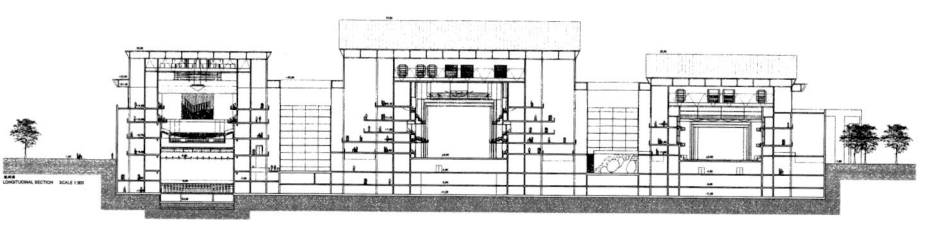

纵剖面
LONGITUDINAL SECTION

180

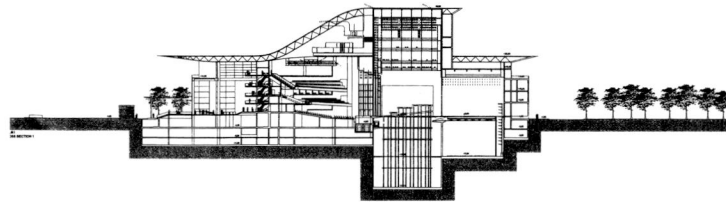

剖面 1
SECTION1

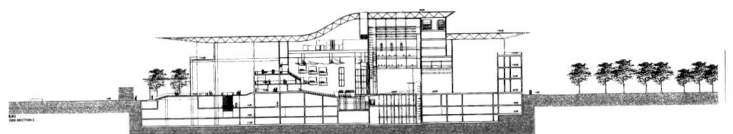

剖面 2
SECTION 2

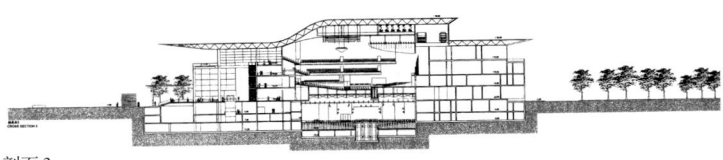

剖面 3
SECTION 3

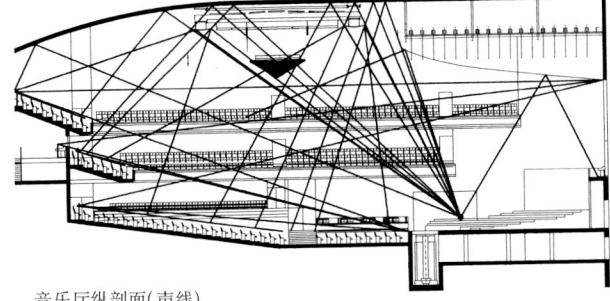

音乐厅纵剖面(声线)
LONGITUDINAL SECTION OF THE CONCERT HALL(ACOUSTICAL ANALYSIS)

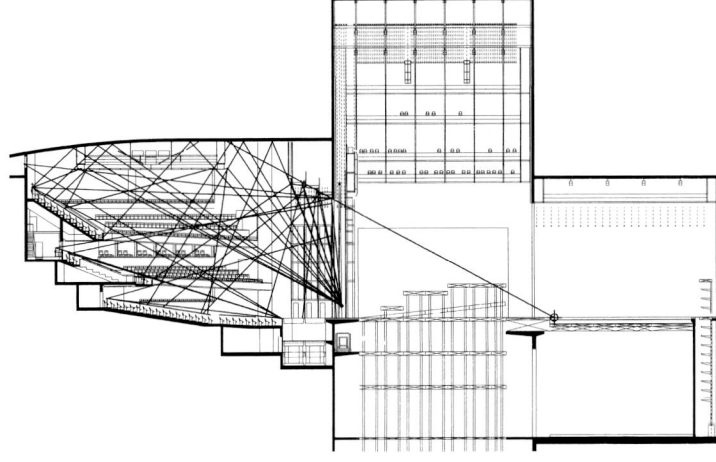

歌剧院纵剖面(声线)
LONGITUDINAL SECTION OF THE OPERA HOUSE(ACOUSTICAL ANALYSIS)

技术经济指标
Technical Economic Index

基地面积(m²) Site Area	总建筑面积(m²) Total Floor Area	建筑高度(m) Building Height	建筑覆盖率 (%) Building Coverage Ratio
38916	地上：83606 地下：88834	23.90/32.90/42.90	91

容积率 Plot Ratio	绿化覆盖率 (%) Green Coverage Ratio	停车数量 Number of Stalls	自行车停车数量 Number of Bicycle Stalls
10		地上：20 地下：578	1100

各观众厅数据
Data For Auditoria

		歌剧院 Opera House	音乐厅 Concert Hall	戏剧院 Theater	小戏场 Mini Theater
	观众厅面积(m²) Floor Area of Auditorium	1920	1770	976	1100
	观众厅体积(m³) Volume of Auditorium	24360	22300	10300	7900
座席数 Seating Capacity	总座席数 Seating Capacity(Seats)	2486	2006	1228	500
	其中：池座 Among Them:Auditorium	1187	1018	896	
	楼座 Floor Seat	1299	988	332	
	休息厅面积(m²) Lounge Floor Area	6490	2915	3680	1325
舞台尺寸(m²) Stage Dimentions	主舞台 Main Stage	920	435	528	110
	左右侧台 Left and Right Side Stages	1245	440	632	256
	后舞台 Back Stage	650		330	
	台仓 Under Stage Storage	1568	195	526	
	升降乐池 Elevating Orchestra Pit	160		205	
	最大视距(m) Max. Visual Distance	44.90	38.05	33.60	21.10
	最大俯角(°) Max. Angle of Depression	23.02	20.14	16.75	9.56

**东南大学建筑设计
研究院(中国)**
Architectural Design &
Research Institute of
Southeast University
(China)

本方案的造型设计力
求在基本比例、尺度和色
彩上与人民大会堂相协
调,在具体空间、体形设
计上设想用几个隐性的大
屋顶轮廓组成在西长安街
上的北立面。用楼电梯间
的顶部升高形成"正吻",
用玻璃的倾斜外墙面形成
"坡屋面",用金黄色的玻
璃高侧窗组成的平屋顶檐
部形成"屋脊"。神似"大
屋顶",又非大屋顶,这就
是现代与传统结合的创
造。

The form design of this
project is to be strict to harmonize with the People's
Great Hall on proportion,
scale and color. In the space
and form design, the north elevation along the West Chang
An Avenue is composed by
several outlines of metaphori-
cal large tiled roof. The
"Zhengwen"is formed by the
lift tower the "ramp roof"is
formed by the glass inclined
external wall, the "roof
edge"is formed by the eaves
that composed by high side
golden glass lights. It is alike
in spirit of "large tiled roof",
but it is not"roof". So this is
truly a creation of combination of modernity and tradition.

北侧模型透视 VIEW FROM NORTH SIDE OF THE MODEL

公共大厅透视 PUBLIC HALL PERSPECTIVE

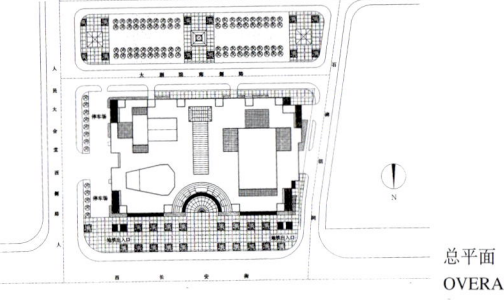

总平面
OVERALL SITE PLAN

观众厅透视 AUDITORIUM PERSPECTIVE

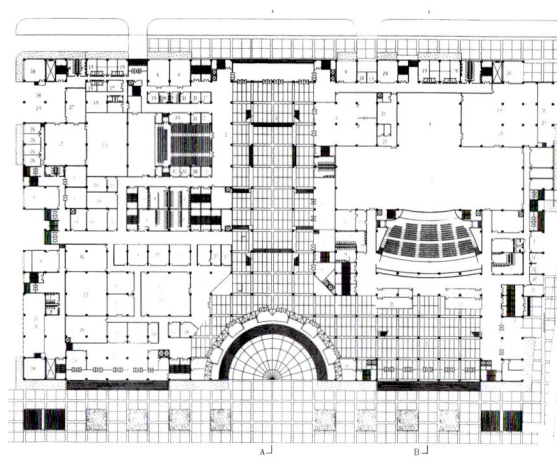

一层平面
GROUND FLOOR PLAN

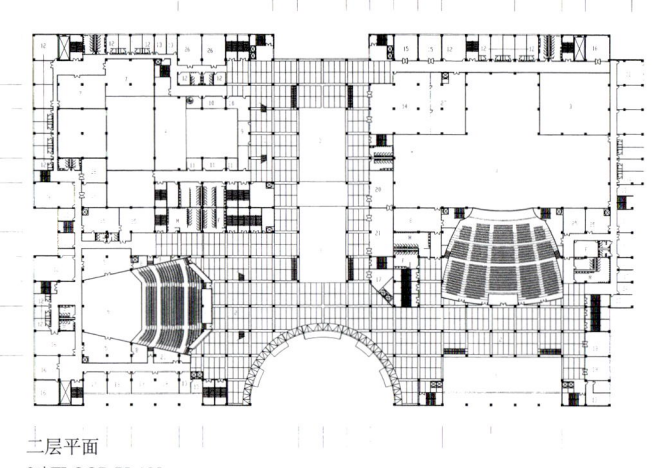

二层平面
2nd FLOOR PLAN

南侧模型透视
VIEW FROM SOUTH SIDE OF THE MODEL

技术经济指标
Technical Economic Index

基地面积（m²） Site Area	总建筑面积（m²） Total Floor Area	建筑高度（m） Building Height	建筑覆盖率（%） Building Coverage Ratio
38900	地上：90026 地下：43404	29.2/最高舞台45	61

容积率 Plot Ratio	绿化覆盖率（%） Green Coverage Ratio	停车数量 Number of Stalls	自行车停车数量 Number of Bicycle Stalls
2.314	6	地上：104 地下：441	地下1200

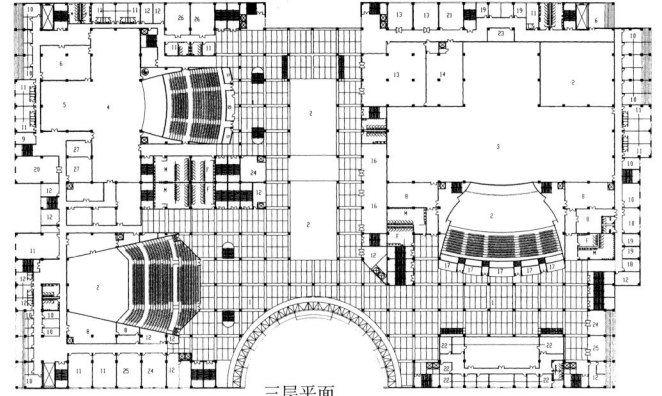

三层平面
3rd FLOOR PLAN

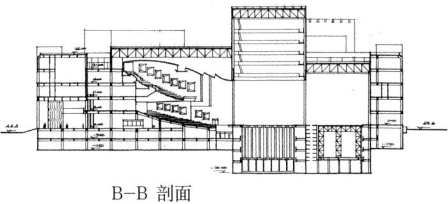

B-B 剖面
SECTION B-B

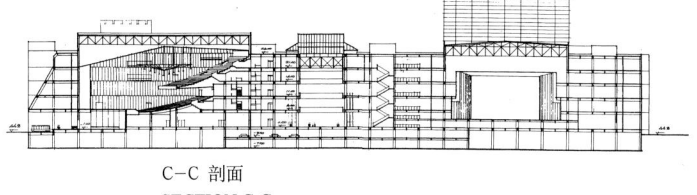

C-C 剖面
SECTION C-C

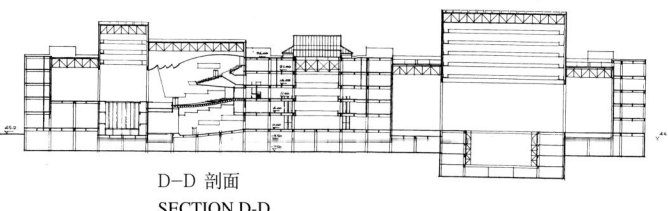

D-D 剖面
SECTION D-D

各观众厅数据
Data For Auditoria

		歌剧院 Opera House	音乐厅 Concert Hall	戏剧院 Theater	小戏场 Mini Theater
观众厅面积（m²） Floor Area of Auditorium		2100	1667	1105	389
观众厅体积（m³） Volume of Auditorium		20140	19242	8044	3143
座席数 Seating Capacity	总座席数 Seating Capacity(Seats)	2579	2021	1205	444
	其中：池座 Among Them:Auditorium	1434	870	630	444
	楼 座 Floor Seat	1145	1151	575	
	休息厅面积（m²） Lounge Floor Area	3564	2349	1296	891
舞台尺寸（m²） Stage Dimentions	主舞台 Main Stage	45 × 27	(18~24) × 19	30 × 18	30 × 12
	左右侧台 Left and Right Side Stages	18 × 27	(9~13) × 18	12 × 18	
	后舞台 Back Stage	27 × 24		18 × 18	
	台 仓 Under Stage Storage	主45 × 27 后27 × 24	18 × 18	30 × 18	18 × 9
	升降乐池 Elevating Orchestra Pit	28 × 5		23 × 8.5	
最大视距（m） Max. Visual Distance		40.9	43.4	35.9	24
最大俯角（°） Max. Angle of Depression		26.76	26.13	22.05	7

格里高里国际公司（意大利）
Gregotti Associati International srl (Italy)

我们的基本设计观念是，把建筑物的硕大屋顶变成公众可以进入的场所。我们在标高+9.00m处设计一个平台，创造出一个新的地平面。这个平台可以从马路上方便地登上，装饰有许多具有戏剧含义的建筑雕塑，从而形成一个神妙的场所。从这里可以俯视下方的城市风光和剧院前厅。平台的高度与"紫禁城"围墙相同：两者具有同样的视野，周围地区交通繁忙，街道宽阔，人们可以轻易地看见这个平台。

Our design is based on the idea of using the large roof of the building as a public space. A podium at level+9.00m creates a new ground level easily accessible from the street and rich in theatrical allusions: architectural sculptures form a magical place from where to observe the city and the theater lobbies below. The podium is as tall as the perimeter wall of the Forbidden City: they share the same horizon of vision, which can be seen also across the wide and heavily used traffic roads of the area.

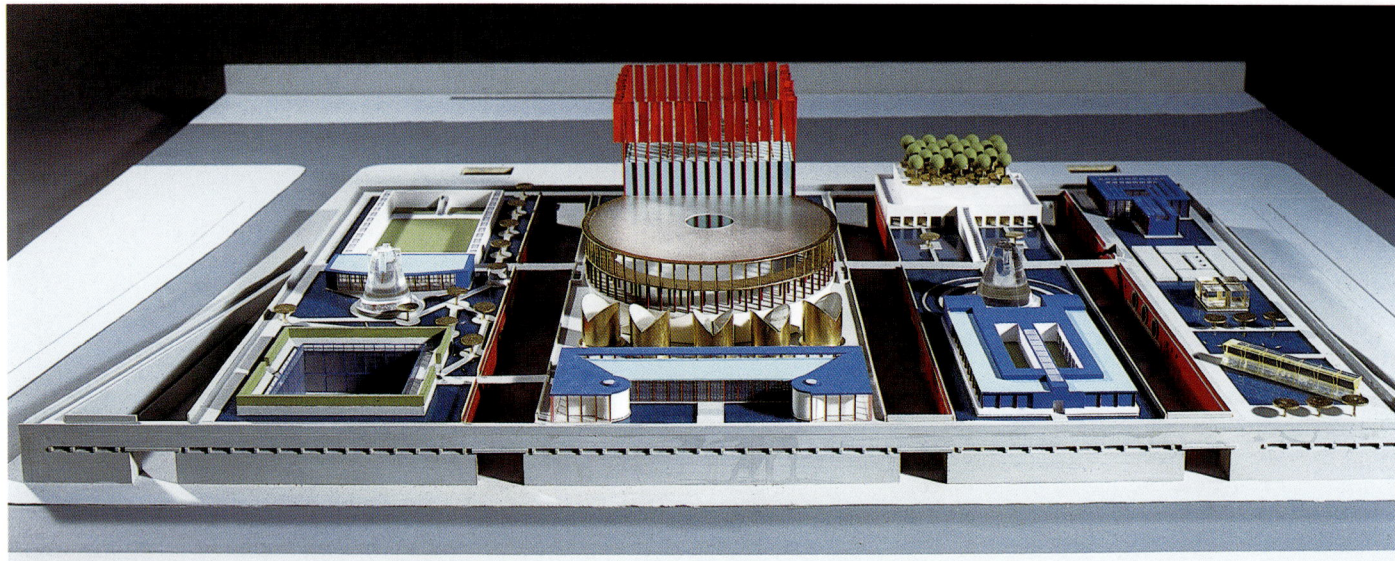

北侧模型透视 VIEW FROM NORTH SIDE OF THE MODEL

南侧模型透视
VIEW FROM SOUTH SIDE OF THE MODEL

观众厅透视
AUDITORIUM PERSPECTIVE

北侧模型透视　PERSPECTIVE OF NORTH ELEVATION

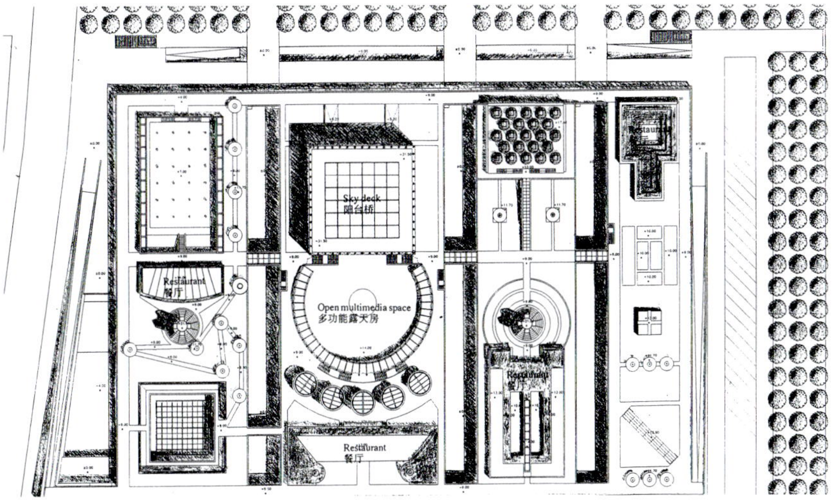

屋顶平台(+9.00m)
ROOF PLATFORM

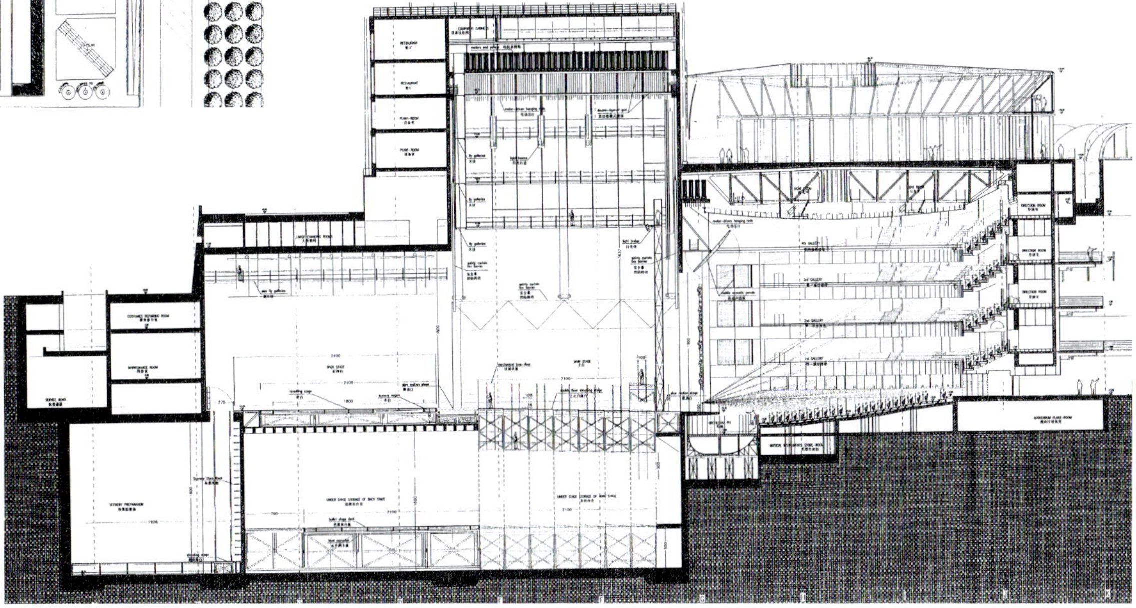

歌剧院剖面
SECTION OF THE OPERA HOUSE

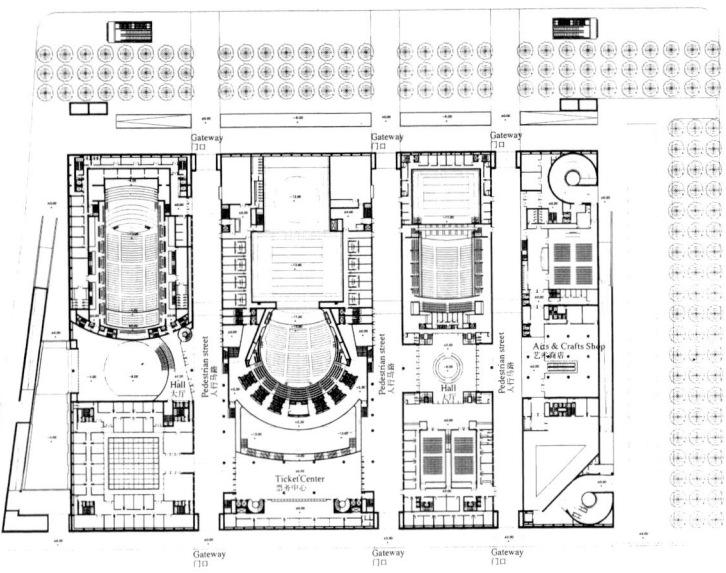

首层平面(±0.00m)
GROUND FLOOR LEVEL(+0.00m)

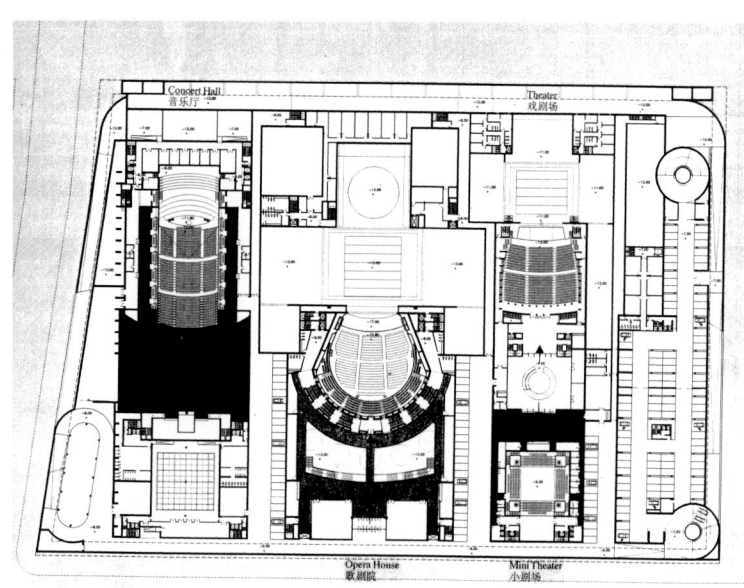

−8.00m平面
GROUND LEVEL(-8.00m)

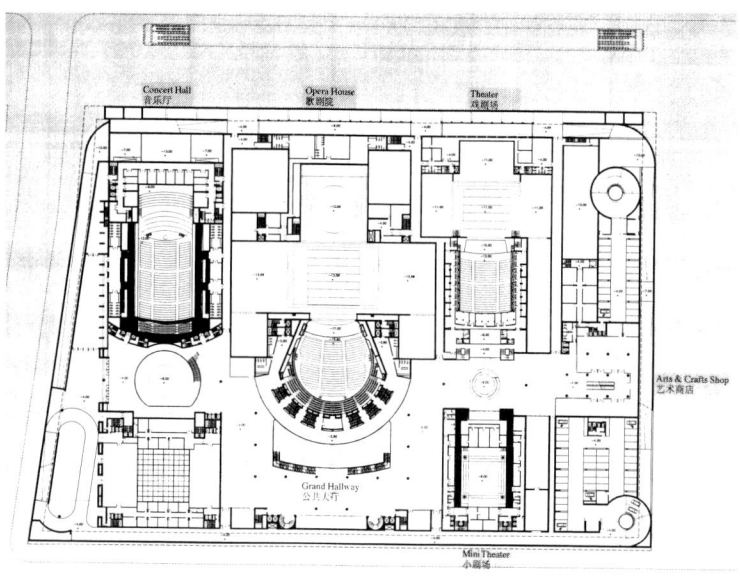

−4.00m平面
GROUND LEVEL(-4.00m)

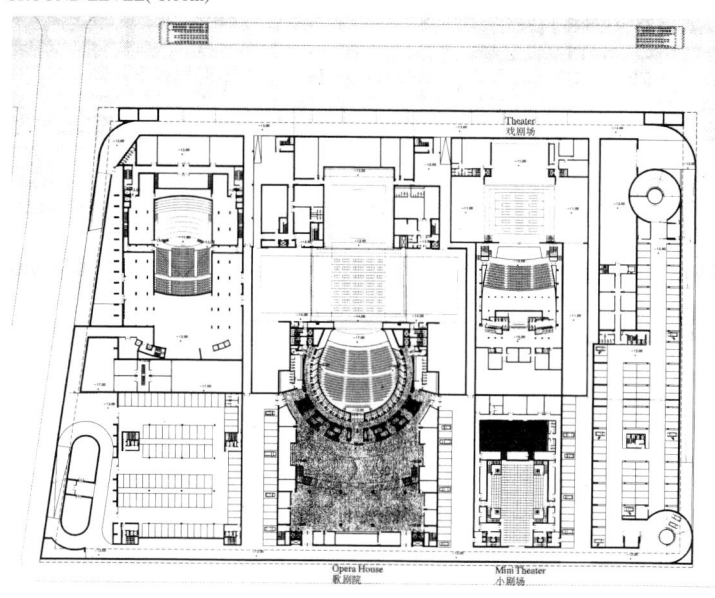

−13.00m平面
GROUND LEVEL(-13.00m)

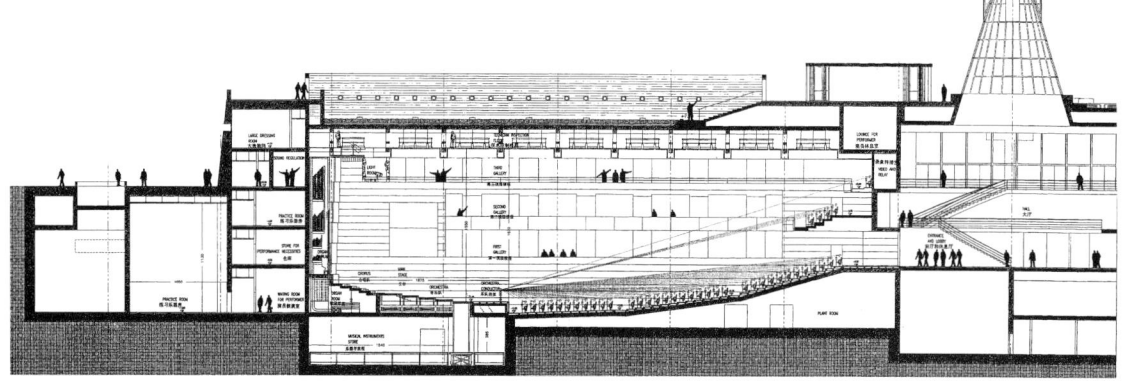

音乐厅剖面
SECTION OF THE CONCERT HALL

186

技术经济指标
Technical Economic Index

基地面积 (m²) Site Area	总建筑面积 (m²) Total Floor Area	建筑高度 (m) Building Height	建筑覆盖率 (%) Building Coverage Ratio
3.2	地上：34500 地下：91600	9/31.5(最高)	69

容积率 Plot Ratio	绿化覆盖率 (%) Green Coverage Ratio	停车数量 Number of Stalls	自行车停车数量 Number of Bicycle Stalls
3.2	63	地上：20 地下：620	1000(地上) 500(地下)

各观众厅数据
Data For Auditoria

		歌剧院 Opera House	音乐厅 Concert Hall	戏剧院 Theater	小戏场 Mini Theater
观众厅面积(m²) Floor Area of Auditorium		2010	2023	1420	515
观众厅体积(m³) Volume of Auditorium		21950	21580	13800	4150
座席数 Seating Capacity	总座席数 Seating Capacity(Seats)	2580	1980	1230	490
	其中：池座 Among Them:Auditorium	760	1408	940	515
	楼 座 Floor Seat	797	1398	900	490
	休息厅面积(m²) Lounge Floor Area	6350	3550	1880	1330
舞台尺寸(m²) Stage Dimentions	主舞台 Main Stage	972		567	145
	左右侧台 Left and Right Side Stages	594	216	316	65
	后舞台 Back Stage	650		330	65
	台 仓 Under Stage Storage	972	192	567	
	升降乐池 Elevating Orchestra Pit	85	288	204	145
最大视距(m) Max. Visual Distance		37	38	31	13
最大俯角（°） Max. Angle of Depression		32	18	20	39

立面 ELEVATION
剖面 SECTIONS

●设立一条清晰的视觉及空间中轴线,从而安排功能及人流组织。

●营造一个在建筑中心的绿化空间(冬季庭院),肩负起空间的组织及导向作用。

●利用公共空间的序列层次处理,从而营造憧憬及寻疑的意境。

●从现代理性结构及营造原理出发,制造出像传统屋檐般的弧形建筑构件形态。

●采取传统建筑中轻巧及凝重对比的视觉艺术,尽量发挥现代材料及构件的不同透明度及纹理特质。

● A clear visual and spatial axis about which functions and circulation are organized.

● A central negative space (Winter courtyard) as the rallying and orientating element in the spatial organization.

● A deliberate sequential arrangement of public space and movement to create anticipation and suspense.

● The creation of curvilinear elements (as in traditional eaves) based on rational structural principle and contemporary construction methodology.

● The play of visual texture and translucency in the architectural elements against mass and solidity.

北侧透视 PERSPECTIVE OF NORTH ELEVATION

南侧模型透视 VIEW FROM SOUTH SIDE OF THE MODEL

大门透视 MAIN ENTRANCE PERSPECTIVE

门廊透视 PORTICO PERSPECTIVE

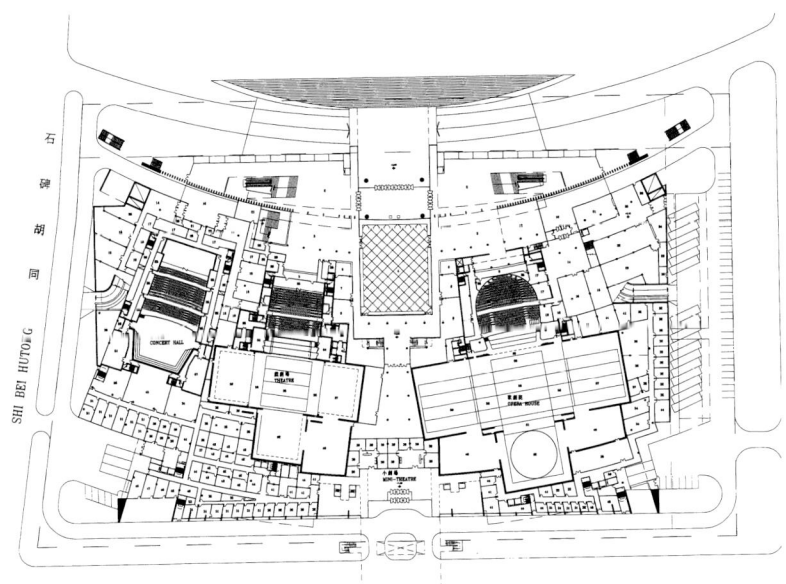

一层平面
GROUND FLOOR PLAN

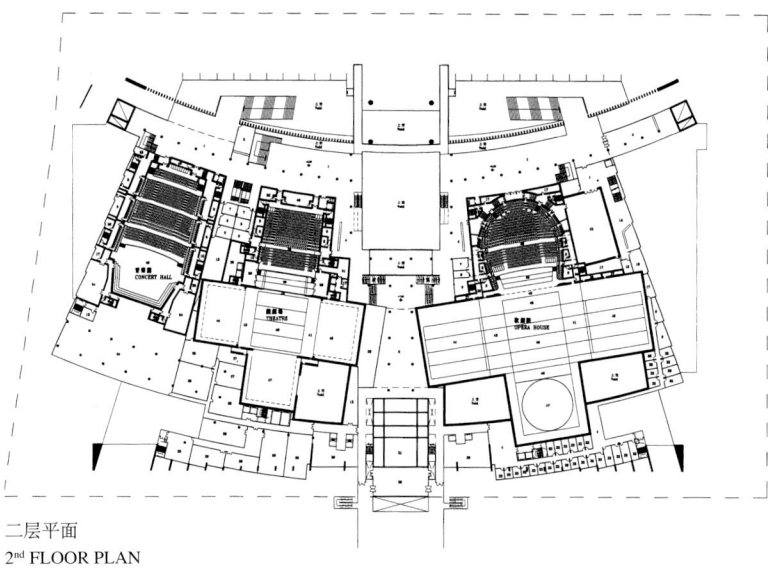

二层平面
2nd FLOOR PLAN

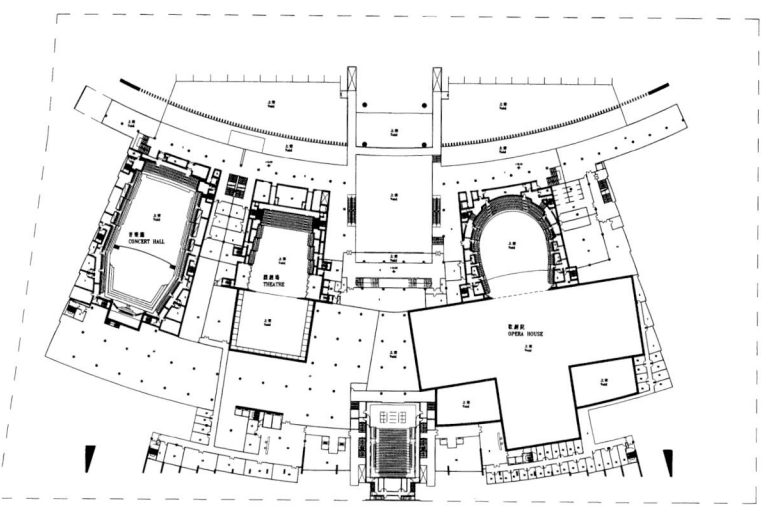

三层平面
3rd FLOOR PLAN

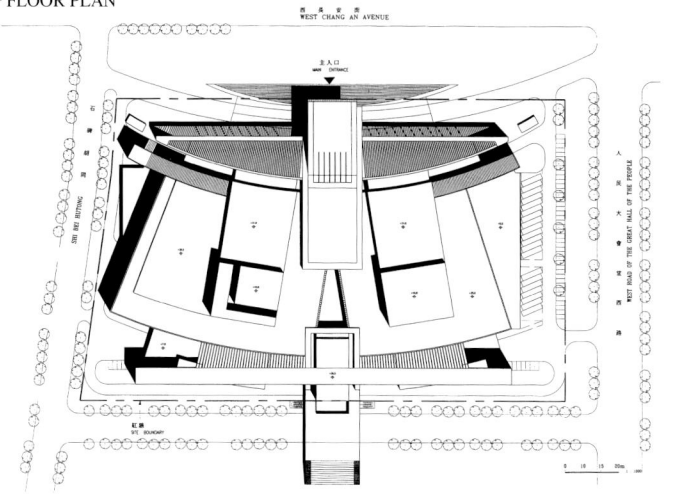

四层平面
4th FLOOR PLAN

总平面
OVERALL SITE PLAN

189

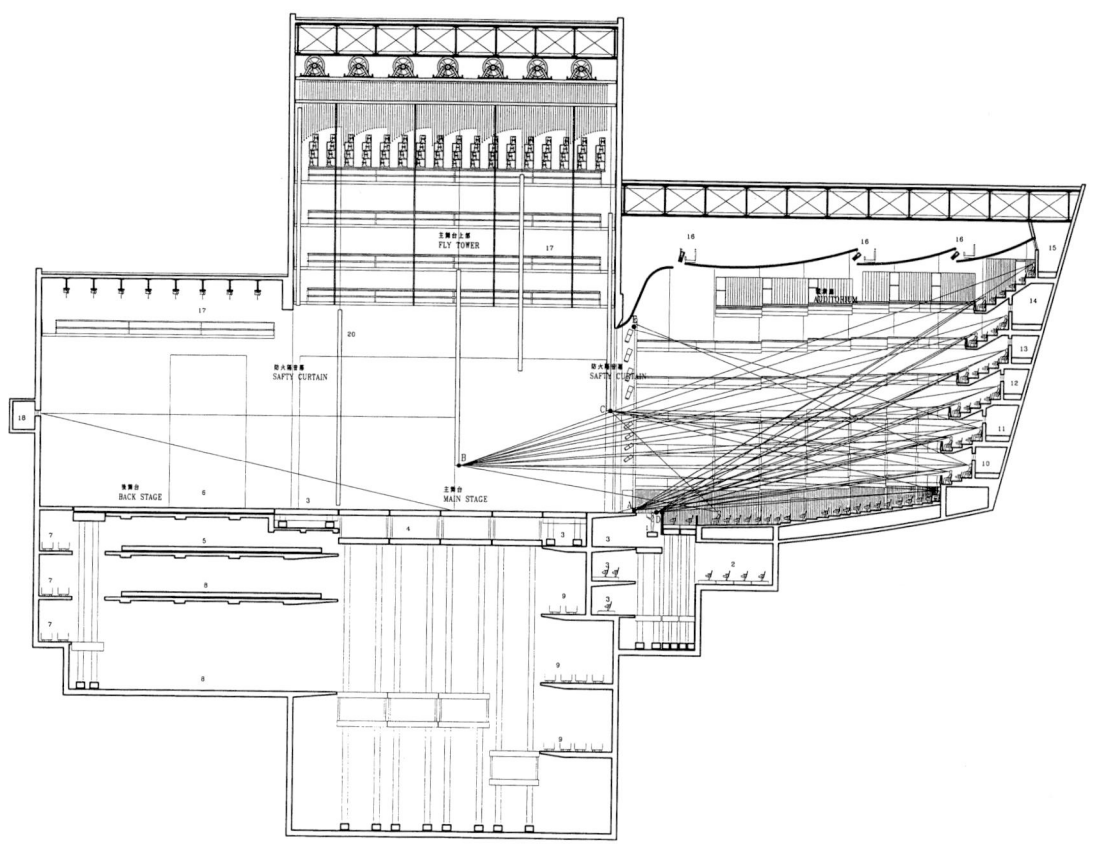

歌剧院视线
VISUAL ANALYSIS OF THE OPERA HOUSE

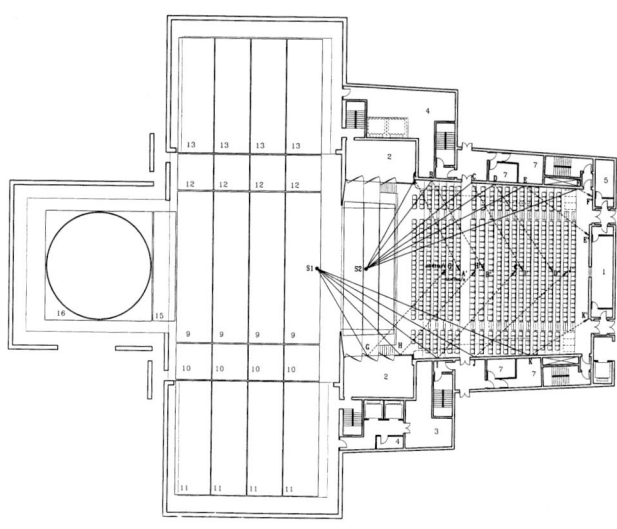

戏剧院视线分析平面
VISUAL ANALYSIS PLAN OF THE THEATER

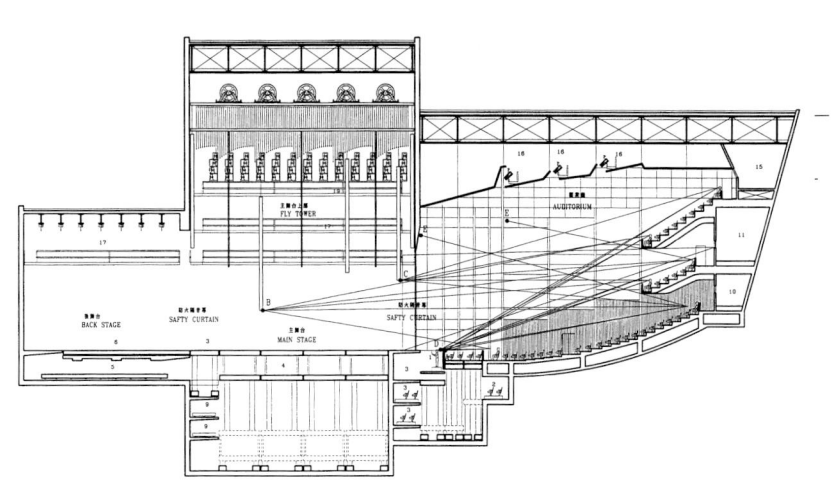

戏剧院视线剖面
VISUAL ANALYSIS SECTION OF THE THEATER

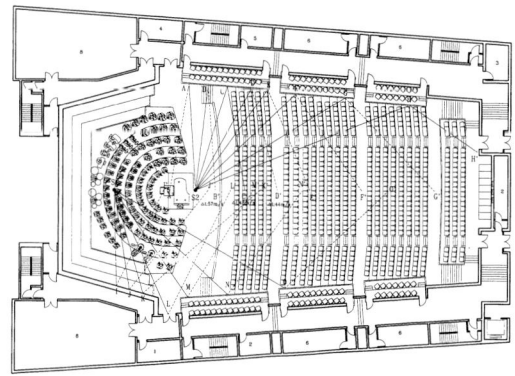

音乐厅平面
CONCERT HALL PLAN

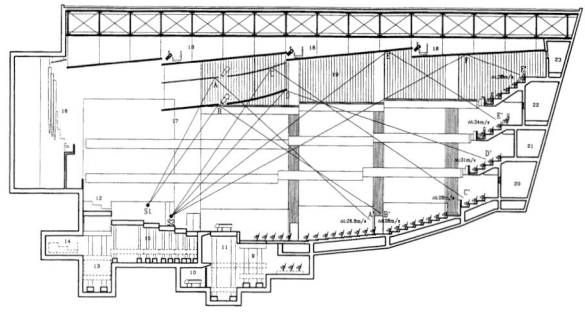

音乐厅声线剖面
VISUAL ANALYSIS SECTION OF THE CONCERT HALL

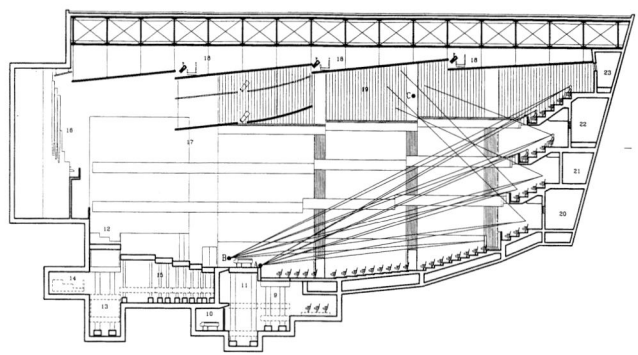

音乐厅视线剖面
ACOUSTICAL ANALYSIS SECTION OF THE CONCERT HALL

技术经济指标
Technical Economic Index

基地面积 (m²) Site Area	总建筑面积 (m²) Total Floor Area	建筑高度(m) Building Height	建筑覆盖率（%) Building Coverage Ratio
43080	地上：79900 地下：51055	26.5/45(局部)	65.8
容积率 Plot Ratio	绿化覆盖率(%) Green Coverage Ratio	停车数量 Number of Stalls	自行车停车数量 Number of Bicycle Stalls
3.04	11.2	地上：70 地下：504	1060

各观众厅数据
Data For Auditoria

		歌剧院 Opera House	音乐厅 Concert Hall	戏剧院 Theater	小戏场 Mini Theater
	观众厅面积(m²) Floor Area of Auditorium	1476	1323	860	450
	观众厅体积(m³) Volume of Auditorium	16460	21200	7400	可变 Varies 3000~3360
座席数 Seating Capacity	总座席数 Seating Capacity(Seats)	636	710	628	可变 Varies 最高 341 人
	其中：池座 Among Them:Auditorium	1873	1299	580	可变 Varies 最高 42 人
	楼 座 Floor Seat	2509	2009	1208	可变 Varies 最高 383 人
	休息厅面积(m²) Lounge Floor Area	2523	2216	2028	370
舞台尺寸 Stage Dimentions (m²)	主舞台 Main Stage	36 宽(W) 27 深(D)	24.75 宽(W) 18 深(D)	27 宽(W) 21 深(D)	可变 Varies
	左右侧台 Left and Right Side Stages	每个27 × 22	每个164	每个15.5 × 21	
	后舞台 Back Stage	26 × 25	7.5 × 12.0	19 × 19	
	台 仓 Under Stage Storage	1899	270	302	215
	升降乐池 Elevating Orchestra Pit	6 × 20	3 × 18	6.2 × 16	可变 Varies
	最大视距(m) Max. Visual Distance	40	37	34	可变 Varies
	最大俯角（°) Max. Angle of Depression	32	28	29	34

汉斯豪莱建筑师+
海兹诺曼建筑师
（奥地利）
Prof. Hans Hollein/
Architect-Heinz
Neumann/Architect
(Austria)

由此带来的是一种新观念，既将所有单独的元素集合起来置于一个屋檐下，供人们自由支配，充分利用，同时使得必要的组织性分割成为可能。

对各种不同元素——剧院、音乐厅等等的布局安排使得从中央的纪念性入口可以直接到达和看见任何一个部分。

主要的入口在长安街。为与人民大会堂的正面成一直线，整个建筑的拐角一块内缩，而这一块内凹区则形成了扁平的入口区。

主要的入口处是通过圆形入口大厅，它是一个倾斜的环行平台，带有通向歌剧院、戏剧场和音乐厅的各自独立的大门。

歌剧院的主入口处位于地面以上3米，由倾斜的圆形广场处进入。

贵宾可驾车驶上环形广场直接通到大门前。

主厅是中国古典园林的精髓的再现，有植物，有流水，有桥、休息处(亭阁)以及其它各种迷人的装饰。

主厅通过楼梯、自动楼梯和升降梯与下面的大厅联系顺畅，而这些大厅又作为通道与地铁和车库连接。

整个建筑群没有背面，无论朝南还是朝北它都有庄重华丽的外表。

This led to a concept of assembling together all the singular elements under one roof giving people free reign

北侧模型透视　VIEW FROM NORTH SIDE OF THE MODEL

to make use of the whole area but also make possible necessary organizational separations.

The arrangement of the different elements of performance-the theaters, auditor concert halls-was thus make in such a way that from a central monumental entrance area all parts can be directly reached and seen.

The main access is from Chang An Avenue. By setting back the corner-block of the building in line with the facade of the Great Hall of the People facade a recessed area is created forming a depressed entrance zone.

The main entry is through the entrance rotunda, an inclined circular platform with separate doors to the Opera House, the Theater and the Concert Hall.

3m above ground-accessible by the inclined rotunda there is the main (Opera) entrance to the building.

VIP will drive up the circular plaza directly in front of the main door.

The Lobby is a modern version of a Chinese garden, also with plants, water bridges, pavilions and other attractions.

There is a fluent connection by stairs, escalators and elevators to the lower part of the lobby which doubles also as passage through the site and access to subway and garages.

This building complex therefore has no real backside-it has prominent fronts towards North and South.

草图 SKETCH

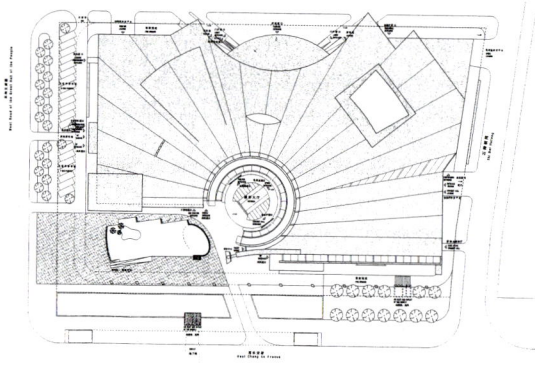

总平面 OVERALL SITE PLAN

模型透视 MODEL PERSPECTIVE

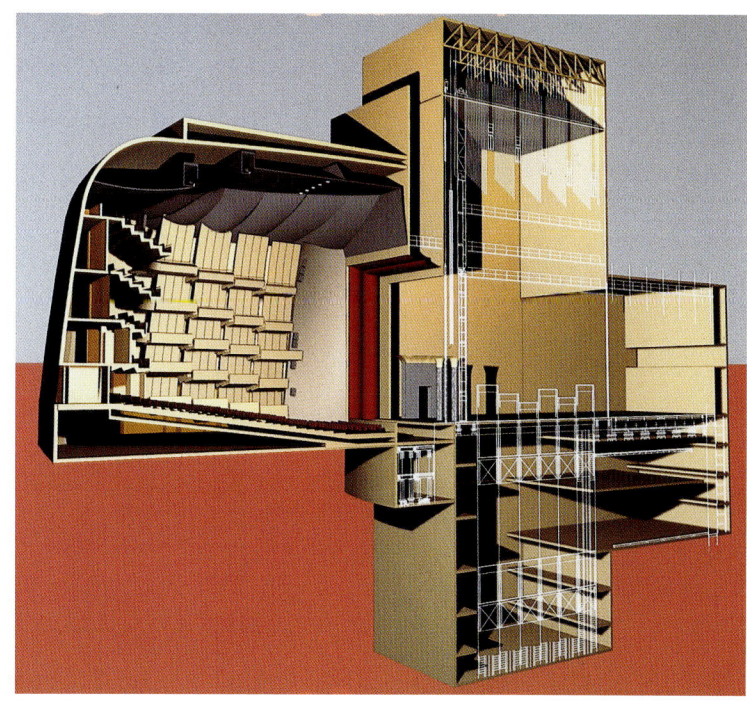

剖面透视 SECTIONS

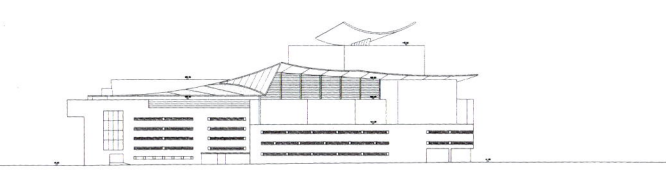

西立面
WEST ELEVATION

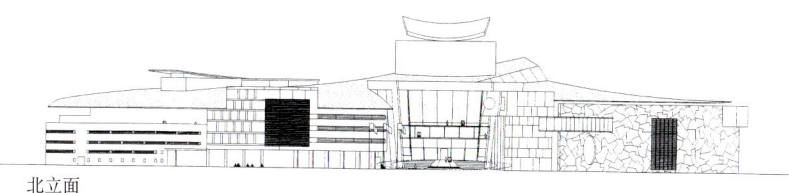

北立面
NORTH ELEVATION

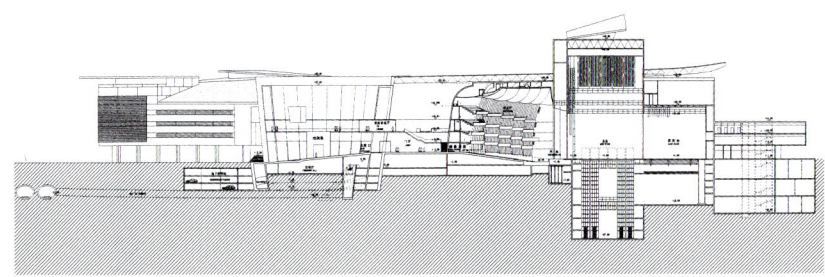

剖面1 SECTIONS(1)

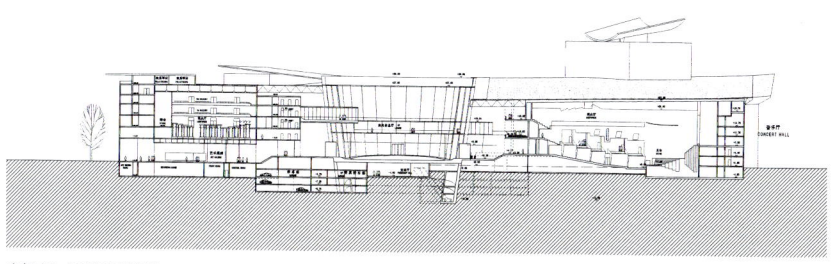

剖面2 SECTION(2)

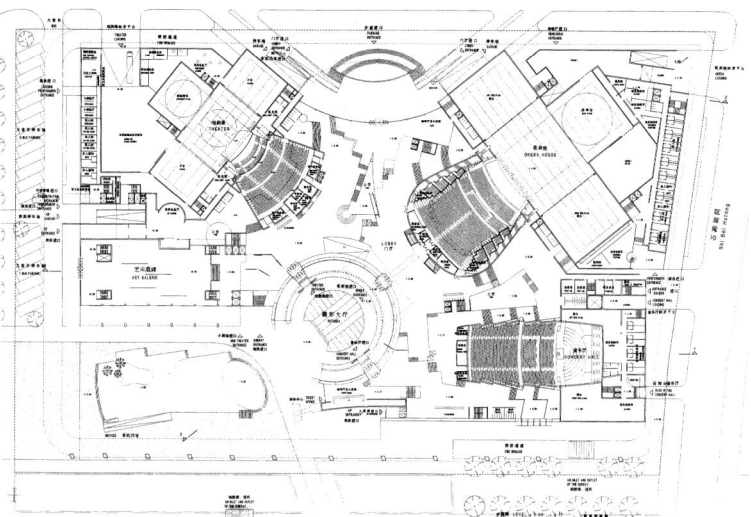

±0.00m平面
GROUND LEVEL(+0.00m)

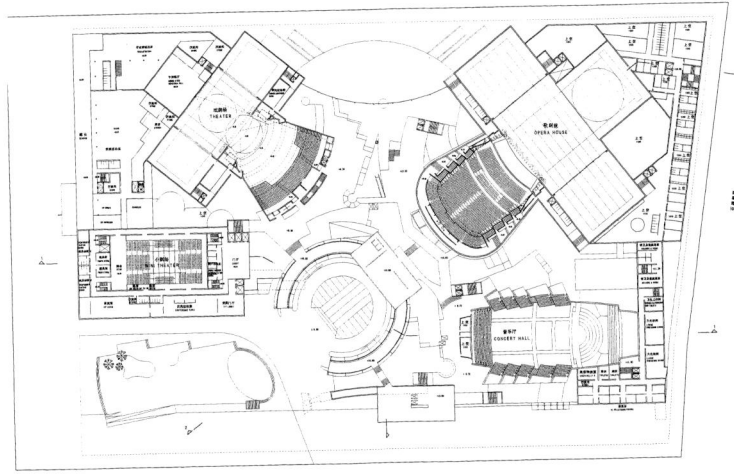

11.20m平面
GROUND LEVEL(+11.20m)

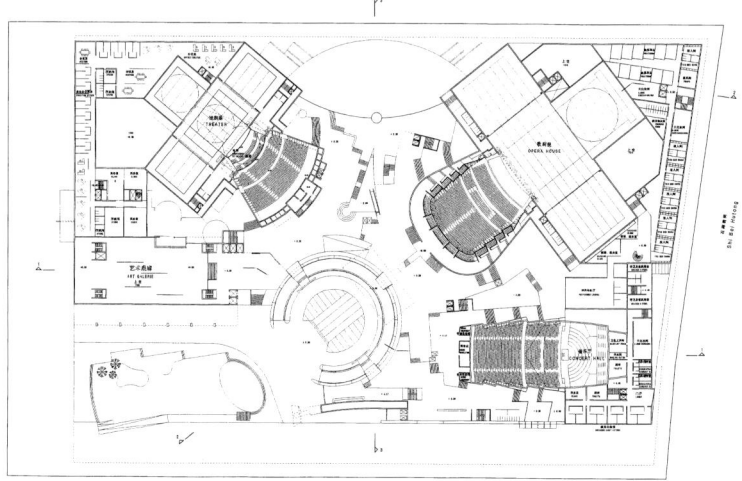

4.17m平面
GROUND LEVEL(+4.17m)

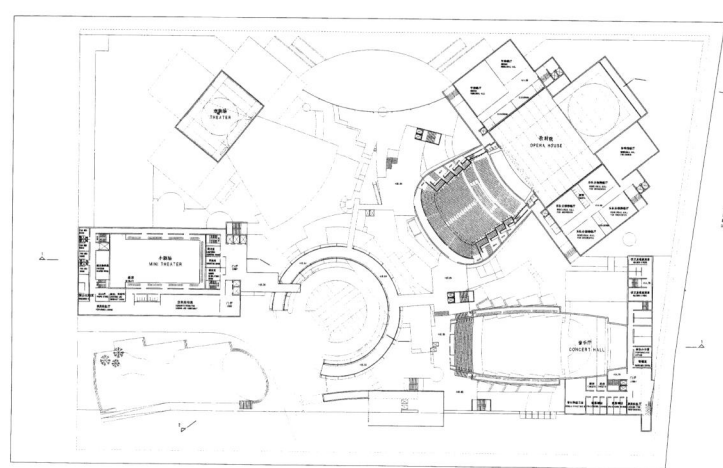

14.00m平面
GROUND LEVEL(+14.00m)

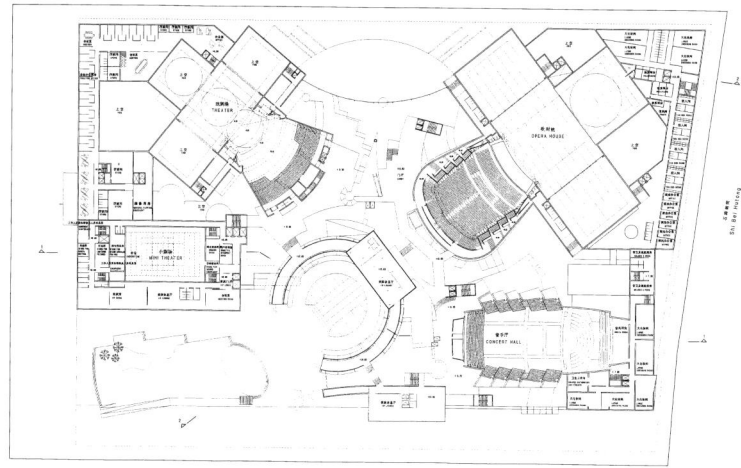

7.80m平面
GROUND LEVEL(+7.80m)

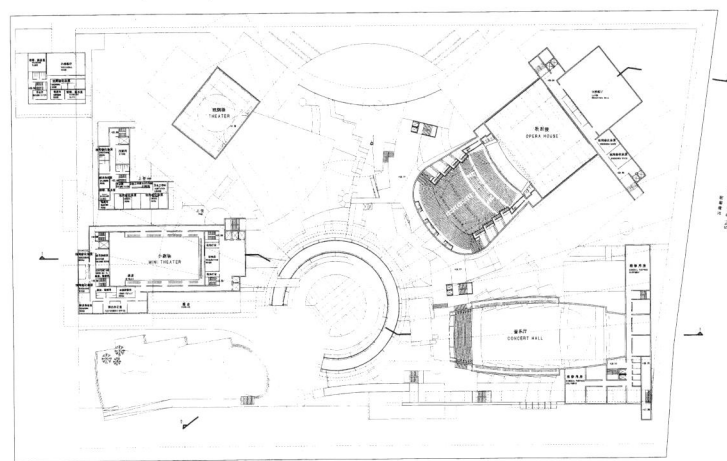

19.77m平面
GROUND LEVEL(+19.77m)

技术经济指标
Technical Economic Index

基地面积 (m²) Site Area	总建筑面积 (m²) Total Floor Area	建筑高度(m) Building Height	建筑覆盖率（%） Building Coverage Ratio
43082	地上：51064 地下：54565	30	58.9

容积率 Plot Ratio	绿化覆盖率 (%) Green Coverage Ratio	停车数量 Number of Stalls	自行车停车数量 Number of Bicycle Stalls
2.45	22.4	地上：20 地下：541	1130

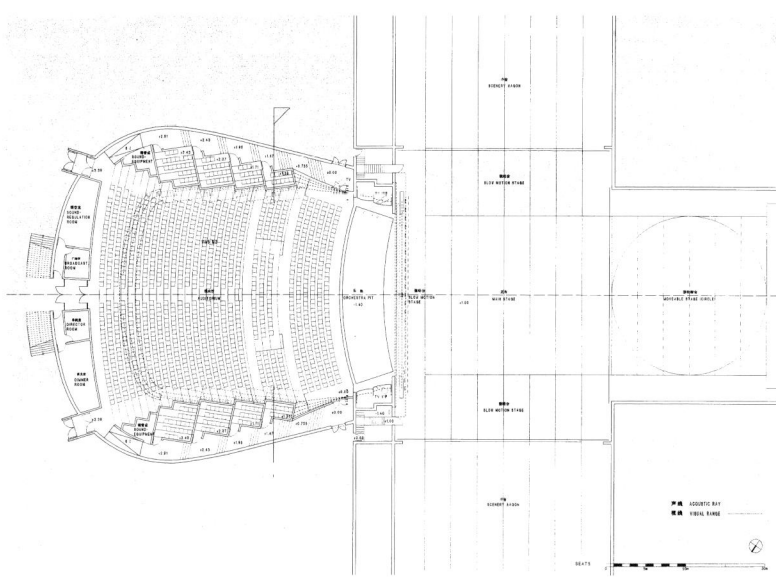

歌剧院声、视平面
OPERA HOUSE ACOUSTIC AND VISVAL ANALYSIS LEVEL

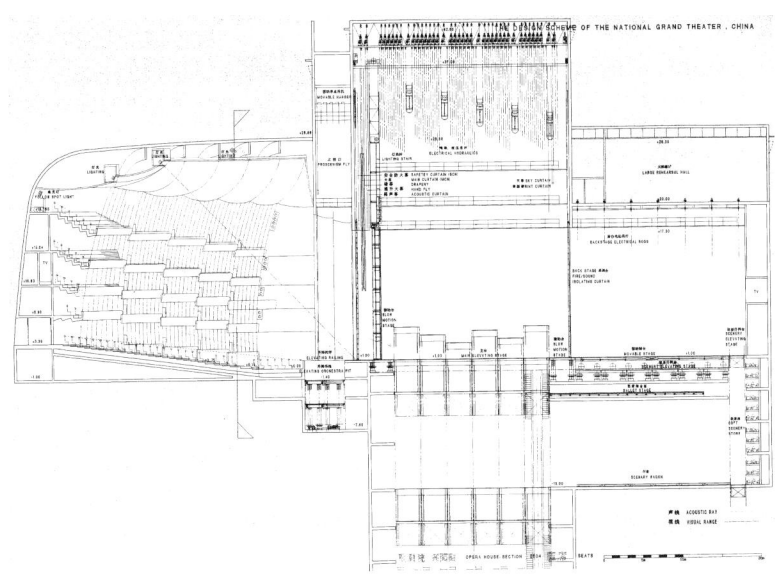

歌剧院声、视线剖面
OPERA HOUSE ACOUSTIC AND VISUAL ANALYSIS BECTION

各观众厅数据
Data For Auditoria

		歌剧院 Opera House	音乐厅 Concert Hall	戏剧院 Theater	小戏场 Mini Theater
	观众厅面积(m²) Floor Area of Auditorium	2242	1374	953	808
	观众厅体积(m³) Volume of Auditorium	20000	22000	7800	7140
座席数 Seating Capacity	总座席数 Seating Capacity(Seats)	2509	2003	1200	300~500
	其中：池座 Among Them:Auditorium				活动式
	楼 座 Floor Seat				活动式
	休息厅面积(m²) Lounge Floor Area	3567	2684	1660	605
舞台尺寸 Stage Dimentions (m²)	主舞台 Main Stage	972	250	526	
	左右侧台 Left and Right Side Stages	2 × 621		2 × 316	
	后舞台 Back Stage	650		330	
	台 仓 Under Stage Storage	1568		526	
	升降乐池 Elevating Orchestra Pit	147	107	204	
	最大视距(m) Max. Visual Distance	43	45	33	20
	最大俯角(°) Max. Angle of Depression	35	44	21	58

HPST 设计事务所(美国)
Hakomori Pali Stenfors Tang (America)

西侧新区的城市空间和建筑风格应与历史性的天安门广场和"紫禁城"有明显的区别。天安门广场凝聚着历史的光辉,而以国家大剧院为代表的新区,则着重于反映时代及展望未来。

长安街是北京城区贯串东西的主干道。它联系着众多的城市标志。相应地,作为与此干道相临的国家大剧院和其南侧的城市绿地自然就成为连接主干道与这个新区之间的门户,或者是一种象征性的连接过去与未来的桥梁。

Furthermore, it was decided that this new district shall be different and autonomous from the neighboring historically significant Tian'anmen Square and Forbidden City. Where Tian'anmen Square derives its significance from its history, the new National Grand Theater and its adjacent urban landscape area should reflect the present and become a beacon to the future.

Chang'an Avenue is the main east/west link for the city of Beijing. It is an important urban artery that connects many of the city's important landmarks. Correspondingly, the National Grand Theater and the urban landscape area becomes the threshold from Chang'an Avenue into the new district. In this we saw the opportunity to design the bridge and mediator between the future and the past.

北侧模型 NORTH SIDE OF THE MODEL

北侧透视 PERSPECTIVE OF NORTH ELEVATION

中央通道 CENTRAL STREET

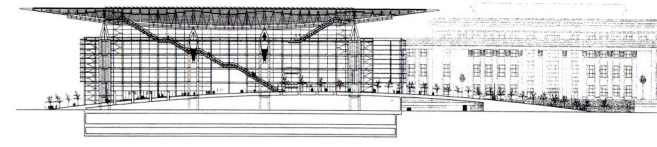

南-北剖面
NORTH-SOUTH SECTION

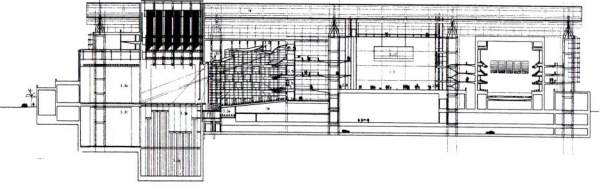

东-西剖面
EAST-WEST SECTION

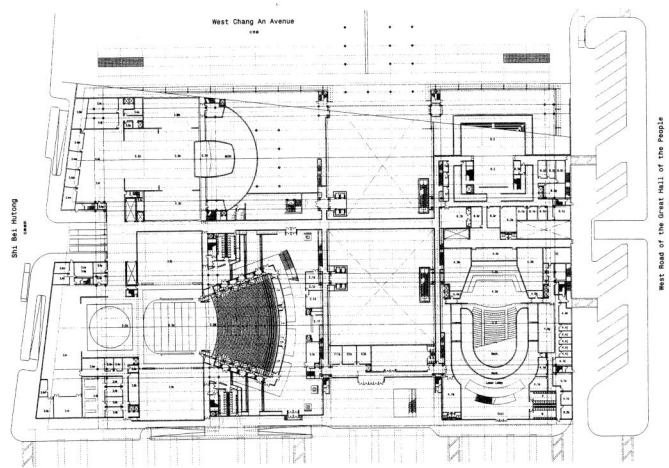

街道平面
STREET PLAN

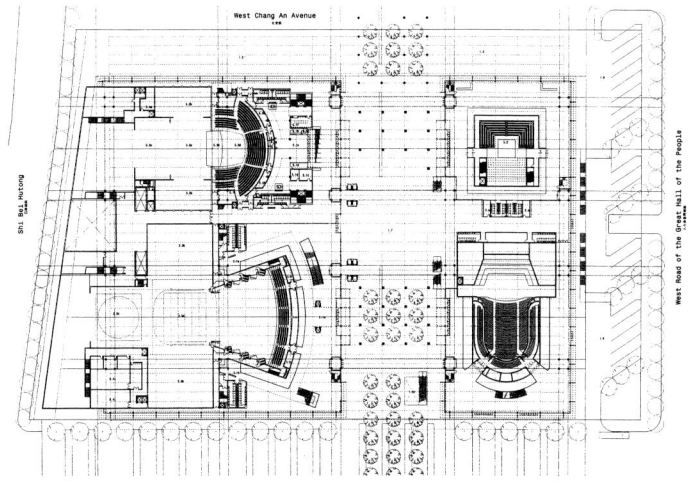

首层平面
GROUND FLOOR PLAN

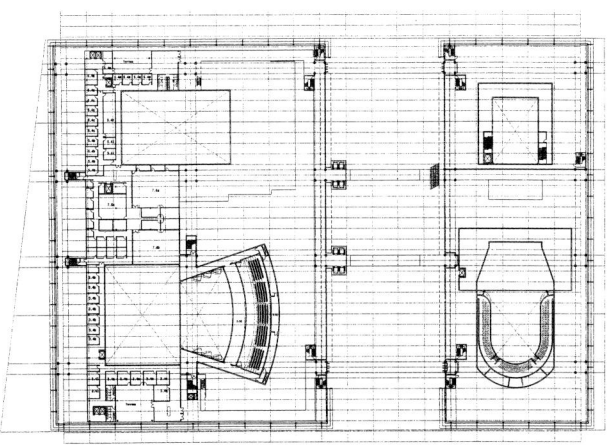

二层平面
2nd FLOOR PLAN

技术经济指标
Technical Economic Index

基地面积（m²） Site Area	总建筑面积（m²） Total Floor Area	建筑高度(m) Building Height	建筑覆盖率（%） Building Coverage Ratio
38916.06	地上：61000 地下：64000	30/45(局部)	82

容积率 Plot Ratio	绿化覆盖率 (%) Green Coverage Ratio	停车数量 Number of Stalls	自行车停车数量 Number of Bicycle Stalls
3.2	5	地上：20 地下：500	150

各观众厅数据
Data For Auditoria

		歌剧院 Opera House	音乐厅 Concert Hall	戏剧院 Theater	小戏场 Mini Theater
观众厅面积(m²) Floor Area of Auditorium		2300	1800	1200	940
观众厅体积(m³) Volume of Auditorium		20000	22000	7800	2500
座席数 Seating Capacity	总座席数 Seating Capacity(Seats)	2500	1823	1200~1345	350~500
	其中：池座 Among Them:Auditorium	1330	977	761~906	
	楼 座 Floor Seat	1170	846	439	
	休息厅面积(m²) Lounge Floor Area	7000	5700	4000	3500
舞台尺寸 Stage Dimentions (m²)	主舞台 Main Stage	1000	500	650~800	75~204
	左右侧台 Left and Right Side Stages	690 × 2	240 × 2	350 × 2	
	后舞台 Back Stage	725		370	
	台 仓 Under Stage Storage	1080		585	270
	升降乐池 Elevating Orchestra Pit	175			
最大视距(m) Max. Visual Distance		32	31	32	15
最大俯角（°） Max. Angle of Depression		35	40	35	35

设计由三个不同但相关的要素组成：第一是剧院的主体，每座剧场都有其独特的流线形的玻璃休息厅和华丽的外墙，所有的演出空间设计都力求达到世界最高水平的声学效果；第二是一个宏伟的中央共享空间，现命名为人民文化艺廊，这一高科技的玻璃大厅是剧院的观众、游客和北京市民进行集会和文化活动的场所；第三是环绕整体的柱廊，作为天安门广场区域文脉的延续以及中国传统建筑的现代诠释。

Our project is composed of three different, yet interrelated elements. First, are the theaters themselves; each has it own curved glass lobby and richly ornamented exterior walls and each performance space is shaped according to the highest standards of acoustic performance. Second, is a large central atrium; the People's Cultural Gallery. This high-tech, glass wrapped gathering place is for theater patrons, tourists, and the people of Beijing. Third, is an encircling colonnade; a solemn completion of the Tian'anmen Square precinct and a contemporary interpretation of traditional Chinese construction.

北侧透视
PERSPECTIVE OF NORTH ELEVATION

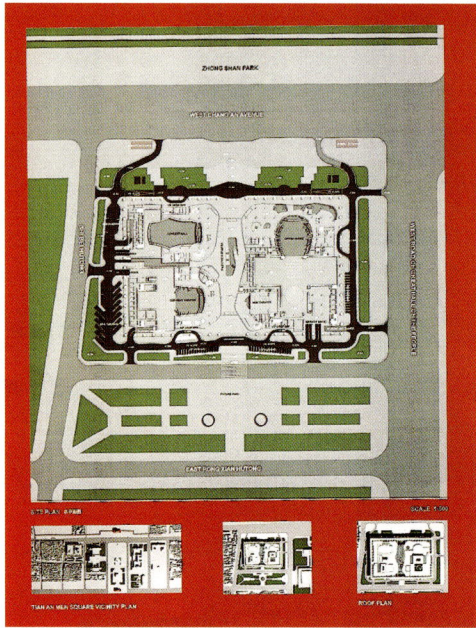

总平面
OVERALL SITE PLAN

北入口透视
PERSPECTIVE OF NORTH ENTRANCE

歌剧院观众厅 THE OPERA HOUSE AUDITORIUM

公共大厅(3) PUBLIC HALL(3)

公共大厅(2) PUBLIC HALL(2)

公共大厅(1) PUBLIC HALl(1)

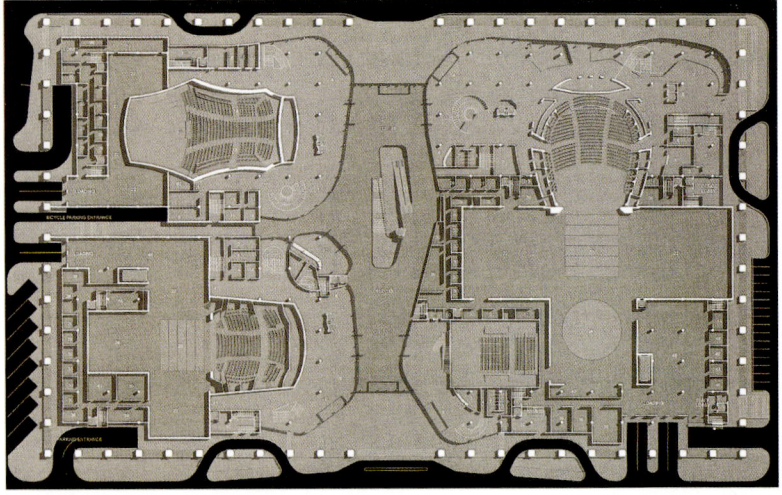

一层平面 GROUND FLOOR PLAN

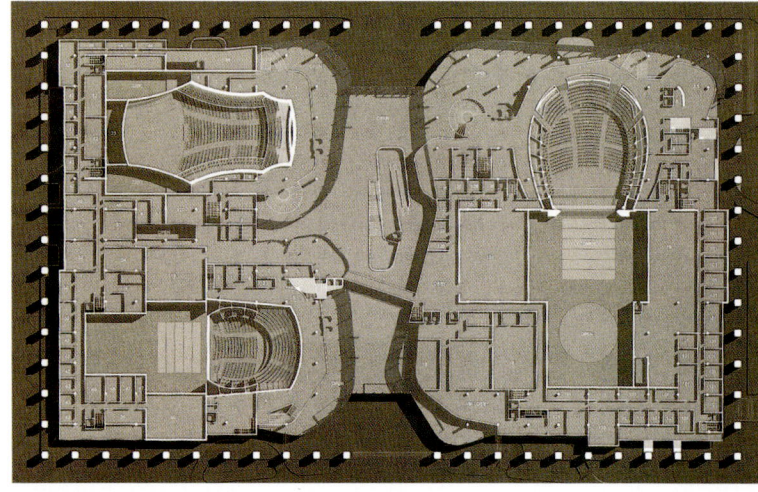

四层平面 4th FLOOR PLAN

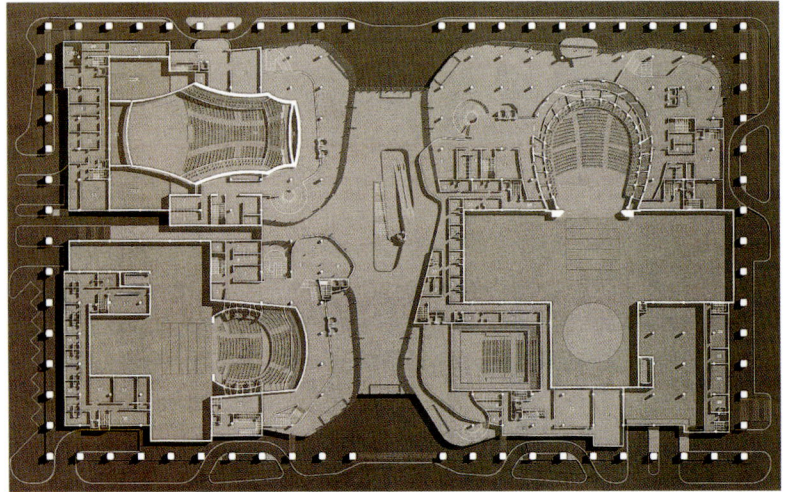

二层平面 2nd FLOOR PLAN

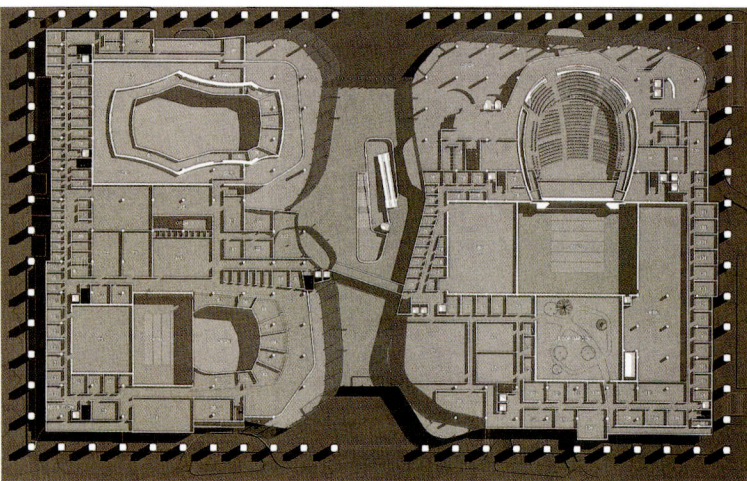

五层平面 5th FLOOR PLAN

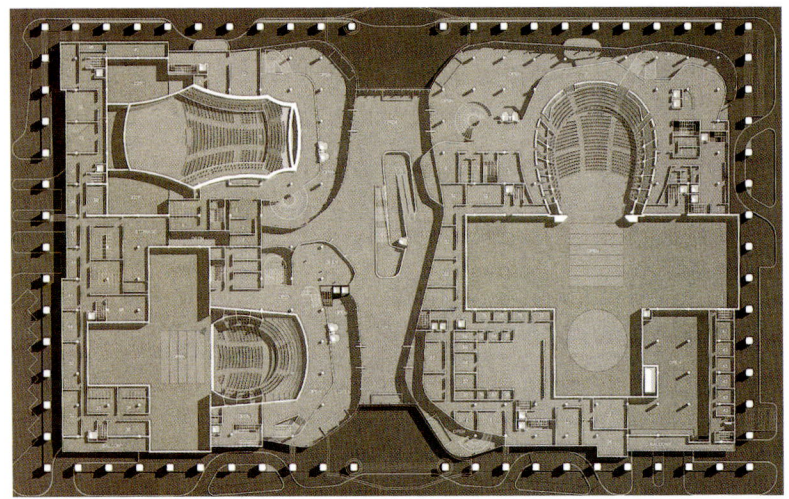

三层平面 3rd FLOOR PLAN

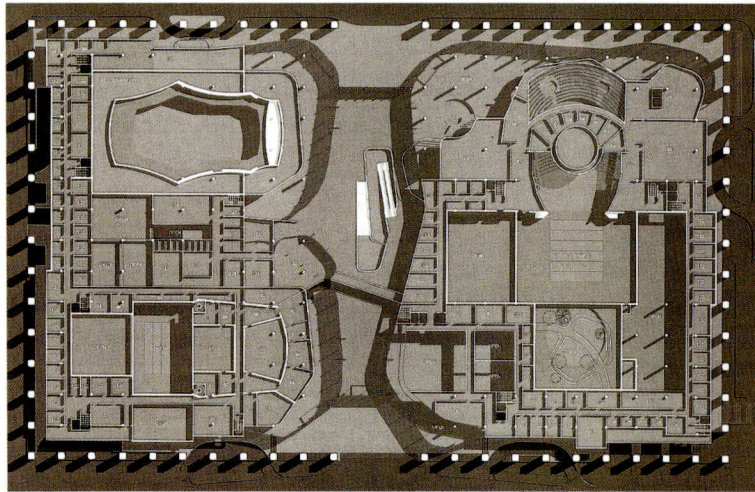

六层平面 6th FLOOR PLAN

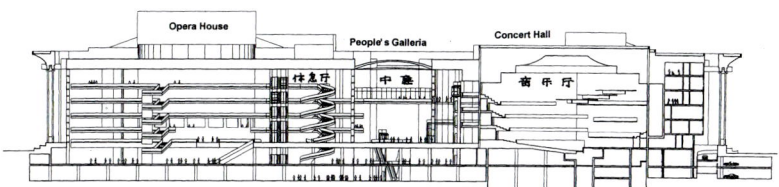

纵剖面 1 LONGITUDINAL SECTION 1

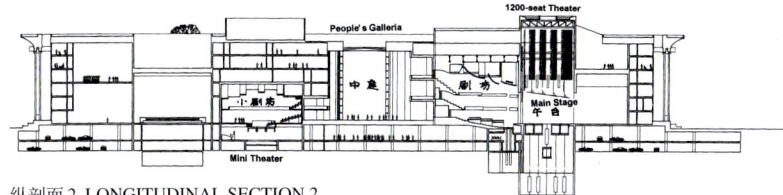

纵剖面 2 LONGITUDINAL SECTION 2

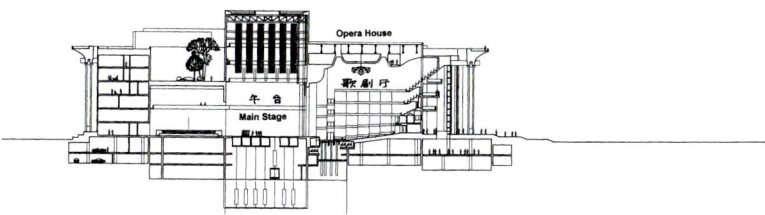

横剖面 CROSS SECTION

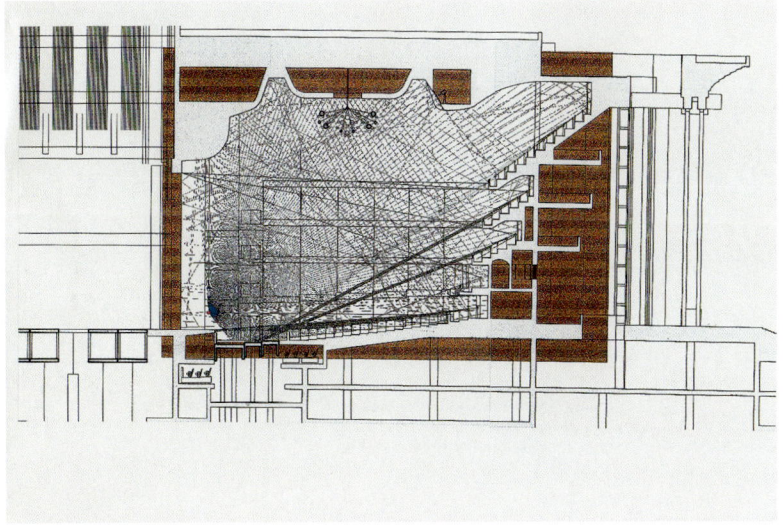

歌剧院声、视线分析
ACOUSTICAL & VISUAL ANALYSIS OF THE OPERA HOUSE

技术经济指标
Technical Economic Index

基地面积 (m²) Site Area	总建筑面积 (m²) Total Floor Area	建筑高度(m) Building Height	建筑覆盖率 (%) Building Coverage Ratio
43083		30/42(局部)	45.3

容积率 Plot Ratio	绿化覆盖率 (%) Green Coverage Ratio	停车数量 Number of Stalls	自行车停车数量 Number of Bicycle Stalls
304	15	地上：50 地下：543	1013

各观众厅数据
Data For Auditoria

			歌剧院 Opera House	音乐厅 Concert Hall	戏剧院 Theater	小戏场 Mini Theater
	观众厅面积(m²) Floor Area of Auditorium		968	1047	617	517
	观众厅体积(m³) Volume of Auditorium		19565	20228	10154	6708
座席数 Seating Capacity	总座席数 Seating Capacity(Seats)		2500	2015	1211	597
	其中：池座 Among Them:Auditorium		1730	1500	880	375
	楼 座 Floor Seat		770	515	333	222
	休息厅面积(m²) Lounge Floor Area		1740	540	420	375
舞台尺寸 (m²) Stage Dimentions	主舞台 Main Stage		972	440	540	240
	左右侧台 Left and Right Side Stages		1296	440	513	
	后舞台 Back Stage		650		324	
	台 仓 Under Stage Storage		972	360	600	320
	升降乐池 Elevating Orchestra Pit		172	150	180	
最大视距(m) Max. Visual Distance			38.3	35.2	30.1	25.0
最大俯角（°） Max. Angle of Depression			36.5	21.0	21.5	27.0

深圳大学建筑设计研究院(中国)

Shenzhen University, The College of Architecture & Civil Engineering (China)

将国家大剧院建筑屋盖和南北立面抽象地表现为"凤"与"凰",并将"凤"与"凰"围着有梭形玻璃顶的中央大厅,而大厅的主轴线朝向东偏北正对着天安门(见总体布局图),以此体现"凤鸣朝阳"的主题,并以此向天安门广场的中心建筑——天安门致意。歌剧院、音乐厅、戏剧场上的表演者既面向本观众厅的观众表演,也隐喻在面向天安门表演,面向全中国表演。

The National Grand Theater roof plan and the south and north facades of the design represent the abstract images of Feng and Huang. The theme of "the singing phoenixes greet the rising sun"is embodied in the orientation of the central atrium enclosed by the Feng Huang, and which is directed diagonally toward Tian'anmen Gate along a northeast axis (see the General Layout Diagram). This arrangement implies that the performances in the Opera House, the Concert Hall and the Theater are not only dedicated to their immediate audiences, but also metaphorically to Tian'anmen and to the country as a whole.

北侧透视 PERSPECTIVE OF NORTH ELEVATION

南侧透视 PERSPECTIVE OF SOUTH ELEVATION

公共大厅(1) PUBLIC HALL(1)

观众厅 AUDITORIUM

公共大厅(2) PUBLIC HALL(2)

局部 SECTIONS

局部 SECTIONS

局部 SECTIONS

局部 SECTIONS

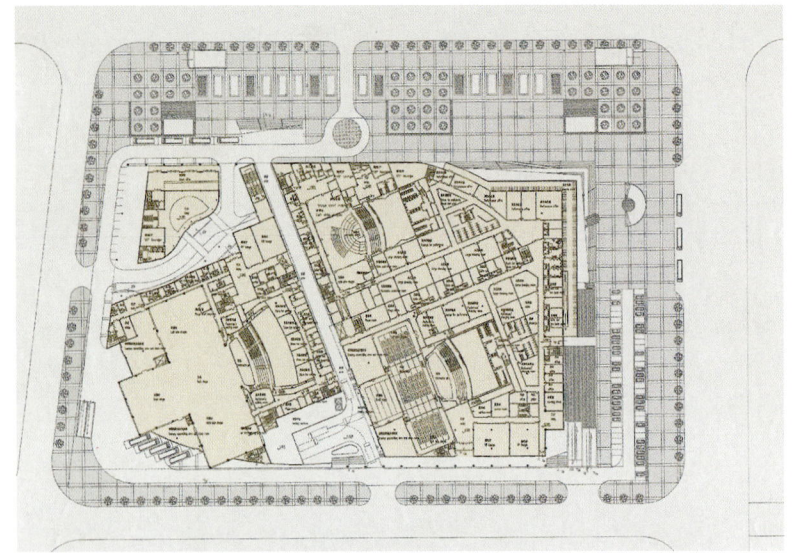

一层平面(±0.00m) GROUND FLOOR PLAN

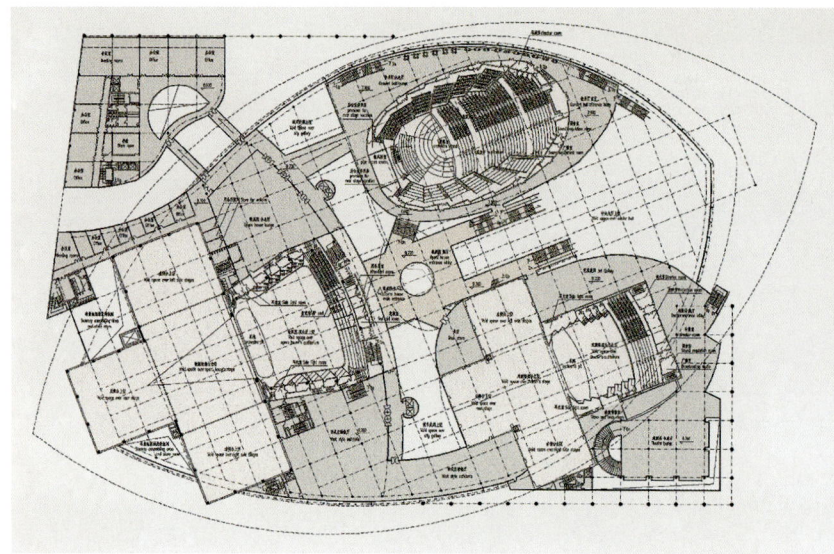

三层平面(9.70m) 3rd FLOOR PLAN

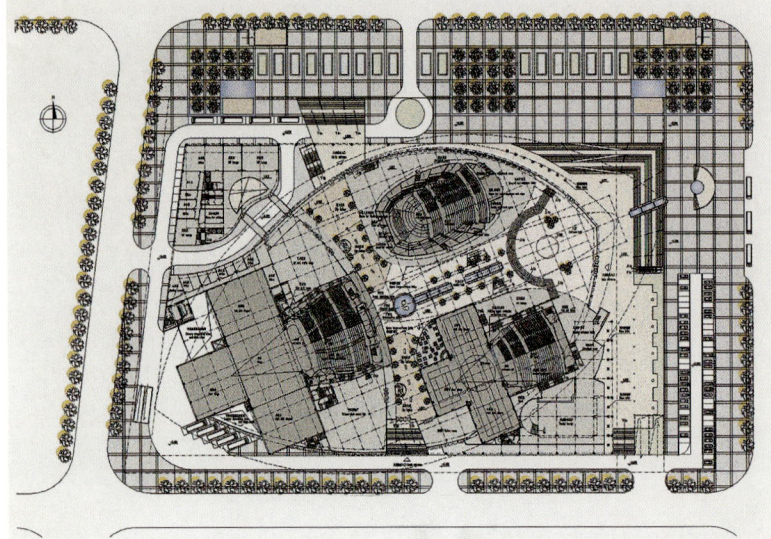

二层平面(4.90m) 2nd FLOOR PLAN

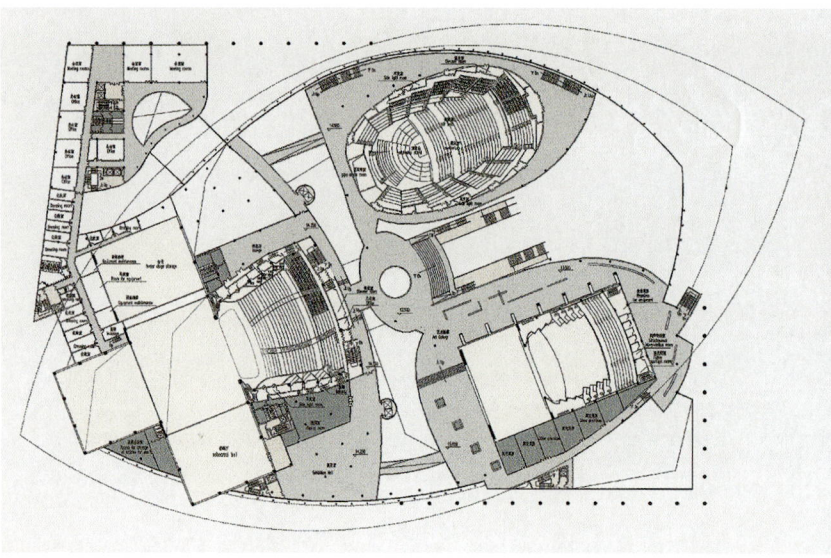

四层平面(13.50m) 4th FLOOR PLAN

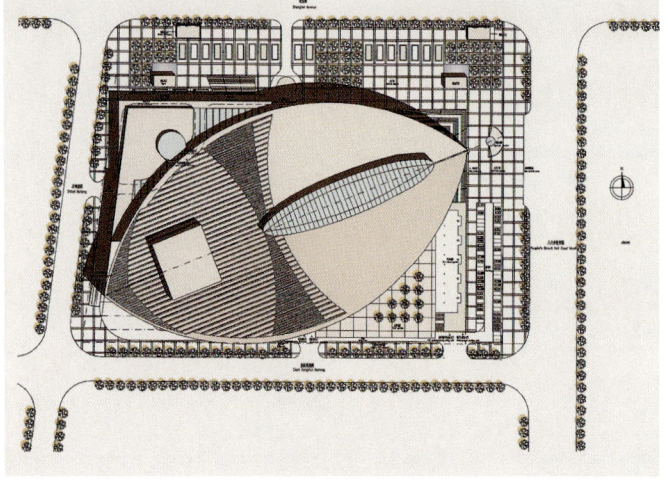

总平面 OVERALL SITE PLAN

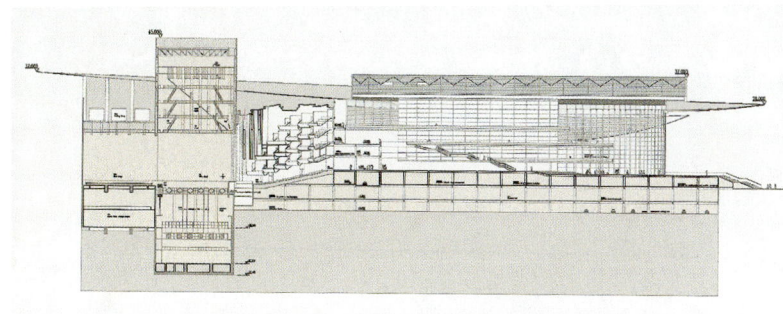

纵剖面 LONGITUDINAL SECTION

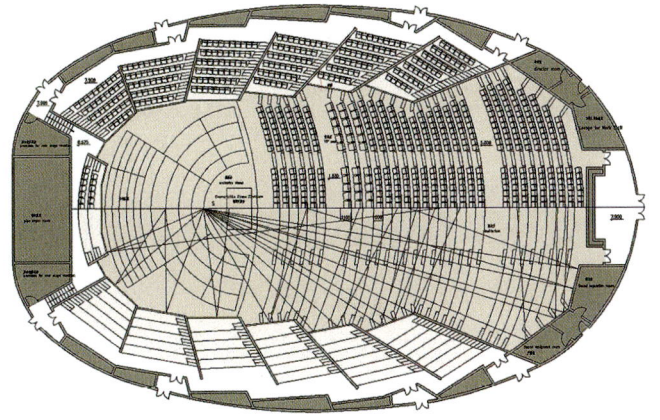

音乐厅平面 PLAN OF THE CONCERT HALL

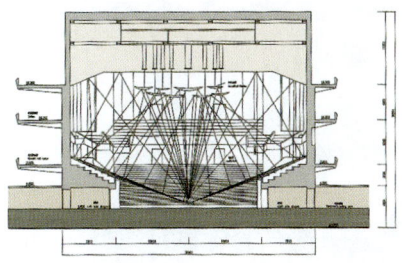

音乐厅剖面(1) SECTION(1) OF THE CONCERT HALL

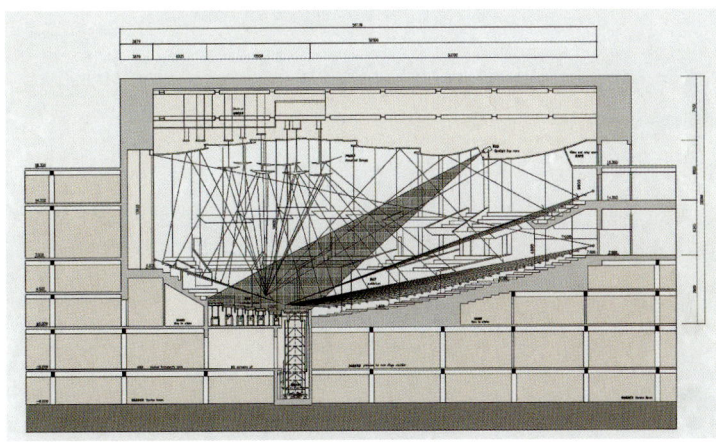

音乐厅剖面(2) SECTION(2) OF THE CONCERT HALL

技术经济指标
Technical Economic Index

基地面积（m²） Site Area	总建筑面积（m²） Total Floor Area	建筑高度(m) Building Height	建筑覆盖率（%） Building Coverage Ratio
36700	地上：72610.7 地下：56497.5	38/45(局部)	64.3

容积率 Plot Ratio	绿化覆盖率（%） Green Coverage Ratio	停车数量 Number of Stalls	自行车停车数量 Number of Bicycle Stalls
1.98	21.2	地上：81 地下：546	1032

各观众厅数据
Data For Auditoria

			歌剧院 Opera House	音乐厅 Concert Hall	戏剧院 Theater	小戏场 Mini Theater
	观众厅面积(m²) Floor Area of Auditorium		2080	1630	972	1202
	观众厅体积(m³) Volume of Auditorium		162800	20400	7260	9908
座席数 Seating Capacity		总座席数 Seating Capacity(Seats)	2524	1964	1204	394~462
		其中：池座 Among Them:Auditorium	1140	1462	852	78
		楼 座 Floor Seat	1384	502	352	316~388
		休息厅面积(m²) Lounge Floor Area	2980	2560	1390	924
舞台尺寸(m²) Stage Dimentions		主舞台 Main Stage	918	440	525.5	152
		左右侧台 Left and Right Side Stages	1188	396	331.5	
		后舞台 Back Stage	650		525.5	83
		台 仓 Under Stage Storage	1568		197	704
		升降乐池 Elevating Orchestra Pit	165			
	最大视距(m) Max. Visual Distance		40	38.5	29	27.2
	最大俯角（°） Max. Angle of Depression		35	楼座33.5	19	26

504

清华大学安地
建筑设计顾问
有限责任公司

Ande Architectural
Consultant Ltd. Co. of
Tsinghua University
(China)

● 时代性
● 民族性
● 开放性
● 纪念性
● 与环境之协调性
● 剧场建筑之个性

● Characteristics of the Times

● Characteristics of the Nation

● Characteristics of Openness

● Characteristic of Commemoration

● Harmony with the Surroundings

● NGT's Architectural Uniqueness

北侧透视　PERSPECTIVE OF NORTH ELEVATION

南侧透视　PERSPECTIVE OF SOUTH ELEVATION

南夜景透视　NIGHT VIEW OF SOUTH SIDE

观众厅透视 AUDITORIUM PERSPECTIVE

中央大厅透视 CENTRAL HALL PERSPECTIVE

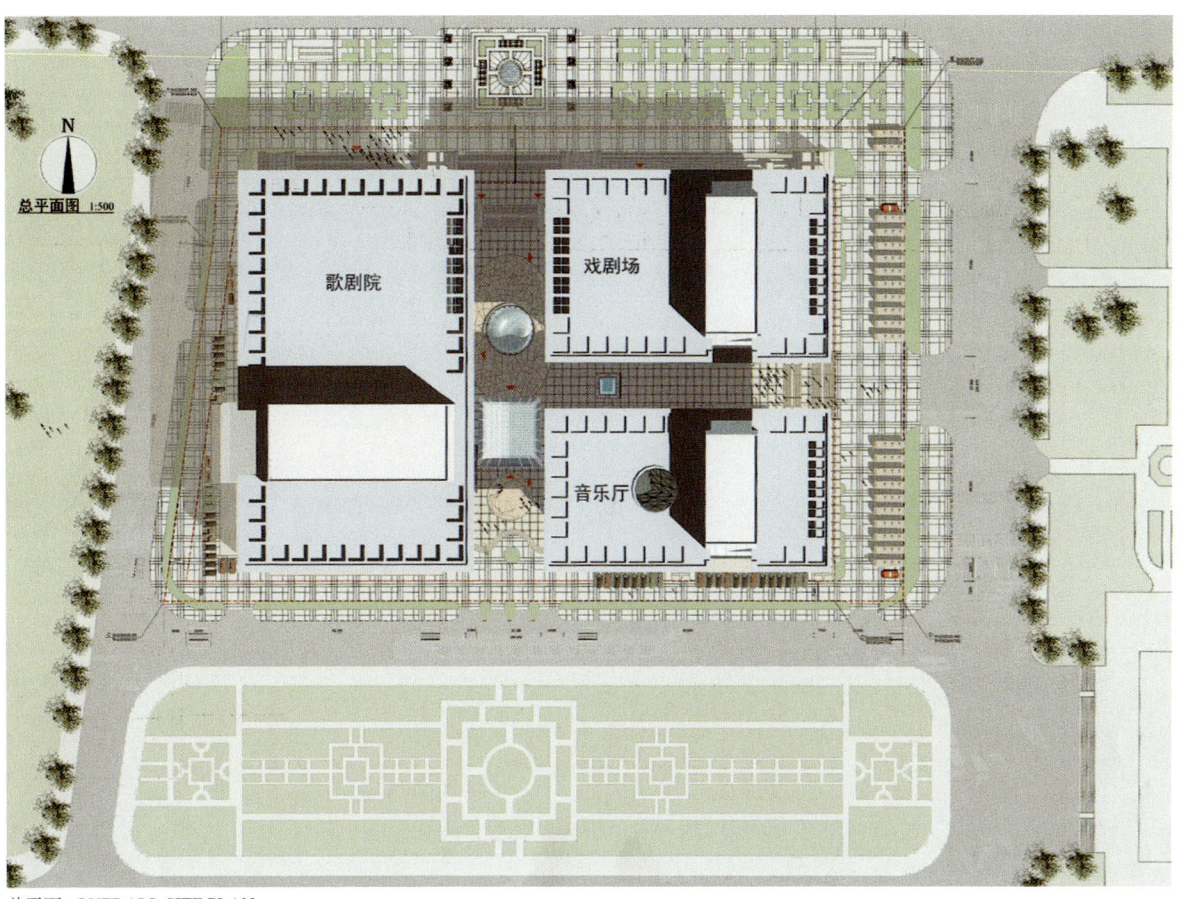

总平面 OVERALL SITE PLAN

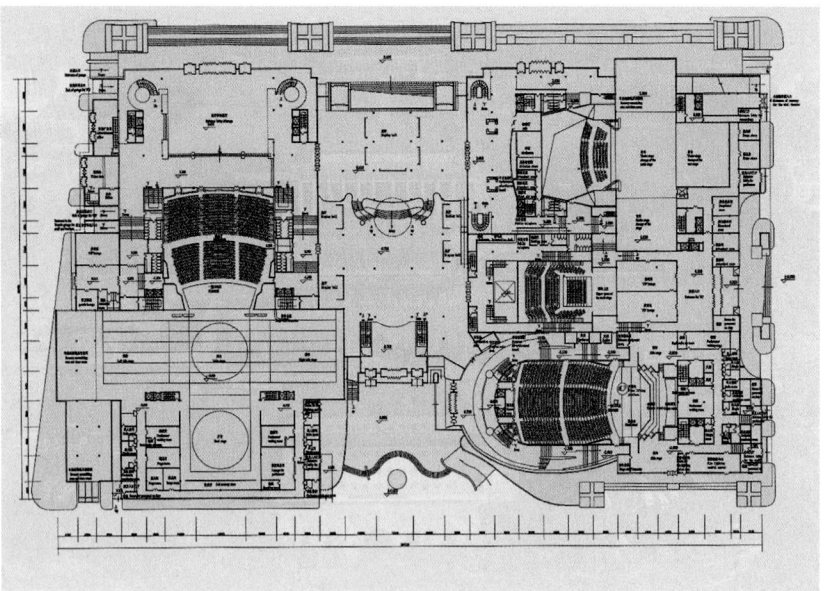

首层平面 GROUND FLOOR PLAN

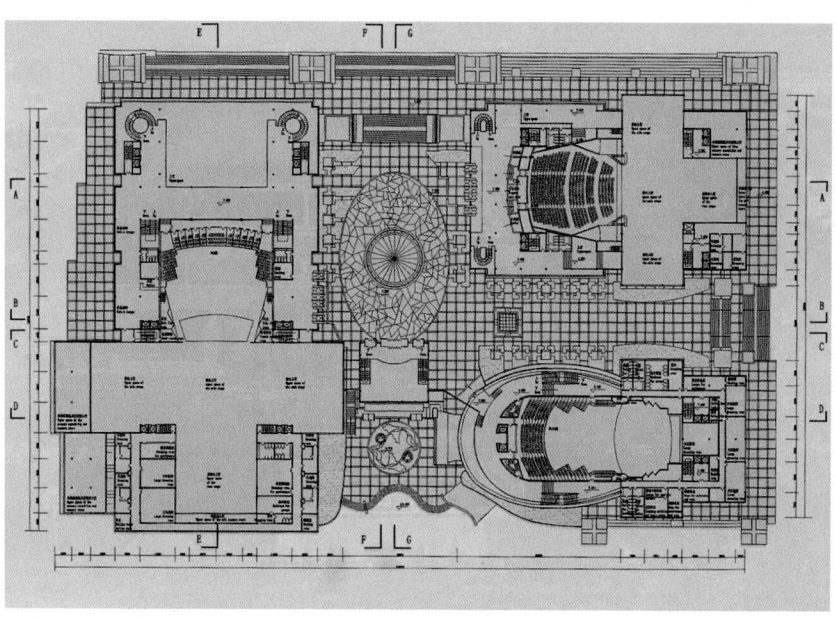

7.00m层平面 LEVEL(+7.00m)

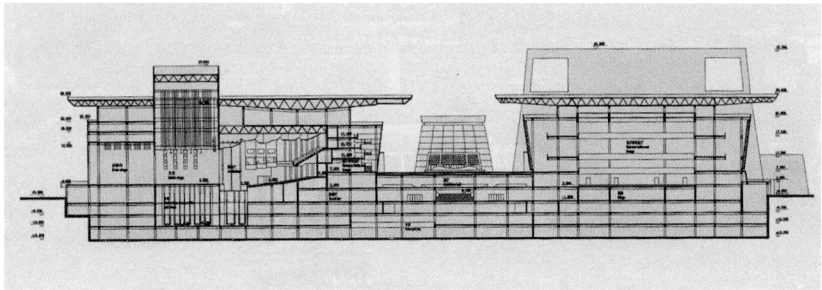

A-A剖面 SECTION A-A

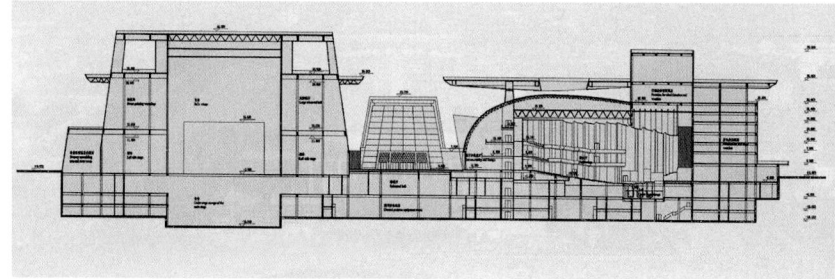

D-D剖面 SECTION D-D

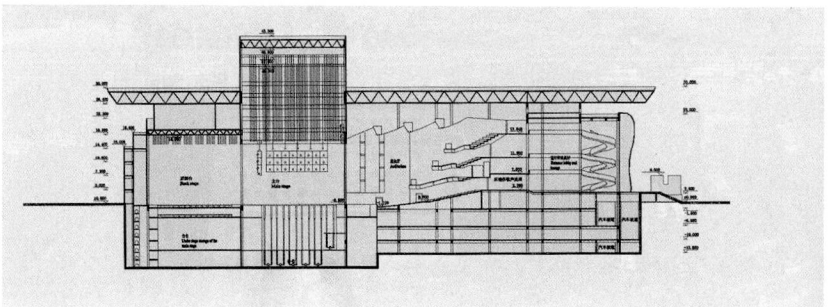

E-E剖面 SECTION E-E

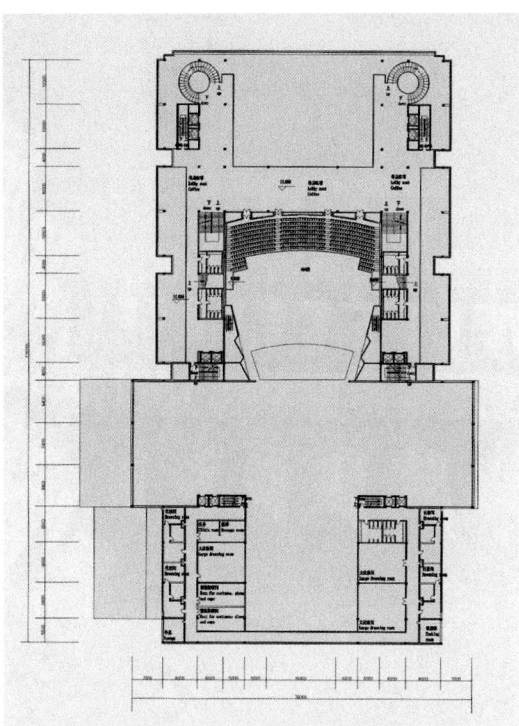

歌剧院10.80m平面　THE OPERA HOUSE LEVEL(+10.80m)

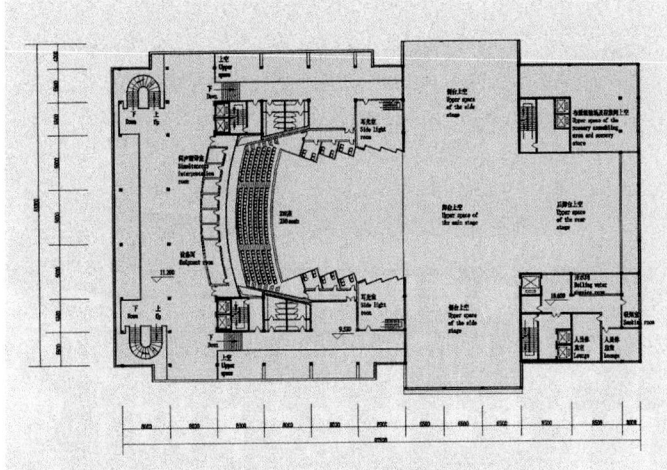

戏剧院10.60m平面　THE THEATER LEVEL(+10.60m)

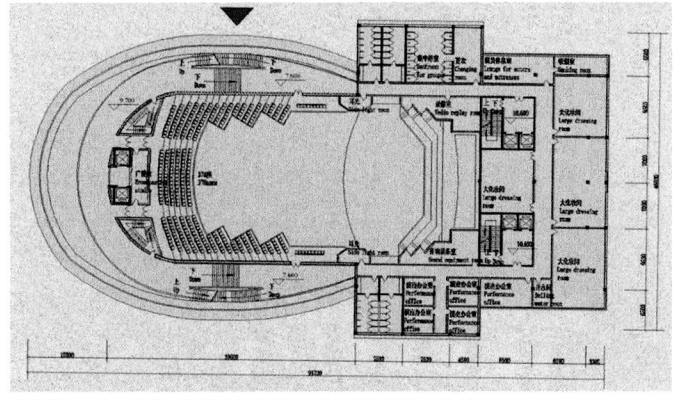

音乐厅10.60m平面　THE CONCERT HALL LEVEL(+10.60m)

技术经济指标
Technical Economic Index

基地面积(m²) Site Area	总建筑面积(m²) Total Floor Area	建筑高度(m) Building Height	建筑覆盖率（%） Building Coverage Ratio
38900	地上：71831 地下：59300	30/43.3(局部)	66.18
容积率 Plot Ratio	绿化覆盖率(%) Green Coverage Ratio	停车数量 Number of Stalls	自行车停车数量 Number of Bicycle Stalls
3.37	16.48	地上：70 地下：501	1017

各观众厅数据
Data For Auditoria

		歌剧院 Opera House	音乐厅 Concert Hall	戏剧院 Theater	小戏场 Mini Theater
	观众厅面积(m²) Floor Area of Auditorium	2180	1520	1096	448
	观众厅体积(m³) Volume of Auditorium	22358	16216	10919	3553
座席数 Seating Capacity	总座席数 Seating Capacity(Seats)	2520	2003	1204	530
	其中：池座 Among Them:Auditorium	1338	1155	844	530
	楼座 Floor Seat	1182	848	360	——
	休息厅面积(m²) Lounge Floor Area	5371	2442	2568	552
舞台尺寸(m²) Stage Dimentions	主舞台 Main Stage	28 × 36	18 × 26	19.5 × 27	8 × 15
	左右侧台 Left and Right Side Stages	28 × 22	15 × 10	19.5 × 16	11 × 8
	后舞台 Back Stage	25 × 26		17 × 20	6.5 × 15
	台仓 Under Stage Storage	1968.24	572	866.5	217.5
	升降乐池 Elevating Orchestra Pit	22 × 8		11 × 4	
	最大视距(m) Max. Visual Distance	40.8	32	31	23
	最大俯角（°） Max. Angle of Depression	26	23	20	16

我们所设计的方案目标是营造一个使人们有激情，有回忆的空间，它是由必然性和偶然性所构成。

我们的目的是希望该设计是一种动力的源点，它既促使了市政建筑与文化建筑共存，同时也促使了历史古城和它自身新的发展共存。

我们的愿望是设计一个建筑实体，它既不会被动地使空间违背或超越自身，也不会强求两者一致，它是完全自由的，被结构和空间注上标记，拥有极高的历史意义，同时也传播和阐述了现实。

This plan aims at building an emotional and memorial place made of certainty and uncertainty, evocations and mysteries.

Our proposal wishes to become the starting point of a dynamic coexistence between government buildings and cultural buildings; but also between the historical city and its new development.

Our desire is to plan an architectonic object which does not oppose or prevail, does not suffer nor conform itself passively to a place which is strongly personalized and marked by buildings and spaces possessing high historical value but interprets reality and transforms it.

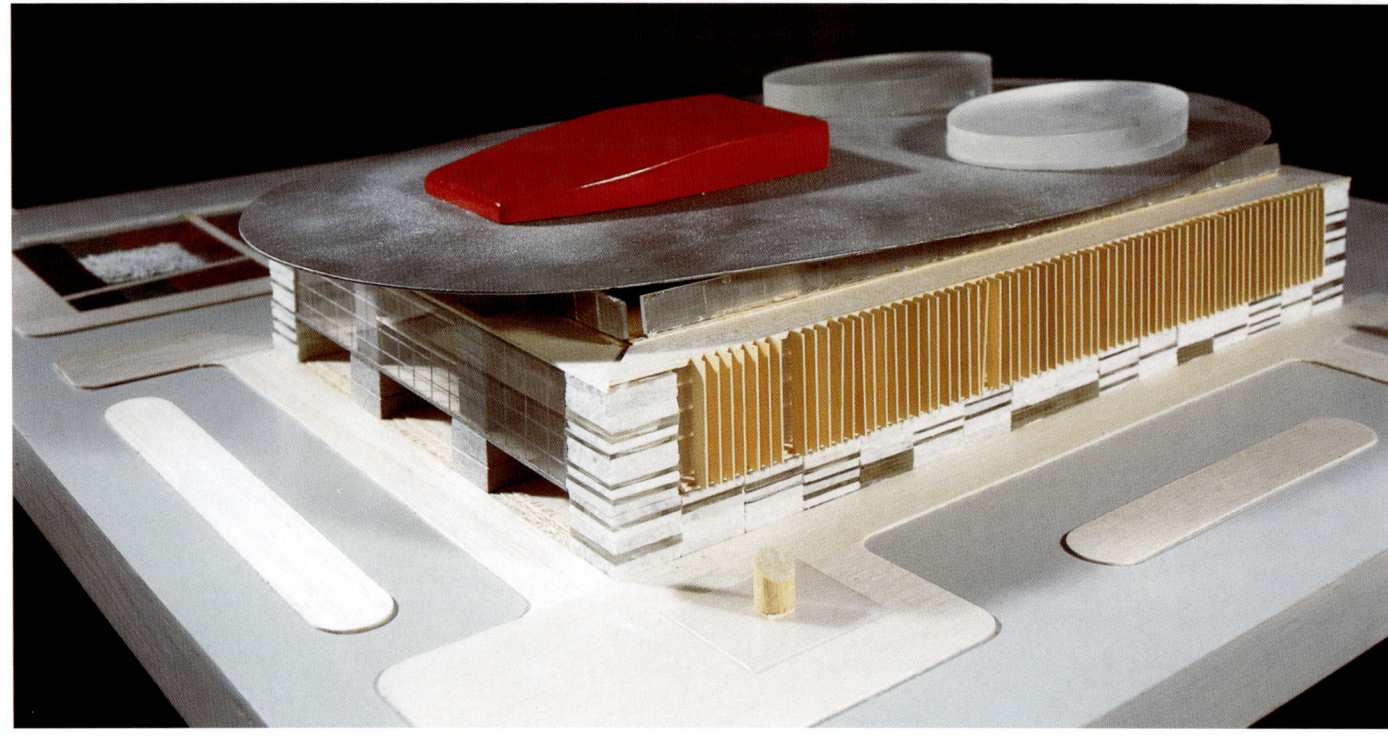

北侧模型透视 VIEW FROM NORTH SIDE OF THE MODEL

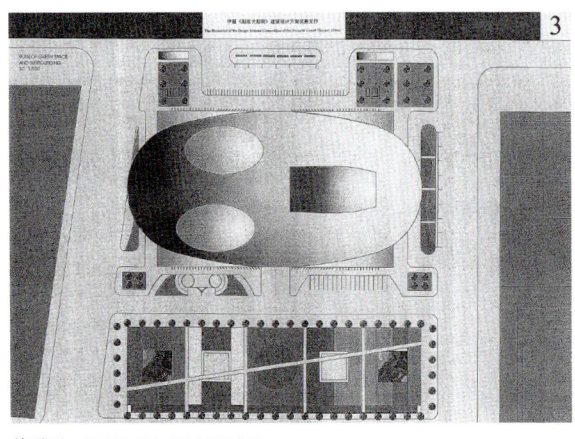

总平面 OVERALL SITE PLAN

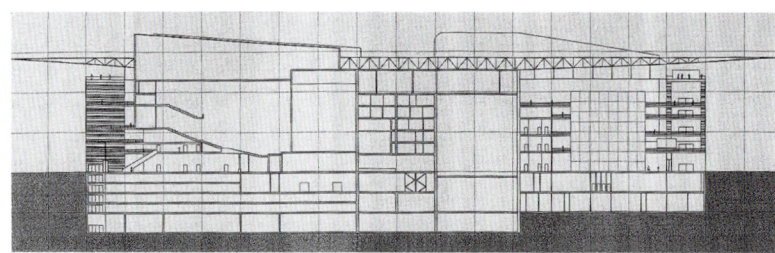

A−A 剖面 SECTION A-A

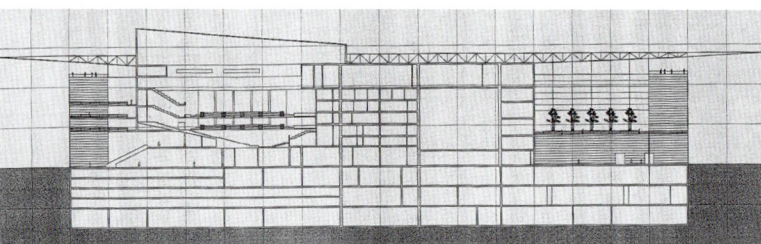

B−B 剖面 SECTION B-B

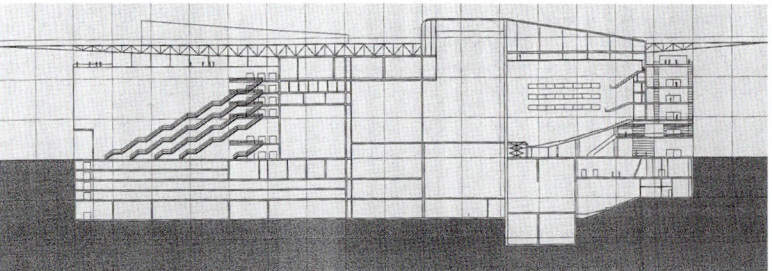

C−C 剖面 SECTION C-C

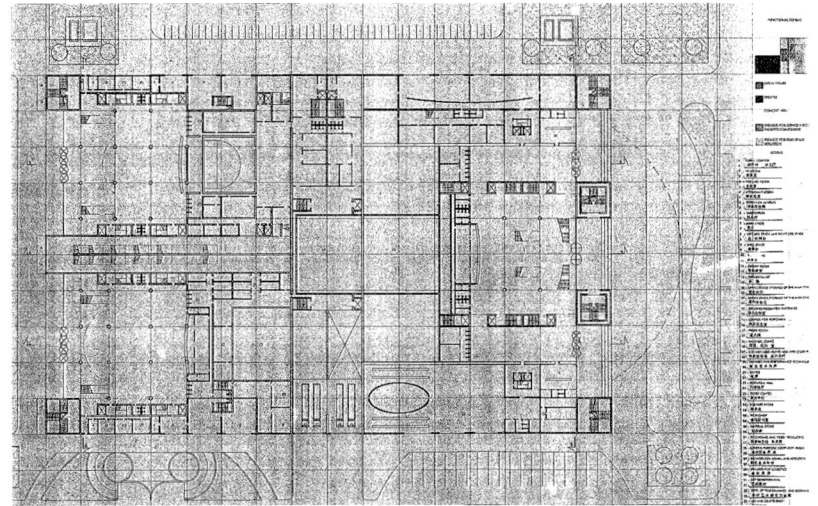

首层平面
GROUND FLOOR PLAN

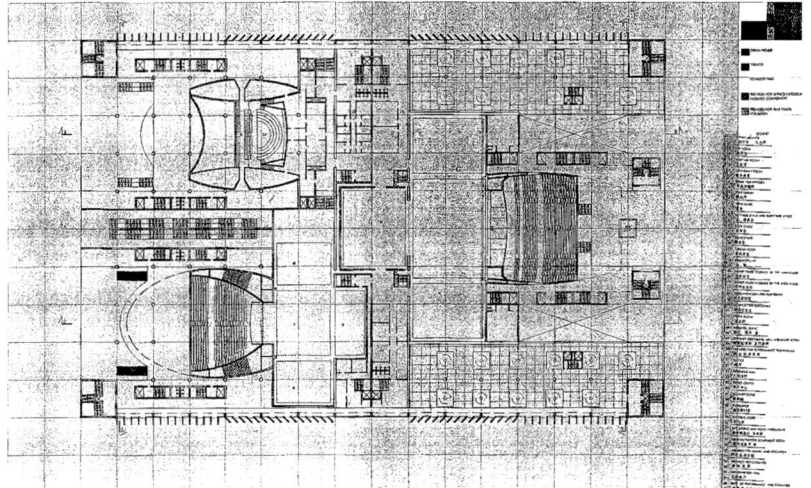

9.75m平面
LEVEL(+9.75m)

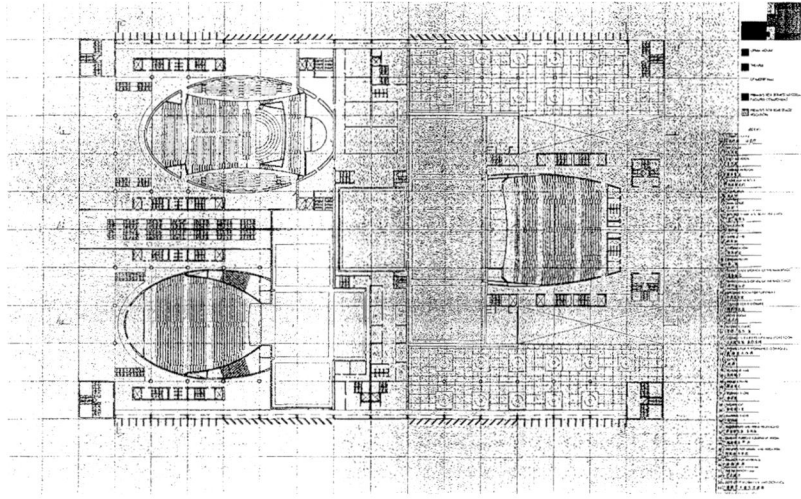

13.65m平面
LEVEL(+13.65m)

技术经济指标
Technical Economic Index

基地面积 (m²) Site Area	总建筑面积 (m²) Total Floor Area	建筑高度(m) Building Height	建筑覆盖率 (%) Building Coverage Ratio
39200	地上：58000 地下：68000	40	1.4
容积率 Plot Ratio	绿化覆盖率 (%) Green Coverage Ratio	停车数量 Number of Stalls	自行车停车数量 Number of Bicycle Stalls
3.2	16	地上：80 地下：490	980

各观众厅数据
Data For Auditoria

		歌剧院 Opera House	音乐厅 Concert Hall	戏剧院 Theater	小戏场 Mini Theater
	观众厅面积(m²) Floor Area of Auditorium	2430	1950	1598	680
	观众厅体积(m³) Volume of Auditorium	20650	27600	7100	2600
座席数 Seating Capacity	总座席数 Seating Capacity(Seats)	2500	1958	1250	480
	其中：池座 Among Them:Auditorium				
	楼　座 Floor Seat	2000	1566	1000	567
	休息厅面积(m²) Lounge Floor Area				
舞台尺寸 (m²) Stage Dimentions	主舞台 Main Stage	980	440	540	326
	左右侧台 Left and Right Side Stages	640	460	660	126
	后舞台 Back Stage	670		350	330
	台　仓 Under Stage Storage	980	210	540	326
	升降乐池 Elevating Orchestra Pit	170		90	
	最大视距(m) Max. Visual Distance	46	42	37	22
	最大俯角（°） Max. Angle of Depression	27.29	24.68	24.08	22.96

北京有色冶金研究
总院（中国）
Central Engineering &
Research Institute for
Non-Ferrous Metallurgical
Industries (China)

方案创意:

建筑方案的构思取材于中国的古琴名曲——"高山流水"意境，以示中国哲学与演艺文化博大精深，并以"天地人和"的民族精神回应"高山流水"，揭示出表演艺术感悟自然，歌颂真善美、鞭挞假恶丑的永恒主题。

The conception of the architectural scheme comes from the mood of a famous piece of ancient Chinese music called "High Mountains and Flowing Waters". It reveals the Chinese philosophy and culture of performing arts at their broadest and most profound. The nation's spirit of "Unity of Heaven, Earth and Humanity"blends with "High Mountains and Flowing Waters",telling an everlasting tale that the performing arts may reflect Nature and extol truth, goodness and beauty.

北侧透视
PERSPECTIVE OF NORTH ELEVATION

南侧模型透视
VIEW FROM SOUTH SIDE OF THE MODEL

中央大厅透视
CENTRAL HALL PERSPECTIVE

戏剧院观众厅透视
PERSPECTIVE OF THE THEATER AUDITORIUM

歌剧院观众厅透视
PERSPECTIVE OF THE OPERA HOUSE AUDITORIUM

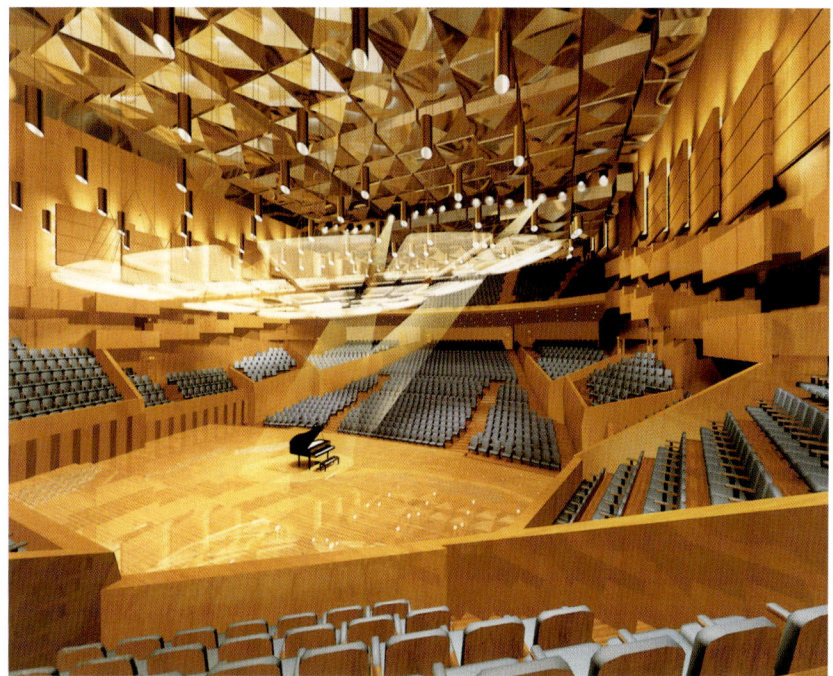

音乐厅观众厅透视
PERSPECTIVE OF THE CONCERT HALL AUDITORIUM

方案创意
CONCEPT CREATION

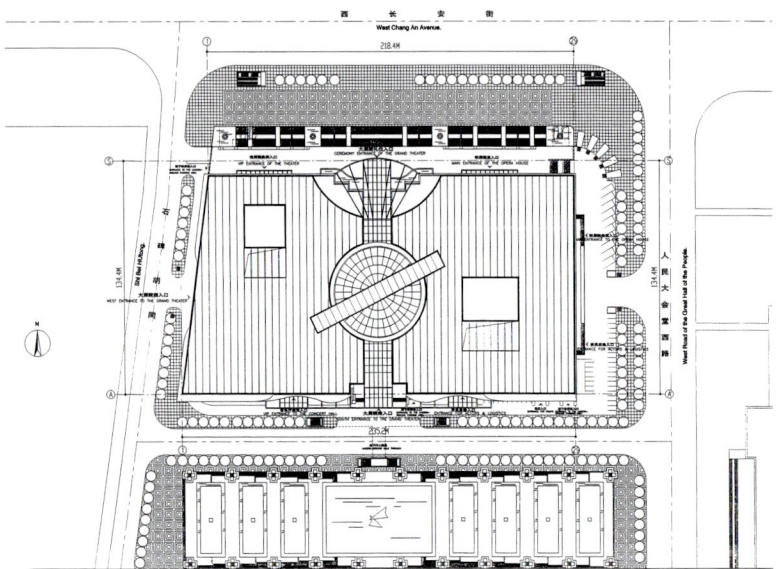

总平面　OVERALL SITE PLAN

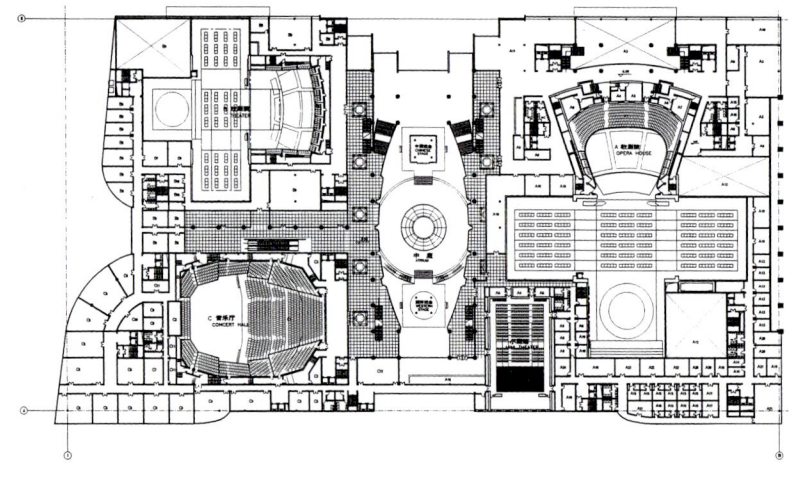

二层平面　2nd FLOOR PLAN

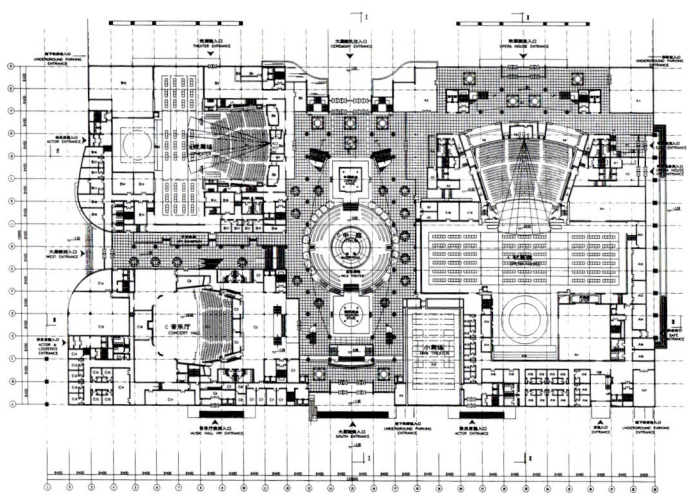

首层平面　GROUND FLOOR PLAN

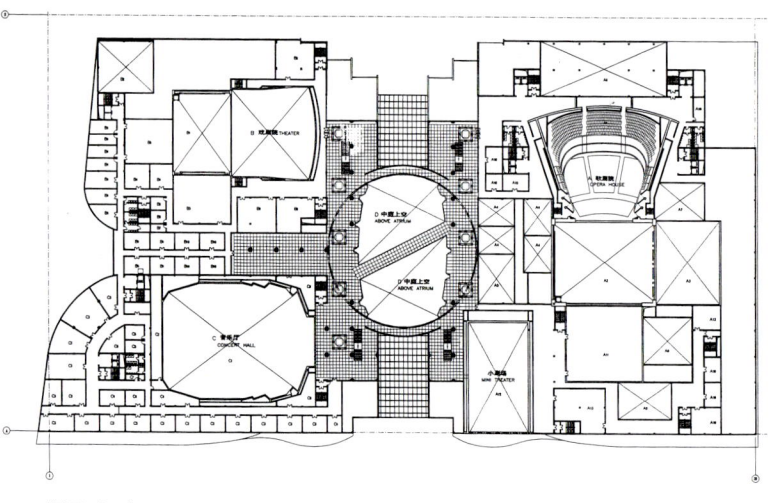

五层平面　5th FLOOR PLAN

214

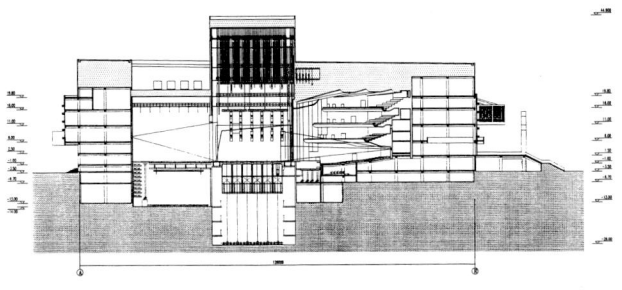

A 剖面
SECTION A

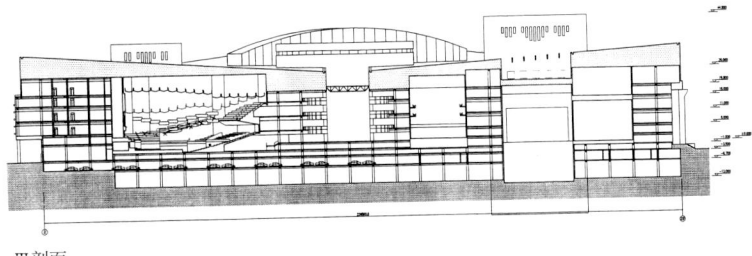

III 剖面
SECTION III

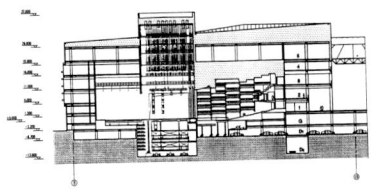

IV 剖面
SECTION IV

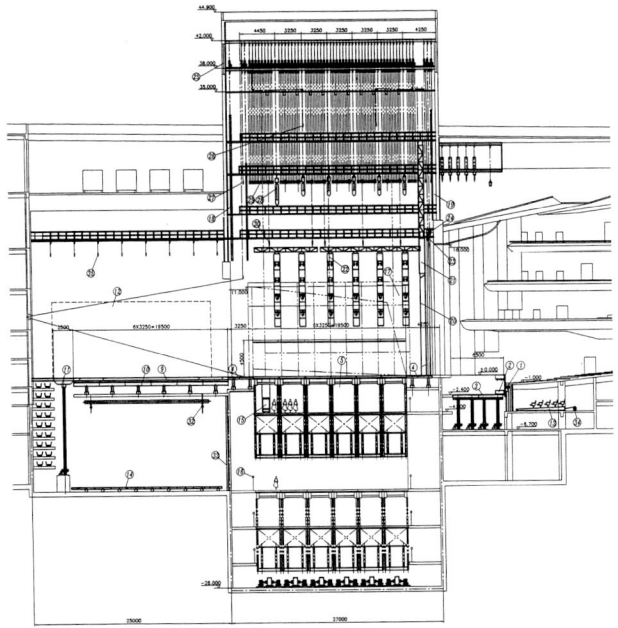

歌剧院舞台机械布置
MECHANICAL ARRANGEMENT OF THE OPERA HOUSE STAGE

技术经济指标
Technical Economic Index

基地面积 (m²) Site Area	总建筑面积 (m²) Total Floor Area	建筑高度(m) Building Height	建筑覆盖率 (%) Building Coverage Ratio
38916	地上：98067 地下：31453	23.20~36.50/44.9(局部)	70

容积率 Plot Ratio	绿化覆盖率 (%) Green Coverage Ratio	停车数量 Number of Stalls	自行车停车数量 Number of Bicycle Stalls
2.5	2	地上：83 地下：532	1100

各观众厅数据
Data For Auditoria

		歌剧院 Opera House	音乐厅 Concert Hall	戏剧院 Theater	小戏场 Mini Theater
	观众厅面积(m²) Floor Area of Auditorium	2050	1633	960	950
	观众厅体积(m³) Volume of Auditorium	17500	20400~29560	9616	8640
座席数 Seating Capacity	总座席数 Seating Capacity(Seats)	2542	2036	1166~1215	302~456
	其中：池座 Among Them:Auditorium	1022	1425	649~786	302~456
	楼 座 Floor Seat	1520	611	429~517	206
	休息厅面积(m²) Lounge Floor Area	1650	980	650	235
舞台尺寸 (m²) Stage Dimentions	主舞台 Main Stage	34 × 27 × 35	430.5	26 × 19.5 × 26	6 × 20
	左右侧台 Left and Right Side Stages	23 × 27 × 11		16.75 × 19.5 × 9	
	后舞台 Back Stage	26 × 25 × 12		16.5 × 20 × 10	
	台 仓 Under Stage Storage	34 × 27 × 26	192	26 × 19.5 × 13	16 × 28
	升降乐池 Elevating Orchestra Pit	145		85	
	最大视距(m) Max. Visual Distance	36.3	28.8	26.9	29
	最大俯角（°） Max. Angle of Depression	29	28.1	29.93	

507
HPP 国际建筑设计
有限公司(德国)
HPP International
Planungsgesellschaft
mbH(Germany)

在一个共同的、透明的、均匀分割的大屋顶下，国家大剧院的所有组成部分被联系在一起。整个建筑综合体以简洁的形式宁静地嵌入城市的肌理，有机地续写着不断滋长的城市平面。建筑本身试图以当代的语汇来解释中国特有的明确清晰、精心分割的建筑格调。传统的建筑元素，特别是园林，反映出一个悠久的文化传统以及民族的个性和特征。这个元素的应用会将建筑和周围环境有机地联系在一起。位于这个开放的、与公众接近的建筑前的水面同时又能创造一个必要的观赏距离。

在大屋顶下由轴线关系组织的歌剧院、戏剧场、音乐厅、小剧场及其它功能部分共同形成一个向公众开放的"文化广场"。整个大屋顶下围合的空间与各个功能单体形成一个"空间中的空间"。

Underneath of a joining, transparent and flat formed roof all elements of the National Grand Theater are combined, same as done in the tradition of the Chinese monumental buildings.

With its simplicity and clearness the project will become an important, unchangeable point of view of the Capital, it highlights the develop-ment of the city, reflects the cultural tradition but also states its own independence and character.

The building design comprises the actual state of architecture and technique but also makes reference to the clear and straight design of the Chinese governmental building architecture. The horizontal base, roof construction on high columns, light base and dark colored superstructure describing clearly the way and the expected result. The incorporated elements of tradition are making reference to the image and also reflect the cultural history of China.

In continuation of the great Chinese history of building construction the surroundings have to be incorporated. The complex of the National Grand Theater opens to the "Chang'an Avenue" and refers to the entrance of the "Forbidden City" with the Zhong Shan Park, without giving up the necessary distance to the actual daily situation.

At the place, art and culture are speaking to the people.

The single places of action-Opera House, Theater, Concert Hall and Mini Theater including all related requirements and facilities are combined underneath of a joining, transparent steel / glass roof construction. The main entrance is situated on the ground floor. All to be seen as a functional center of art, designed as a "house in house" structure.

北侧模型透视
VIEW FROM NORTH SIDE OF THE MODEL

南侧模型透视
VIEW FROM SOUTH SIDE OF THE MODEL

公共大厅透视
PUBLIC HALL PERSPECTIVE

观众厅透视
AUDITORIUM PERSPECTIVE

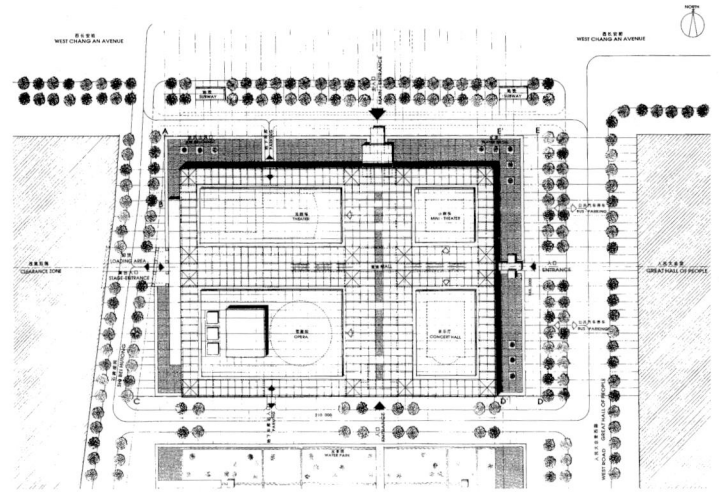

总平面 OVERALL SITE PLAN

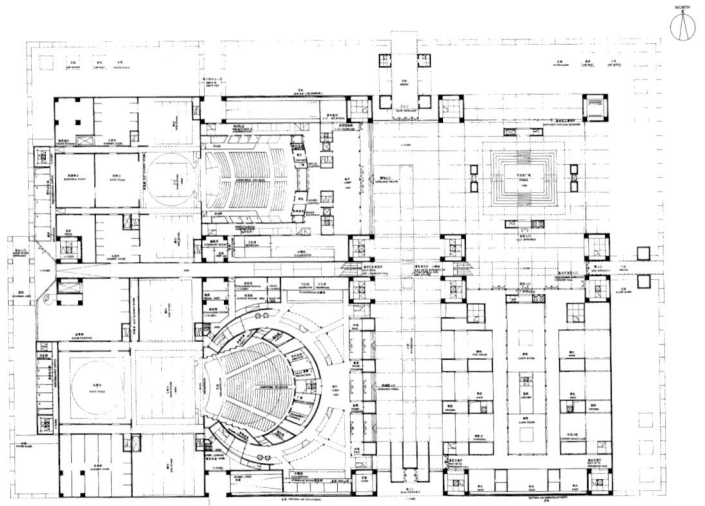

一层平面 GROUND FLOOR PLAN

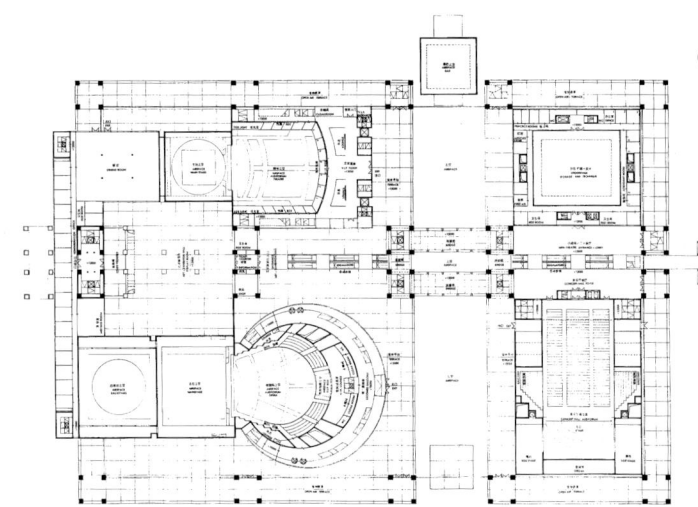

四层平面 4th FLOOR PLAN

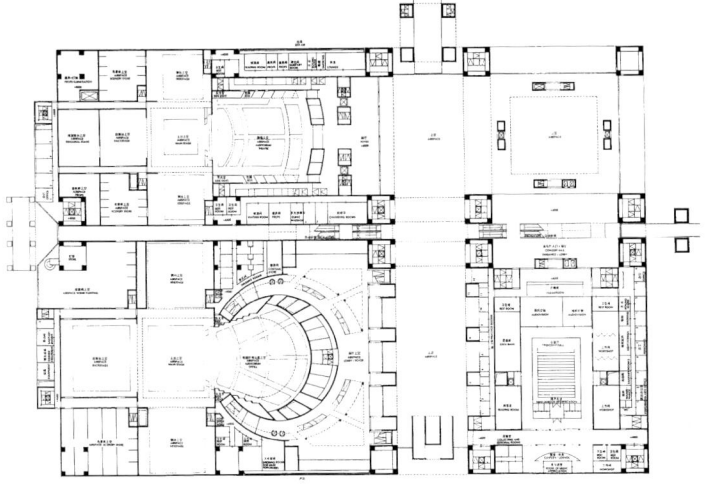

二层平面 2nd FLOOR PLAN

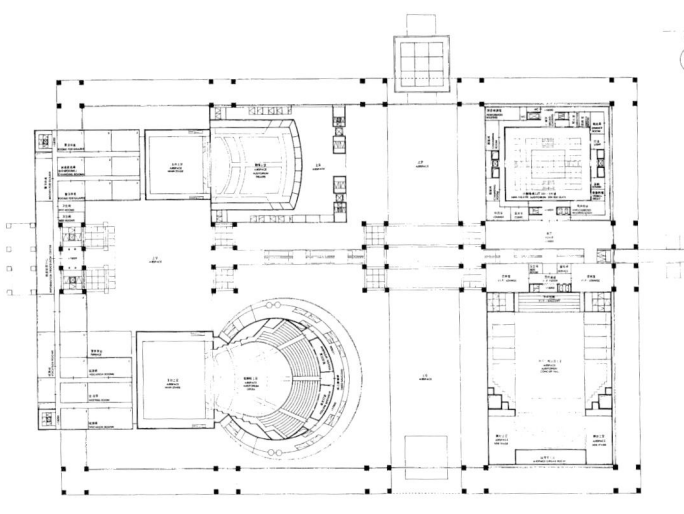

六层平面 6th FLOOR PLAN

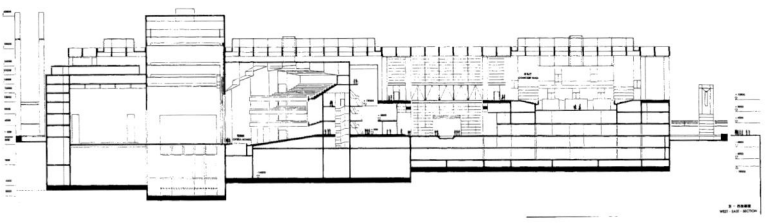

歌剧院东－西剖面
EAST-WEST SECTION OF THE OPERA HOUSE

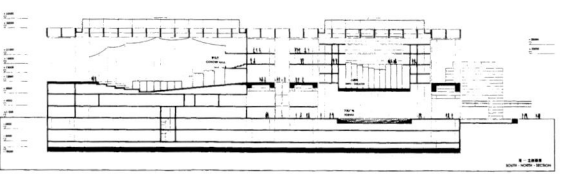

南－北剖面
SOUTH-NORTH SECTION

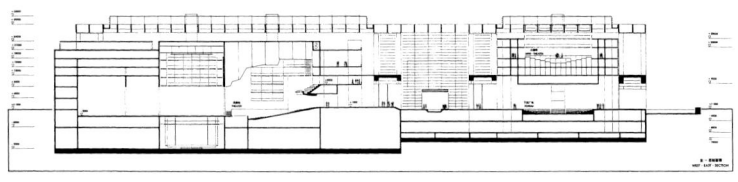

戏剧院东－西剖面
EAST-WEST SECTION OF THE THEATER

技术经济指标
Technical Economic Index

基地面积 (m²) Site Area	总建筑面积 (m²) Total Floor Area	建筑高度(m) Building Height	建筑覆盖率（%） Building Coverage Ratio
43082	147.628	30/42(局部)	73

容积率 Plot Ratio	绿化覆盖率 (%) Green Coverage Ratio	停车数量 Number of Stalls	自行车停车数量 Number of Bicycle Stalls
3.43	31	地上：102 地下：559	902

各观众厅数据
Data For Auditoria

		歌剧院 Opera House	音乐厅 Concert Hall	戏剧院 Theater	小戏场 Mini Theater
观众厅面积(m²) Floor Area of Auditorium		957	1170	875	750
观众厅体积(m³) Volume of Auditorium		20000	23000	11500	7500
座席数 Seating Capacity	总座席数 Seating Capacity(Seats)	2514	2262	1246	300~500
	其中：池座 Among Them:Auditorium	1111	1628	874	300~500
	楼座 Floor Seat	1403	634	372	60
	休息厅面积(m²) Lounge Floor Area	3000	1600	2320	900
舞台尺寸 Stage Dimentions (m²)	主舞台 Main Stage	891	325	598	750
	左右侧台 Left and Right Side Stages	2 × 564	2 × 148	2 × 379	
	后舞台 Back Stage	650		310	
	台仓 Under Stage Storage	2105	≈ 160	1356	750
	升降乐池 Elevating Orchestra Pit	110 (160m² intotal)	≈ 160	90 (160m² intotal)	根据舞台需要定
最大视距(m) Max. Visual Distance		32	38	32	根据舞台需要定

北京三磊建筑设计
有限公司(中国)
Sunlight Architects &
Engineers(China)
法国建筑家
协会(法国)
Partenaires Architects
(France)

●巨大的曲面屋顶及檐廊表达出传统精神，根据内部功能需要，屋顶由东北向西南层层叠高，表达出音乐的韵律节奏和传统的重檐意念。

●国家大剧院西立面的实体界面与人民大会堂北墙保持在同一水平线上，其实面效果与天安门城墙及人民大会堂的实体形成呼应。

●国家大剧院北立面由水平飞檐及19开间柱廊组成，其柱廊的比例、节奏与人民大会堂柱廊有一致性，并传达出国家大剧院作为"国家最高艺术殿堂"的象征意义。

● Huge and curved roof and eaves gallery demonstrate the traditional spirit. According to the requirements of internal functions, the roof is raised layer by layer from northeast to southwest and expresses the music melody and rhythm and traditional idea of double eaves roof.

● The entity interface of west elevation of the National Grand Theater keeps the same horizon with the north wall of the Great Hall of the People. Its real effect works in concert with the Tian'anmen wall and

北侧夜景透视
NIGHT VIEW OF NORTH SIDE

南侧透视
PERSPECTIVE OF SOUTH ELEVATION

公共大厅透视
PUBLIC HALL PERSPECTIVE

歌剧院观众厅透视
PERSPECTIVE OF THE OPERA HOUSE AUDITORIUM

entity of the Great Hall of the People.

●The north elevation of the National Grand Theater is composed of horizontal up-turned eaves and the gallery of 19 bay colonnade, the proportion and rhyme of which are identical with colonnade of the Great Hall of the People and represent the symbolic significance of National Grand Theater as "National supreme art palace".

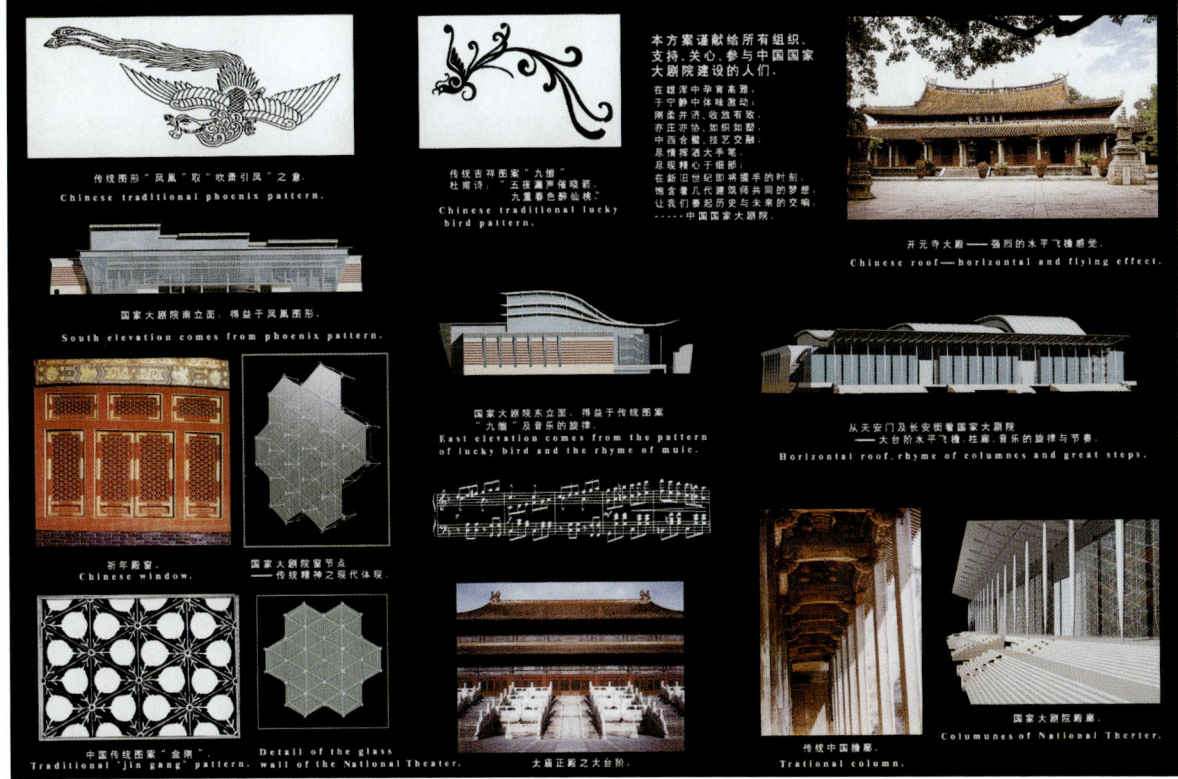

概念及分析图
CONCEPT & ANALYSIS

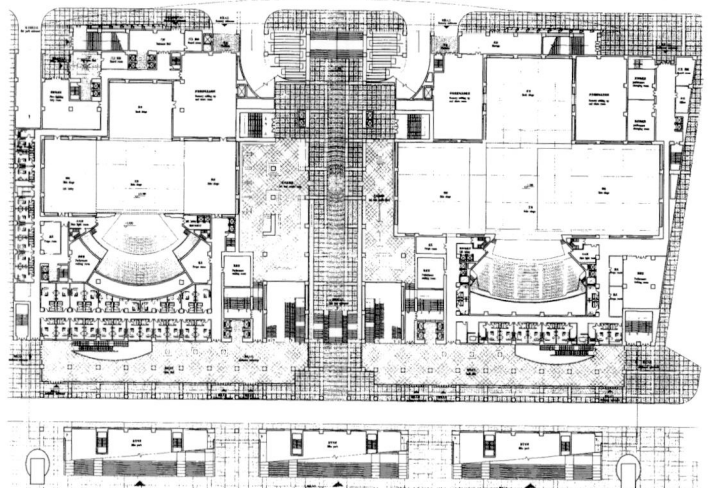

首层平面　GROUND FLOOR PLAN

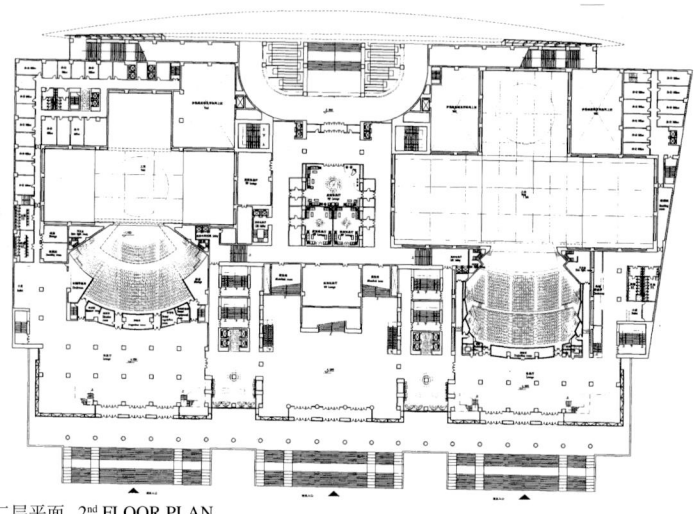

二层平面　2nd FLOOR PLAN

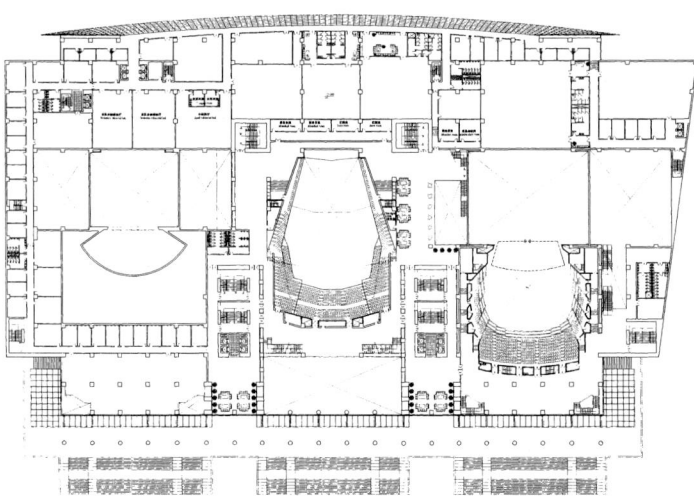

五层平面　5th FLOOR PLAN

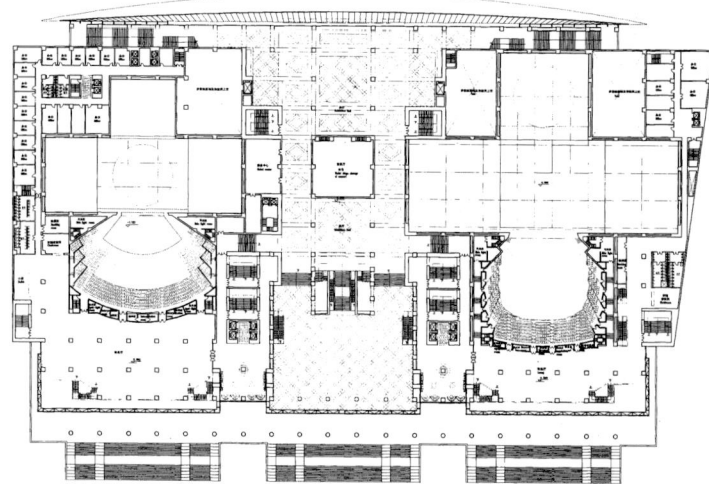

三层平面　3rd FLOOR PLAN

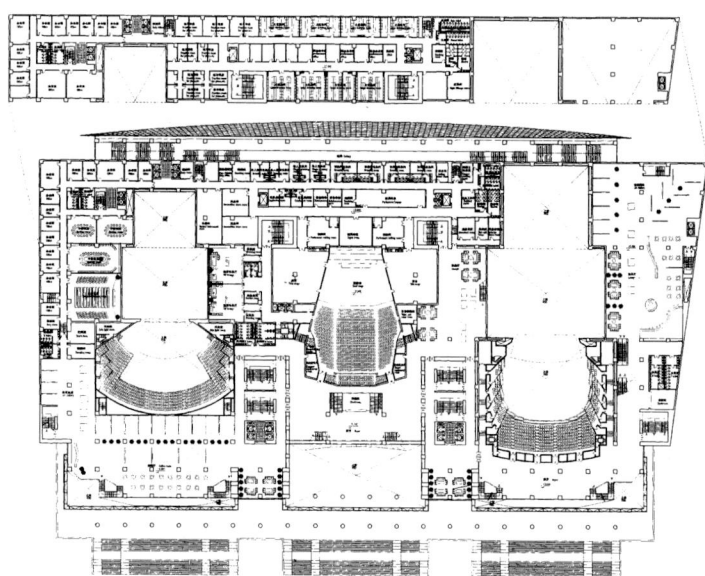

四层平面　4th FLOOR PLAN

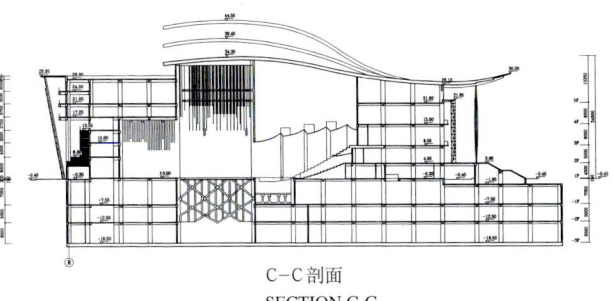

A−A剖面
SECTION A-A

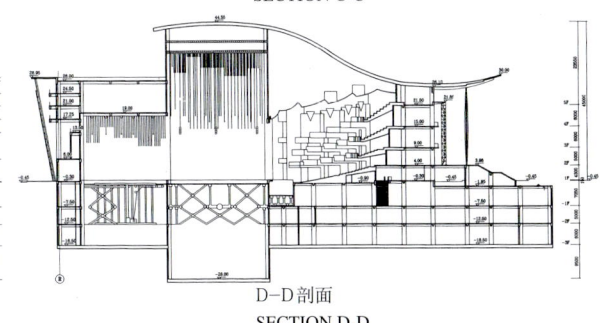

C−C剖面
SECTION C-C

D−D剖面
SECTION D-D

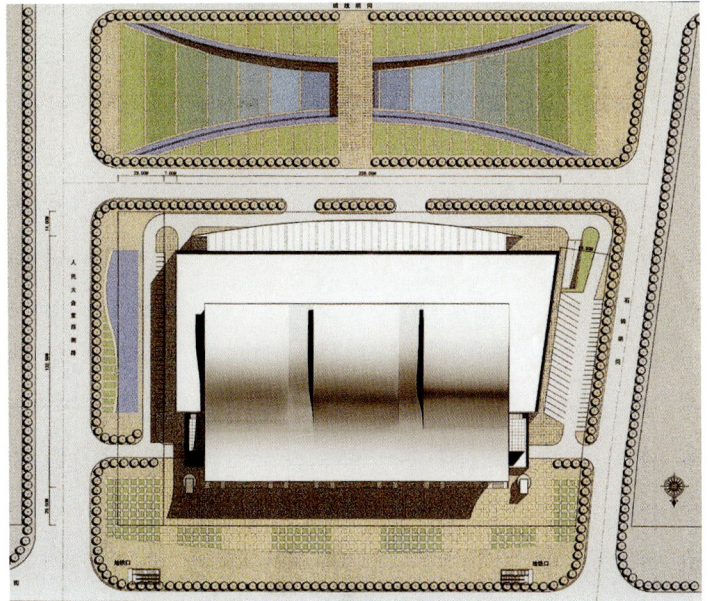

总平面 OVERALL SITE PLAN

技术经济指标
Technical Economic Index

基地面积 (m²) Site Area	总建筑面积 (m²) Total Floor Area	建筑高度(m) Building Height	建筑覆盖率 (%) Building Coverage Ratio
38900	地上：59574 地下：62232	44.55	56

容积率 Plot Ratio	绿化覆盖率 (%) Green Coverage Ratio	停车数量 Number of Stalls	自行车停车数量 Number of Bicycle Stalls
1.52	28	地上：83 地下：612	1500

各观众厅数据
Data For Auditoria

			歌剧院 Opera House	音乐厅 Concert Hall	戏剧院 Theater	小戏场 Mini Theater
		观众厅面积(m²) Floor Area of Auditorium	2083.9	1713.2	810.6	420.3
		观众厅体积(m³) Volume of Auditorium	17750	21000	7200	可调
座席数 Seating Capacity		总座席数 Seating Capacity(Seats)	2492	1992	1198	462
		其中：池座 Among Them:Auditorium	984	968	558	462
		楼 座 Floor Seat	1508	1024	640	
		休息厅面积(m²) Lounge Floor Area	4800	3980	2360	2380
舞台尺寸 Stage Dimentions (m²)		主舞台 Main Stage	36 × 27 × 36	373.6	27 × 21 × 25	
		左右侧台 Left and Right Side Stages	22 × 27 × 12	258.8 × 2	15 × 21 × 9	
		后舞台 Back Stage	26 × 25 × 18		19.3 × 17 × 15	
		台 仓 Under Stage Storage	36 × 27 × 28	17 × 17 × 7.5	27 × 21 × 12	34 × 17 × 56
		升降乐池 Elevating Orchestra Pit	103.3		197.2	
		最大视距(m) Max. Visual Distance	35.6	34.4	28.8	
		最大俯角（°） Max. Angle of Depression	32	20	28	

**尼克拉斯设计师
事务所（希腊）**
Nicolas Petropoulos,
Architect Designer
(Greece)

设计之基础是以一个"迷人城市"的概念作为"紫禁城"的对比。因此一堵围墙用以保护里面为献给艺术而建的建筑物。离开热闹的街道，游人就能走在里面的小路上或庭院里。这片宁静的地方，就好像是由一系列不规则的地方所组成的，使人产生在此散步流连之意。

庭院里也有可供露天表演之场所，同时隔声性极强，隔绝从外来之声响。

兴建围墙和里面剧场、剧院的目的是要给整座综合建筑一种关闭的、明确的结构，以作为文化中心的总体表现。

Basis for the design is the idea of an "Enchanted City" as an antithesis to the "Forbidden City". So a surrounding wall protects the buildings inside, dedicated to the arts. Leaving the lively streets, visitors find themselves in an internal pathway or courtyard, a zone of calmness, formed like a sequence of irregular places, inviting to walk and linger.

The internal courtyard is also a space for open air performances, acoustically well shielded from environmental noise.

The intention to provide a surrounding wall with halls and theaters positioned inside, is to give the complex a closed, clear organization and an overall appearance as a cultural center.

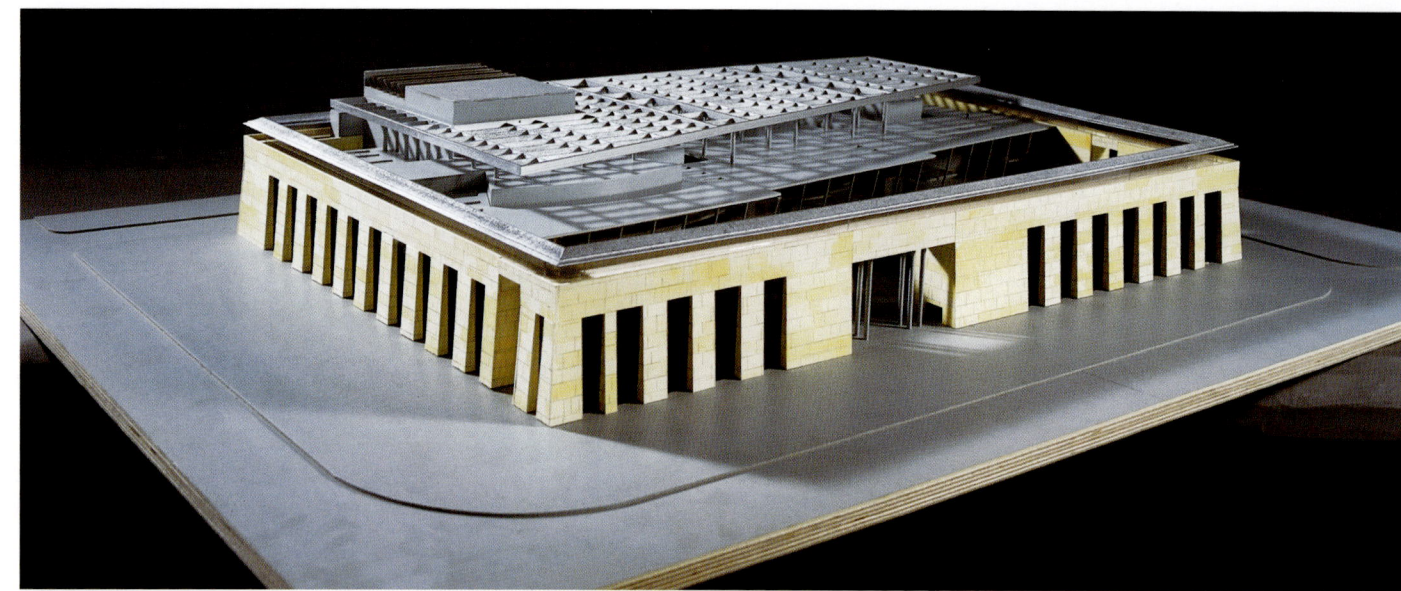

北侧模型透视
VIEW FROM NORTH SIDE OF THE MODEL

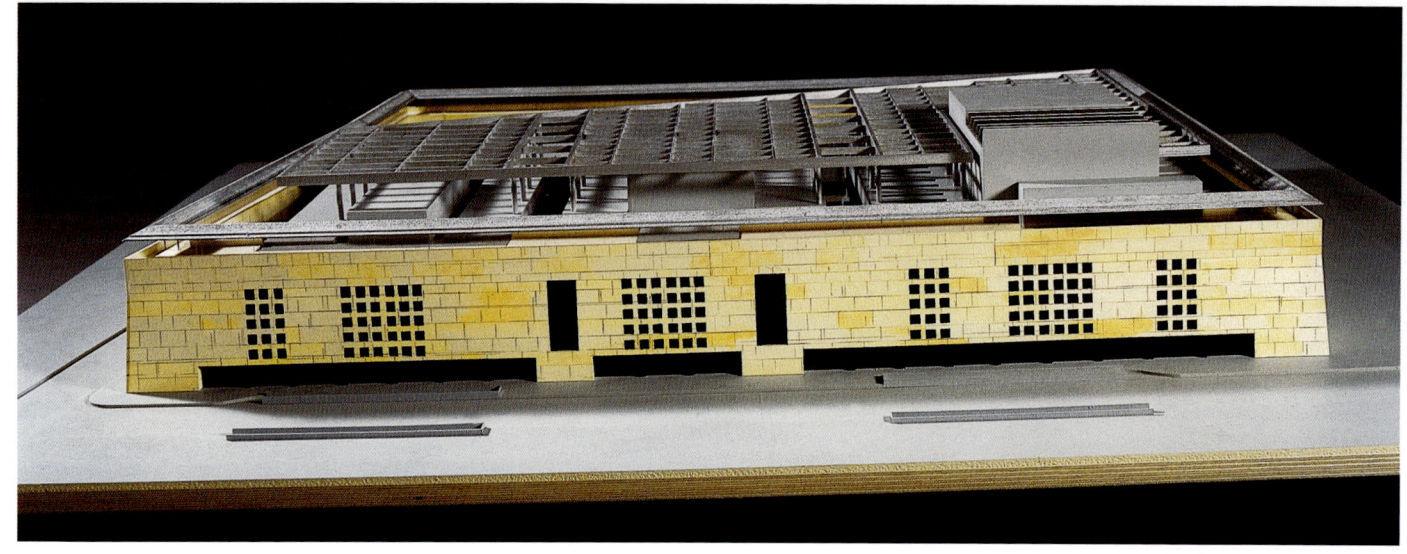

南侧模型透视
VIEW FROM SOUTH SIDE OF THE MODEL

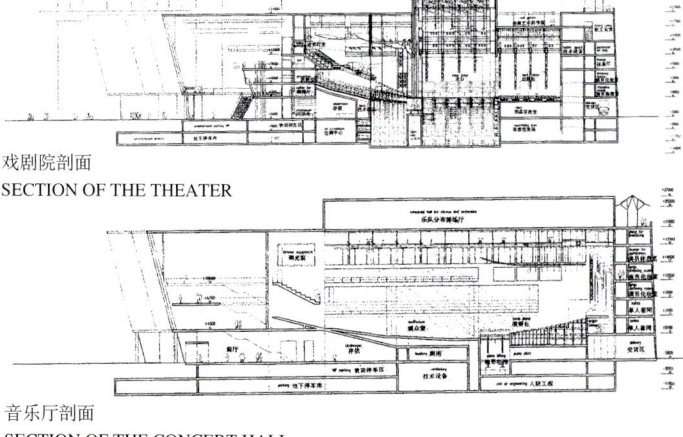

戏剧院剖面
SECTION OF THE THEATER

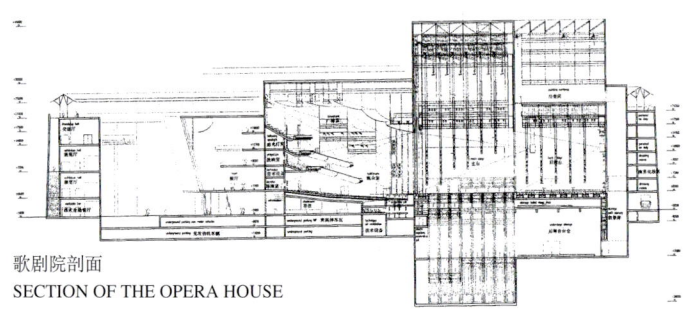

歌剧院剖面
SECTION OF THE OPERA HOUSE

音乐厅剖面
SECTION OF THE CONCERT HALL

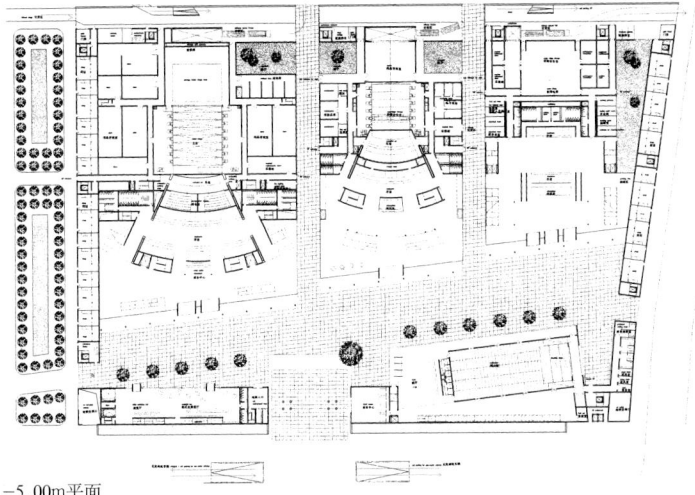

−5.00m平面
LEVEL(-5.00m)

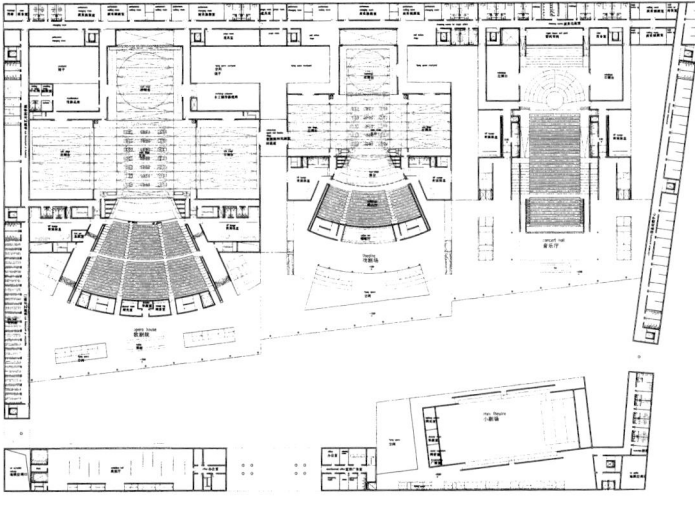

±0.00m平面
LEVEL(±0.00m)

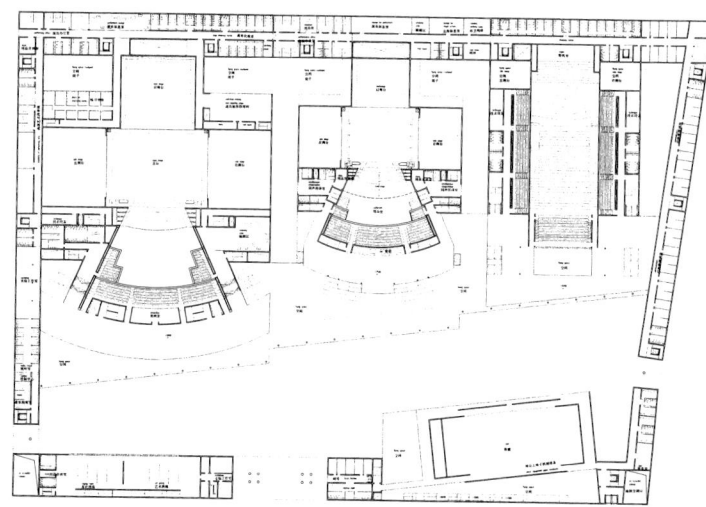

7.00m平面
LEVEL(+7.00m)

技术经济指标
Technical Economic Index

基地面积 (m²) Site Area	总建筑面积 (m²) Total Floor Area	建筑高度 (m) Building Height	建筑覆盖率 (%) Building Coverage Ratio
38900	147500	49	0.74

容积率 Plot Ratio	绿化覆盖率 (%) Green Coverage Ratio	停车数量 Number of Stalls	自行车停车数量 Number of Bicycle Stalls
2.62	0.26	地上：20 地下：551	1020

各观众厅数据
Data For Auditoria

		歌剧院 Opera House	音乐厅 Concert Hall	戏剧院 Theater	小戏场 Mini Theater
	观众厅面积(m²) Floor Area of Auditorium	2157	1518	864~1054	720
	观众厅体积(m³) Volume of Auditorium	23850	29874	9000	11520
座席数 Seating Capacity	总座席数 Seating Capacity(Seats)	2635	2209	1120	575
	其中：池座 Among Them:Auditorium	1156	949	472	
	楼　座 Floor Seat	1479	1260	643	576
	休息厅面积(m²) Lounge Floor Area	7281	6089	3966	2264
舞台尺寸(m²) Stage Dimentions	主舞台 Main Stage	972	528	567	240
	左右侧台 Left and Right Side Stages	1232		630	
	后舞台 Back Stage	650		330	
	台　仓 Under Stage Storage	2272	550	567	960
	升降乐池 Elevating Orchestra Pit	176		190	
	最大视距(m) Max. Visual Distance	44	40	30	36
	最大俯角（°） Max. Angle of Depression	25		23	

鉴于对该建筑的这种要求和其所处的条件，建筑设计应追求一种特性，即应避免采用那些众多过时的（哪怕是最优秀的）东、西方建筑风格，并且不应受那些明显是90年代建筑风格的影响。应将一种时间的永恒性体现在设计中以保证其在新世纪乃至于更为久远的岁月里获得赞叹。我们尝试着运用来自于中国哲学和诗歌的灵感，使建筑造型既具有曲线美和戏剧性，又富有和谐宁静的特征。该建筑物以其庭园和山水之特征在喧嚣的闹市中营造出一个世外桃源，恬静安祥一如"紫禁城"墙内一般。

Studio Valle Progettazioni
Percy Thomas Partnership
SGA Design Studio

A building in this context and for this purpose must, therefore, achieve a quality of architecture that avoids the passing fashions of so many of even the best western and eastern hemispheres' architecture. The building must be free of stylistic forms which would place it too obviously as a building of the 1990's. A timeless quality must ensure that it will be applauded in the century to come and beyond. We have tried to draw some of our inspiration from Chinese philosophy and poetry and to create forms, which are both curvilinear and dramatic in silhouette while being harmonious and tranquil. The building with its gardens and water features may create a haven in the bustling city center, as do the walled spaces in the Forbidden City.

北侧透视
PERSPECTIVE OF NORTH ELEVATION

南侧透视
PERSPECTIVE OF SOUTH ELEVATIONL

局部模型
SECTIONAL VIEW OF THE MODEL

歌剧院观众厅透视
PERSPECTIVE OF THE OPERA HOUSE AUDITORIUM

公共大厅透视　PUBLIC HALL PERSPECTIVE

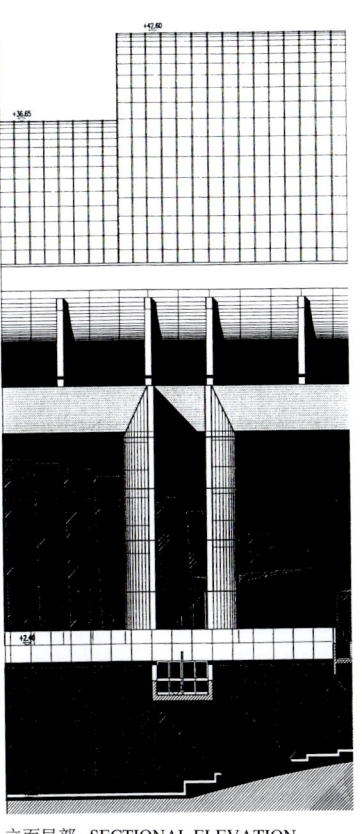

立面局部　SECTIONAL ELEVATION

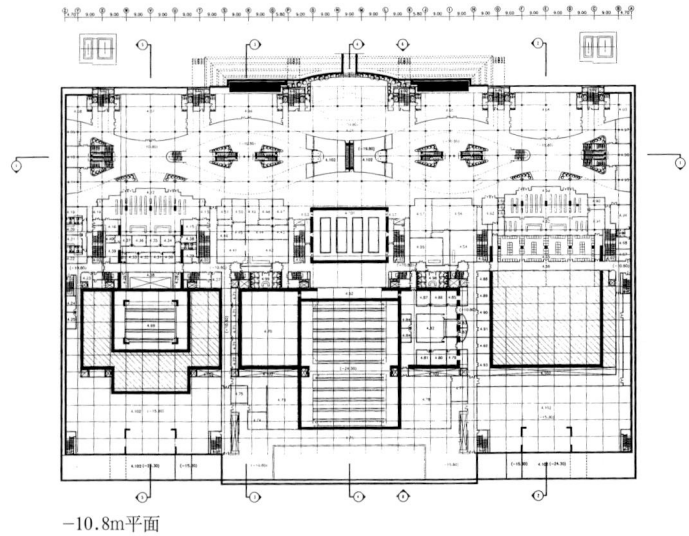

−10.8m平面
GROUND LEVEL(-10.80m)

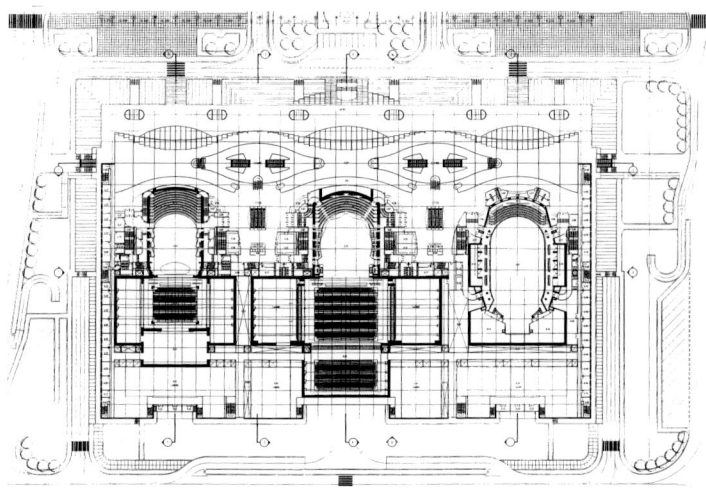

7.20m平面
GROUND LEVEL(+7.20m)

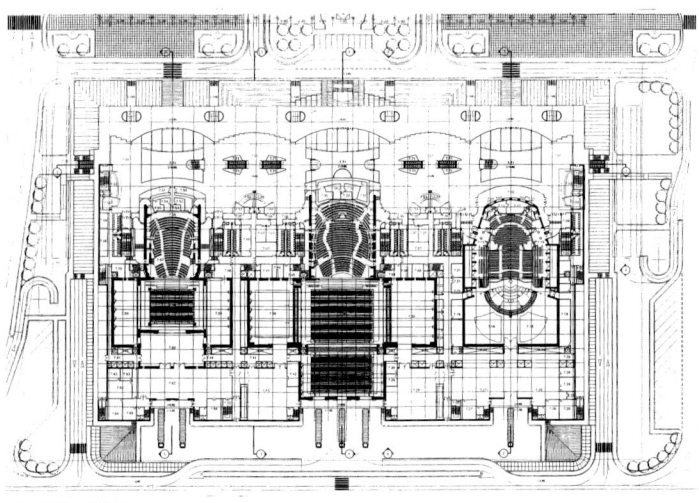

±0.00m平面
GROUND LEVEL(+0.00m)

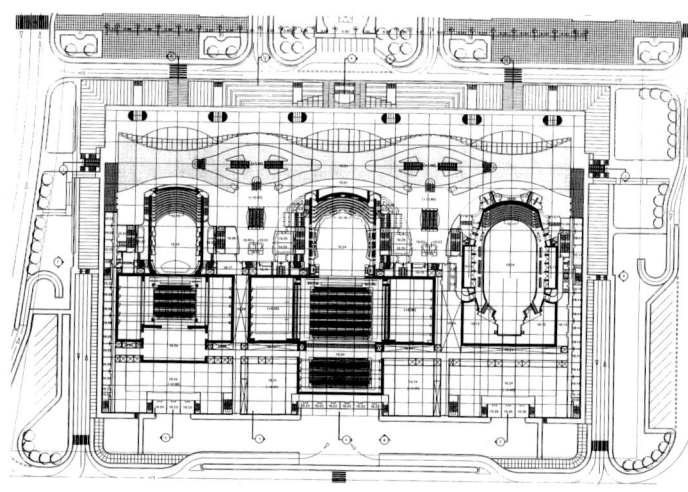

10.80m平面
GROUND LEVEL(+10.80m)

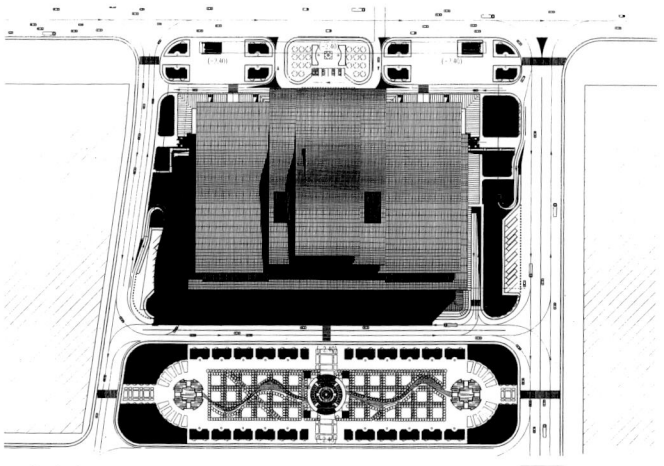

总平面
OVERALL SITE PLAN

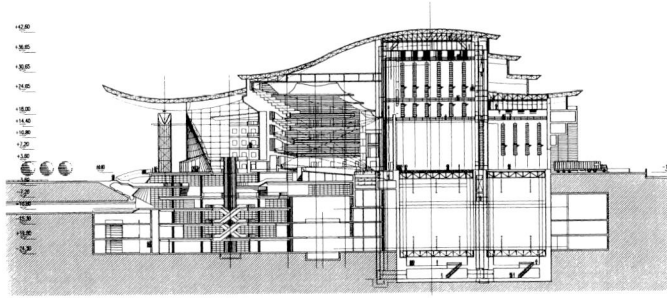

歌剧院剖面
SECTION OF THE OPERA HOUSE

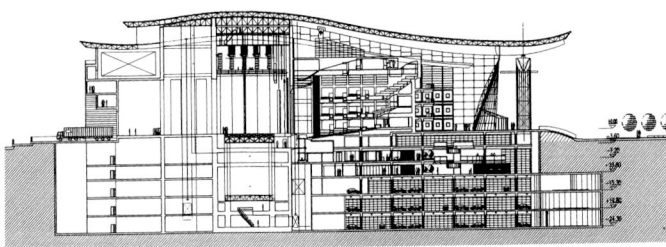

戏剧院剖面
SECTION OF THE THEATER

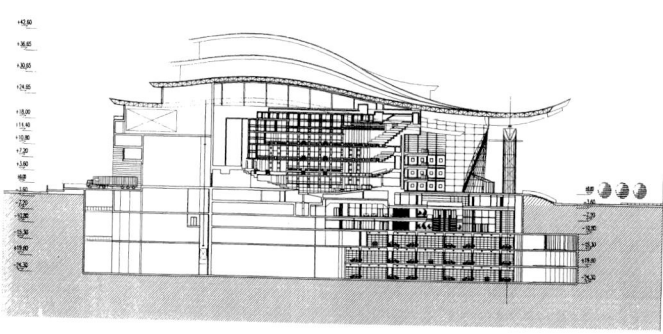

音乐厅剖面
SECTION OF THE CONCERT HALL

技术经济指标
Technical Economic Index

基地面积（m²） Site Area	总建筑面积（m²） Total Floor Area	建筑高度(m) Building Height	建筑覆盖率（%） Building Coverage Ratio
38844	地上：50335 地下：80750	26/27（局部）	70

容积率 Plot Ratio	绿化覆盖率(%) Green Coverage Ratio	停车数量 Number of Stalls	自行车停车数量 Number of Bicycle Stalls
1.30	15	地上：72 地下：686	1000

各观众厅数据
Data For Auditoria

		歌剧院 Opera House	音乐厅 Concert Hall	戏剧院 Theater	小戏场 Mini Theater
	观众厅面积(m²) Floor Area of Auditorium	2320	2400	1100	1525
	观众厅体积(m³) Volume of Auditorium	20700	25000~35000	11428	6700
座席数 Seating Capacity	总座席数 Seating Capacity(Seats)	2527	2186	1200	500
	其中：池座 Among Them:Auditorium	939	831	382	340
	楼座 Floor Seat	1588	1355	820	160
	休息厅面积(m²) Lounge Floor Area	3450	2615	1460	690
舞台尺寸 Stage Dimentions (m²)	主舞台 Main Stage	918	410	530	87
	左右侧台 Left and Right Side Stages	625 × 2	660	316 × 2	
	后舞台 Back Stage	650		330	
	台仓 Under Stage Storage	918	180	530	400
	升降乐池 Elevating Orchestra Pit	123	65	(35)80	87
	最大视距(m) Max. Visual Distance	44.5	40.8	37.4	11.3
	最大俯角（°） Max. Angle of Depression	34.50	28.84	22.50	25.60

同济大学建筑设计研究院(中国)

The Architectural Design & Research Institute of Tongji University (China)

由于基地已有一个范围很大较深的基坑,于是将戏剧场置于音乐厅之下,将小剧场置于歌剧院之下,以空间换平面,使原来基地上应当安置的四个演出厅转化成同一平面上安置两个演出厅,使得平面空间开敞、舒展。

大剧院采用了上部有水平线条束、蓬勃向上、挺拔刚健有力、不积灰尘的斗形顶部,置于内缩的玻璃体量上,再置于大台基座上的这一中国传统建筑具形方式,将整体空间体量组合成庄重、典雅、简洁的体块。汉白玉色大理石台基、浅米黄色大沿口墙的主调,在大面积色彩上使大剧院与人民大会堂紧密呼应,成为一个有机的整体;同时简洁体块的大剧院与壮丽丰富的、高低错落的大会堂空间体量形成一定的差异,表现出人民大会堂为主、大剧院为辅,不喧宾夺主的关系。

There exists a large and deep open pit, so we put the Theater underneath the Concert Hall, and the Mini-theater underneath the Opera House. Four auditoriums become two on the site, it feels stretch and open.

Use the fofrmation of Chinese traditional architecture, a vigorous and energetic cream colored compact succinct without dust gathering bucket-shape block with horizontal strips is supported on a large white marble platform, to create a solemn, elegant and compact form. Then the same color could be echo each other from a short distance as a whole of the People's Great Hall and the National Grand Theater. The compact abundant magnificent interlocked the Great Hall of the people plays a supporting role clearly.

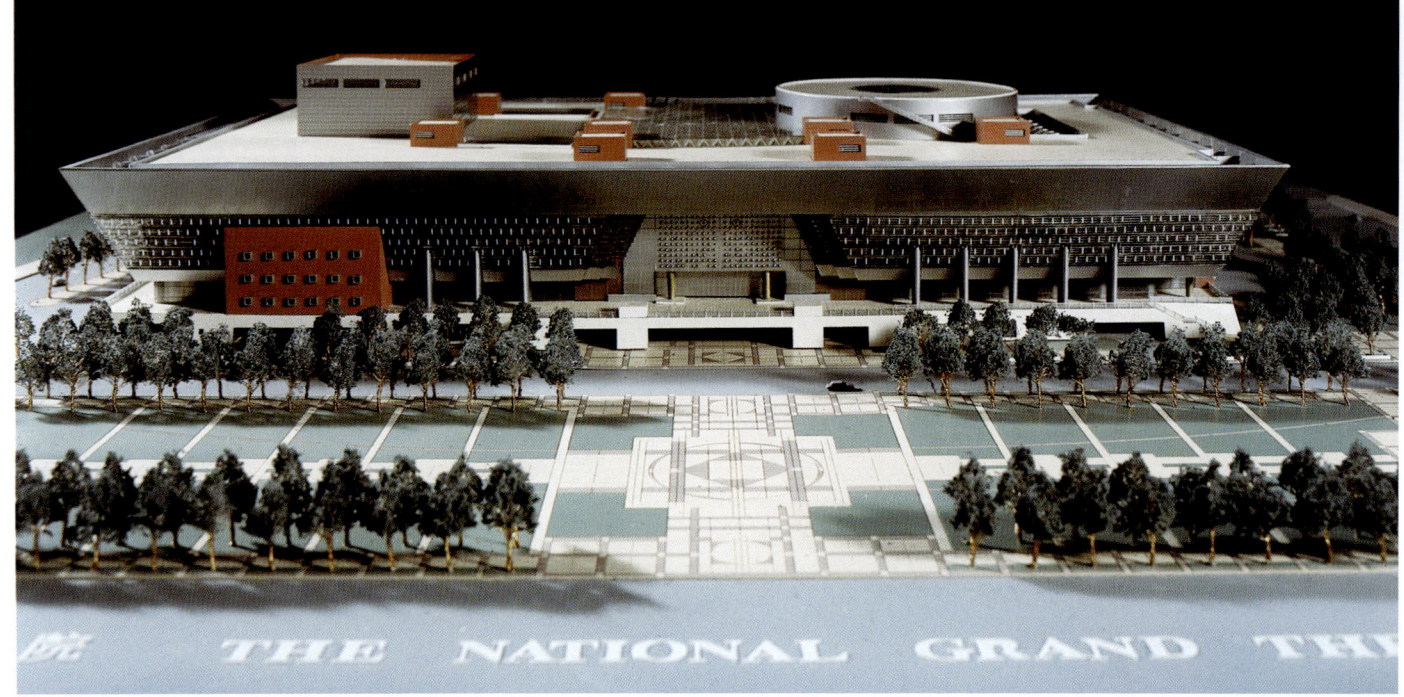

北侧夜景透视
NIGHT VIEW OF NORTH SIDE

南侧模型透视
VIEW FROM SOUTH SIDE OF THE MODEL

公共大厅透视
PUBLIC HALL PERSPECTIVE

歌剧院观众厅透视
PERSPECTIVE OF THE OPERA HOUSE AUDITORIUM

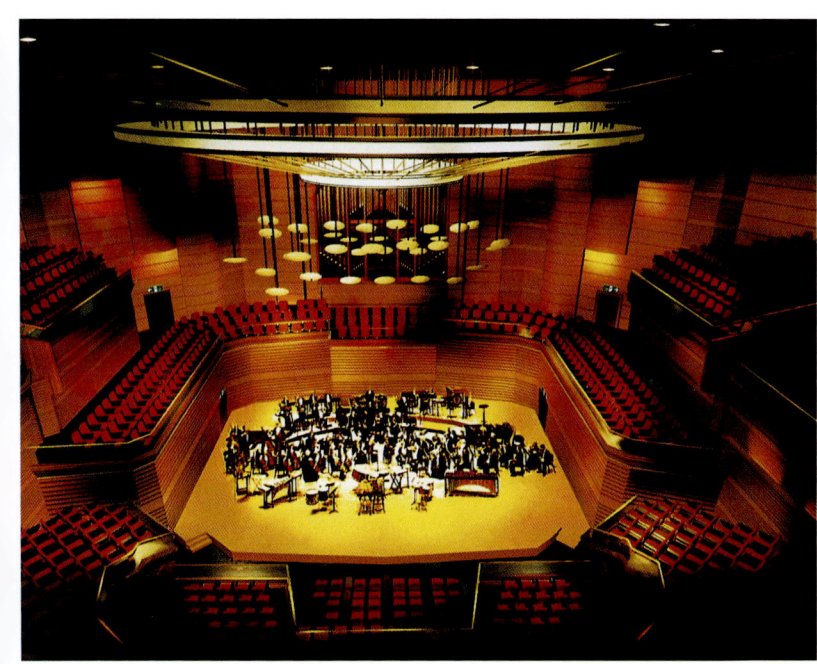

音乐厅观众厅
PERSPECTIVE OF THE CONCERT HALL AUDITORIUM

戏剧院观众厅
PERSPECTIVE OF THE THEATER AUDITORIUM

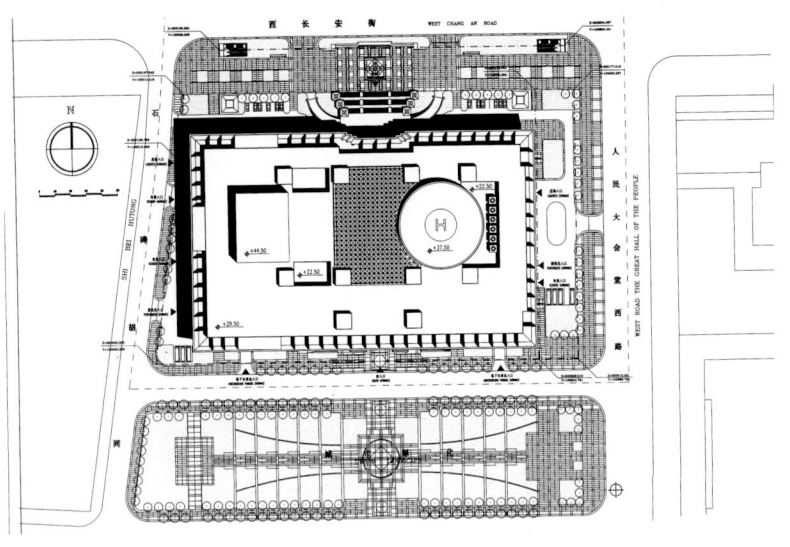

总平面
OVERALL SITE PLAN

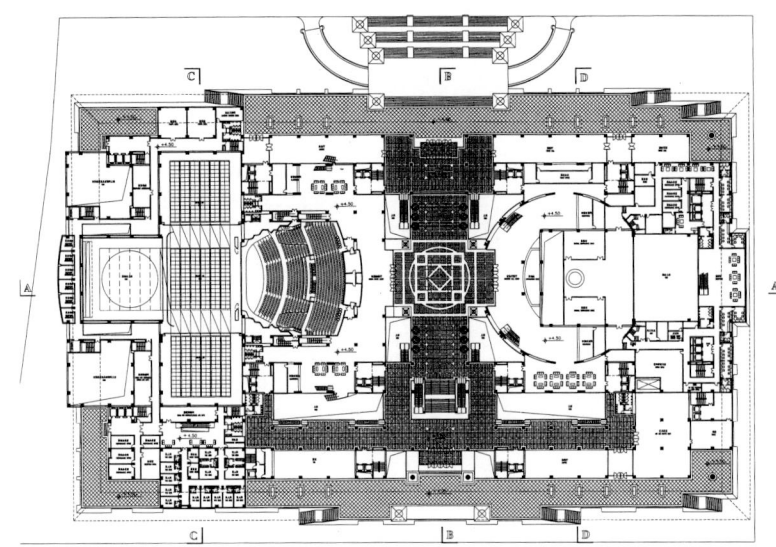

4.50m平面
GROUND LEVEL(+4.5m)

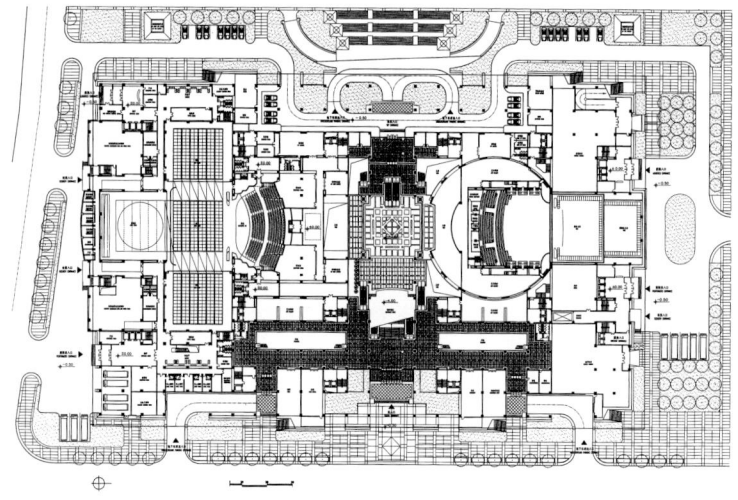

±0.00m平面
GROUND LEVEL(+0.00m)

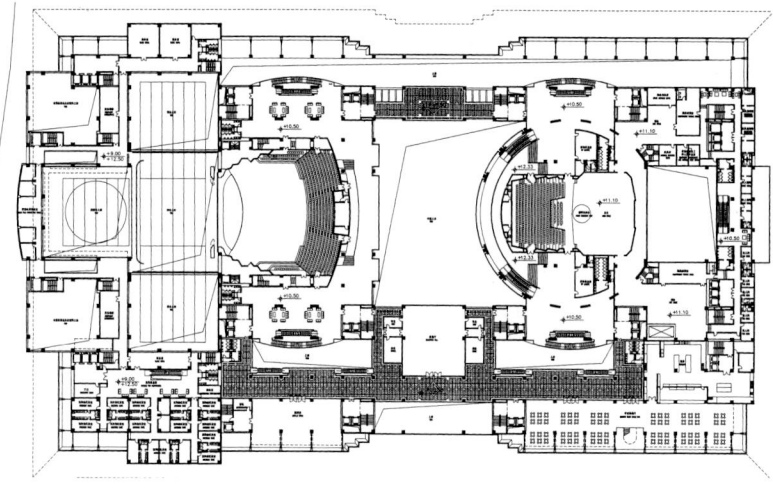

10.50m平面
GROUND LEVEL(+10.5m)

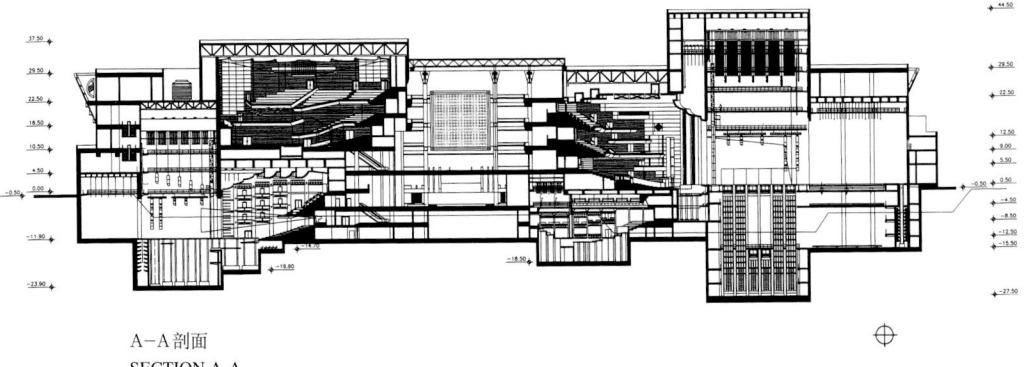

A－A剖面
SECTION A-A

232

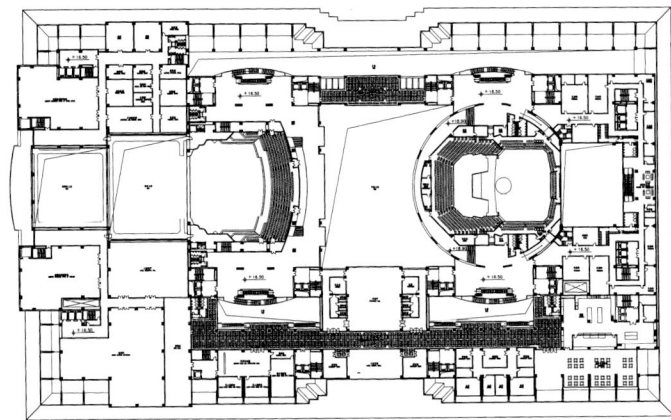

16.50m平面
GROUND LEVEL(+16.5m)

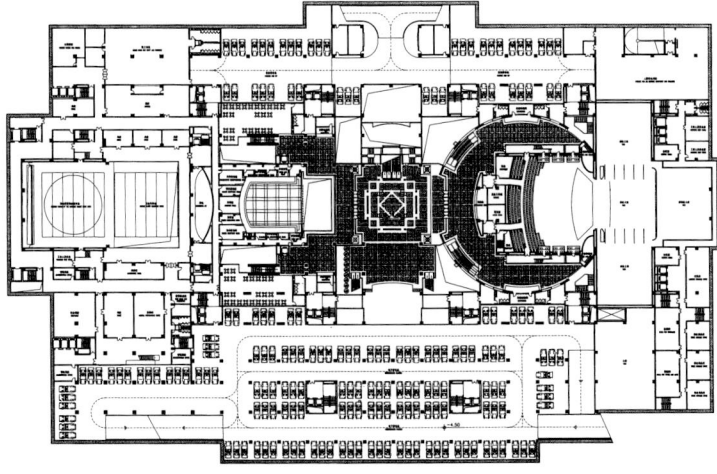

−4.50m平面
GROUND LEVEL(-4.5m)

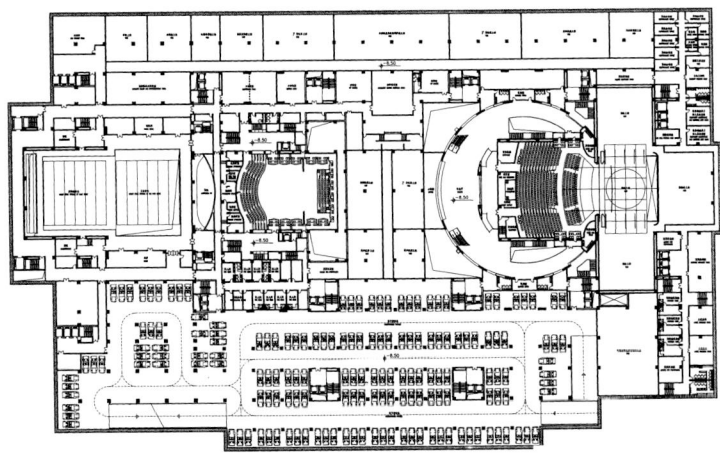

−8.50m平面
GROUND LEVEL(-8.5m)

技术经济指标
Technical Economic Index

基地面积（m²） Site Area	总建筑面积（m²） Total Floor Area	建筑高度(m) Building Height	建筑覆盖率（%） Building Coverage Ratio
38900	地上：72630 地下：61080	30(沿街)/45(局部)	47.12

容积率 Plot Ratio	绿化覆盖率(%) Green Coverage Ratio	停车数量 Number of Stalls	自行车停车数量 Number of Bicycle Stalls
1.87	19.04	地上：30 地下：526	1160

各观众厅数据
Data For Auditoria

		歌剧院 Opera House	音乐厅 Concert Hall	戏剧院 Theater	小戏场 Mini Theater
	观众厅面积(m²) Floor Area of Auditorium	1986	15900	1072	550
	观众厅体积(m³) Volume of Auditorium	20147	20500	8800	3880
座席数 Seating Capacity	总座席数 Seating Capacity(Seats)	2432	2020	1272	499
	其中：池座 Among Them:Auditorium	1254	464	759	276
	楼座 Floor Seat	1178	1556	513	223
	休息厅面积(m²) Lounge Floor Area	3420	2956	1826	800
舞台尺寸 (m²) Stage Dimentions	主舞台 Main Stage	36 × 27 × 36	23.8×18.9×21.0	27 × 21 × 25	24 × 12 × 17
	左右侧台 Left and Right Side Stages	22 × 27 × 12	19.5×11.4×6.0	18 × 18 × 10.5	9 × 12 × 8
	后舞台 Back Stage	25.5 × 26 × 18		22 × 15 × 13.4	
	台仓 Under Stage Storage	36 × 27 × 28 25.5 × 26 × 16		27 × 21 × 12	24 × 24 × 6
	升降乐池 Elevating Orchestra Pit	160	15	80	50
	最大视距(m) Max. Visual Distance	36.3	25.8	28.9	20.5
	最大俯角（°） Max. Angle of Depression	27.95	30	29.30	18.00

233

中国国家大剧院——向所有人开放的21世纪的城市建筑

中华文化的内在气质——雄浑、辽阔、精致、洒脱

技术文明的外在表现——轻盈、透明、简洁、幻化

信息时代的城市建筑——融入环境、崇尚多元

古都北京的城市意境——绿树丛中、屋顶隐现。

The National Grand Theater-an urban architecture for the 21st century open to the public.

The temperament of Chinese culture: Powerful, grand, delicate, free and easy.

The expression of technological advancement-light, transparent, succinct and ever-changing.

An artistic manipulation valued by Chinese art-dialectical understanding and well controlled play between the dense and the loose, between the complicate and the straight-forward.

Urban Architecture in the Age of information-dissolving into the surrounding and providing variety of choices.

Urban image of an ideal Beijing-gray, the color of its roofs shimmering above and among the green.

北侧模型透视
VIEW FROM NORTH SIDE OF THE MODEL

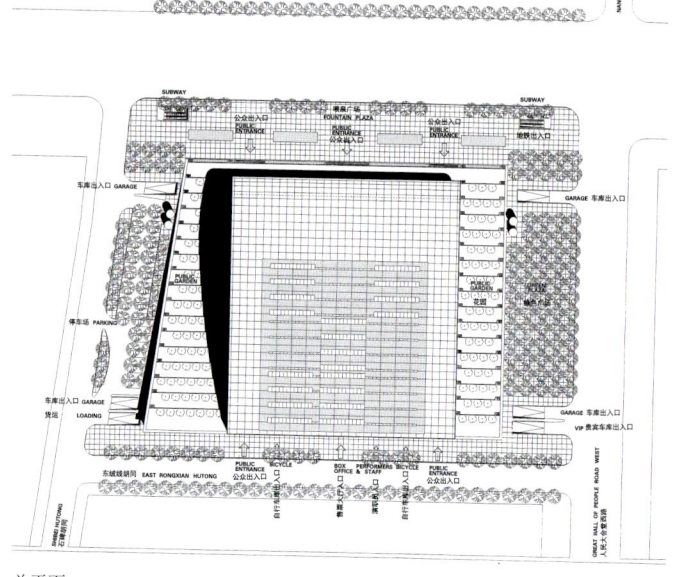

总平面
OVERALL SITE PLAN

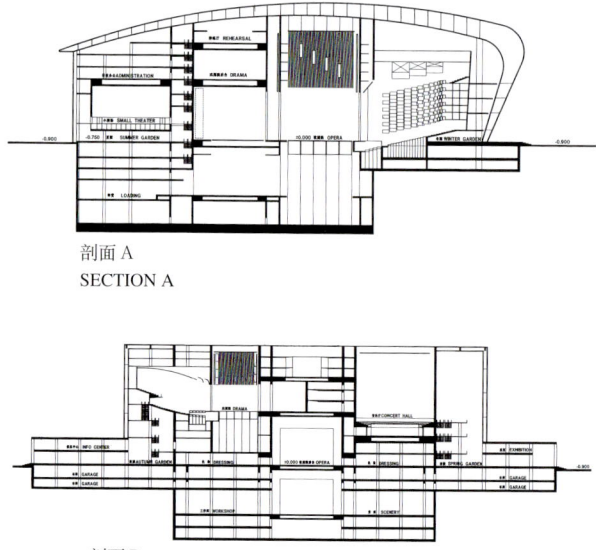

剖面 A
SECTION A

剖面 B
SECTION B

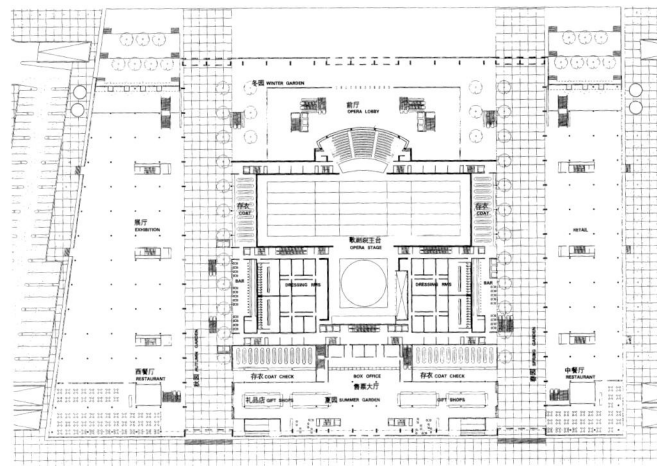

±0.00m平面
GROUND LEVEL(±0.00m)

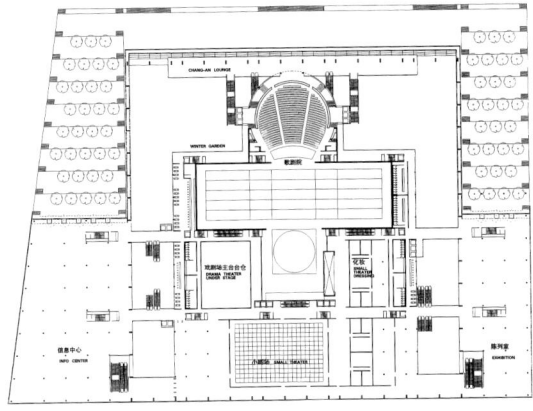

6.00m平面
GROUND LEVEL(+6.00m)

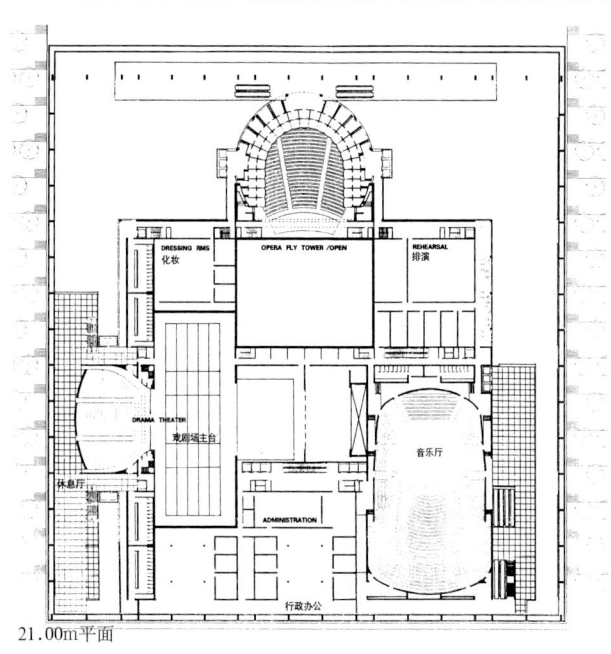

21.00m平面
GROUND LEVEL(+21.00m)

技术经济指标
Technical Economic Index

基地面积（m²） Site Area	总建筑面积（m²） Total Floor Area	建筑高度(m) Building Height	建筑覆盖率（%） Building Coverage Ratio
38916	地上：77040 地下：52700	30/45(局部)	68.7
容积率 Plot Ratio	绿化覆盖率 (%) Green Coverage Ratio	停车数量 Number of Stalls	自行车停车数量 Number of Bicycle Stalls
4.85	43.0	地上：52 地下：660	840

各观众厅数据
Data For Auditoria

		歌剧院 Opera House	音乐厅 Concert Hall	戏剧院 Theater	小戏场 Mini Theater
	观众厅面积(m²) Floor Area of Auditorium	2630	1560	980	2188
	观众厅体积(m³) Volume of Auditorium	65750	38530	14700	32700
座席数 Seating Capacity	总座席数 Seating Capacity(Seats)	2320	2020	1248	300~500
	其中：池座 Among Them:Auditorium	1293	1318	710	
	楼座 Floor Seat	1237	702	538	
	休息厅面积(m²) Lounge Floor Area	4455	2430	1720	1300
舞台尺寸(m²) Stage Dimentions	主舞台 Main Stage	27×36	22×27	22×27	27×40.5
	左右侧台 Left and Right Side Stages	20×27	各220	各364	
	后舞台 Back Stage	27×27		18×24	
	台仓 Under Stage Storage	2795	200	576	1094
	升降乐池 Elevating Orchestra Pit	162	594	220	
	最大视距(m) Max. Visual Distance	34.5	27	28	24
	最大俯角（°） Max. Angle of Depression	31	25	21	可调 Varies

第二轮竞赛方案(14项)
Schemes of the 2nd Round Competition(Fourteen)

01 法国巴黎机场公司（法国）
ADP Aeroports de Paris(France)

02 矶崎新建筑师事务所（日本）
Arata Isozaki & Associates (Japan)

03 建设部建筑设计院(中国)
Architectural Design Institute of Ministry of
Construction(China)

04 HPP 国际建筑设计有限公司（德国）
HPP International Planungsgesellschaft mbH(Germany)

05 塔瑞·法若建筑师事务所(英国)
Terry Farrell & Partners(UK)

06 北京市建筑设计研究院（中国）
Beijing Institute of Architectural Design &
Research(China)

07 深圳大学建筑设计研究院（中国）
Shenzhen University,The College of Architecture & Civil
Engineering(China)

08 清华大学建筑设计研究院（中国）
Architectural Design & Research Institute of Tsinghua
University(China)

09 王欧阳(香港)有限公司（中国香港）
Wong & Ouyang(HK) Ltd (China Hong Kong)

10 法国建筑工作室（法国）
Architecture · Studio(France)

11 卡洛斯·奥特建筑师事务所（加拿大）
Carlos Ott & Associates, Architects(Canada)

12 汉斯豪莱建筑师＋海兹诺曼建筑师（奥地利）
Design Group:Prof. Hans Hollein/Architect-Heinz
Neumann/Architect Planungsgemeinschaft(Austria)

13 山崎实建筑师事务所（美国）
Minoru Yamasaki Associates,Inc(USA)

14 欧博迈亚设计咨询有限公司＋戴尔曼教授建筑
设计事务所（德国）
Design Group: Obermeyer+Deilmann (Germany)

注：因01,05,11,方案进入到下一轮修改，部分图纸是通
用的，故上述三个方案连同修改方案在下一部分发表。

第二轮竞赛方案的设计构思及改进

从"凸起"到"反翘"

我们提出的中国国家大剧院设计方案，是在继承了天安门广场的历史和空间的连续性的前提下，作为新中国的国家象征，拥有全新的造型。这里，我们吸取了天安门广场的台座、列柱、屋顶的三个空间特征要素，运用计算机技术将其变形为现代现象。三维立体空间的大型波状屋顶，在天安门金黄耀眼的大屋顶和人民大会堂多层屋顶的水平线的延长线上，构成了天安门广场的第三个屋顶。这里，故宫博物院黄金瓦屋顶的波纹，将在国家大剧院屋顶的新轮廓上得到延伸。由曲面构成的屋顶可以说是故宫、天坛等中国传统建筑物中经常可以看到的高高向上挑起的大屋檐的现代翻版。这一屋顶结构通过薄型壳膜的空间解析，将铁板作为模板现场浇灌混凝土制作。铁板熔接采用了造船技术。这一造型设计并非只是单纯依靠计算机的演算能力创造出来的形状，而是在彻底追求合理的结构，并拥有充分的施工技术保障的前提下产生的。

在第二轮竞赛方案中，我们在继续贯彻第一轮竞赛方案中所提出的主要设计构思的基础上，对大剧院屋顶微妙的形状变化的修正及其美学效果等多方面，进行了反复的论证探讨。在第一轮竞赛方案评审报告书中，我们提出的大屋顶的形状，从其与周围环境的特殊性上考虑，未能得到大多数评委的赞同。当时此屋顶形状形成的主要原因在于，与天安门及故宫等中国传统建筑屋顶的轻快反翘形状相比，考虑到为了覆盖歌剧院舞台上部的挑架而高高凸起，从而使在整体上形成了带有"凸起"的建筑样式。与我们提出的带有向上方"凸起"的屋顶相比较，传统的中国建筑屋顶通过带有弦状"反翘"的曲线结构，使屋顶整体形成了向下方"凸起"的曲面。

Design Concept and Further Development in the Second Round Scheme Grand Roof, Curving Outwards or Inwards

Our proposal for the National Grand Theater extends the historical and spatial continuity with Tian An Men Square. On this basis, a completely new architectural form is created to symbolize the contemporary China. Three major architectural elements which characterize the visual impression of Tian An Men Square are extracted:Plinth,Colonnade and Roof. By virtue of the advancing computer technology, the three elements are transformed into contemporary figures. The Grand Roof dynamically undulates in a three dimensional field. It becomes the third roof of Tian An Men Square,following the predecessors of the yellow roof of Tian An Men Rostrum and the layers of cornice lines of the Great Hall of the People. The waves of the yellow rooftiles of the Forbidden City repeat in the new silhouette of the National Grand Theater.This curved surface is the contemporary re-interpretation of large curved roofs of Chinese traditional architecture well represented by the monuments of the Forbidden City and Tian Tan Park. Structurally the curved roof is modeled and analyzed as a shell. And it is realized in cast-in-place reinforced concrete with steel plate form works. Shipbuilding technology is applied in the welding of the forms. The complex shape of the roof was not only made possible by the utilization of computing power,but a thorough persuasion for the structural rationality assures the construction technology.

In the design development at the second round scheme, this major achitectural concept is maintained. On this basis, aethetic modification and refinement related to the subtle form of the roof was further considered. As mentioned in the jury report for the first scheme design, the proposed silhouette of the grand roof was not able to acomplish a general consent in view of the specific context of the site.The reason is analyzed that the grand roof aroused the impression of the roof with a dome resulting from its convex surface covering the fly tower over the opera house. In other words, the surface was curving outwards. By contrast,the rafter of the traditional Chinese architecture is given a downwards curve to form a chord shape and therefore the surface of the entire roof creates a hollow profile. The dome has played a significant role as the symbol of urban unity in the traditional European cities. In China, the roof of the large-scale wooden buildings represented by the series of the monuments in the Forbidden City has played this role. The dynamically curving roof creates the ridge at the top. Also at the bottom, it makes possible the deep and horizontal projection of eaves. Often such eaves are multiply layered and create impressive profiles. In summary, the inwards-curved roof, the ridge, and the multi-layered eaves are the major charactaristics of Chinese architecture. Consequently such silhouette of the roof has determined the characteristics of the urban space in China.

北侧模型透视
VIEW FROM NORTH SIDE OF THE
MODEL

南侧模型透视
VIEW FROM SOUTH SIDE OF THE
MODEL

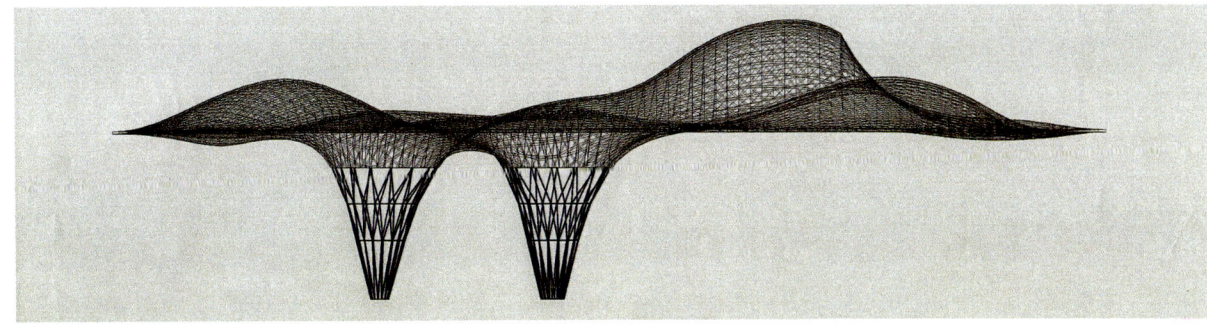

屋顶透视
ROOF PERSPECTIVE

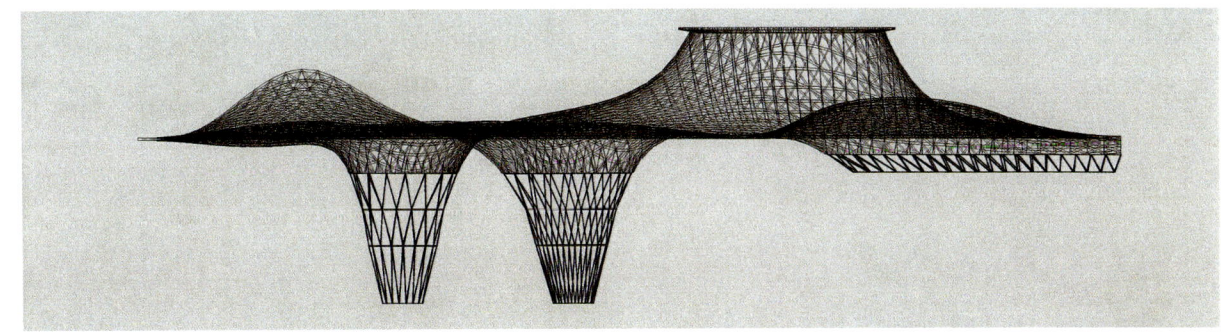

屋顶轮廓
ROOF OUTLINE

239

技术人员活动区
Technical Staff Zone

演员活动区
Performers' Zone

歌剧院、戏剧场后台入口
Opera House and Theater
Backstage Entry

小剧场—音乐厅后台入口
Mini-Theater and Concert Hall
Backstage Entry

地下二层平面
SECOND BASEMENT PLAN

贵宾用车道
VIP Car Access

贵宾上下车处
VIP Drop Off

贵宾专用电梯通往歌剧院
VIP Elevator for Opera House

贵宾专用通道
VIP Path

贵宾专用电梯通往戏剧场
小剧场、音乐厅
VIP Elevator to Theater,
Mini-Theater and Concert Hall

技术人员活动区
Technical Staff Zone

演员活动区
Performers' Zone

歌剧院货运通道
Opera House Loading

服务坪
Service Yard

戏剧场货运通道
Theater Loading

音乐厅演职活动区
Performers' Zone for Concert

一层平面
GROUND FLOOR PLAN

五层平面
5th FLOOR PLAN

歌剧院贵宾休息室
Opera House
VIP Lounge

中式快餐厅
Chinese Style
Snack Bar

表演艺术交流部
Department of Performance, Arts
Research and Exchange

带顶平台
Roof Terrace

带顶平台
Roof Terrace

戏剧场贵宾休息室
Theater
VIP Lounge

小剧场贵宾休息室
Mini-Theater
VIP Lounge

音乐厅贵宾休息室
Concert Hall
VIP Lounge

带顶平台
Roof Terrace

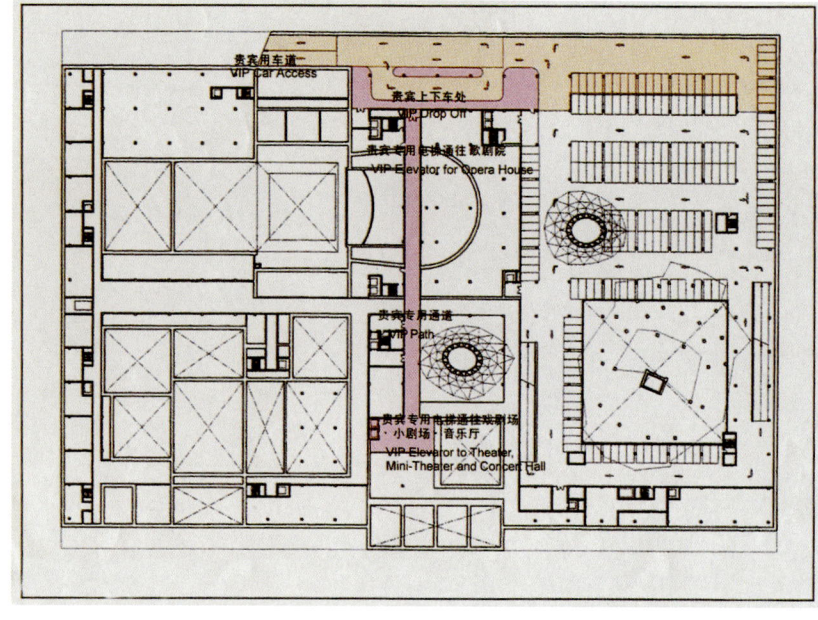

北立面
PERSPECTIVE OF NORTH ELEVATION

南立面
PERSPECTIVE OF SOUTH ELEVATION

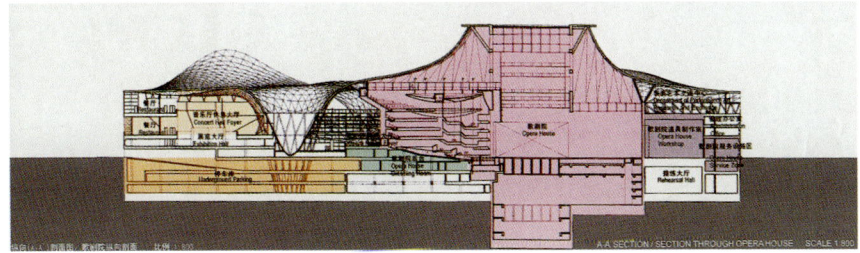

A-A 剖面
SECTION A-A

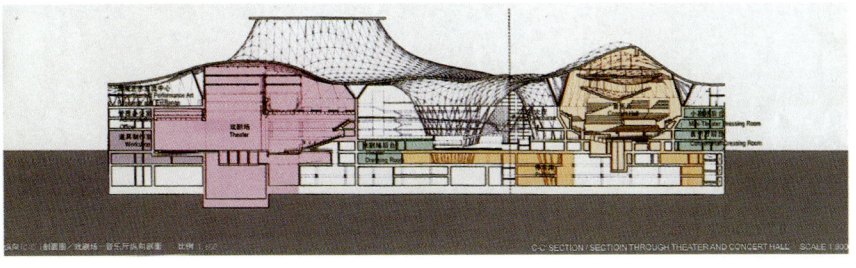

C-C 剖面
SECTION C-C

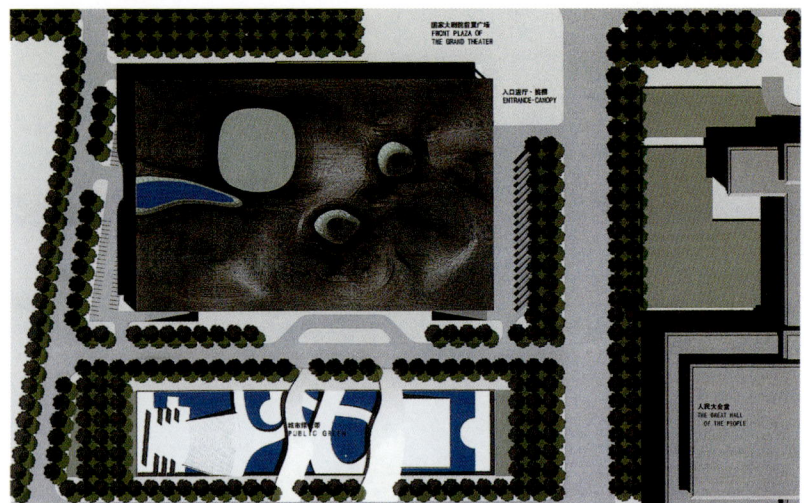

总平面
OVERALL SITE PLAN

戏剧场观众厅
PERSPECTIVE OF THE THEATER AUDITORIUM

音乐厅观众厅
PERSPECTIVE OF THE CONCERT HALL AUDITORIUM

技术经济指标
Technical Economic Index

基地面积(m²) Site Area	总建筑面积 (m²) Total Floor Area	建筑高度(m) Building Height	建筑覆盖率 （%） Building Coverage Ratio
38916	地上：69830 地下：57660	45	74.58

容积率 Plot Ratio	绿化覆盖率 (%) Green Coverage Ratio	停车数量 Number of Stalls	自行车停车数量 Number of Bicycle Stalls
3.3	19.4	地上：72 地下：496	1000

各观众厅数据
Data For Auditoria

		歌剧院 Opera House	音乐厅 Concert Hall	戏剧院 Theater	小戏场 Mini Theater
	观众厅面积(m²) Floor Area of Auditorium	1973	1930	1102	678.2
	观众厅体积(m³) Volume of Auditorium	20300	32800	8440	6782
座席数 Seating Capacity	总座席数 Seating Capacity(Seats)	2520	2050	1252	550
	其中：池座 Among Them:Auditorium	782	2050	804	500
	楼 座 Floor Seat	1738		448	50
	休息厅面积(m²) Lounge Floor Area	5208	3861	2593	1543
舞台尺寸 (m²) Stage Dimentions	主舞台 Main Stage	36 × 27	24 × 18	27 × 21	30 × 15
	左右侧台 Left and Right Side Stages	22 × 27	L:12.5 × 11 R:8.5 × 13	15 × 21	
	后舞台 Back Stage	26 × 25		18 × 18	
	台 仓 Under Stage Storage	36 × 27	24 × 12.5	27 × 21	30 × 15
	升降乐池 Elevating Orchestra Pit	23 × 6.5		21 × 10.5	
	最大视距(m) Max. Visual Distance	44	35	28	
	最大俯角（°) Max. Angle of Depression	27	22	21	

●国家——金色的琴弦意象讴歌中国源远流长的、灿烂的民族艺术。

●首都——波动的屋顶形象表达对作为中国古典都市与建筑杰作的北京的敬意与融合。

●天安门广场——作为配角的大剧院形象与广场建筑群风格上既有对话又有创新。

●纸窗、藻井——现代科技的玻璃幕墙传达传统的审美取向，洋溢着高雅的东方情调。

●戏剧意象——北侧柱廊象征徐徐拉开人幕的台口，向长安街展现新世纪的辉煌。

●红墙意向——天安门的红墙引入现代巨构之中，既表达对历史的尊重又是共和国旗帜之象征。

●竹的意象——中国文化赋予了竹以高贵、清白、高风亮节等含义，中国传统音乐又素以"丝竹"传世；竹荫大道于华贵中开辟出一片宁静的世外桃源。

●社火意象——腾空飞架的坡道象征龙灯飞舞，用以表现大剧院的市民性。

● State--The image of the golden string eulogizes the brilliance and long history of Chinese national arts.

● Captial--The undulating roof expressos its respect to and its harmony with Beijing--the Chinese classical capital and the masterpieces of architecture.

● Tian An Men Square--As the supporting role, the Grand Theatre has both coversation and creativeness in relation to the architecure group of Tian An Men Square.

● Paper Window, Coffer--The glass curtain wall shows traditional aethetic view with mordern high technology and permeates gracegful eastern style.

●Opera--The colonnade at the north side symbolizes an opening stage curtain, showing the splendour of the new era to Chang An Street.

●Red Wall--The red wall of Tian An Men Tower is introduced into this modern building to show respect to history and to signify the national flag.

●Bamboo--In Chinese culture, bamboo means gracs, pure and integrity. It is also an important element in traditionl Chinese music. The bamboo avenue at the center of the Grand Theatre creates a peaceful Shaugrila among the overall brilliance.

● Traditional Festival Performance--The slopway high above the courtyard is just like the flying dragon which represent the public spirit of the Grand Theatre.

北侧透视
PERSPECTIVE OF NORTH ELEVATION

南侧透视
PERSPECTIVE OF SOUTH ELEVATION

竹荫大道
BAMBOO AVENUE

歌剧院观众厅
PERSPECTIVE OF THE
OPERA HOUSE
AUDITORIUM

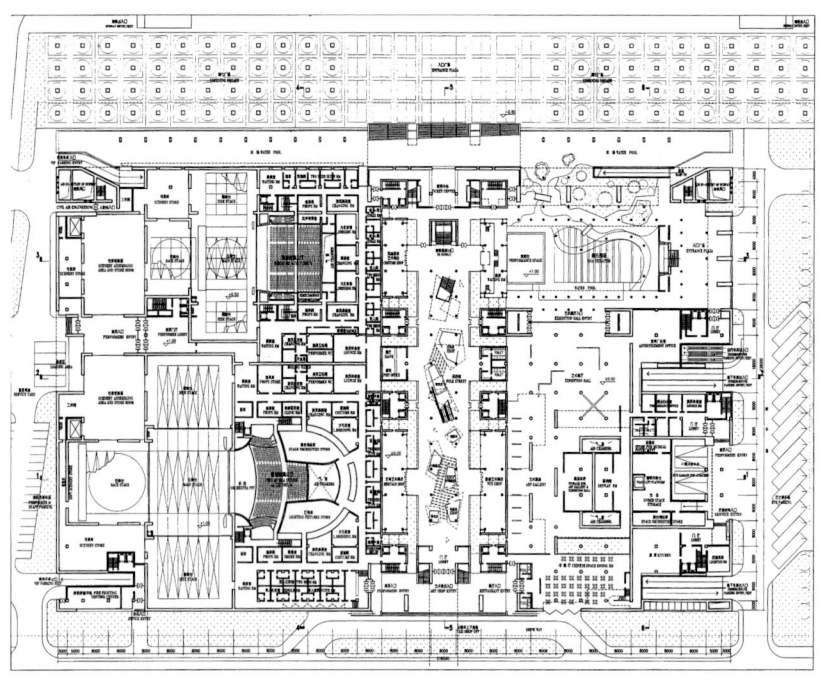

一层平面
GROUND FLOOR PLAN

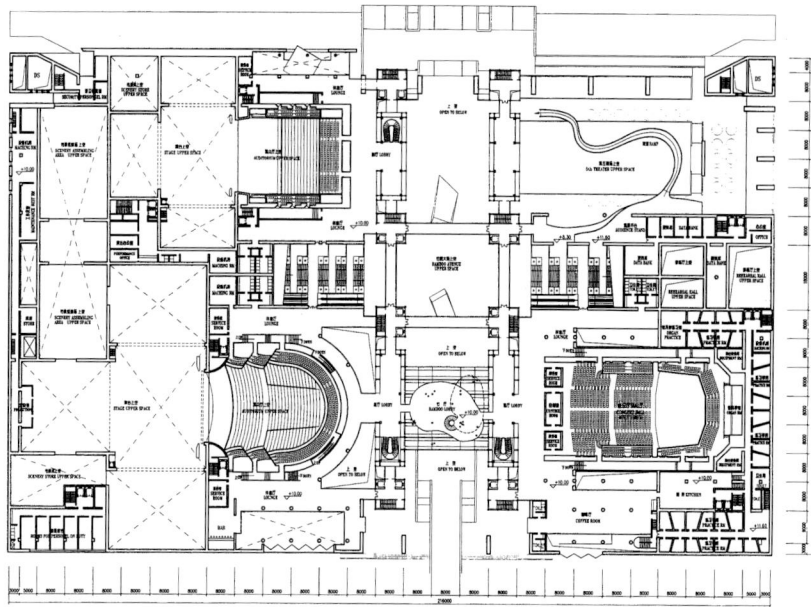

三层平面
3rd FLOOR PLAN

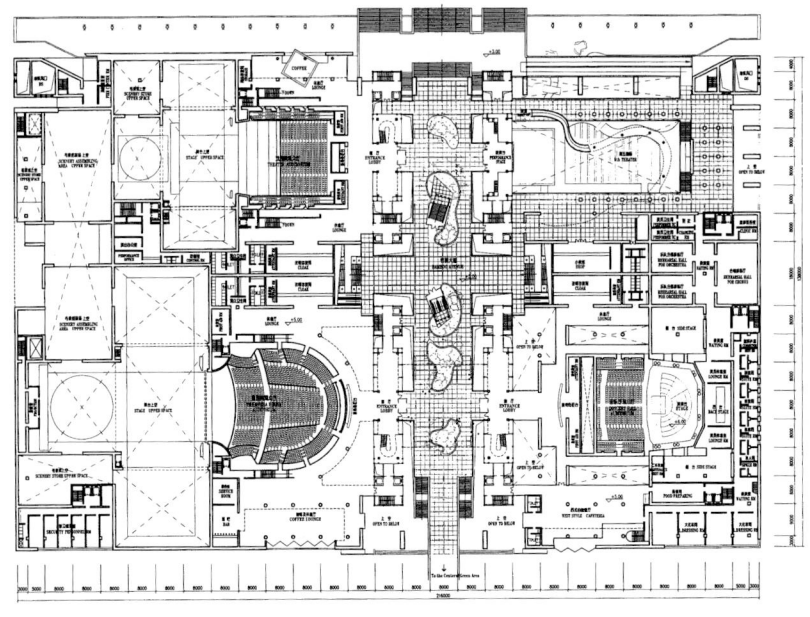

二层平面
2nd FLOOR PLAN

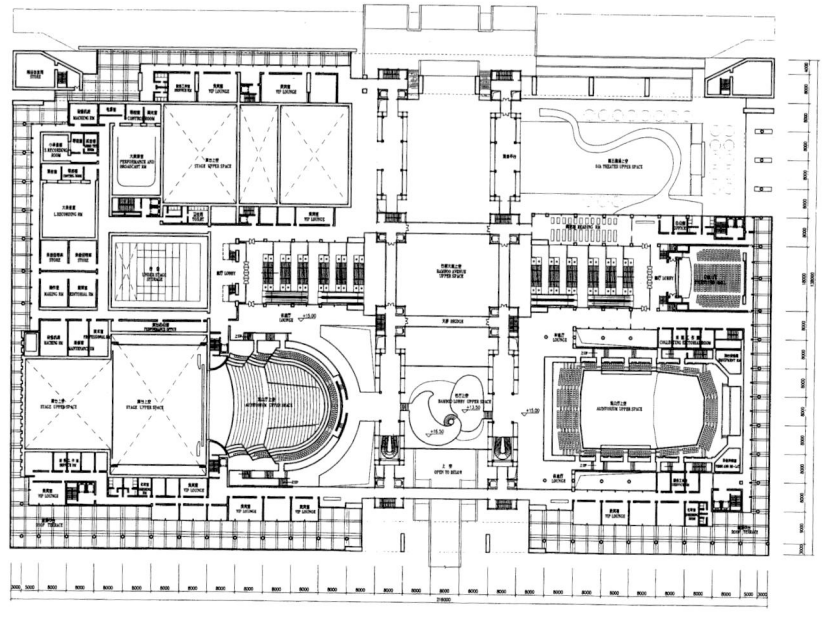

四层平面
4th FLOOR PLAN

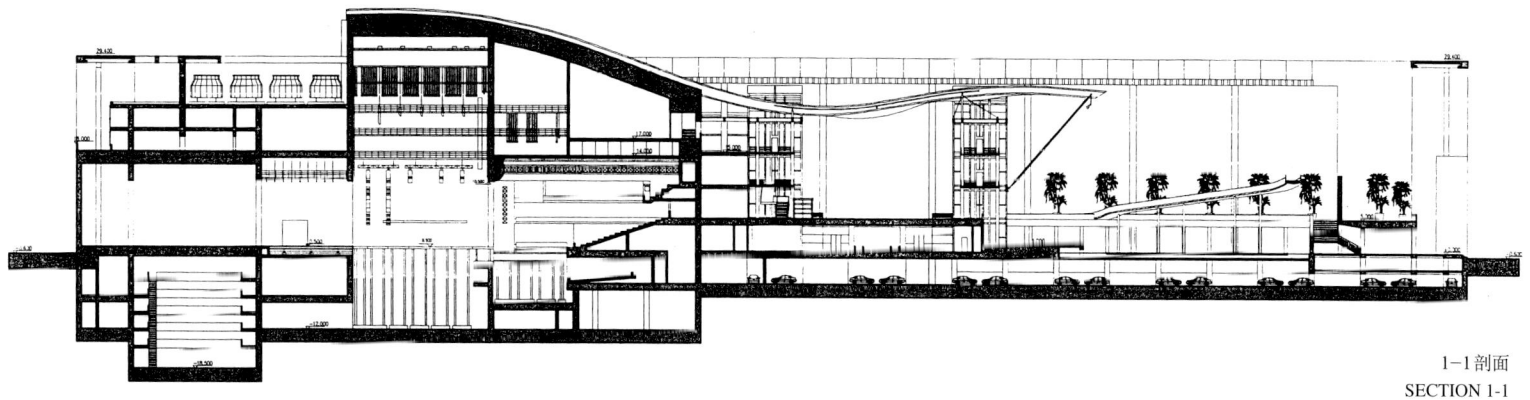

1—1剖面
SECTION 1-1

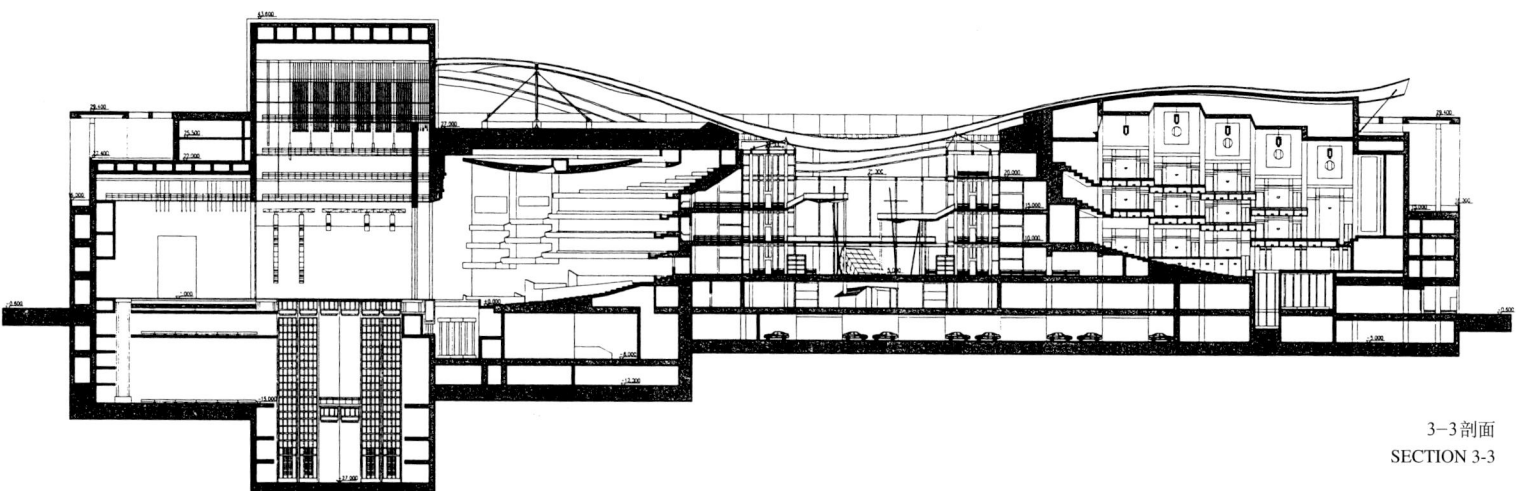

3—3剖面
SECTION 3-3

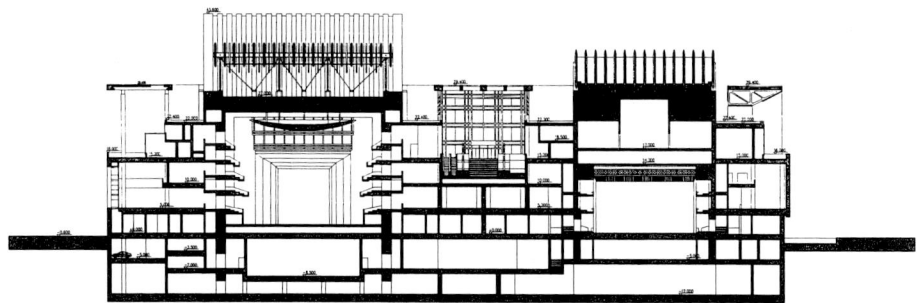

4—4剖面
SECTION4-4

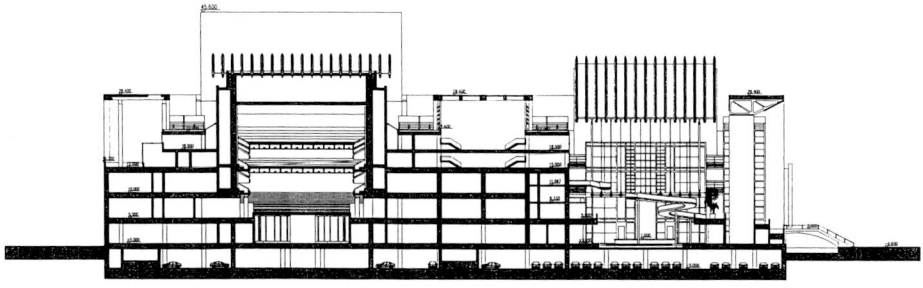

6—6剖面
SECTION6-6

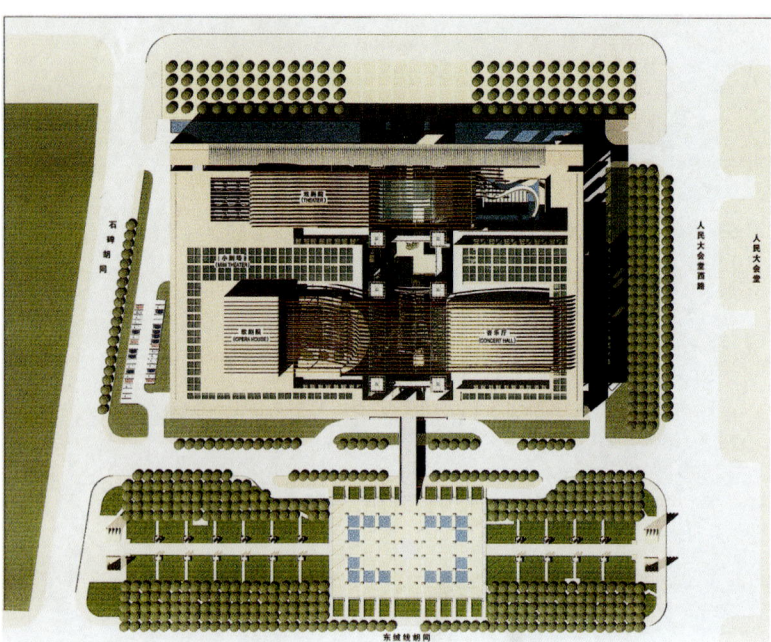

总平面
OVERALL SITE PLAN

模型透视
MODEL PERSPECTIVE

技术经济指标
Technical Economic Index

基地面积(m²) Site Area	总建筑面积 (m²) Total Floor Area	建筑高度(m) Building Height	建筑覆盖率（%） Building Coverage Ratio
38900	地上：75675 地下：40822	30/45(局部)	71
容积率 Plot Ratio	绿化覆盖率 (%) Green Coverage Ratio	停车数量 Number of Stalls	自行车停车数量 Number of Bicycle Stalls
1.95	12.8	地上：85 地下：513	1260

各观众厅数据
Data For Auditoria

		歌剧院 Opera House	音乐厅 Concert Hall	戏剧院 Theater	小戏场 Mini Theater
	观众厅面积(m²) Floor Area of Auditorium	1985	1980	1102	712
	观众厅体积(m³) Volume of Auditorium	1150	1070	640	544
座席数 Seating Capacity	总座席数 Seating Capacity(Seats)	2540	1992	1236	380~572
	其中：池座 Among Them:Auditorium	1222	926	514	260~452
	楼座 Floor Seat	1318	1066	556	120
	休息厅面积(m²) Lounge Floor Area	5100	4850	3240	1450
舞台尺寸(m²) Stage Dimentions	主舞台 Main Stage	34 × 27 × 39	22 × 16 × 21	27 × 21 × 29	64~192
	左右侧台 Left and Right Side Stages	23 × 27 × 12	550	15 × 21 × 12 10 × 21 × 12	
	后舞台 Back Stage	26 × 25 × 18		23 × 15 × 16	
	台仓 Under Stage Storage	34 × 27 × 16 26 × 25 × 16	12 × 17 × 6	27 × 21 × 12	24 × 16 × 4
	升降乐池 Elevating Orchestra Pit	20 × 7.6		20 × 6	
	最大视距(m) Max. Visual Distance	43.2	31	29	
	最大俯角（°） Max. Angle of Depression	25	25	25	

对于设计构想和创意的深化思维

1.中国国家大剧院建筑综合体应该融于其所处的城市大环境、历史渊源和文化氛围之中,同时又必须具有自己的独特风貌,并反映出时代精神。

2.国家大剧院建筑将北京拥有上千年历史和文化传统的基本城市结构体系沿袭演化为具有现代特征的建筑语汇,是本设计的一个十分重要的基本原则,同时亦在城市的发展和建设中具有广泛而深刻的意义。

3.北京典型的建筑传统——自然、绿化和所建成的人工环境间的紧密联系——应该在城市规划和建筑设计的创作中得到更多的重视。

4.国家大剧院建筑综合体的基本体系——"房中房"应更为明确地体现在其外部造型之中,使之真正成为在一个大屋顶下的小型文化城。她面向市民,并对所有来访者展开欢迎之臂,她将融于北京的日常生活之中,即便在单个功能区处在非使用状态的情况下,仍然由于其开放性,以及设置其内的多种多样的商业服务设施而充满生机。

5.该建筑综合体向公众开放,并对公众的日常生活提供多种多样的服务及活动空间,不仅对于其能否获得人们的接受是十分重要的,同时还能大大提高其经济效益。在本方案的设计中,建筑师力图在文化生活和商业效益之间找到一个完美的契合点。

6.在深入设计阶段中,建筑师在保持本方案在竞赛第一阶段中简洁严谨的建筑形体的基础上,力图使其在造型、功能和色彩上更为明快开放。这与评委们在竞赛第一阶段所提出的建议和要求是完全一致的。

7.中国国家大剧院的设计,应是中国悠久历史和灿烂文化的沿袭光大,现代科学技术的集中表现,人类即将进入第三千纪和通向未来的灯塔,因而必须在设计中避免自我表现,或追逐时髦。设计任务的严肃性要求建筑自身具有超越时代的风格和特征。

8.中国国家大剧院所处之地,拥有上千年灿烂的文化传统,今后这里将是世界各民族优秀文化相互交流之地,因此,其建筑风格既要能包容这丰富多彩的大千世界于一体,又扎根于生之于斯的土壤之中。

9.在当今瞬息万变的世界上,人们对日益增长的混乱已习以为常,在本方案的设计中,建筑师力图创造一个与之相对应的场所,在这里人们能获得宁静、思索和均衡。

Additional Throughts of Task

1.The building should be incorporated in the future town planning aspects as well as in the already existing cultural surroundings.

2.The basics of all further developements of the metropolis Bejing should be well organized and planned to harmonize the historically grown-up, cleary structured layout of the old town quarters and the new town planning elements.

3.The changes between free spaces and built-on areas are to be a characteristic of the city of Beijing, and have to be more considered and adhered to in the future.

4.The simple and clear internal features of architecture in the National Grand Theatre with the "house in house" concept has to be more transferred to the external building appearance as well.

Underneath of a common light, transparent roof structure someone will find a fully functional little township making reference to cultural happenings, open for the public but as well outlooking to the pulsating atmosphere of the nearby Beijing city.

5.In terms of use and function the building has to be more open, more accountable for visitors and occupants, day by day. This target should be achieved by offering a large deal of service,shopping-and relaxing components which make the whole complex more acceptable and therefore will create a desirable economic status. For sure this a small grade between culture and commerce, not easy to establish but most essential for this undertaking.

6.As suggested by the Jury,the architecture of the National Grand Theatre has to be open and cheerful without softening the clearness of form and structure.

These suggestions are fully right and now are adopted in this revision to the Competition Design Concept with all regards to form, function and colour schemes.

7.The new building should bridge the grown-up national traditions via the spirit of age up to the needs of the next millennium. This requirement does not allow any kind of self-satisfaction features and modernistic gags. The seriousness of this big task requires a long standing consideration.

8.At this place, where the Chinese cultural tradition was grown-up over general millenniums and now will meet with the various traditions and styles of art celebrations of other nations,a very special location for the "coming together" has to be created.

This place has to be free of any reservations, full of interest but at the same time someone should not fail to recognize the roots of the beginning.

9.Growing chaos, which partly receives components of esthetic consideration all over the world, is nowadays ruling our live. Therefore a counterweight of silence, reflection and harmony will be established at this location.

北侧透视
PERSPECTIVE OF NORTH ELEVATION

南侧透视
PERSPECTIVE OF SOUTH ELEVATION

东侧透视
PERSPECTIVE OF EAST ELEVATION

西侧透视
PERSPECTIVE OF WEST ELEVATION

中央通道
CENTRAL THOROUGHFARE

入口局部
ENTRANCE SECTION

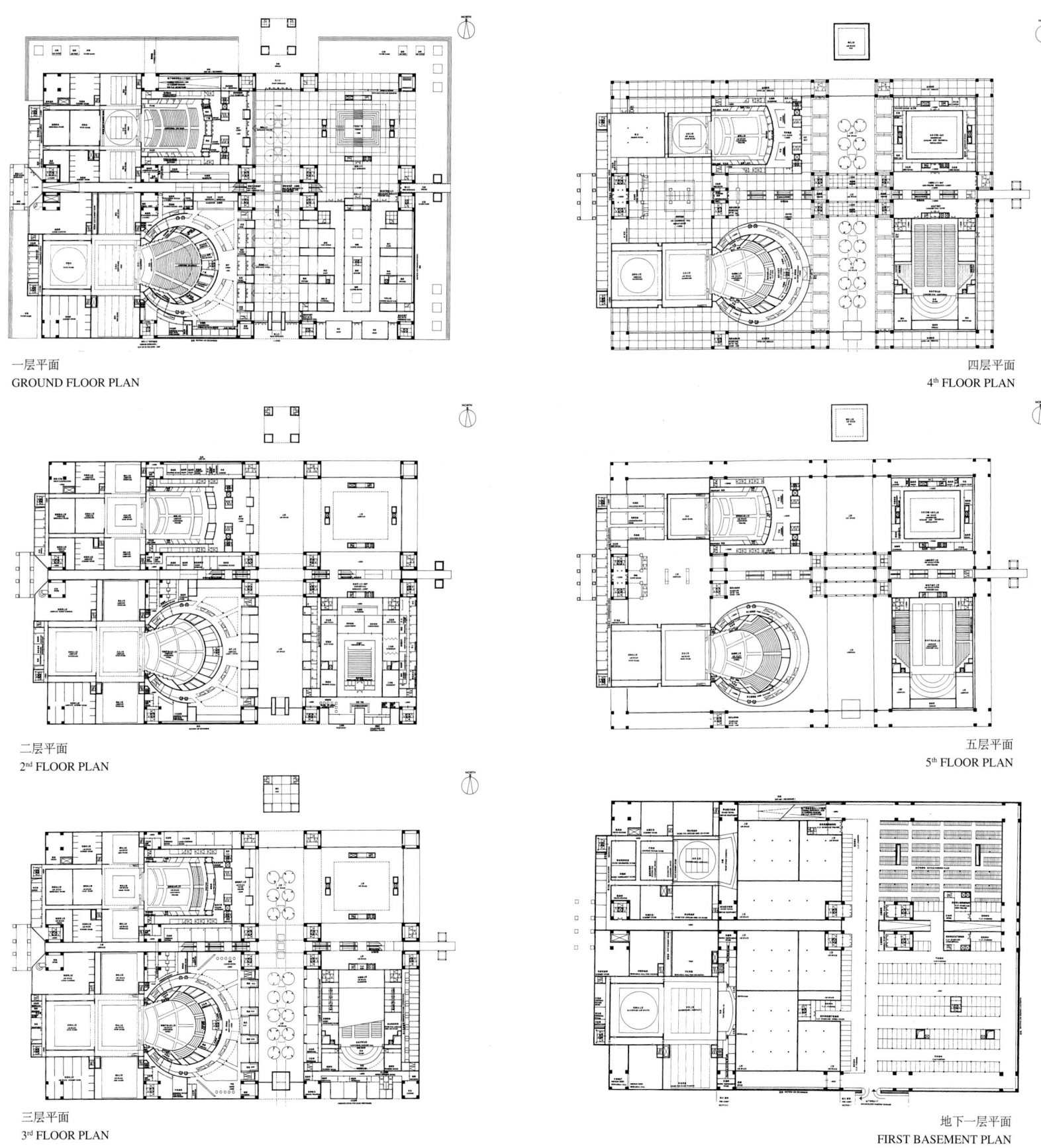

一层平面
GROUND FLOOR PLAN

四层平面
4th FLOOR PLAN

二层平面
2nd FLOOR PLAN

五层平面
5th FLOOR PLAN

三层平面
3rd FLOOR PLAN

地下一层平面
FIRST BASEMENT PLAN

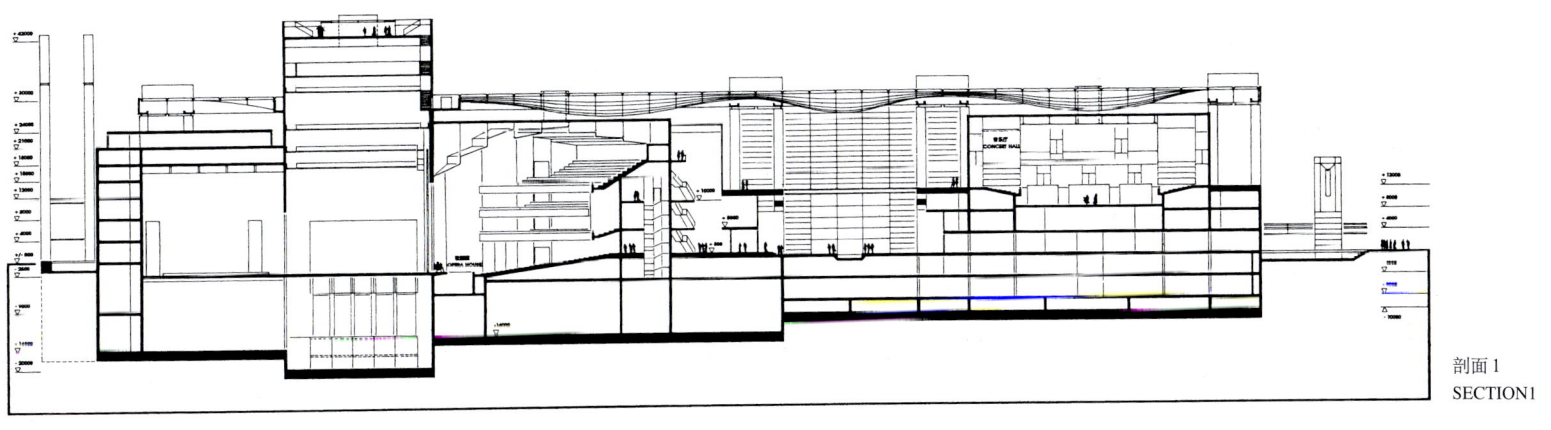

剖面 1
SECTION1

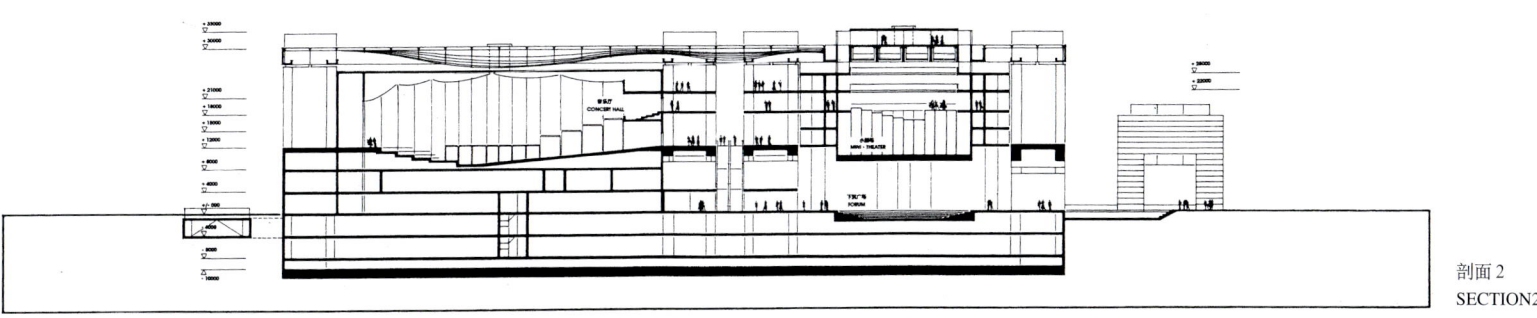

剖面 2
SECTION2

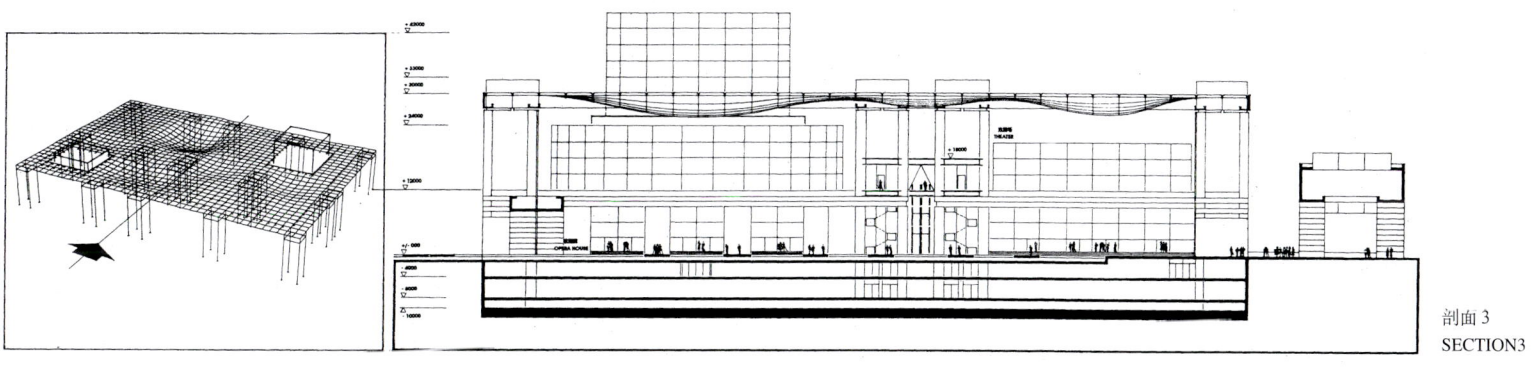

剖面 3
SECTION3

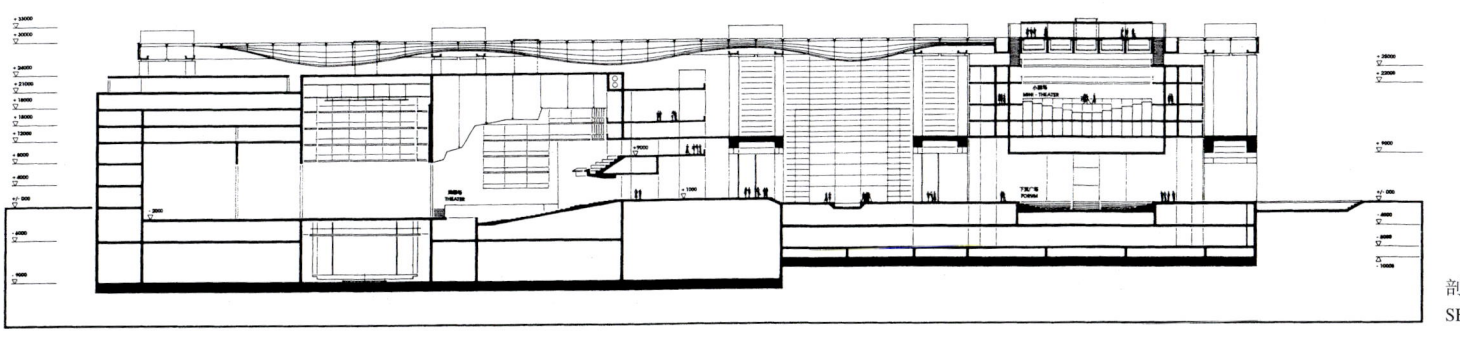

剖面 4
SECTION4

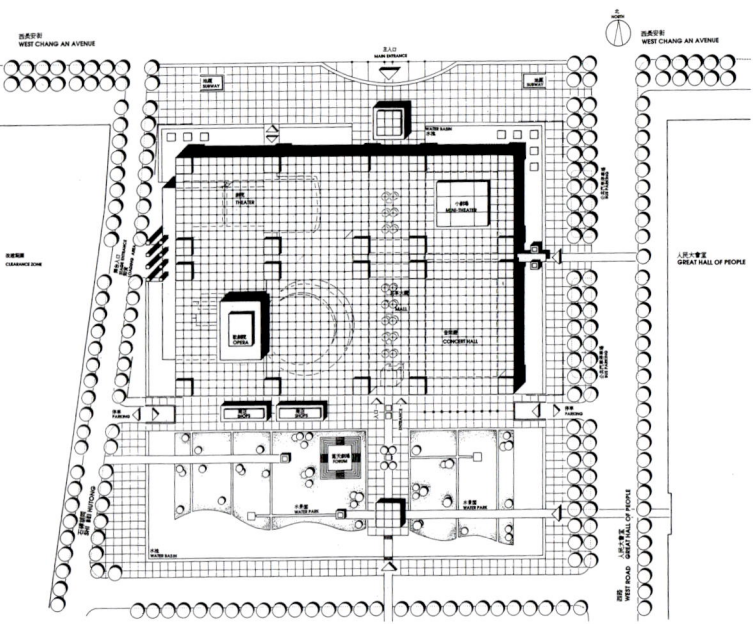

总平面
OVERALL SITE PLAN

模型透视
MODEL PERSPECTIVE

技术经济指标
Technical Economic Index

基地面积(m²) Site Area	总建筑面积 (m²) Total Floor Area	建筑高度(m) Building Height	建筑覆盖率 (%) Building Coverage Ratio
43082	地上：92876 地下：54725	30/42(局部)	73

容积率 Plot Ratio	绿化覆盖率 (%) Green Coverage Ratio	停车数量 Number of Stalls	自行车停车数量 Number of Bicycle Stalls
3.43	31	地上：102 地下：559	902

各观众厅数据
Data For Auditoria

		歌剧院 Opera House	音乐厅 Concert Hall	戏剧院 Theater	小戏场 Mini Theater
	观众厅面积(m²) Floor Area of Auditorium				
	观众厅体积(m³) Volume of Auditorium	957	1.170	875	750
座席数 Seating Capacity	总座席数 Seating Capacity(Seats)	20.000	23.000	11.500	7.500
	其中：池座 Among Them:Auditorium	2514	2262	1246	300~500
	楼 座 Floor Seat	1111	1628	874	300~500
	休息厅面积(m²) Lounge Floor Area	1403	634	372	60
舞台尺寸(m²) Stage Dimentions	主舞台 Main Stage	3000	1600	2320	900
	左右侧台 Left and Right Side Stages	891	325	598	750
	后舞台 Back Stage	2 × 564	2 × 148	2 × 379	
	台 仓 Under Stage Storage	650		310	
	升降乐池 Elevating Orchestra Pit	2.105	160	1.356	750
	最大视距(m) Max. Visual Distance	160	160	160	
	最大俯角（°） Max. Angle of Depression	32	38	32	根据舞台情况定

国家大剧院地处天安门广场人民大会堂西侧,用地南北约166m,东西约224m～245m,用地北临长安街,南侧为60m×250m的城市集中绿地。根据规划设计要求,本方案布局考虑的基本原则是:

(1)大剧院的主要入口应朝长安街,因为这是观众来剧院的主要方向,正入门偏向天安门一侧,没有采用中轴对称,以表示本建筑物对天安门广场的依盼和呼应,也避免了与人大会堂并列、重复。

(2)充分考虑本建筑对南侧集中绿地的连续性,入口向南通往绿地,用半室外共享空间贯穿南北,把长安街和南边绿地沟通,共享空间为三个观众厅及前厅围合而成,充分体现"屋围绕院","院勾通屋"的中国建筑空间构成的传统思想。同时,这一丰富的半室外空间也提供了市民交往、活动的场所,成为一个名副其实的市民广场,弥补了北面缺少广场的不足。

共享空间东北角留一豁口,对景将是天安门之英姿。

共享空间和歌剧、话剧院前厅以上之屋顶,将沿一圆弧线坡向绿地,又形成外形对绿地的感应。同时在北侧经过屋脊转折,又向北面开口处做成坡顶,这样就使人联想到中国大屋顶,也就是用现代手法体现出传统的神韵。

(3)建筑形式及风格

本建筑在人大会堂一侧,因此,应与天安门广场建筑群协调是不言而喻的,但是协调不是简单重复,雷同只能削弱主体。本建筑东侧采用柱廊取得与人大会堂立面之对应,而总体趋向完整,用总体之完整,外表简洁取得广场周围公建的一致和共性,与其不争不挤,相得益彰,相映成趣,烘托天安门广场建筑的庄严、肃穆感,同时也用丰富、自然的收敛、内向,锋芒含而慎露的"内秀"完成国家级演出建筑的个性,室内丰富的空间构图,使人充分领略到建筑艺术与音乐、戏剧之交相辉映,将体现出浓郁的文化性和跨世纪的时代精神,使之成为首都具有标志性的现代化建筑。

本方案平面布局为以贯通南北的半室外共享空间为交通中心,西侧为歌剧院和戏剧场,东侧为椭圆形音乐厅,总之,使三个剧场有分有合,便于管理,而小剧场设在西部五层。半室外共享空间将成为从城市进入各剧场的过渡空间,也将成为供城市居民享受的文化广场和缓冲。将充分体现最高艺术殿堂的标志性、文化性和群众性。

The National Grand Theatre is situated to the west of the Great Hall of the People at Tian'anmen Square.The project site is 166m wide from north to south and 224-255m long from east to west with Chang'an Avenue on its north and a 60m×250m central green area of the city on its south. In conformity with the planning design requirements the schematic design is made with the following principles taken into consideration:

(1)The main entrance of the Grand Theatre is to face Chang'an Avenue to welcome visitors coming mainly from this direction and it is not positioned in the centre but to the east toward Tian'anmen Tower. The Project is designed as a subordinate part of tian'an men Square. The National Theatre will relate to the Square and no parallelism or competition with the Great Hall of the People will appear by the construction of the Theatre.

(2)To facilitate accessibility to the central green area on its south, the south entrance faces the green area and the semi-outdoor promenade runs through the site connecting Chang'an Avenue on the north and green area on the south. The promenade is enclosed with the three auditoriums and entrance lobbies, expressing "court enclosed with houses" and "houses linked with courts",which is the traditional concept of Chinese architectural space composition. This magnificent semi-outdoor space will provide a place for citizens for interactions and events,forming a new real people's square and making up the inadequacy of square areas on the north. On the north east corner of the semi-outdoor space there is an opening toward Tian'anmen Tower, that will be a contrast scene. The roofs of the common space and the entrance lobbies of the Opera House and the Theatre incline toward the green area along a curved line to relate the green. On the north side of the ridge the roofs incline toward the north opening of the common apace. All this would make people think of Chinese traditional roofs.Modern methods are used here to show the spirit and rhyme of traditional architecture.

(3)Building form and style

The project is to the west of the Great Hall of the People. The building form and style should coordinate with buildings around Tian'anmen Square. Repetition or duplication should be avoided. Colonnade is designed on the east side of the Project to relate to the Great Hall of the People. Consistency and identity are achieved by means of simplicity. The new building of the National Theatre will set off the solemnity and majesty of the buildings around the Square by contrast.The rich internal spaces and exteriors show architectural characters of a national centre of performing arts.Architecture art blends with music and performing arts.The National Grand Theatre will have strong cultural characters and express the spirit of times at the threshold of the new century. It will become a symbolic modern building of the capital.

In the design the semi-outdoor space running through the site from north to south is treated as the circulation centre of the layout plan with the Opera House and the Theatre on the west side and the elliptical Concert Hall on east and the Mini-theatre on the 4th level of the west. The three buildings are separate but united as a whole for ease of management. The semi-outdoor common space will form a transit area for people coming from different directions to enter the buildings and become an enjoyment venue for citizens to have cultural events. The national performing arts centre will be a popular cultural centre and a land mark as well.

北侧透视
PERSPECTIVE OF NORTH ELEVATION

南侧透视
PERSPECTIVE OF SOUTH ELEVATION

公共大厅
PUBLIC HALL

入口局部
ENTRANCE SECTION

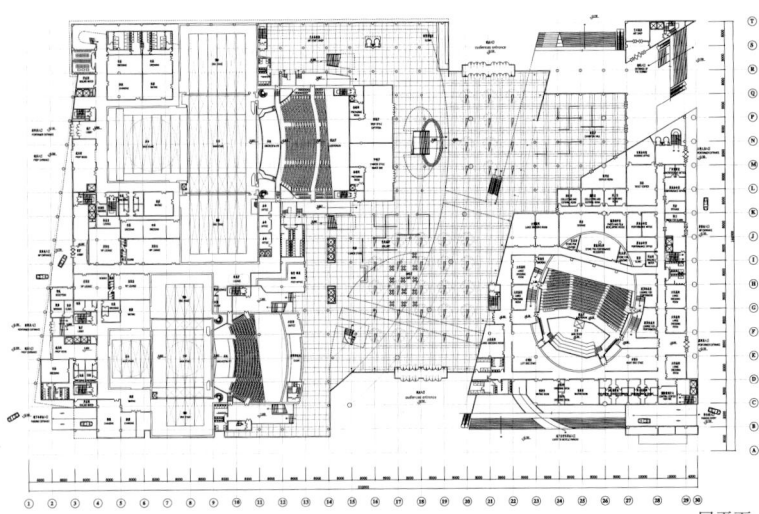

一层平面
GROUND FLOOR PLAN

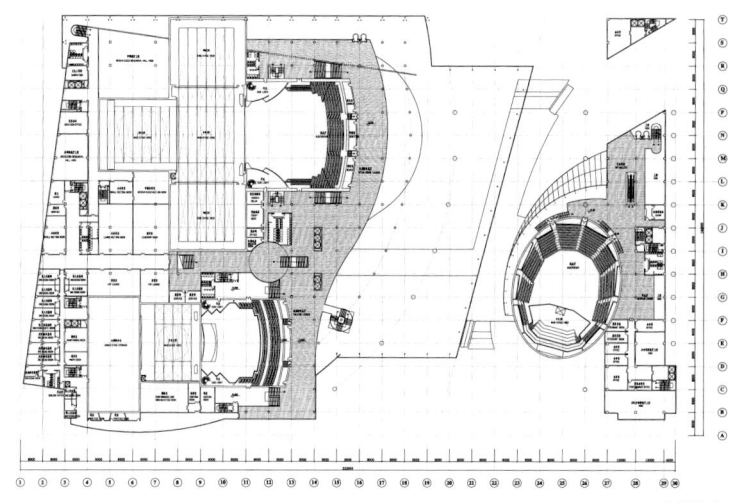

四层平面
4th FLOOR PLAN

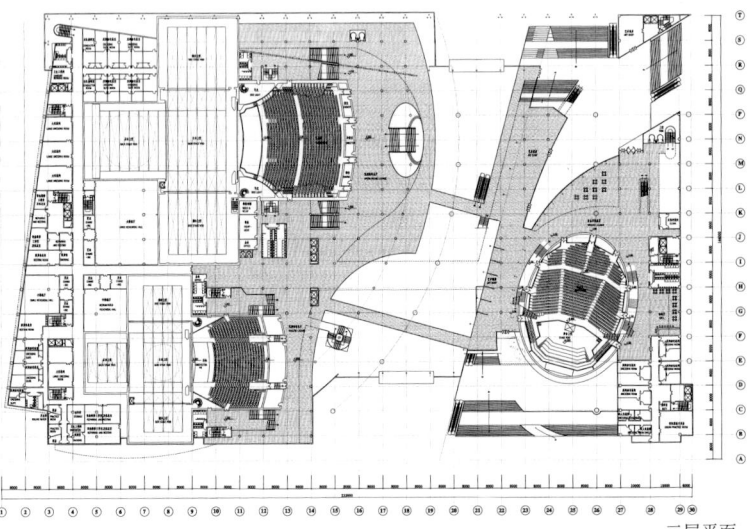

二层平面
2nd FLOOR PLAN

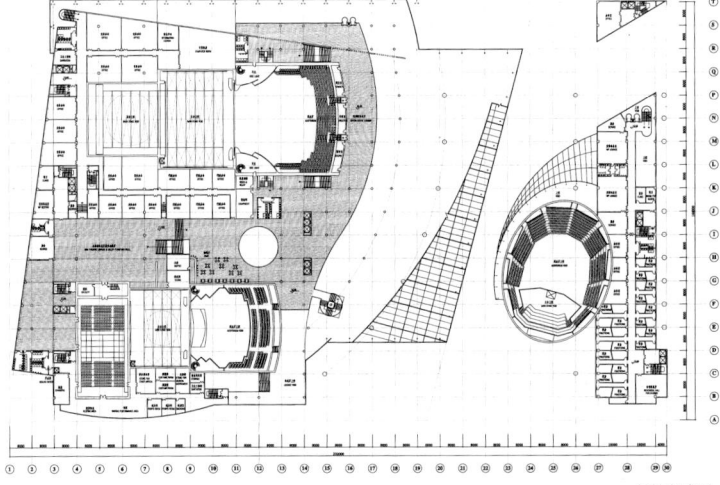

五层平面
5th FLOOR PLAN

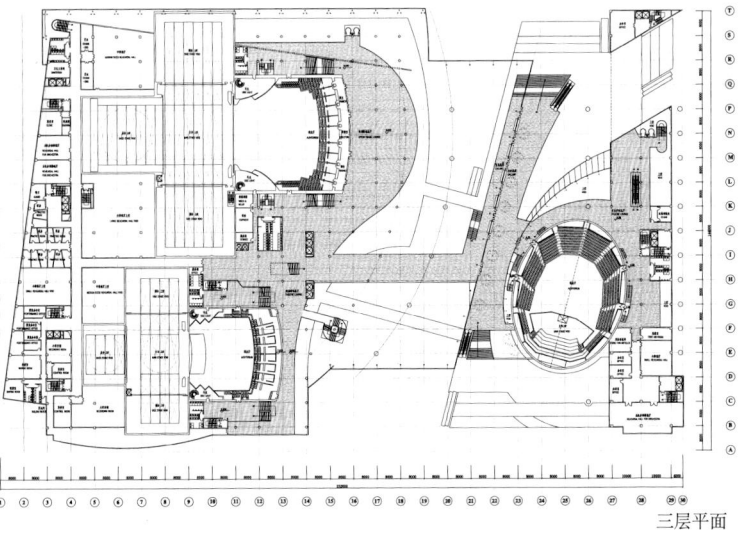

三层平面
3rd FLOOR PLAN

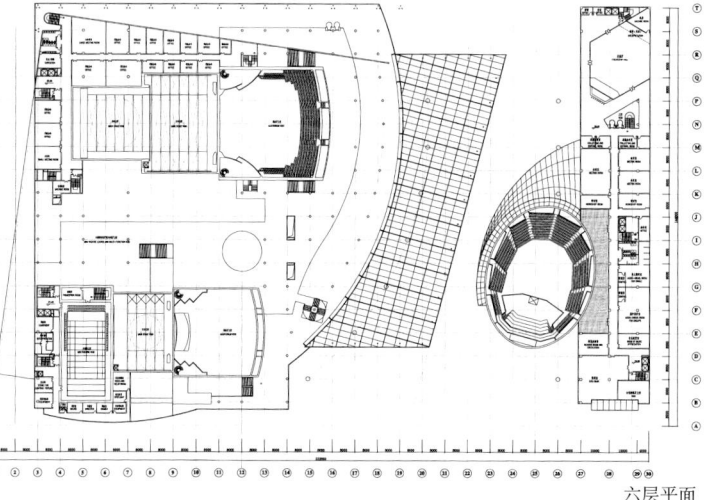

六层平面
6th FLOOR PLAN

环境草图
ENVIRONMENT SKETCH

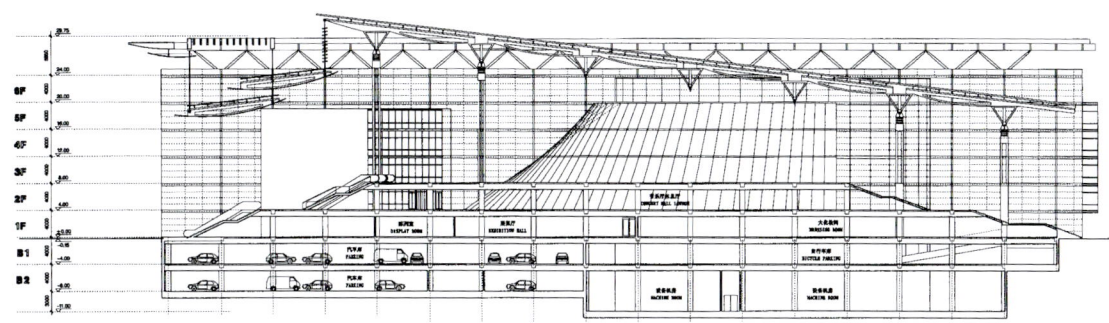

A—A 剖面
SECTIONA-A

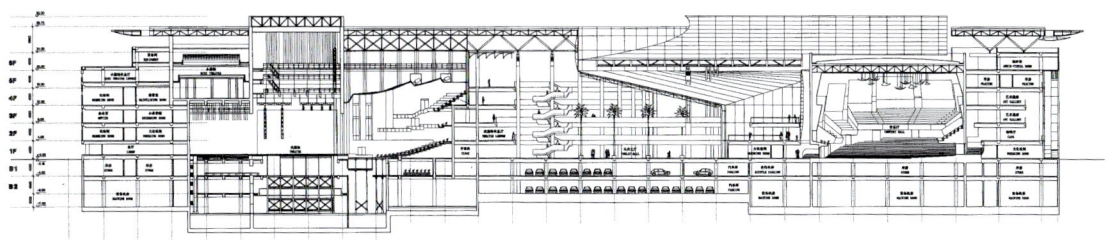

B—B 剖面
SECTIONB-B

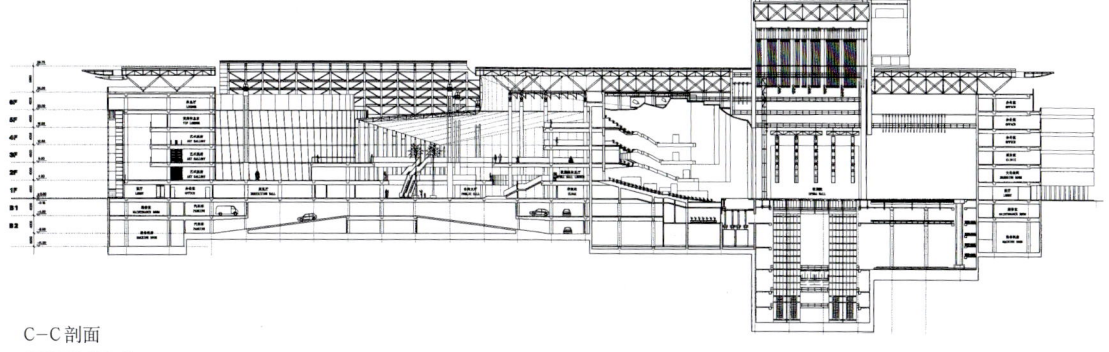

C—C 剖面
SECTIONC-C

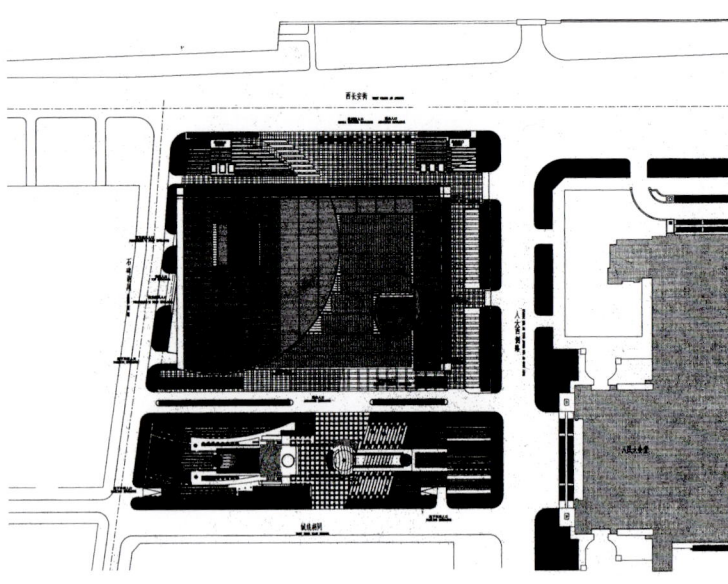

总平面
OVERALL SITE PLAN

技术经济指标
Technical Economic Index

基地面积(m²) Site Area	总建筑面积 (m²) Total Floor Area	建筑高度(m) Building Height	建筑覆盖率 （%） Building Coverage Ratio
38900	地上：752000 地下：54600	30	73.5
容积率 Plot Ratio	绿化覆盖率 (%) Green Coverage Ratio	停车数量 Number of Stalls	自行车停车数量 Number of Bicycle Stalls
1.94	3.0	地上：64 地下：502	1000

模型透视
MODEL PERSPECTIVE

各观众厅数据
Data For Auditoria

		歌剧院 Opera House	音乐厅 Concert Hall	戏剧院 Theater	小戏场 Mini Theater
	观众厅面积(m²) Floor Area of Auditorium	2175	1242	730	591
	观众厅体积(m³) Volume of Auditorium	20900	19860	9485	4435
座席数 Seating Capacity	总座席数 Seating Capacity(Seats)	2515	2027	1376	476
	其中：池座 Among Them:Auditorium	1372	902	794	396
	楼 座 Floor Seat	1143	1125	582	80
	休息厅面积(m²) Lounge Floor Area	9501	4291	3354	1303
舞台尺寸 Stage Dimentions (m²)	主舞台 Main Stage	34 × 27	23.8 × 16.9	27 × 22	15 × 8
	左右侧台 Left and Right Side Stages	23 × 27		22 × 14.6	
	后舞台 Back Stage	25 × 25		19.5 × 14.5	
	台 仓 Under Stage Storage	34 × 27 × 28		27×22×12.6	34.6×17.6×4
	升降乐池 Elevating Orchestra Pit	6.5			
	最大视距(m) Max. Visual Distance	35.45	34.2	29.2	21.5
	最大俯角（°） Max. Angle of Depression	29.11	19	25	13~23

07
深圳大学建筑设计
研究院(中国)
Shenzhen University,
The College of
Architecture & Civil
Engineering(China)

"凤鸣朝阳"——国家大剧院方案设计构思主题

天安门广场是首都政治、文化中心的象征,天安门本身是该广场最重要的建筑,它在天安门广场规划中具有唯一的、真正的中心地位。因此,国家大剧院总体布局中使大剧院与天安门"对话"并尊重天安门的中心地位是本方案构思时最重要的出发点。

作为国家最高艺术殿堂,国家大剧院在建筑艺术上应当既有国家的象征也有文化艺术的象征,隐喻(metaphor)是古今中外常见的建筑艺术表现手段,在本方案设计上,采用"凤凰"作为国家大剧院的隐喻和象征。"凤凰"和"龙"一样,是家喻户晓的国家最高等级的象征,也是中华民族吉祥的象征。早在2000多年前,中国最早的一部诗集中,就有"凤鸣朝阳"的描述。《诗经·大雅·卷阿》:"凤凰鸣矣,于彼高岗;梧桐生矣,于彼朝阳"。诗意是"凤凰在太阳初升时的鸣叫,比喻稀有的吉兆"。在本方案中,也象征着中华民族下世纪的腾飞。

凤凰是古代传说中的鸟王。雄的叫"凤",雌的叫"凰",通称为凤或"凤凰"。凤凰不仅有国家级的象征,也有文化艺术的象征。"凤凰来仪"、"凤歌鸾舞"、"百鸟朝凤"等成语和乐曲名说明凤凰也是音乐舞蹈等表演艺术上的象征。

如果国家大剧院需要象征或隐喻的话,无论是从中国的普通大众还是从历史文献记载看,以凤凰来象征国家最高级的艺术殿堂最为恰当。本方案构思正是从这一点出发构思,将国家大剧院建筑屋盖和南北立面抽象地表现为"凤凰",而大厅的主轴线面向东方,偏向天安门(见总体布局图),以此体现"凤鸣朝阳"的主题,并以此向天安门广场的中心建筑——天安门致意。

The Singing Phoenix Greets the Rising Sun-Principal Design Concept for the National Grand Theater

Tian'anmen Square symbolizes the capital city's status as a political and cultural center. Tian'anmen Gate is the most significant landmark on the Square;it has long been the true and only center in the planning of Tian'anmen Square.Hence,to dialogue with and honor the central position of Tian'anmen Gate became the most important point of departure during the concptual design phase.

As the most prominent performance arts center in the country,the architectural design of the Grand Theater is to symbolize the nation as well as art and culture;The image of phoenix is used here as a metaphor and symbol. Like the dragon, the phoenix has been one of the symbols of the Chinese nation;it is also a symbol for the blessed prosperity of Chinese people. The image of the singing phoenix greeting the rising sun first appeared two thousand ycars ago in the Book of Odes-the first Chinesc poetry anthology. In The Crawling Height from Major Odes in the second part of the anthology, the anonymous poet describes the beauty of the phoenix's landscape:the sky at dawn,a mountain of the wutong trees (phoenix trees),and the phonenix's triumphant voice raised in song to greet the rising sun. This is believed to be a rare sign of good fortune. The metaphor is adopted here in the design to symbolize the rising of Chinese people in the coming century.

According to ancient Chinese Legends, Feng Huang(female and male phoenixes) are the king and queen of all birds.Feng is the female phoenix and Huang is the male one. Phonenixes are generally referred to as Feng of Feng Huang. They are not only symbols of the nation,but also symbols of art and culture. This is illustrated by the many idioms and titles of traditional dances, songs and music pieces, which constantly refer to the phoenixes.

It is highly appropriate to use the Phoenix to symbolize the most prominent performance arts center in China, whether from a general public point of view or from a cultural,historical point of view. Hence, the shape of the roof and the south and north facades in our design reflect the abstract images of phoenix.The theme of "the singing phoenix greets the rising sun" is embodied in the orientation of the central atrium,which is directed diagonally toward Tian'anmen Gate (see the General Layout Diagram)and pays respect to this central landmark in the square.

凤鸣朝阳
THE SINGING
PHOENIX GREETS
THE RISING SUN

北侧透视
PERSPECTIVE OF NORTH ELEVATION

南侧透视
PERSPECTIVE OF SOUTH ELEVATION

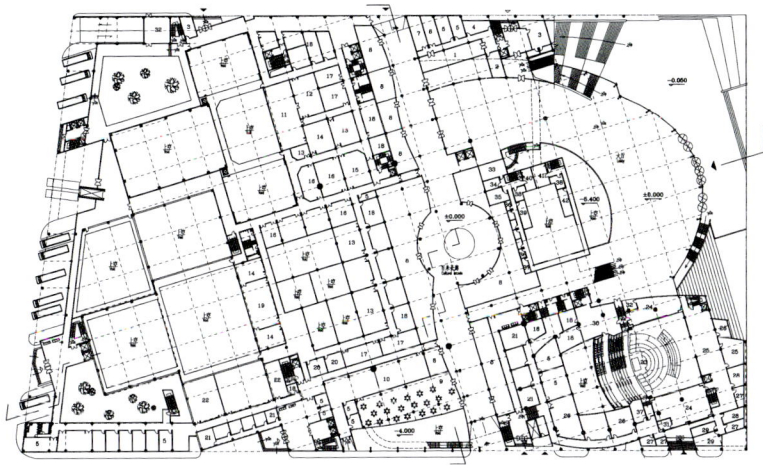

一层平面
GROUND FLOOR PLAN

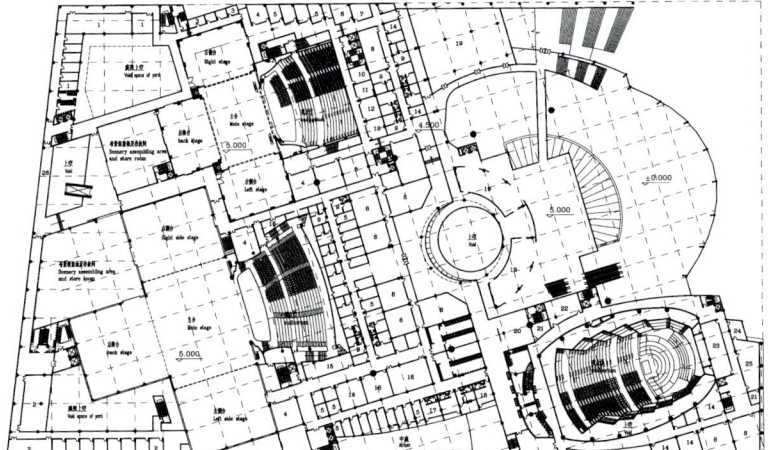

二层平面
2nd FLOOR PLAN

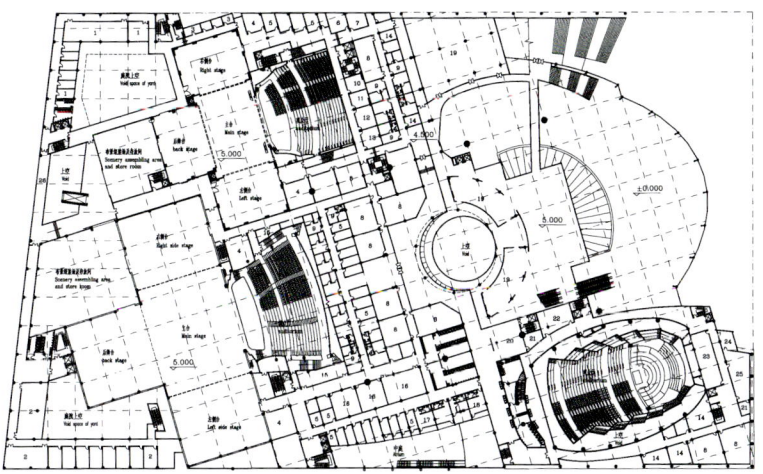

三层平面
3rd FLOOR PLAN

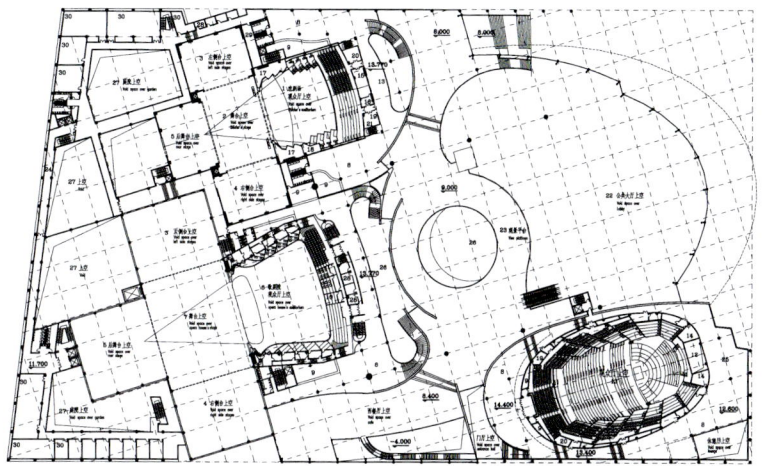

四层平面
4th FLOOR PLAN

公共大厅
PUBLIC HALL

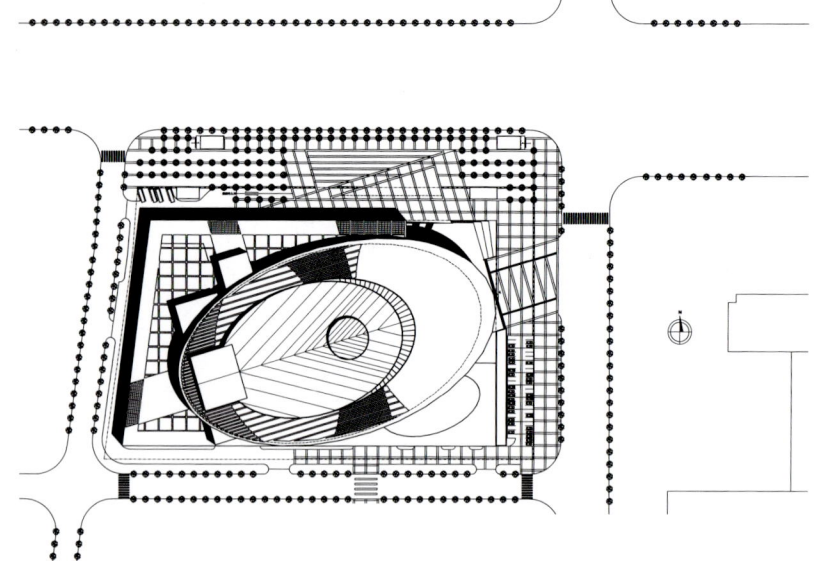

总平面
OVERALL SITE PLAN

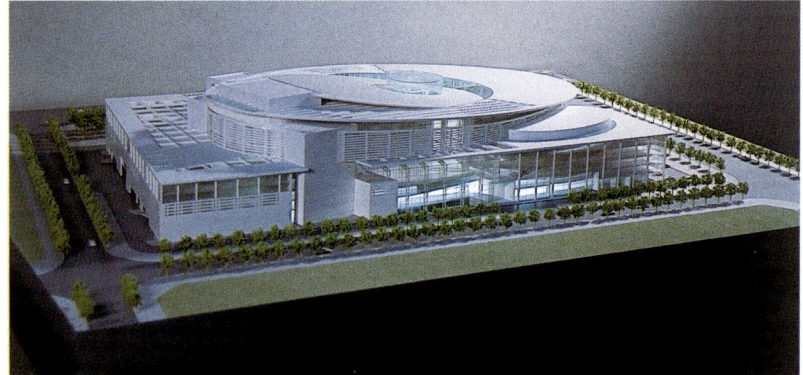

模型透视
MODEL PERSPECTIVE

技术经济指标
Technical Economic Index

基地面积(m²) Site Area	总建筑面积(m²) Total Floor Area	建筑高度(m) Building Height	建筑覆盖率（%） Building Coverage Ratio
36700	地上：72853 地下：59213	23.5～38/45(局部)	56.4

容积率 Plot Ratio	绿化覆盖率(%) Green Coverage Ratio	停车数量 Number of Stalls	自行车停车数量 Number of Bicycle Stalls
1.99	21.6	地上：72 地下：509	1095

各观众厅数据
Data For Auditoria

		歌剧院 Opera House	音乐厅 Concert Hall	戏剧院 Theater	小戏场 Mini Theater
	观众厅面积(m²) Floor Area of Auditorium	2080	1630	972	960
	观众厅体积(m³) Volume of Auditorium	162800	20400	7260	6807
座席数 Seating Capacity	总座席数 Seating Capacity(Seats)	2524	1986	1192	314~482
	其中：池座 Among Them:Auditorium	1140	1462	836	409
	楼座 Floor Seat	1384	524	356	73
	休息厅面积(m²) Lounge Floor Area	2980	2560	1296	624
舞台尺寸 Stage Dimentions (m²)	主舞台 Main Stage	918	360	525.5	143
	左右侧台 Left and Right Side Stages	1188	396	322	
	后舞台 Back Stage	650		328	
	台仓 Under Stage Storage	1568		333	525
	升降乐池 Elevating Orchestra Pit	165		196	
	最大视距(m) Max. Visual Distance	40	38.5	29	18.5
	最大俯角（°） Max. Angle of Depression	35	33.5	19	32

08
清华大学建筑
设计研究院
（中国）
Architectural Design &
Research Institute
of Tsinghua
University (China)

本方案总体设计构思基于下述六条理念和原则：

a.本设计方案在建筑整体设计上大胆采取东半部全面开敞，西半部为建筑实体的处理，使得地段东半部南北完全贯通，为北京市中心区提供了一片新的，有丰富文化内涵的绿色空间。在天安门广场之外，形成了文化性广场，既丰富和美化了天安门广场周围建筑群的外部空间，又为硬质的城市中心区创造出一片"绿色城市肺"，有利于城市的可持续发展。

b.本方案建筑形体构思充分关注建筑与整个城市环境的关联性。设计采用了中国传统的"九宫格"模式和设计母体，与北京内在的城市肌理有机地融合为一个整体，并与中国、北京、天安门广场的深层历史文化内涵相匹配。

c.本方案设计面向未来，运用新理念、新布局、新技术和新形式创造出一种21世纪的崭新大剧院模式。本方案设计采用开放式的平面布局，通过建筑和绿化、空中和地面、内部和外部、地上和地下穿插结合的空间和造型，运用高科技手段，创造出一种开放的、公共的、内涵丰富的新型剧院模式。

d.本方案设计强调国家大剧院的人民性和开放性。国家大剧院观演设施的视听条件、舞台设备、电气设备、通信技术设备等应是世界一流水准的，在确保这一目标的基础上，其也应是面向人民大众的，应具有广泛的人民性和开放性，要兼顾专业性和大众性的结合。本方案除了安排歌剧院、戏剧场、音乐厅、小剧场等专业性极强的四个演出场所外，还在整体设计中增设了露天剧场、水幕电影、传统式室外戏台、室外激光映示信息墙等四项演示场所，使得一个国家大剧院内包含有"8个剧场"。呈九宫格型设置的空中文化艺术管廊与上述设施一起，可以全天候、全方位向公众开放。

e.本方案在建筑外观造型设计上致力于表现中国特色。在形体设计中采用中国传统的"九宫格"模式，在色彩、材料、建筑细部处理等方面借鉴和采用了典型的中国传统建筑处理手法，诸如圆与方、红与金、虚与实，以及入口处白色"金水桥"皇家风范的菱花窗三交六碗图案、舞台高出部分的三重檐处理及汉唐铜镜的图案变异等等，并在绿色开放空间内设置一座中国传统的戏台楼阁，增强了建筑自身的戏剧性矛盾。上述设计手法展现出强烈的中国传统特色，而它们又都与简洁、开放、虚实对比、具有时代感的建筑体型相结合，使得大剧院设计既具有传统意味，又具有新世纪前卫建筑的风采。

f.本方案设计了较大面积的面向公众的开放型空中文化艺术管廊，其中布置有空中艺术展廊、艺术商店、艺术书店、空中咖啡廊等高品位的文化艺术休闲设施，并安排有可直接对外服务的地下商场、地下艺术展厅等商业文化设施，有利于大剧院建成后的管理运营。

The general layout of this scheme is based on the following conception and principles.

a.The theater is the people's arts palace which design is in pursuit of opening, Publicity,peace and modesty.Such layout is the best way to respect the special place of the Tian'anmen Square and to contribute to the whole environment of the city. In architectural design,the east part of the theater is absolutely open to the city and the west part is designed with solid form, which makes the east part penetrated from north to south and provides an open green space with new cultural meaning to Beijing central district. The open green space acts as the Chinese cultural square besides the political one of Tian'anmen Square. It is not only enrich the surrounding environment of Tian'anmen Square, but also provide the hard city center with a soft 'green city lung' which benefits the sustainable development of the city.

b.The historical, cultural and national intention within the site of the Tian'anmen Square is well inherited. The connection between the theater and urban surroundings is paid closed attention to, in order that a monumental cultural facility with harmonious relations with China, Beijing City and the Tian'anmen Square is created out. The ancient traditional Chinese 'Nine Grid System' pattern is adapted as the design modular, which makes the theater as an organic part of the Beijing City.

c.The scheme is designed with new conception,new layout,new technology and new form towards the future, in order that a new pattern of the 21st century theater is created out. The ordinary theater usually takes the shape of a big,close and solie appearance, while ,in this proposal,an unique open character is produced through the combination of solid form and green space, sky part and ground one,interior and exterior,aboveground and under-ground. An open, public, meaningful grand theater is designed with high technology.

d.The National Grand Theater of China is not only geared to the needs of the future,but also set up with good function and security.As the national highest arts center,the whole Chinese people will be proud of it.The National Grand Theater of China should be one of the best theaters in the world with the most advanced stage equipment and the best quality of acoustic effects.On the other hand, the theater must be open to the public including professional experts and ordinary people. Besides the four professional performance facilities of opera house, theater,concert hall and mini theater are set up rationally, four other performance places of open theater,water curtain film Chinese traditional exterior stage and high-tech laser information wall are also created out.'Eight theaters' in one grand theater and the sky structure facilities of cultural arts are open to the public at the whole time.

e.The theater is designed to embody the new figure of reforming and opening China, to embody the spirit of times and nationality. It will be the symbol of modernized China in the 21st century. The architectural appearance design tries to embody the Chinese character.The Chinese ancient traditional 'Nine Grid System' is adapted as the design modular,and some design methods of typical traditional Chinese architecture are used for reference in the aspects of colors, materials and details,such as square and circle,red and gold,void and solid,white 'golden water bridge' at the entrances,one of the typical patterns of Chinese royal style windows,the three layer roof of the highest part of opera stage and the transition of the pattern of ancient Chinese copper mirror in Han and Tang dynasty,etceteras.A Chinese ancient traditional stage is set up on the open green square at the east part of the theater,which shows the dramatic contradiction of the grand theater and adds the rich historical and cultural meaning to it.The above parts are well designed and integrated with the simple,open,modern whole.It means that the grand theater belongs to China,and also belongs to the world,with traditionalmeaning and unique style of the new century.

f.The open public sky cultural facilities are designed characteristically.The high quality cultural entertainment facilities,such as sky art gallery, exhibition hall,art shop,art book store,sky coffee and snake bar etceteras,are arranged there.Meanwhile,many cultural and commercial facilities are arranged underground.They are very helpful for the management of the theater and provides with good social effects and economic returns.

北侧透视
PERSPECTIVE OF NORTH ELEVATION

南侧透视
PERSPECTIVE OF SOUTH ELEVATION

歌剧院观众厅
OPERA HOUSE AUDITORIUM

南北通道
NORTH-SOUTH PASSAGEWAY

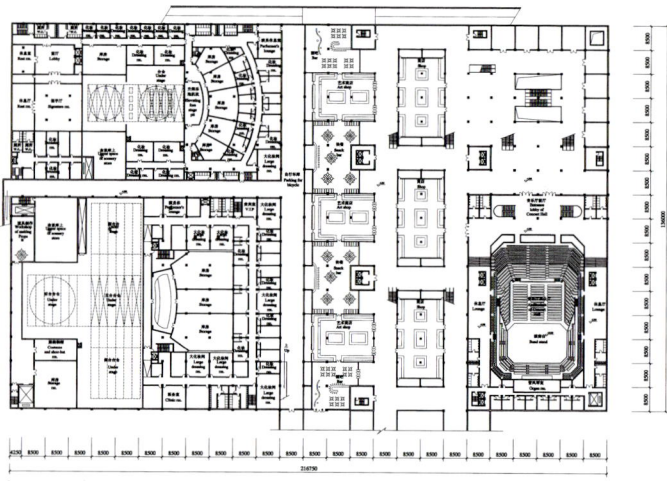

地下一层平面
FIRST BASEMENT PLAN

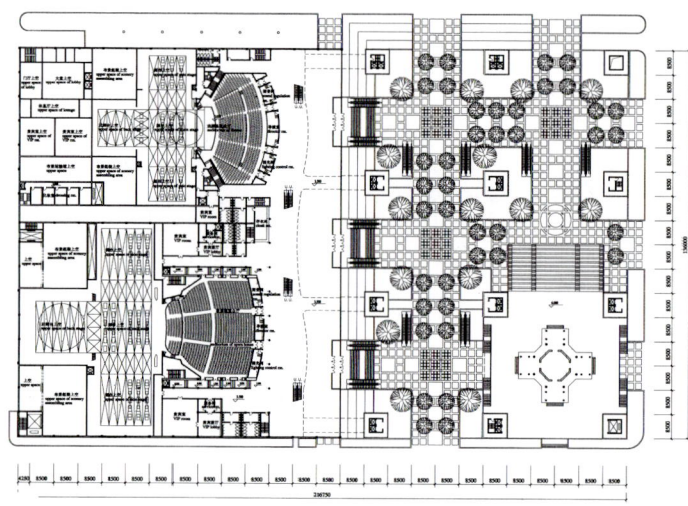

二层平面
2nd FLOOR PLAN

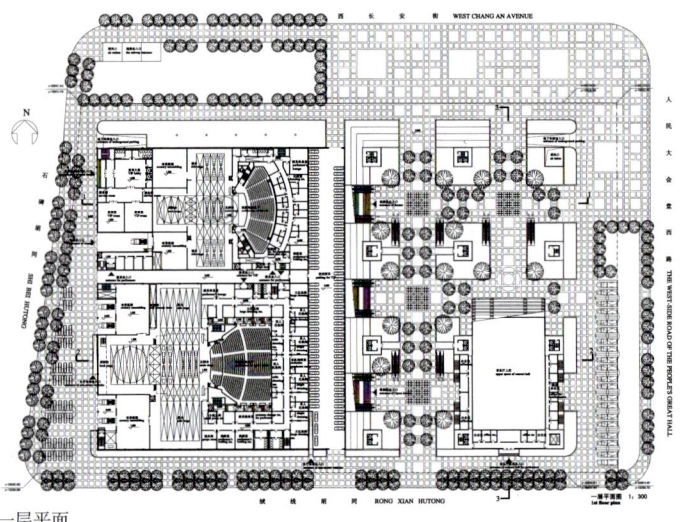

一层平面
GROUND FLOOR PLAN

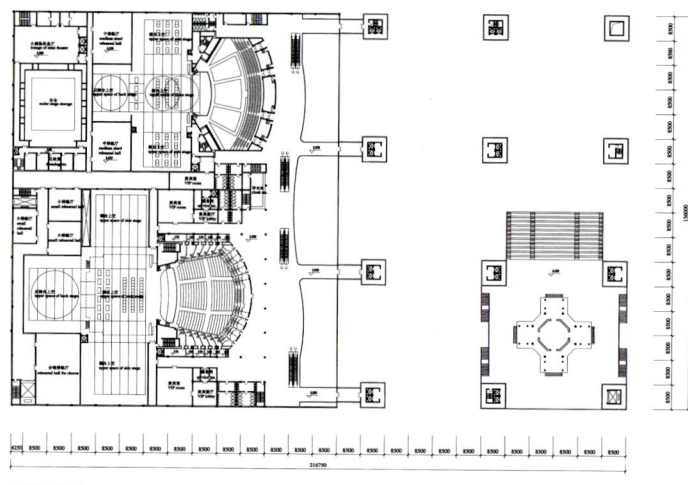

三层平面
3rd FLOOR PLAN

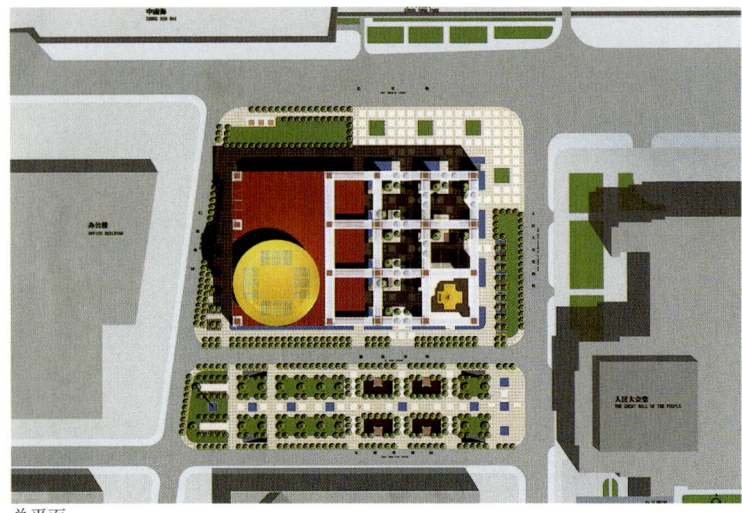

总平面
OVERALL SITE PLAN

城市开放空间
CIVIC OPEN SPACE

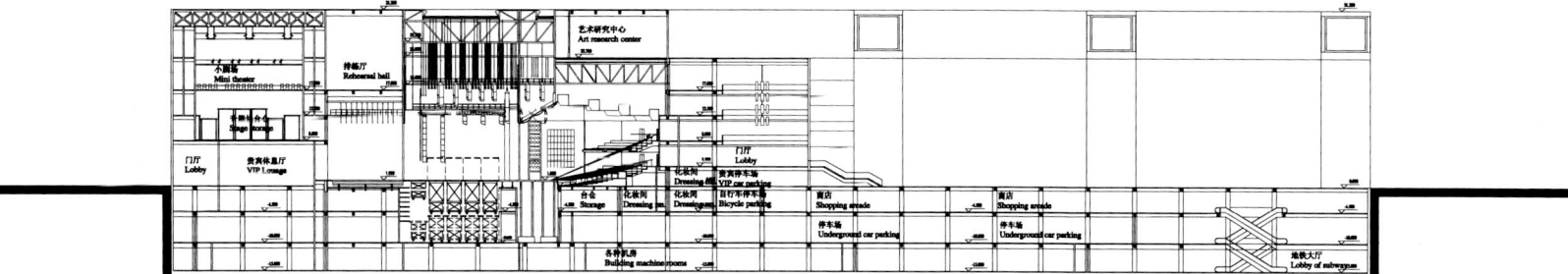

艺术研究中心
Art research center

小剧场
Mini theater

排练厅
Rehearsal hall

门厅
Lobby

贵宾休息厅
VIP Lounge

台仓
Storage

化妆间
Dressing room

化妆间
Dressing room

门厅
Lobby

贵宾停车场
VIP car parking

自行车停车场
Bicycle parking

商店
Shopping arcade

商店
Shopping arcade

停车场
Underground car parking

停车场
Underground car parking

各种机房
Building machine rooms

地铁大厅
Lobby of subway

1—1剖面
SECTION 1-1

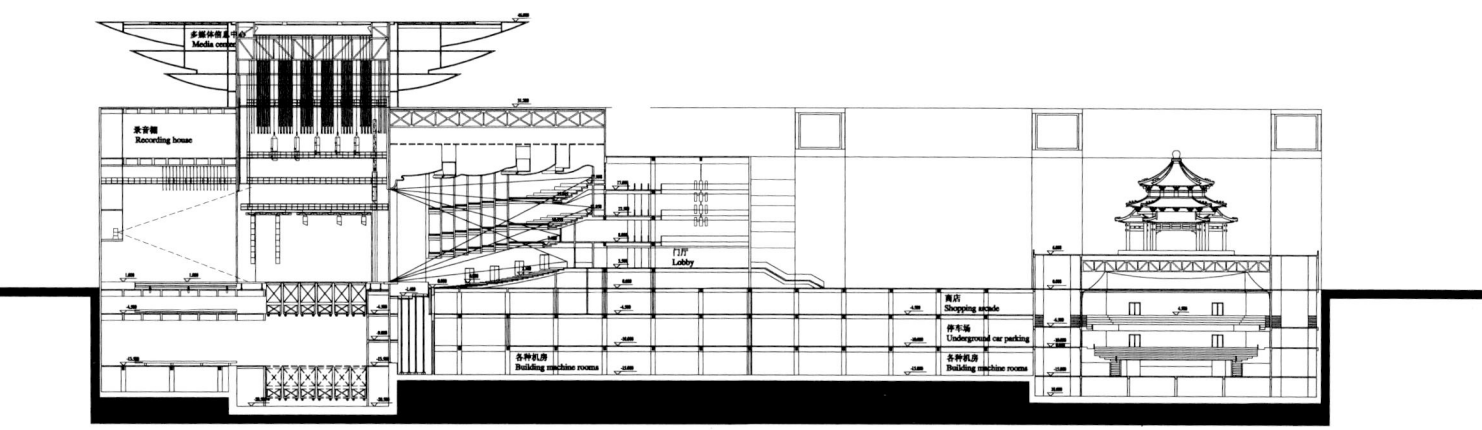

多媒体信息中心
Media center

录音棚
Recording house

门厅
Lobby

商店
Shopping arcade

停车场
Underground car parking

各种机房
Building machine rooms

各种机房
Building machine rooms

2—2剖面
SECTION 2-2

商店
Shopping arcade

停车场
Underground car parking

激光水幕电影位置
Laser water screen cinema position

音乐厅门厅
Concert hall lobby

地下停车库出入道
Entrance of underground car parking

地铁大厅
Lobby of subway

地铁出入口
Entrance of subway

3—3剖面
SECTION 3-3

基地面积(m²) Site Area	总建筑面积 (m²) Total Floor Area	建筑高度(m) Building Height	建筑覆盖率（%） Building Coverage Ratio
43300	地上：47230 地下：71960	31.2/45(局部)	42
容积率 Plot Ratio	绿化覆盖率 (%) Green Coverage Ratio	停车数量 Number of Stalls	自行车停车数量 Number of Bicycle Stalls
3.51	32	地上：156 地下：915	800

开放空间 A
OPEN SPACE A

开放空间 B
OPEN SPACE B

各观众厅数据
Data For Auditoria

		歌剧院 Opera House	音乐厅 Concert Hall	戏剧院 Theater	小戏场 Mini Theater
	观众厅面积(m²) Floor Area of Auditorium	1970	1850	1007	673
	观众厅体积(m³) Volume of Auditorium	18609	21800	5995	2059
座席数 Seating Capacity	总座席数 Seating Capacity(Seats)	2621	2011	1249	222~406
	其中：池座 Among Them:Auditorium	1580	976	892	368
	楼 座 Floor Seat	1041	1035	357	38
	休息厅面积(m²) Lounge Floor Area	6929	5173	4578	560
舞台尺寸 Stage Dimentions (m²)	主舞台 Main Stage	33 × 27	21.56 × 14	27 × 21	18 × 15
	左右侧台 Left and Right Side Stages	20 × 24		14.5 × 21	
	后舞台 Back Stage	27 × 24		20 × 16	
	台 仓 Under Stage Storage	33 × 27 27 × 24	8.3 × 5.8	27 × 21 20 × 16	18 × 15
	升降乐池 Elevating Orchestra Pit	19 × 6		15 × 6	
	最大视距(m) Max. Visual Distance	37.4	33.1	18.8-25.5	
	最大俯角（°） Max. Angle of Depression	25	18	20-27	

鸟瞰图
BIRD'S-EYE VIEW

国家大剧院选址在天安门广场，人民人会堂西侧。

天安门广场和故宫是北京城的中心，南北中轴线布局是广场的特点。人民大会堂等建筑物是天安门广场的重点建筑。国家大剧院是整个建筑群的一个组成。

国家大剧院的主要入口面向西长安街，各种交通工具包括地下铁均可到达。20m 高的中央入口大厅，是整个建筑群的焦点、中心，引领观众通向不同的演出大厅及公共空间。各演出厅前厅设在不同高度，形成室内丰富的空间变化。每个演出厅及前厅，除具有本身特点外，又有其共同的属性，和谐、统一。无论白天或晚上，大剧院的歌剧院、音乐厅、戏剧场透过大片玻璃幕墙，展现于人们面前。

与中央入口大厅相连的公众大厅设置了展览厅、营业厅、餐厅，并引领人们通往南面共享绿化园林。园林设置了露天剧场，节日欢庆广场等，它是大剧院艺术空间的延续。

为使各个演出厅的设计更为完善，我们研究了世界上各个成功的实例，用最新的科技加以提升、改善。

歌剧院观众厅采用马蹄形平面设计，音乐厅采用长方形及端部圆形平面设计，利用混响室调节音质。戏剧场采用扇形平面，镜框式及可升降伸出式舞台设计。小剧场采用卵形与矩形平面结合，观众厅及舞台可灵活地转换。

大剧院所采用的流线形屋面及建筑造型设计，灵感是来自吸取了中国传统书法、绘画艺术的抽象手法，注重神似而非形同，并结合现代建筑高科技所形成的。

高科技建筑技术提供了可建造 20m 高玻璃幕墙，大跨度室内空间及更加轻盈、现代感的屋面造型，使大剧院建筑能超越时空，将成为 21 世纪国际性重要文化建筑。

The Chinese government has finally decided to carry out the construction of a world class National Grand Theater on a privileged sire west of the Great Hall of the people in Tian An Men Square a project that has undergone more than 40 years of planning .Due to the rapid economic growth and successful open policy in CHina,a world class performance palace is required to host international performing art exchange events. It will be a landmark building that symbolizes China entering into the 21st Century.

The Tian'anmen Square and the axial entry into the Forbidden City is the center of the city planning of Beijing,with the Great Hall of the People and other buildings on the Square as the main focus .The massing and layout of the National Grand Theater shall compliment the Tian An Men ensemble, and not detract from the key buildings of this political and historical center.

Main entry to the Grand Theater will be from west Chang An Avenue,also taking advantage of the MTR connection.The 20m tall Grand Foyer will act as the central focal point of the complex, channeling the public to the various venues, Each of the performing venue and its supporting public space has its independent character,with one common theme.The public space of each element, i.e. the Opera, the Concert Hall and the Theater, are transparent through the glass wall, ready to be seen from the outside offering a key attraction to passers- by day and night.

A pedestrian concourse supported by exhibition space and retail runs through the National Grand Theater, integrating it with the large open space and outdoor performing area to the south of the site, giving the development a further dimension.

For the design of the individual venues, we have analyzed numerous, contemporary and traditional, world renowned examples of the different venues, and we have studied improvements on these models with modern technology and state of the art equipment.

For the Opera House we adopt a 'horse-shoe' auditorium form so successful in the Opera Houses of Milan, Vienna and Paris.

For the Concert Hall, reverberation chambers are designed to wrap around the auditorium in order to achieve appropriate reverberation time for different musical works.

The fan-shaped Theater has a proscenium stage and an elevating thrust stage to cater for the different theatrical requirements.

The oval form Mini Theater is equipped with multiple possibilities of theater staging and seating arrangement essential for the experimental nature of the theater.

The major architectural characteristics of our design proposal, the free flowing curve of the roof line and the transparency of the exterior walls, aim at the spirit and not the form in the same manner as the Chinese tradition of ink painting and poems. Integrated with a high technology approach, the transparency of 20 meter glass wall seems to extend the buildng towards the Chang An Avenue, and the lightness of the roof profiles gives the National Grand Theater the sense of timelessness and marks it as a major building of the 21st Century.

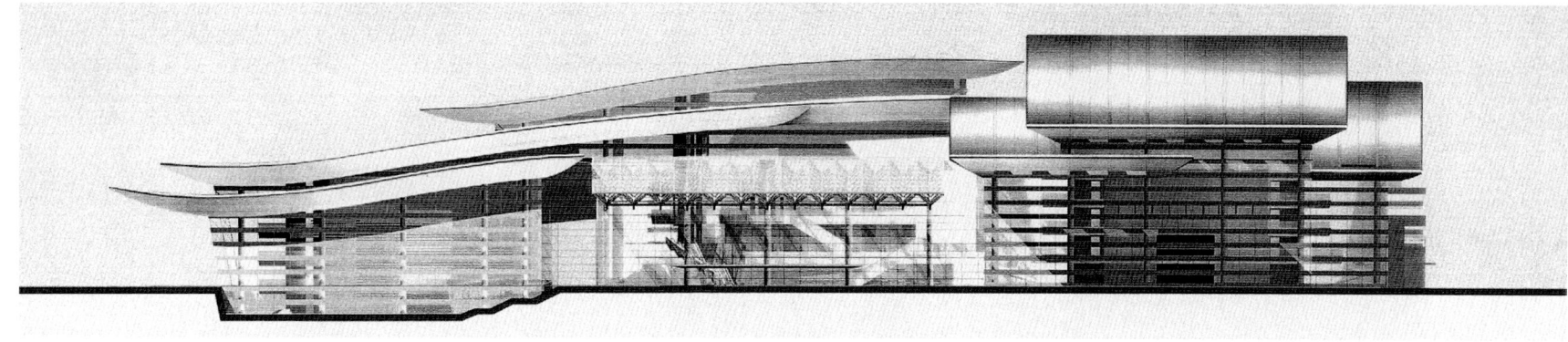

北立面
PERSPECTIVE OF NORTH ELEVATION

南立面
PERSPECTIVE OF SOUTH ELEVATION

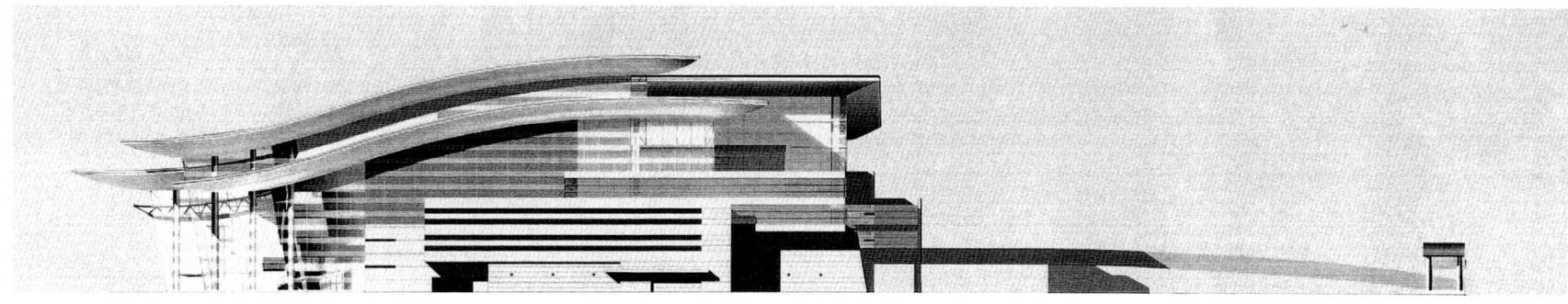

西立面
PERSPECTIVE OF WEST ELEVATION

东立面
PERSPECTIVE OF EAST ELEVATION

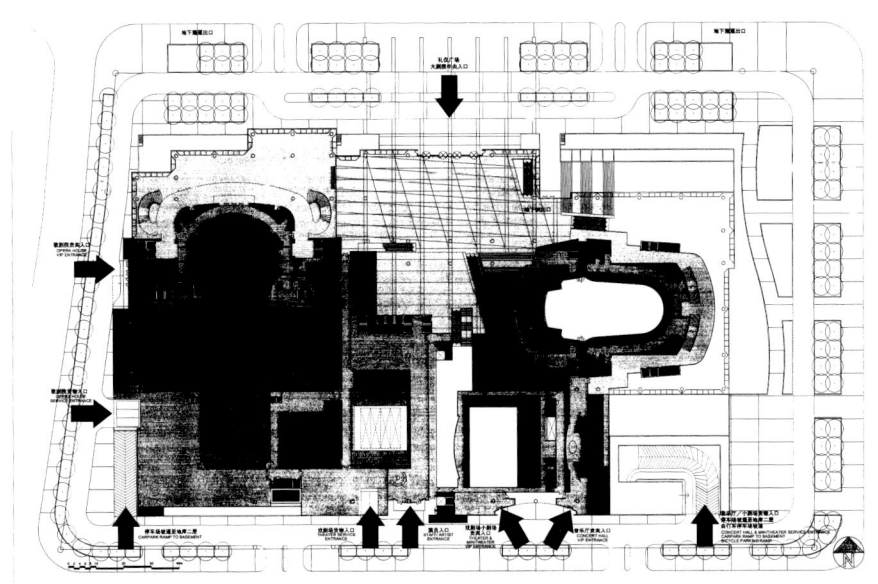

交通示意图
TRAFFIC SKETCH MAP

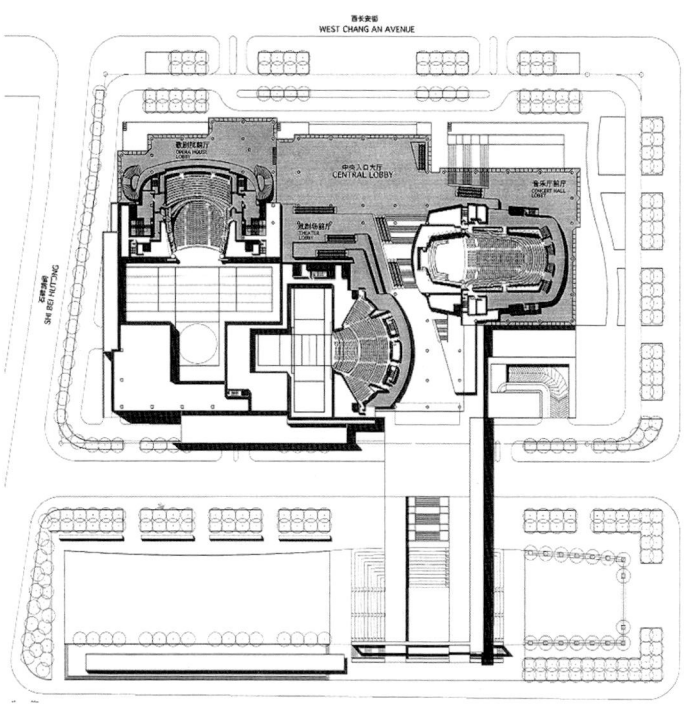

中央大厅与各前厅关系图
INTER CONNECTION OF THE CENTRAL LOBBY AND OTHER
ENTRANCE LOBBIES

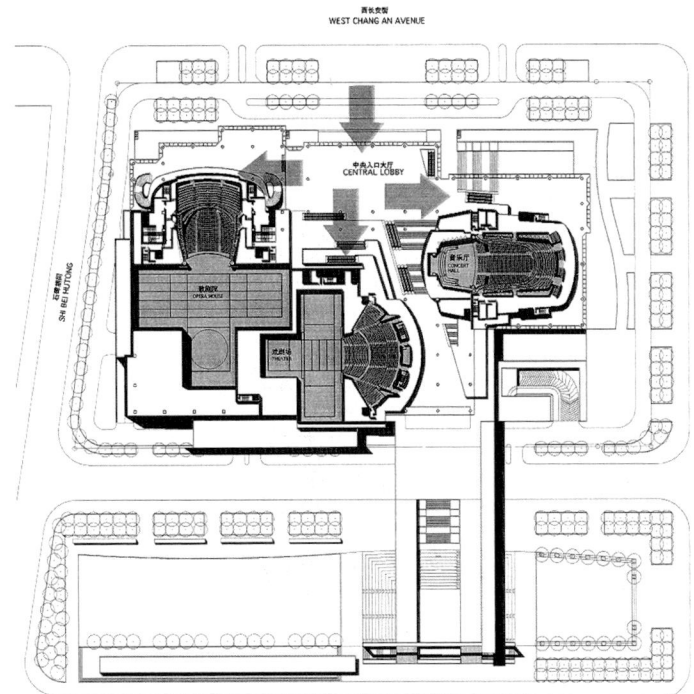

观众厅人流示意图
SKETCH MAP OF AUDITORIUMS' STREAMS
OF PEOPLE

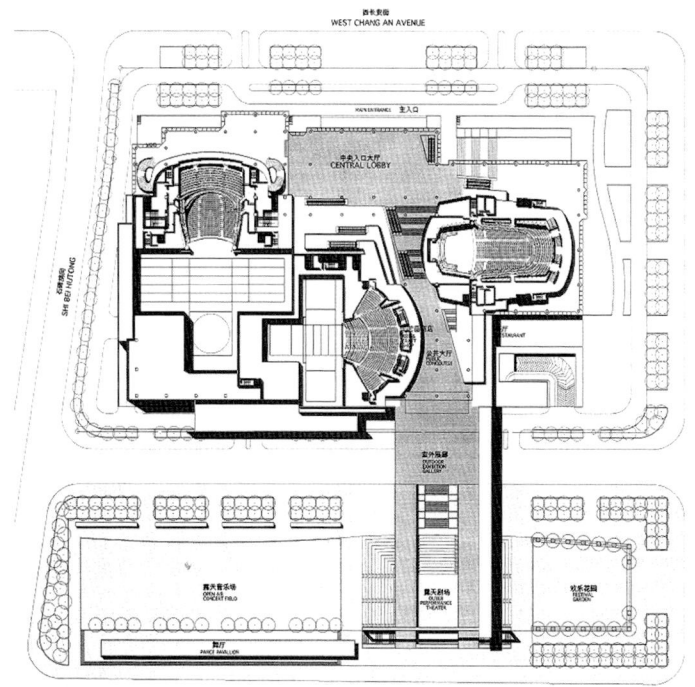

中央大厅，公众大厅与南面共享绿化园林关系图
IUTER CONNECTION OF THE CENTRAL LOBBY,
THE PUBLIC CONCOUSE AND THE SOUTHERN PUBLIC GARDEN

歌剧院前厅
OPERA HOUSE LOBBY

音乐厅前厅
CONCERT HALL LOBBY

274

戏剧场前厅
THEATER LOBBY

公众大厅望向戏剧场
INTERIOR PERSPECTIVE OF PUBLIC CONCOURSE

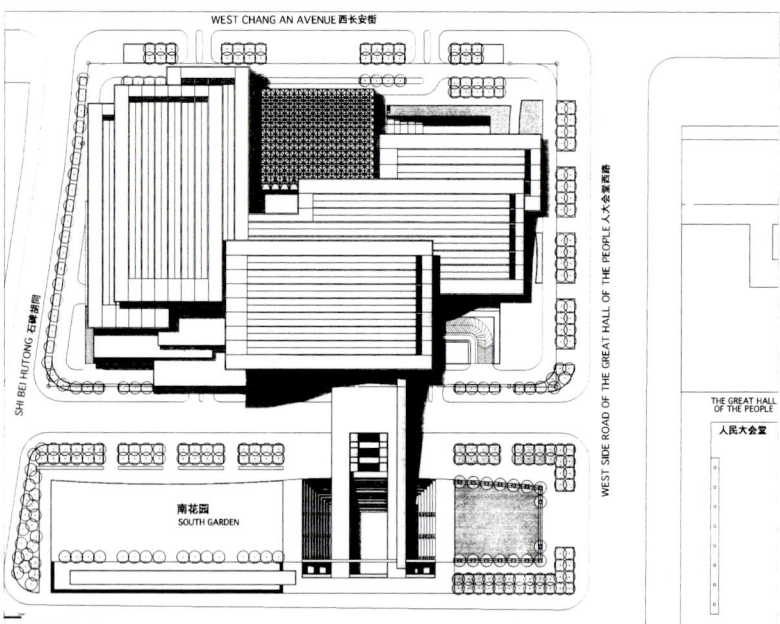

总平面
OVERALL SITE PLAN

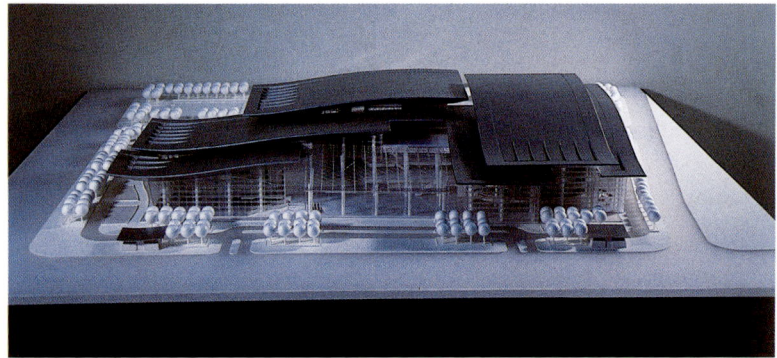

北侧模型透视
VIEW FROM NORTH SIDE OF THE MODEL

南侧模型透视
VIEW FROM SOUTH SIDE OF THE MODEL

技术经济指标
Technical Economic Index

基地面积(m²) Site Area	总建筑面积 (m²) Total Floor Area	建筑高度(m) Building Height	建筑覆盖率（%） Building Coverage Ratio
38900	地上：68850 地下：56150	30/45(局部)	78
容积率 Plot Ratio	绿化覆盖率 (%) Green Coverage Ratio	停车数量 Number of Stalls	自行车停车数量 Number of Bicycle Stalls
3	41	地上：80 地下：500	1000

各观众厅数据
Data For Auditoria

		歌剧院 Opera House	音乐厅 Concert Hall	戏剧院 Theater	小戏场 Mini Theater
	观众厅面积(m²) Floor Area of Auditorium	2175	1600	960	520
	观众厅体积(m³) Volume of Auditorium	21500	21200	16800	5663
座席数 Seating Capacity	总座席数 Seating Capacity(Seats)	2502	2006	1200	498
	其中：池座 Among Them:Auditorium	1075	1066	938	320
	楼座 Floor Seat	1427	940	262(378)	178
	休息厅面积(m²) Lounge Floor Area	3600	2800	1600	600
舞台尺寸 Stage Dimentions (m²)	主舞台 Main Stage	916	432	526	
	左右侧台 Left and Right Side Stages	1242	432	632	
	后舞台 Back Stage	650		356	
	台仓 Under Stage Storage	1566	290	526	
	升降乐池 Elevating Orchestra Pit	160		205	
	最大视距(m) Max. Visual Distance	38	39	26.5	21.5
	最大俯角（°） Max. Angle of Depression	35	23	31	22

我们最初的方案技术及功能方面的构想,特别是其建筑风格设计上的高质量,新颖独特,在简练中创造出在代表着新世纪的中国的鲜明形象,为之博得了好评。尽管如此,我们仍然得到了一些有关建筑表现形式的诸因素的意见：

——建筑各个组成部分之间的层次不够分明,如歌剧院相对其它两个剧院的规模及主观众厅与附属观众厅之间的关系等,屋顶的处理过于单一；

——各立面、屋顶及台基的透明度过高,不适应气候条件及周围的建筑环境,因而使整体建筑略显呆板,缺乏"可读性"。

对于这些言之有理的意见,我们的回答是这样的：

在代表大剧院鲜明特征的三座"金桥"中,最重要的歌剧院在建筑物中心,其规模是构图的主要因素。在屋顶坡面的衬托下,歌剧院规模一直延伸至空地大台阶并于此开设主入口,因而显示出国家大剧院不同于长安街上其它建筑物的特色。

处于建筑物中轴线上的宏伟的门厅构成国家大剧院的主要入口,并连通所有的建筑物内部通道。但是,仍然可能通过南面广场独立地出入音乐厅和话剧院。

屋顶朝向长安街倾斜以呈现歌剧院单纯而动态的曲线,并显示着,且重合于宏伟的内部大台阶的坡面。

在歌剧院的处理上,该单体同样十分突出,使得侧翼与中心部分区别分明。坡面屋顶则构成一个更为现代的表现形式。

屋顶坡面起着分界面的作用：

——与外界的分界面：铝金属表面反映天空,从而使屋顶随时间而变换色彩和外观。

——与内部的分界面：屋顶天窗在白天将自然光引入内部空间,而在夜晚则成为光亮的照明灯。

在方案中,该坡面象征着历史与现代的关系。

屋顶坡面被构想为一层"膜",一方面反作用于自然环境,另一方面与国家大剧院的文化活动相呼应。

外部的两个开阔的绿化庭院在石质大台阶的最高部分穿透屋顶,引入建筑物内并为人们提供又一个观赏的景致。

玻璃幕墙立面的面积由于屋顶的坡面而减小。

建筑台基的立面几乎不透光,只开设有方形小窗。空间体积及各立面上的修改使得国家大剧院与天安门广场周围建筑环境更加和谐,并更好地适应北京地区的气候条件。

Although our initial project was appreciated for the quality of its technical and functional design,but especially for the quality of its architectural design,both new and original,which in simple forms was able to create a strong and recognizable image capable of incarnating the People's Republic of China of the new millennium,a certain number of observations were made reagrding the formal expression of some of these elements.

a The insufficient hierarchical relationship of the different components of the building, such as the volume of the Opera with respect to the 2 other auditoriums,the main lobby compared to the secondary lobbies, and a roof which was too uniformly treated;

b Facades,the roof and base which were too transparent with respect to the climatic conditions of the built-up environment, endowing the whole with a slightly static expression lacking in legibility.

We will thus respond to these perfectly justified remarks,in the following way:

The volume of the Opera, the largest of the 3 "golden bridges" which characterize the architecture of the Great National Theater,is affirmed in the center of the building as a major element of its composition. Revealed by the incline of the roof, it advances to the steps of the open space in front to close a majestic entrance way hence specifying the image of the National Grand Theater with respect to the other buildings of chang An avenve.

The monumental lobby in the building's axis constitutes the main access to the National Grand Theater and separates all the interior passageways. It is possible,however, to access the Concert Hall and Theater independently via the public square to the south.

The roof slopes toward Chang An avenue to uncover the curved,pure and dynamic lines of the Opera and even reveal the inside monumental steps whose grade it reflects.

There is also a hierarchy in its handling, clearly distinguishing the lateral parts from the central part. With respect to the initial proposal, which in a horizontal plane evoked the geometry of traditional frameworks, the incline now imposes a much more contemporary expression.

The inclined plane of the roof acts like an interface:

a Interface with outside:The aluminum surface reflects the sky giving the roof an aspect which changes according to the time of day and the weather.

b Interface with inside:The pierced openings lead the natural light inwards during the day, allowing it to escape to form a luminescent lantern at night.

As such, within the project it incarnates the relation between history and modernity.

The inclined plane of the roof has been designed an a "membrane" which reacts to the natural environment on one hand, and in resonance with the cultural programming of the National Grand Theater on the other hand.

Two great outdoor patios planted with vegetation pierce the roof in the upper part of the stone steps, accentuating the presence of Nature at the heart of the building and providing additional pleasure for users.

The surfaces of the glazed facades has been decreased by the incline of the roof which overhangs each facade like an awing.

The building's base offers nearly blind stone facades, pierced with small square openings. The changes to the volumes and facades improves the inscription of the National Grand Theater in the context of the nearby buildings on Tian An Men square and better meet the climatic constraints of the region of Beijing.

北侧透视
PERSPECTIVE OF NORTH
ELEVATION

南侧透视
PERSPECTIVE OF SOUTH
ELEVATION

公共大厅
PUBLIC HALL

总平面
OVERALL SITE PLAN

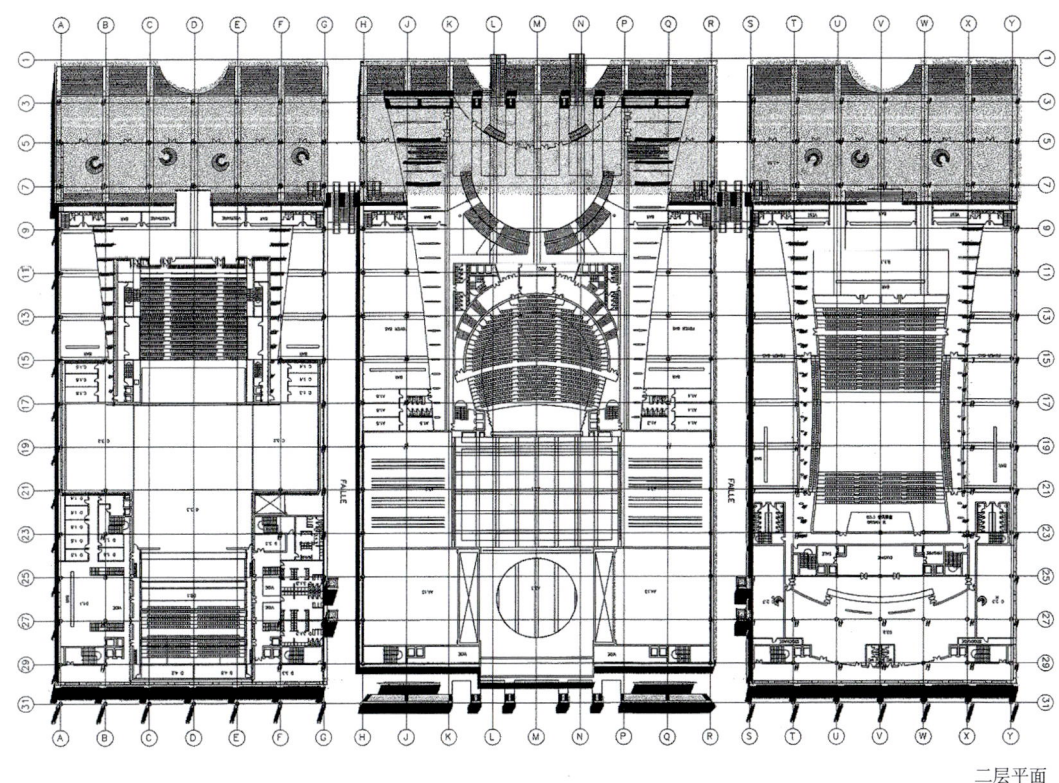

二层平面
2nd FLOOR PLAN

七层平面
7th FLOOR PLAN

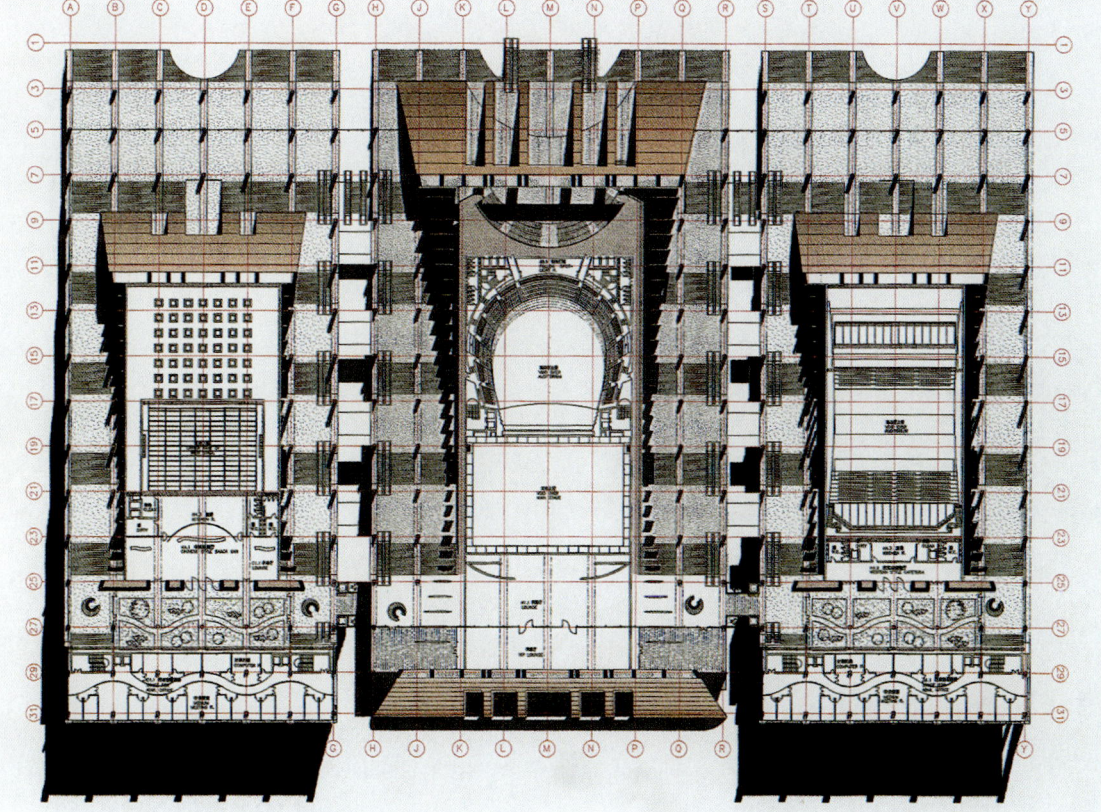

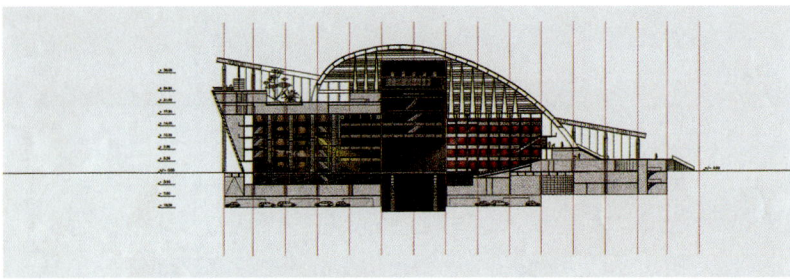

戏剧场剖面
SECTION OF THE THEATER

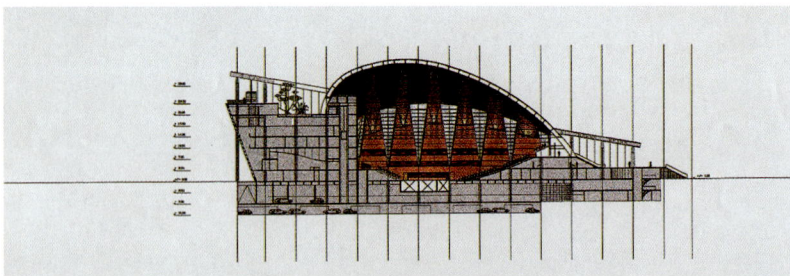

音乐厅剖面
SECTION OF THE CONCERT HALL

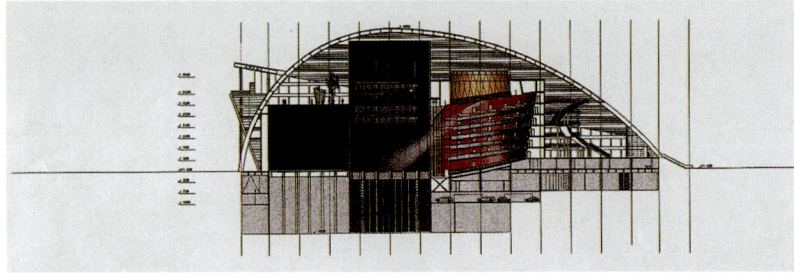

歌剧院剖面
SECTION OF THE OPERA HOUSE

模型透视
MODEL PERSPECTIVE

技术经济指标
Technical Economic Index

基地面积(m²) Site Area	总建筑面积(m²) Total Floor Area	建筑高度(m) Building Height	建筑覆盖率（%） Building Coverage Ratio
38900	地上：78952 地下：52324	30/45(局部)	76

容积率 Plot Ratio	绿化覆盖率(%) Green Coverage Ratio	停车数量 Number of Stalls	自行车停车数量 Number of Bicycle Stalls
3.37	12	地上：50 地下：570	1060

各观众厅数据
Data For Auditoria

		歌剧院 Opera House	音乐厅 Concert Hall	戏剧院 Theater	小戏场 Mini Theater
	观众厅面积(m²) Floor Area of Auditorium	1840	1857	867	788
	观众厅体积(m³) Volume of Auditorium	19610	22340	12950	11850
座席数 Seating Capacity	总座席数 Seating Capacity(Seats)	2506	2020	1203	456
	其中：池座 Among Them:Auditorium	1064		851	
	楼座 Floor Seat	1442		352	
	休息厅面积(m²) Lounge Floor Area	5005	3037	2950	1595
舞台尺寸(m²) Stage Dimentions	主舞台 Main Stage	955	400	522	
	左右侧台 Left and Right Side Stages	2 × 542	2 × 278	2 × 300	
	后舞台 Back Stage	637		353	
	台仓 Under Stage Storage	1710	330	522	1153
	升降乐池 Elevating Orchestra Pit	144		195	
	最大视距(m) Max. Visual Distance	33	27	29	25
	最大俯角（°） Max. Angle of Depression	45	20	26	可变的

12

汉斯豪莱建筑师＋
海兹诺曼建筑师
（奥地利）
Prof.Hans Hollein/
Architect-Heinz
Neumann/Architect
(Austria)

第二轮方案的设计思想是对第一轮方案的进一步完善,特别是在处理与周围现存历史建筑物的关系上——自然法则与文化方面，还存在一些值得商榷的地方。

在保持原方案基本不变的条件下,重点在以上这些方面对原方案进行了进一步思考、探索与改进。我们在这个方案中融入了一些更加强化建筑与城市概念的内容。

该建筑设计思想的出发点在于它在城市文化中所扮演的角色——它理应与周围的空间场所与文化环境相融合。第一轮方案中的有关这一概念的说明在此仍然适用。

为使该建筑与周围城市环境相融合,第二轮方案中采取的一个重要方法就是在入口广场处引入柱廊的概念,它不仅保留了周围建筑物的特征,并把这种特点引伸至国家大剧院的方案中。这种文化氛围反映了"和而不同"的思想——这一思想显然与中国的一些哲学思想不谋而合。

正规与非正规的法则的辨证关系一直存在于当代建筑的布局当中。

在处理复杂的不同部分的关系时,引入了中国园林的原则和概念———一种独一无二的建筑思想,这种原则和概念仍有益于现代建筑的空间构成。

中国古老的"天圆地方"的宇宙观包含了"方"与"圆"、"地"与"天"的辨证关系。这种概念也被引入了此方案的设计中。像入口的圆形大厅(现已围合以延伸门廊)具有特殊的仪式性场所意义及重要性。它不仅为人们提供了一个开放的空间,同时，高耸的柱廊、宽敞的圆形大厅,使人们置身其中时更有一种"天人合一"的感觉。

建筑物的轮廓与屋顶一直是思考的一个要求,作为人民大会堂的紧邻,它的建筑特点透过柱廊渗入到新的建筑物之中,这种由此及彼的过渡极为重要。

世界上没有任何一个地方的建筑像中国建筑那样,赋予大屋顶以特殊的意义。用当代的语言来说,这一特征仍被认为充满活力而经久不衰。与此同时,大屋顶以一种纪念性的方式将它所庇护的各个部分融合在一起。——就这一点而言,一个平屋顶是无论如何也无法满足此多元与巨大的建筑。

基于建筑的永恒的原则以及中国古典建筑的特征与空间思想,进一步探索当代建筑的特点,这座建筑像一座在方阵规矩中跳跃的音符，以一种时代精神展望未来。

The idea for the second round constitutes an elaboration and improvement on the project of the first round, especially in terms of integration into the surrounding environment.

No specific criticism of the jury had been forwarded to us-however,information transmitted mentioned that functionally this project was considered satisfactory but that there were questions about the integration into the existing historic substance-both physically and cultur-ally.

We therefore concentrated on these aspects by further developing primarily some specific points in the original design which we basically kept unaltered. Consequently we asked-and were allowed-to present in the second phase again the project of the first phase with a few strategic changes and additions-elements to strengthen the architectural and urbanistic concept.

The idea of this building is based on its urbanistic and cultural role-its integration in the spatial and cultural environment. The statements in the description of the concept of the first round project are still valid.

Integration into the urban situation has been taken one important step further by introducing an element of colonnades around the entrance square which carries and continues existing characteristic featrures of the surrounding into the new complex of the National Grand Theater. This cultural precinct reflects"unity within difference"-its philosophy is consciously based on some characteristic elements of Chinese thought.

The dialectic between a formal and informal order has been carried over into a contemporary situation.

The complex heterogenity of the task and its program has been handled on the principles and concept of the Chinese garden-a unique attitude to architectural thought but continuously valid for modern space-making.

Primeval archaic elements as the square and the circle have been introduced to give some parts-as the entrance rotunda(which now is enclosed to extend the lobby)-specific meaning and importance.Openess has been searched for as specific quality-provoking passage into and through the building.

The silhouette,the roof has been an important concern. From the immediate neighbourhood of the Great Hall of the People-its architectural attitude carried over into the new building via the motive of the colonnades-the transition to the other borders was thought to be important.

No other world architecture is so dominated by a specific idea of the roof as the Chinese architecture and it was considered as an appropriate dynamic element of continuity-translated into a contemporary language. At the same time this all-encovering element integrates in a monumental way the variety of elements it encloses-a task which a flat roof would not fulfill.

Based on eternal principles of architecture, on characteristics and elements of historic Chinese architectural and spatial thought developing further contemporary attitudes, this building-in the spirit of its time-is looking into the future.

北侧透视
PERSPECTIVE OF NORTH ELEVATION

模型透视
MODEL PERSPECTIVE

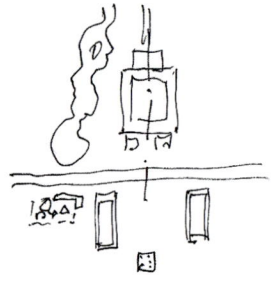

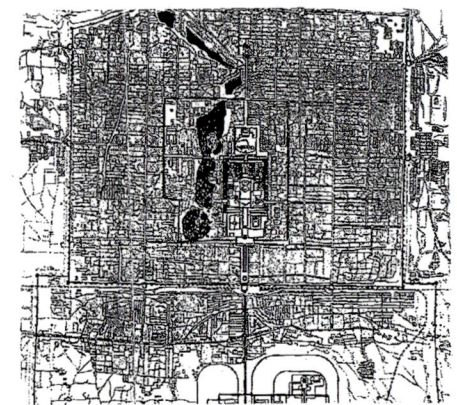

正规与非正规的法则
FORMAL AND INFORMAL ORDER

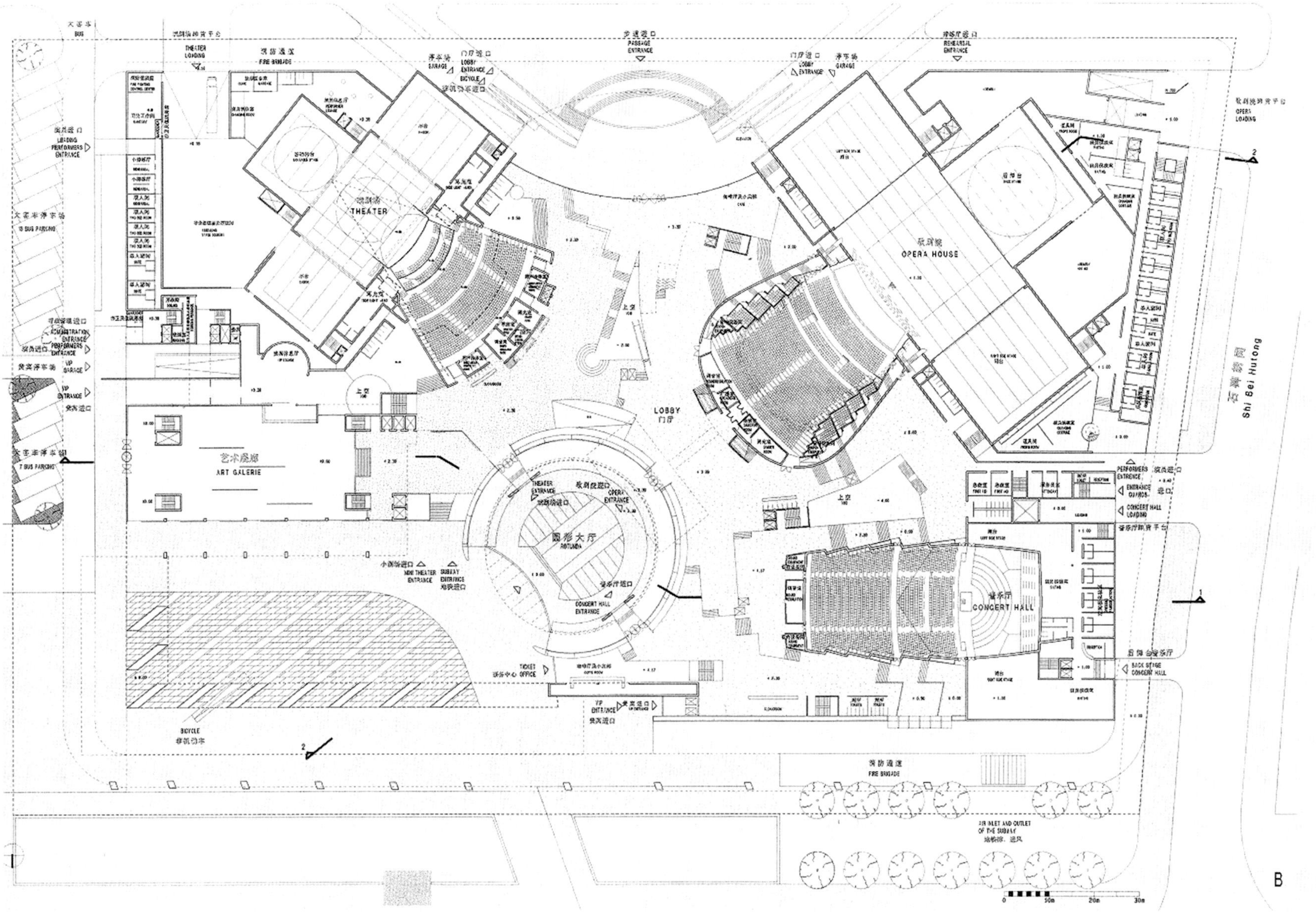

THEATER

OPERA HOUSE

LOBBY

ART GALERIE

ROTUNDA

CONCERT HALL

THEATER ENTRANCE
OPERA ENTRANCE

MINI THEATER ENTRANCE
SUBWAY ENTRANCE
CONCERT HALL ENTRANCE

TICKET OFFICE

VIP ENTRANCE

BICYCLE

PERFORMERS ENTRANCE
ENTRANCE GUARDS
CONCERT HALL LOADING

BACK STAGE CONCERT HALL

FIRE BRIGADE

AIR INLET AND OUTLET OF THE SUBWAY

BUS

THEATER LOADING
FIRE BRIGADE
GARAGE
BICYCLE

PASSAGE ENTRANCE

LOBBY ENTRANCE
GARAGE

REHEARSAL ENTRANCE

OPERA LOADING

LOADING PERFORMERS ENTRANCE

BUS PARKING

ADMINISTRATION ENTRANCE
PERFORMERS ENTRANCE
UP GARAGE

VIP ENTRANCE

7 BUS PARKING

PERFORMERS ENTRANCE

Shi Bei Hutong

B

一层平面
GROUND FLOOR PLAN

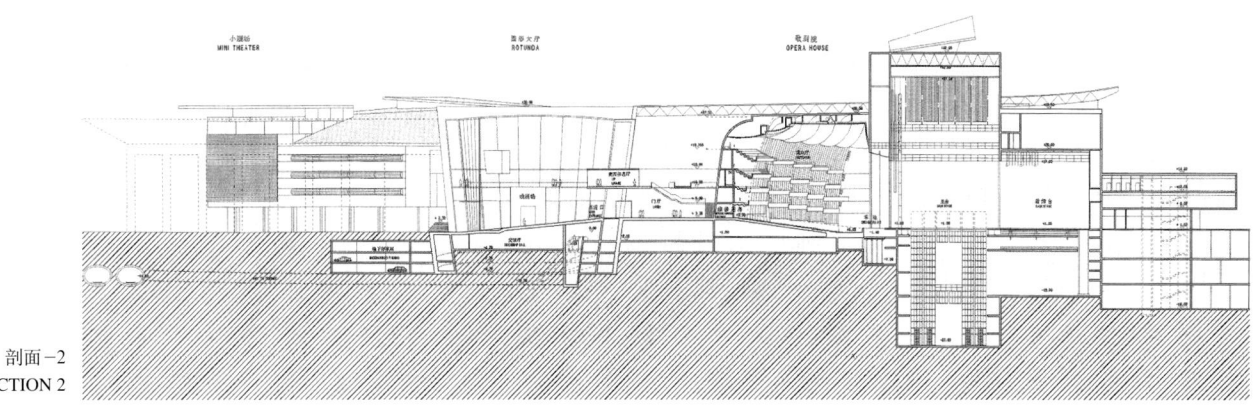

MINI THEATER
ROTUNDA
OPERA HOUSE

剖面-2
SECTION 2

283

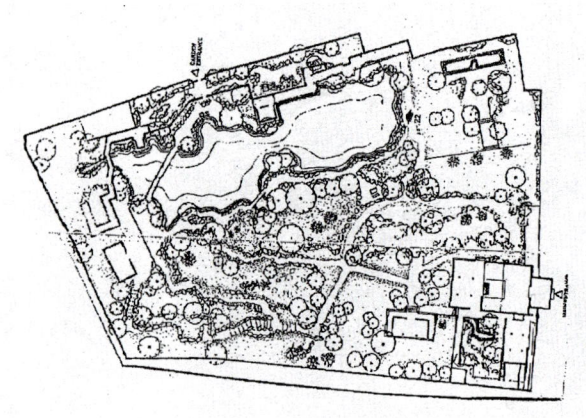

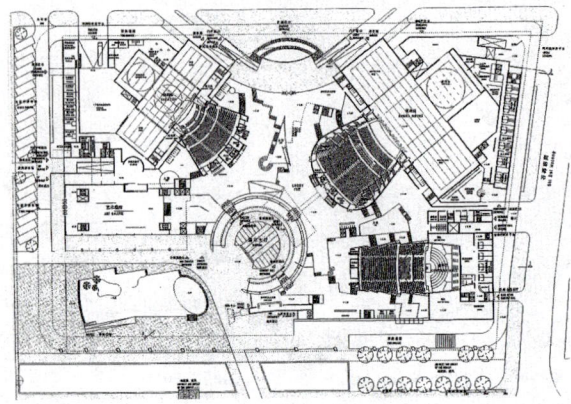

中国园林的空间概念是大剧院的构成原则
THE SPATIAL CONCEPT OF THE CHINESE GARDEN-STRUCTURE ORDER OF THE
NATIONAL GRAND THEATER

技术经济指标
Technical Economic Index

基地面积(m²) Site Area	总建筑面积 (m²) Total Floor Area	建筑高度(m) Building Height	建筑覆盖率（%） Building Coverage Ratio
43082	地上：54064 地下：64565	30	61.2
容积率 Plot Ratio	绿化覆盖率 (%) Green Coverage Ratio	停车数量 Number of Stalls	自行车停车数量 Number of Bicycle Stalls
2.75	20.1	地上：20 地下：541	1130

各观众厅数据
Data For Auditoria

		歌剧院 Opera House	音乐厅 Concert Hall	戏剧院 Theater	小戏场 Mini Theater
	观众厅面积(m²) Floor Area of Auditorium	2242	1374	953	808
	观众厅体积(m³) Volume of Auditorium	20000	22000	7800	7140
座席数 Seating Capacity	总座席数 Seating Capacity(Seats)	2509	2003	1200	300~500
	其中：池座 Among Them:Auditorium	1152	911	392	活动式
	楼 座 Floor Seat	1352	1092	808	活动式
	休息厅面积(m²) Lounge Floor Area	3567	2684	1660	605
舞台尺寸 Stage Dimentions (m²)	主舞台 Main Stage	972	250	526	
	左右侧台 Left and Right Side Stages	2 × 621		2 × 316	
	后舞台 Back Stage	650		330	
	台 仓 Under Stage Storage	1568		526	
	升降乐池 Elevating Orchestra Pit	147	107	204	
	最大视距(m) Max. Visual Distance	43	45	33	20
	最大俯角（°） Max. Angle of Depression	35	44	21	58

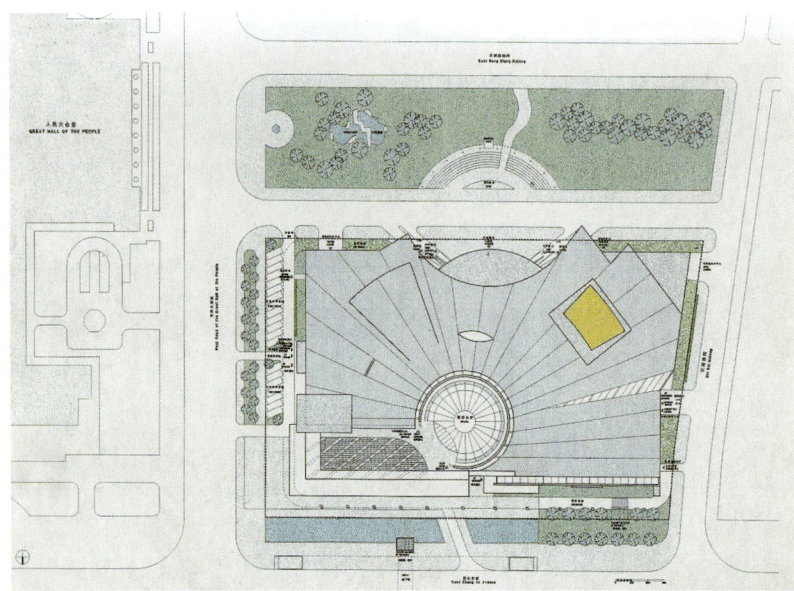

总平面
OVERALL SITE PLAN

中国国家大剧院将成为联系中国的过去、现在与未来的纽带，促进人民生活品质的提高和国际文化交流。它将成为新世纪的文化里程碑。

要想体现这样丰富的内涵，仅仅从建筑形式的表达上去寻求答案是远远不够的。我们的设计构想旨在寻求一种丰富和多元化的建筑精神，更希望能藉此来体现中国人民的精神。我们的设计目标在于为北京的城市和中国人民创造一崭新而充满活力的公共空间，使其恰当融入北京特有的都市景观和文化背景，并提供具有世界一流水准的表演场所。遵循这样的目标，并在严谨的设计框架指导下，我们融合中国的传统建筑风格于现代的建筑形式之中，并试图通过材料的运用来传达一种博爱的人文主义精神。我们的设计基于对人的建筑体验的思考和关怀。

设计由三个不同但相关的要素组成：第一是剧院的主体，每座剧场都有其独特的流线形的玻璃休息厅和华丽的外墙，所有的演出空间设计都力求达到世界最高水平的声学效果；第二是一个宏伟的中央共享空间，现命名为人民文化艺廊，这一高科技的玻璃大厅是剧院的观众、游客和北京市民进行集会和文化活动的场所；第三是环绕整体的柱廊，作为天安门广场区域文脉的延续以及中国传统建筑的现代诠释。总体而言，我们的剧院设计体现了合理的组织以及令人振奋而又内涵丰富的空间感受。我们利用多年来在国际设计和建筑界的经验，通过经济、合理、可行的方式塑造了丰富的建筑与空间形式。

The National Grand Theater becomes a nexus of China past and present, of public enrichment, and of a shared international culture. it will be a landmark of culture for the next century.

It is impossible to. contain such broad aspirations within a single, one-dimensional, architectural expression. Our project for the National Grand Theater seeks a rich and multi-faceted architecture that is first and most importantly a celebration of the people of China. It creates a vital mew public space for the city of Beijing and the people of China, it addresses unique features of its urban context, and it provides world class performance halls. Within the framework of a rigorous design, we have incorporated both traditional motifs and modern architectural

expressions. Throughout we have used materials that convey warmth and invite touch.Our project is conceived at a human experiential scale.

Our projectis composed of three different, yet interrelated, elements. First, are the theaters themselves; each has its own curved glass lobby and richly ornamented exterior walls and each performance space is shaped according to the highest standards of acoustic performance. Second, is a large central atrium; the People's Cultural Galleria. This high-tech,glass wrapped gathering place is for theater patrons, tourists, and the people of Beijing. Third, is an encircling colonnade; a solemn completion of the Tian An Men Square precinct and a contemporary interpretation of traditional Chinese construction. As an ensemble our Theater design presents both a logical arrangement and an exciting, sophisticated experience. To achieve this we have applied our long experience with international design and construction to create a rich spatial effect with means that are economical, rational, and fully feasible.

北侧夜景透视
NIGHT VIEW OF NORTH SIDE

艺廊接待大厅
PERSPECTIVE OF GALLERIA LOBBY

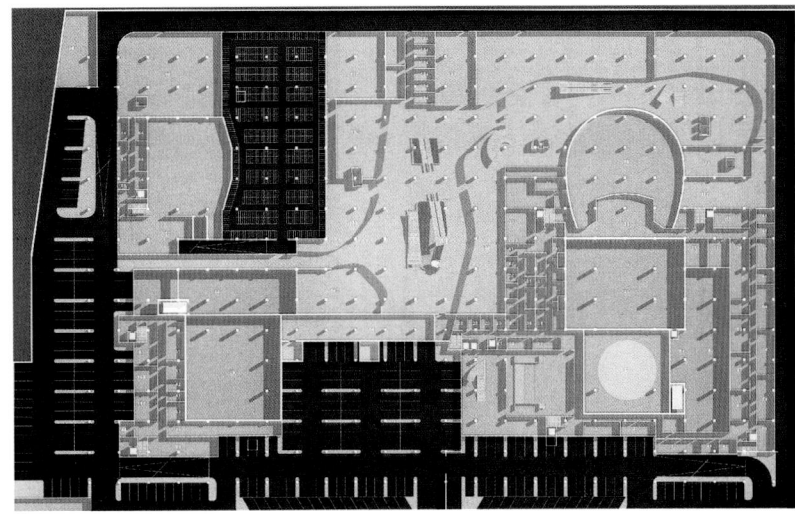

地下一层
FIRST BASEMENT PLAN

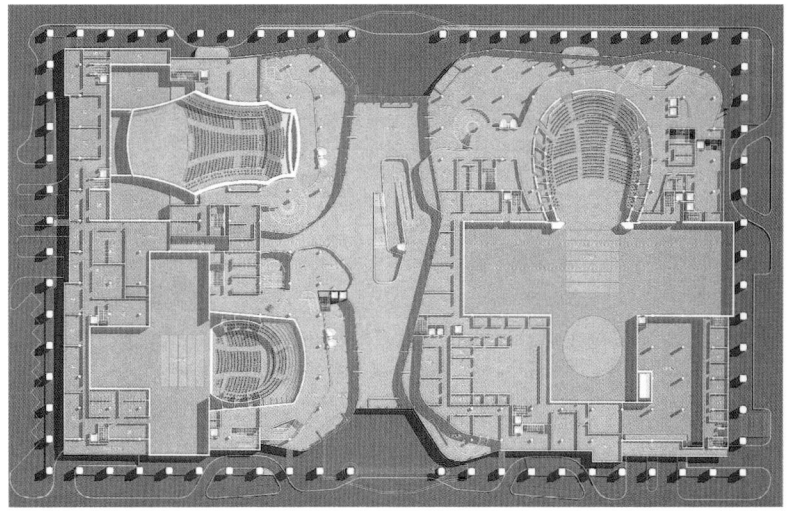

三层平面
3rd FLOOR PLAN

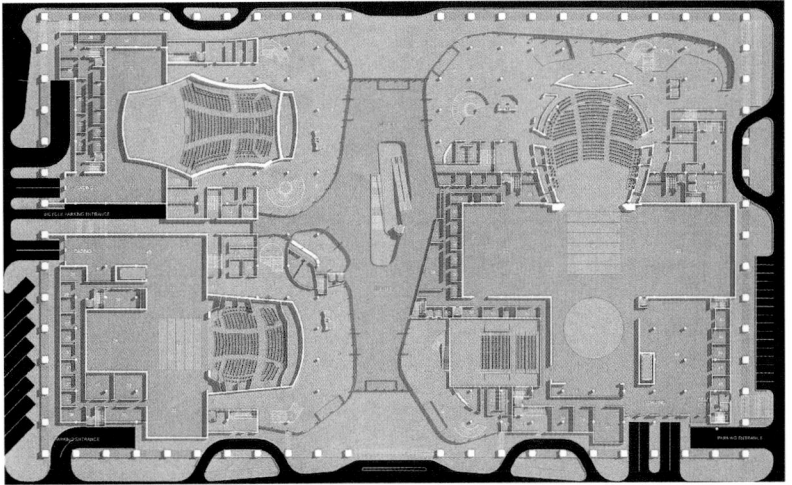

一层平面
GROUND FLOOR PLAN

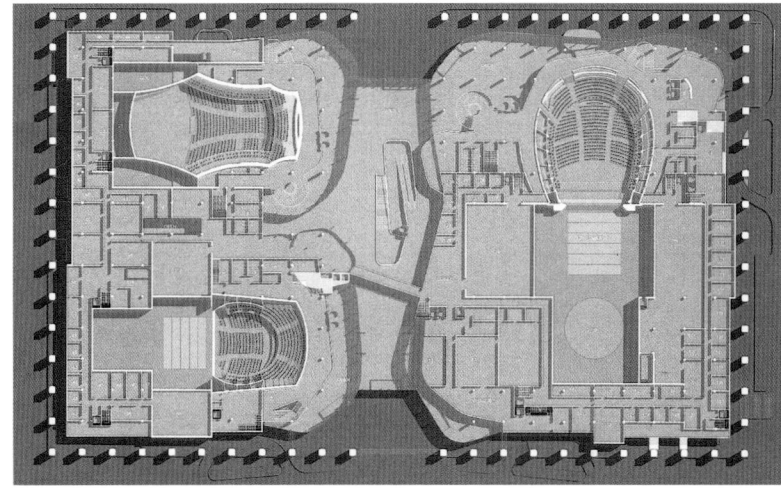

四层平面
4th FLOOR PLAN

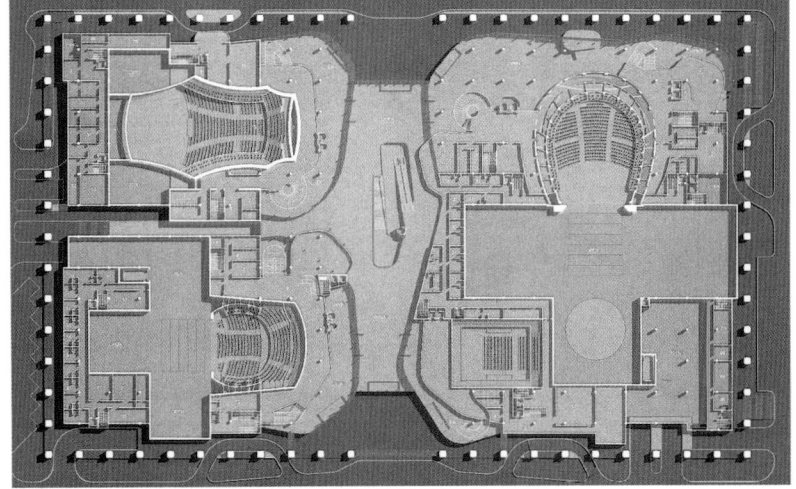

二层平面
2nd FLOOR PLAN

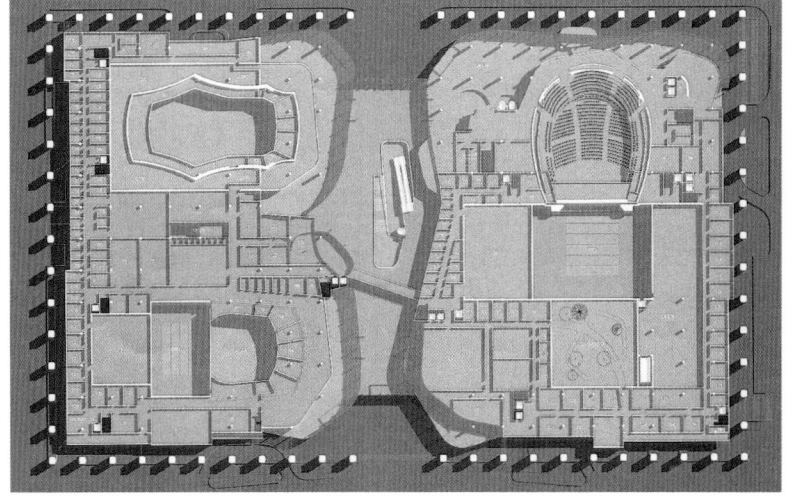

五层平面
5th FLOOR PLAN

287

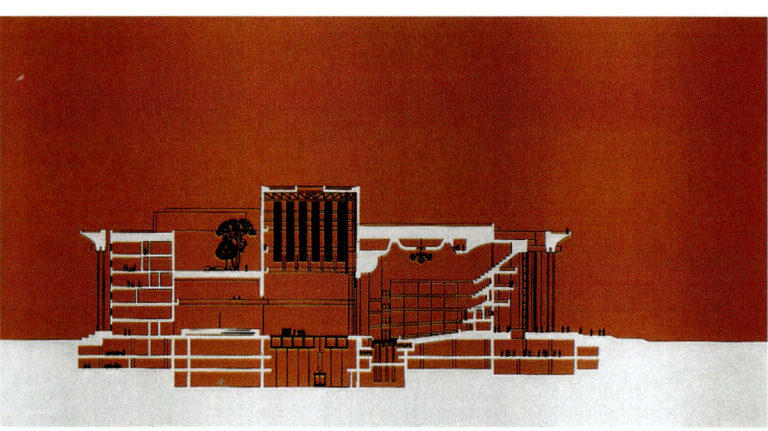

横剖面
CROSS SECTION

技术经济指标
Technical Economic Index

基地面积(m²) Site Area	总建筑面积(m²) Total Floor Area	建筑高度(m) Building Height	建筑覆盖率(%) Building Coverage Ratio
43083	地上：66327 地下：64582	30/42（局部）	45.3

容积率 Plot Ratio	绿化覆盖率(%) Green Coverage Ratio	停车数量 Number of Stalls	自行车停车数量 Number of Bicycle Stalls
3	15	地上：50 地下：543	1013

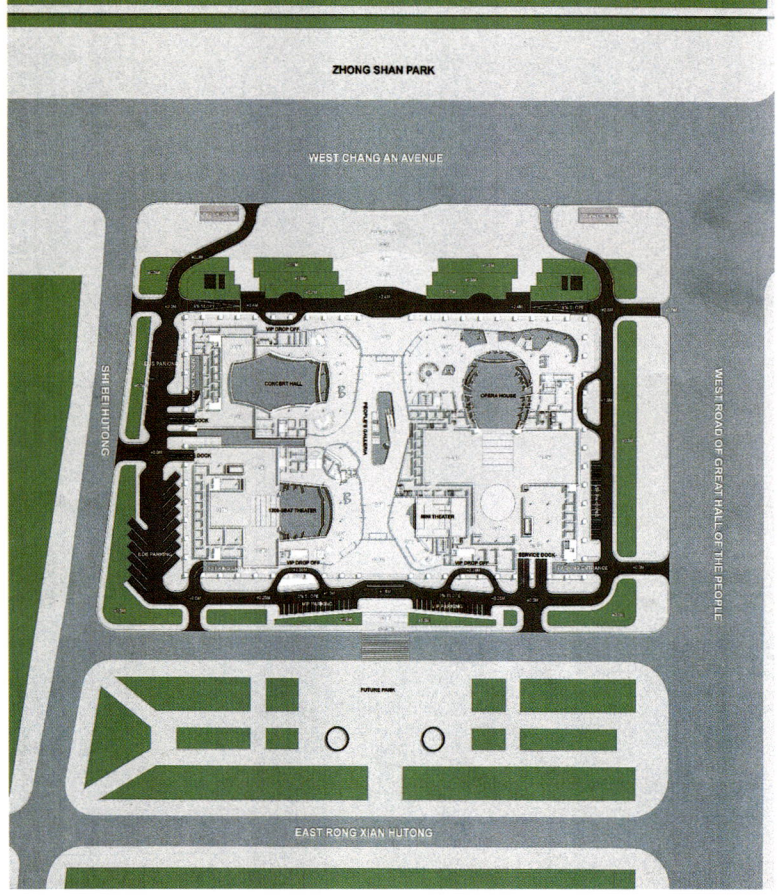

总平面
OVERALL SITE PLAN

各观众厅数据
Data For Auditoria

		歌剧院 Opera House	音乐厅 Concert Hall	戏剧院 Theater	小戏场 Mini Theater
	观众厅面积(m²) Floor Area of Auditorium	968	1047	617	517
	观众厅体积(m³) Volume of Auditorium	19565	20228	10154	6708
座席数 Seating Capacity	总座席数 Seating Capacity(Seats)	2500	2015	1211	597
	其中：池座 Among Them:Auditorium	1730	1500	880	375
	楼座 Floor Seat	770	515	333	222
	休息厅面积(m²) Lounge Floor Area	1740	540	420	375
舞台尺寸(m²) Stage Dimentions	主舞台 Main Stage	972	440	540	240
	左右侧台 Left and Right Side Stages	1296	440	513	
	后舞台 Back Stage	650		324	
	台仓 Under Stage Storage	972	360	600	320
	升降乐池 Elevating Orchestra Pit	172	150	180	
	最大视距(m) Max. Visual Distance	38.3	35.2	30.1	25.0
	最大俯角(°) Max. Angle of Depression	36.5	21.0	21.5	27.0

14

欧博迈亚设计咨询
有限公司＋
戴尔曼教授建筑
设计事务所(德国)
Design Group:
Obermeyer+Deilmann
(Germany)

本方案设计者认为：剧院对于舞台操作的顺利进行具有很高的要求，不仅要给观众振奋、赏心悦目之感，而且要方便演员和舞台工作人员的工作。在功能上，不仅要满足歌剧、芭蕾、戏剧、音乐会和室内剧等不同层面上的要求，还要满足中国民间戏剧的演出条件和国际剧团的演出需要。此外，在建筑学的表现上，室内的环境质量上，要体现地理位置的特殊性及这座建筑物的特征；既宏伟壮观又通透明亮，既有典雅的艺术性又有鲜明的时代感。使之成为中国标志性的建筑物之一。

国家大剧院应具有中华民族传统的特点，但也应具有现代化建筑的特色。一味追求传统形式，将无法反映社会时代的发展。应当在继承传统的义务和反映时代精神责任之间找到合理的结合点。

另外，也应考虑建筑地理位置的重要性和与其有关的城市规划的要求。以天安门广场为轴线，位于西长安街的国家大剧院建筑用地同位于东长安街的公安部大楼要相互对应。这样，国家大剧院的主要朝向也就明确定为朝北方向，朝向主要交通干道。然而，朝南也是作为重要出入口，即有直接的通道与城市绿化带连接。

本方案设计既庄重又典雅，既与环境相配又透射出现代建筑的气息，看上去像一个极有吸引力的"艺术殿堂"。共两层的购物层与南北轴线的长廊自然、紧密的连接，有利于人们的联系、交流。

建筑物结构布局的特点是按其主要功能厅(歌剧院、戏剧场、音乐厅和小剧院)进行划分。内设一条长廊，中心为一汇聚点——圆形中庭将不同的功能厅有机地联系在一起。明亮的圆形中庭和入口大厅内外呼应，红色的圆柱廊点缀出现代化大剧院内在的中国情调。通透的长廊也将不同的楼层在水平和垂直方向连接起来，使它们在使用上能够相互联通、配合，并且使西长安街和拟建的绿化地带之间能够相互联通。在功能的分配上，将用途相同的区域相对集中在一起，其顺序依次为：

●长廊与店铺通道、车道边和入口；
●入口前厅、休息厅及观众厅。

The planners naturally assume that the theatre operation will place high demands on the smooth functioning of the stage presentation, and that the demands of the players and the stage staff are not to be taken any less seriously than those of the public which is in search of personal edification. From the functional aspect, the problem is rooted in the many-sided aspects of the expectations and demands.This not only involves the different requirements for opera, ballet, drama, concerts and chamber theatre. Both the presentation demands for Chinese folk opera and the production conditions for international theatre ensembles must be equally met in the same efficient manner, In its architectural expression, and in the quality of the interiors, the demands on the siting and purpose of the building must be convincingly reflected: dignity and majesty as well as brightness, openness, clearness, elegance, artistry in addition to the spirits of the times.

The National Grand Theatre should be characterised by Chinese national tradition, but should also possess the features of today's modern architecture. A strict adherence to traditional forms would be in contradiction to the present social developments. A synthesis is to be sought between the commitment to tradition and the obligations to the present and the future.

No less important in this regard is the geographic location and its associated urban planning requirements. In relationship to the Tian An Men Square axis, the site chosen for the National Grand Theatre on West Chang An Avenue requires that it be analogous with the office building of the Ministry of Public Security on East Chang An Avenue. The principal orientation for the National Grand Theatre is thus plainly specified: to the north, towards the major avenue. But nevertheless there are entrances as well as exits also from/to the south, offering direct access to the planned greenbelt.

In its unified overall architectural form-which, however, also makes comprehensible the differing functions to be found there-the National Grand Theatre as proposed in this design represents an inviting 'Temple of the muses' with its 'Urb an two -story mall',the great inner communication axis leading from north to south.

Characteristic for the structure of the complex is its interior division in accordance with its main auditoriums: opera house, concert hall, theatre and mini theatre. One internal mall with a brilliant rotunda in the centre organically joins these four main components. The rotunda and the entrance hall echo each other whereas the red pillars arranged in a circle present a unique Chinese touch to the modern building complex. The mall also connects the various levels, horizontally and vertically, making variations in usage possible and providing a through way between West Chang An Avenue and the new planned park. This functional arrangement connects zones of similar usage together at the following stages:

●the stroll-about mall with the shopping passage, entrances and vehicle approaches

●the entrance lobbies and lounges with the auditoriums

The different usage areas are clearly comprehensible in the architectural design.

北侧透视
PERSPECTIVE VIEW OF NORTH ELEVATION

南侧透视
PERSPECTIVE VIEW OF SOUTH ELEVATION

圆型大厅
VIEW OF ROTUNDA

公共大厅
PUBLIC HALL

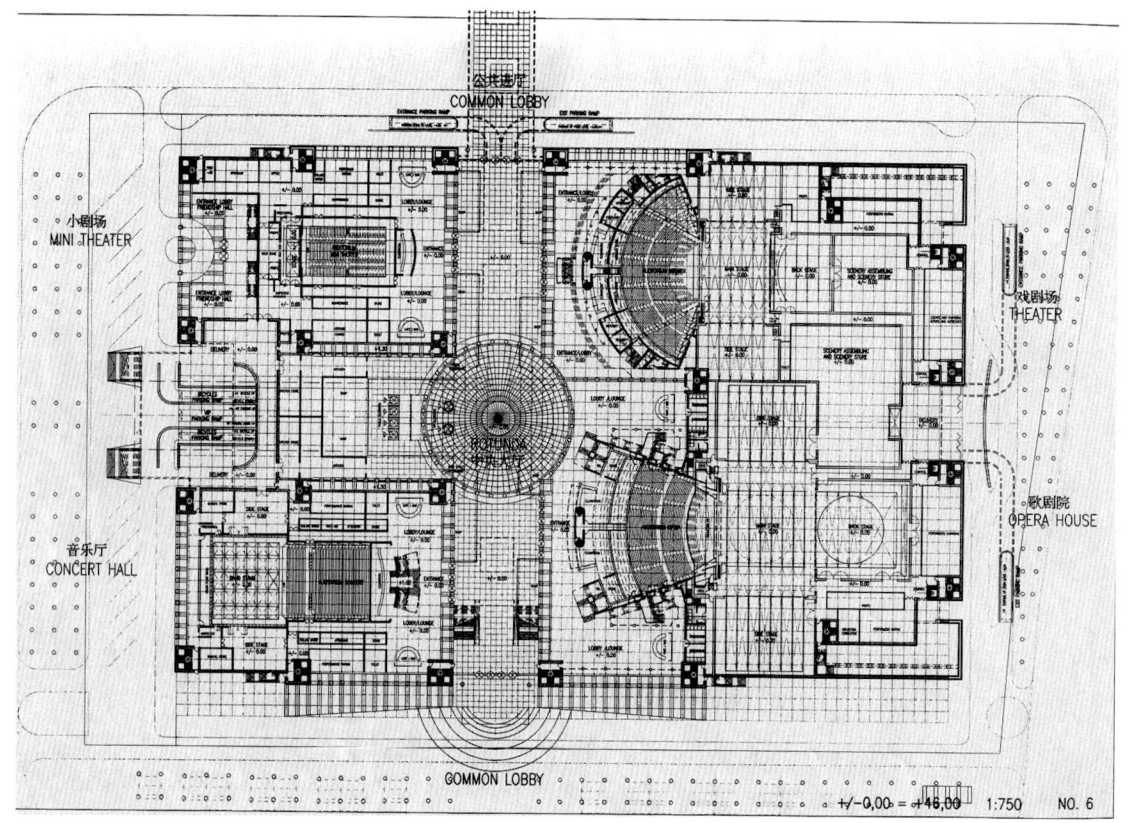

一层平面
GROUND FLOOR PLAN

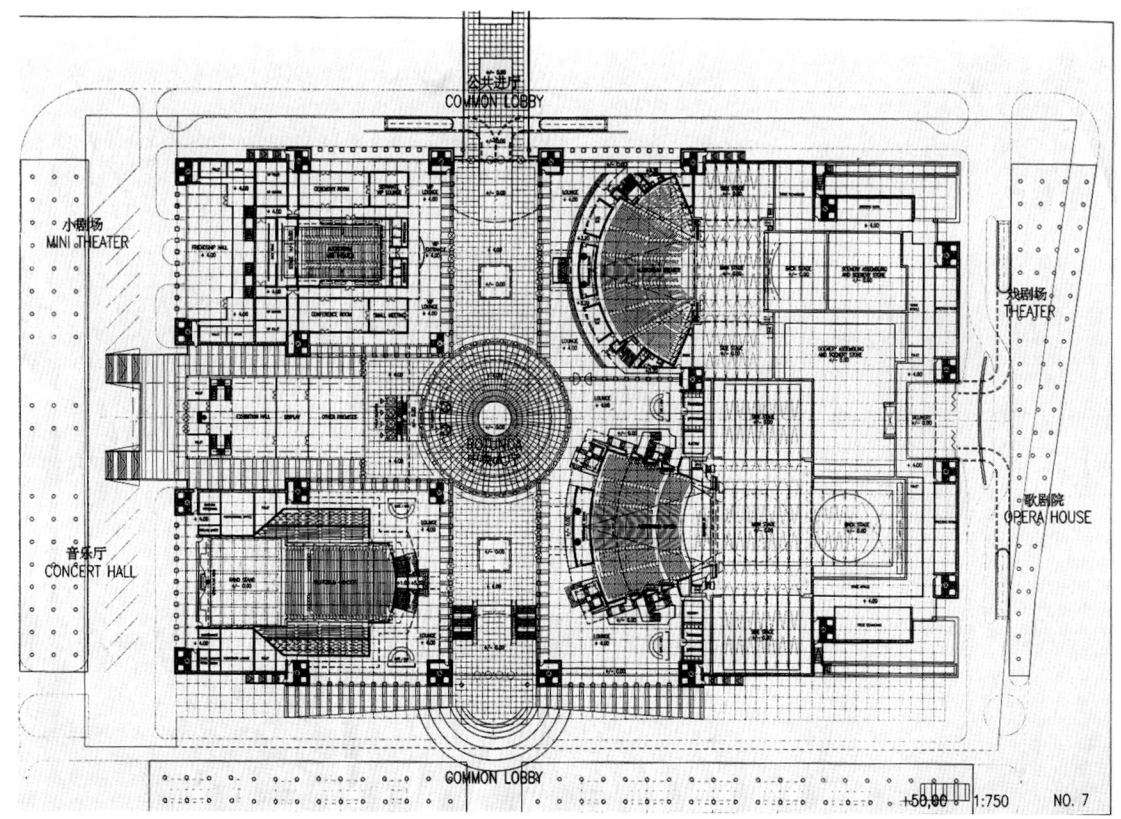

二层平面
2nd FLOOR PLAN

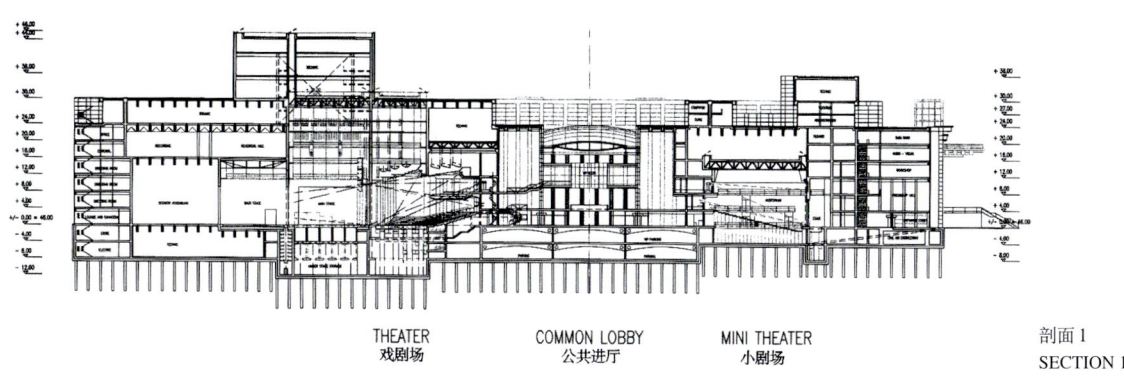

THEATER
戏剧场
COMMON LOBBY
公共进厅
MINI THEATER
小剧场

剖面 1
SECTION 1

CONCERT HALL
音乐厅
COMMON LOBBY
公共进厅
OPERA HOUSE
歌剧院

剖面 2
SECTION 2

北侧鸟瞰
BIRD'S-EYE VIEW OF NORTH SIDE

音乐厅
歌剧院
歌剧院
戏剧场
音乐厅
小剧场
艺术画廊

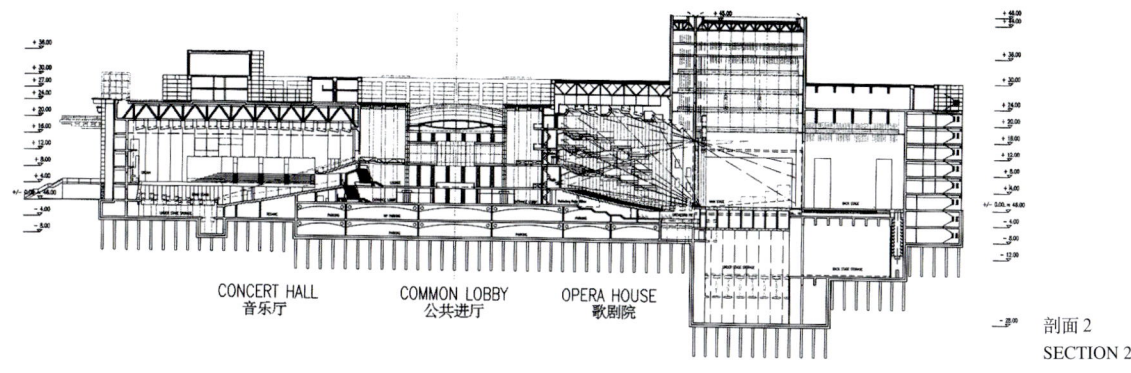

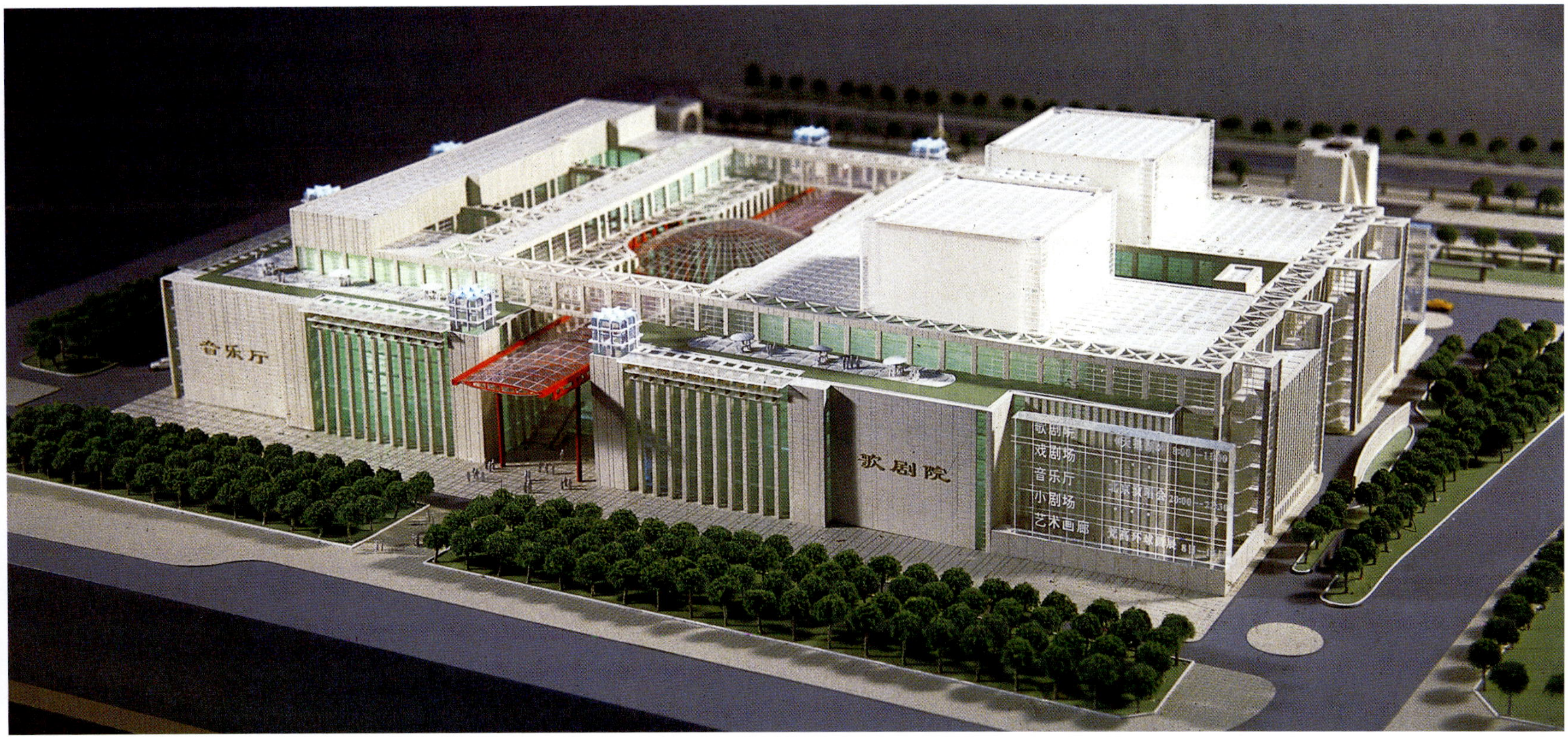

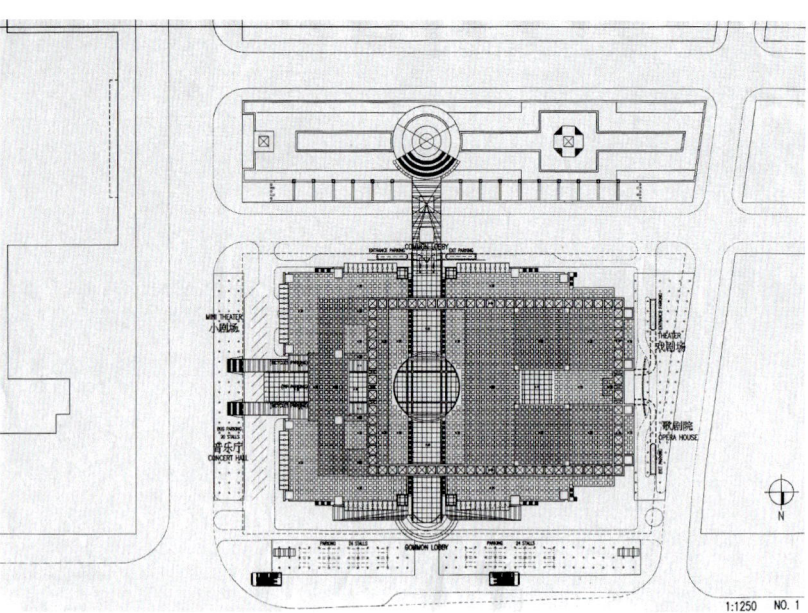

总平面
OVERALL SITE PLAN

技术经济指标
Technical Economic Index

基地面积(m²) Site Area	总建筑面积 (m²) Total Floor Area	建筑高度(m) Building Height	建筑覆盖率 (%) Building Coverage Ratio
38916.0	地上：85383 地下：44085	24/45(局部)	73

容积率 Plot Ratio	绿化覆盖率 (%) Green Coverage Ratio	停车数量 Number of Stalls	自行车停车数量 Number of Bicycle Stalls
2.2	6	地上：74 地下：595	1200

各观众厅数据
Data For Auditoria

		歌剧院 Opera House	音乐厅 Concert Hall	戏剧院 Theater	小戏场 Mini Theater
	观众厅面积(m²) Floor Area of Auditorium	1866	1428	920	1783
	观众厅体积(m³) Volume of Auditorium	25022	25162	15776	4837
座席数 Seating Capacity	总座席数 Seating Capacity(Seats)	2518	1972	1210	500
	其中：池座 Among Them:Auditorium	1258	990	968	440
	楼 座 Floor Seat	1260	982	242	60
	休息厅面积(m²) Lounge Floor Area	3018	2452	1329	621
舞台尺寸 (m²) Stage Dimentions	主舞台 Main Stage	972	437	540	100
	左右侧台 Left and Right Side Stages	1175	472	627	
	后舞台 Back Stage	650		330	
	台 仓 Under Stage Storage	954	195	490	225
	升降乐池 Elevating Orchestra Pit	159		200	
	最大视距(m) Max. Visual Distance	37	35.5	27	25
	最大俯角（°） Max. Angle of Depression	29.5	17	23.5	11.5

模型透视
MODEL PERSPECTIVE

第二轮竞赛第一次修改后的方案（4 项）

Part of the Schemes of the 2nd Round Competition and Their First Modifications (four items)

01　法国巴黎机场公司(法国)
ADP Aeroports de Paris(France)
清华大学(中国) 协作
Joint Collaborator: Tsinghua University(China)

05　塔瑞 · 法若建筑师事务所(英国)
Terry Farrell & Partners(UK)
北京市建筑设计研究院(中国)
Beijing Institute of Architectural Design &
Research(China)

11　卡洛斯 · 奥特建筑师事务所(加拿大)
Carlos Ott & Associates, Architects(Canada)
建设部建筑设计院(中国)
Architectural Design Institute of Ministry of
Construction(China)

清华大学(中国)
Tsinghua University(China)
法国巴黎机场公司(法国) 协作
Joint Collaborator:ADP Aeroports de Paris(France)

01
法国巴黎机场
公司（法国）
ADP Aeroports de
Paris(France)
清华大学协助
（中国）
Joint Collaborator:
Tsinghua University
(China)

第二轮说明

根据我们的最初设想，北京国家大剧院应首先成为一个"都市剧院"。这座紧靠天安门广场和"紫禁城"的象征性建筑的功能性设计包括两个独特的入口，一个在长安街上，另一个位于人民大会堂一侧的花园内。首先，这两个入口的设计不同，但同样重要的入口将形成在建筑物内部穿流的人群。其次，在歌剧厅顶部舞台升降机周围设计一个可俯瞰大部分城区的休息室。为体现大剧院属于所有北京市民的构思，我们把它的主立面象征性地设计成一座"舞台"，向城市开放。我们认为，应进一步突出这一构思，把大剧院建成一个舒适的步行场所，一个时刻充满活力的戏剧之城。为此，我们第二阶段的工作重点之一是设计大剧院内部道路和体现戏剧的神秘魔幻意境。此外，附属的商业和娱乐设施也将增加其对外吸引力，使"都市剧院"成为真正的"戏剧之城"。修改方案扩大了人行面积，重新设计了向公众开放的商场和餐厅区。为了突出开放感，我们除了在两侧主立面采用半透明的玻璃墙外，还特别构思了一个钢和玻璃结构的张开的天篷。它通过自然光或灯光造成了一种独特的氛围：整个顶篷变成了一个巨型的"歌剧厅分枝烛台"。除了大厅里的观众之外，所有行人都可一睹其神韵。大剧院外观和会议厅的设计也在营造一种更为亲切活泼的氛围。灰色的观演厅，会议厅墙面与被休息厅点缀得金碧辉煌的歌剧厅墙面曲直各异，相映成趣。这里的每一幢建筑物都像是一道道开放的幕帘，逐渐把人引向神秘的心脏。各入口处的白色大幕日夜都可通过投影预告剧目，介绍剧情概要，或打出绚丽变幻的艺术灯光。这时，闪闪发光的"枝形烛台"重新呈现出象征长安街外部"舞台"的银色。大厅内地面为素雅的灰色，嵌以发光的美人蕉果图案。其中的光线白天如同博物馆那样柔和均一，夜晚则变得欢快热烈，将观众从容不迫地引入舞台梦境。

The National Grand Theater of Beijing is, first and foremost, "a theater in the city." From the outset we regarded the special location of the site-nearTian'anmen Square and the Forbidden City,opening on one side onto a grand avenue and on the other onto a garden, and adjoining the land around the Great Hall of the People-as necessitating an original functional layout:firstly,two entrances of equal importance but different in character, and consequently the existence of interior circulations that traverse the building; secondly,an upper lobby for the opera above the stage shaft providing a commanding view of an extensive part of the city.

Because this theater is to be the theater of all the people of Beijing, we symbolically designed the main building fronts in the form of a stage opening onto the city and then we pushed this important idea of openness even further by making the theater complex into an inviting place for people to stroll through, a thriving and lively city site around the clock.

This is why we focused a significant amount of work during the second stage on the layout of interior circulations, and on seeking an architectural design that would convey their character in relation to the theatrical world, its mystery and magic while at the same time putting them at the disposal of the widest public possible and making them attractive by adding secondary shopping and entertainment functions. In this way'a theater in the city' has become a kind of 'theater city' as well. We have enlarged the circulation areas, redefined the placement of shops and services open to the public, and opened the view onto the outside through side walls in glass.

Most significantly,for the ceiling we have devised a steel and opalescent glass structure that yields an original radiant atmosphere in the sunlight during the day and the artificial light at night. This roof structure assumes the form of a gigantic chandelier, 'the opera's chandelier',but instead of being enclosed inside the opera house and reserved to theater-goers, it is offered to the sight of one and all, spectators and passersby.

The design of the envelopes of the theater and auditorium was determined by our desire to create an area that is always lively and inviting. These two halls have gray vertical walls that stand out against the gilt curved trim of the opera house wbich now picks up and brings inside the color of the roof covering the main upper lobby. The walls are perceived not so much as the sides of a closed box but as a series of screens to be crossed so as to penetrate into the core of the mystery. Each entrance is equipped with a huge white screen which can be used for projections to announce current or upcoming performances,to show excerpts from previous ones,or even simply as a site for temporary artworks.

The radiant 'chandelier',lit from end to end,will pick up the silvery color of the building's outer 'stage' walls. The floor will be gray and very calm, with small recessed lights at regular intervals. The quality of the lighting during the day will be uniform and subdued as in a museum, while at night it will be lively and cheerful, unhastily leading throngs of spectators into the show world of fiction and dreams.

二轮修改后说明

这本金皮书在不改变建筑物原有的风格和主要布局的前提下，从三个方面对原设计进行了修改。其中两项有关外貌和象征意义的主要改动涉及外主立面和歌剧厅休息厅的穹顶。我们接受了把国家大剧院的主立面的尺寸比例与近旁的人民大会堂协调起来的宝贵意见，以使这两座建筑都能充分展现其象征意义。这一点与我们强调的国家大剧院应与历史、政治和文化相呼应的想法完全一致。为了更好地完成这一平衡和协调，经过研究，我们认为可在不影响入口庄严外观的前提下，将其尺寸略为缩小，再加上把两侧玻璃的北顶端向主立面方向翻转过来。这样，位于长安街上的主立面既能保持原有的力度，又显得更为轻盈，突出了象征意义。在休息厅的改造中，我们本着不影响象征性意义和使用功能的宗旨，设计了一个新颖、低扁的方形金顶，使整体设计得到更新。虽然我们不能断言这样做是否更具有中国特色，但从外表来看，它充满了现代化气息和个性，夜间熠熠发光的金顶像一盏明灯那样，给人带来同样的神奇的感觉。

Neither the principles nor the main layouts have been modified here. Further specifications have been brought on three points, two of which are essential insofar as they have an impact on the building's outer appearance and on how it will be perceived. The modificationg in the design of the building't two main fronts and the roof over the opera house lounge both come in response to comments that were made to us concerning the symbolic perception of the building. We regard these comments as precious. The need to create a balanced relationship between the scale of the Grand Theater's main fronts and that of the Great Hall of the People so as to bring out the symbolic import of each building is something that we hoped to achieve by underscoring the importance of the site in terms of the triple relationship between history, power and culture. Thanks to a comment that was made, we then realized that we could achieve an even better balance by slightly reducing the size of the main entrances without in any way diminishing their force and by extending the side walls in glass around both corners. The Chang An Avenue front is now lighter but just as intense, and its sense is clearer to all. Furthermore, by taking into consideration the comments made about the symbolic reading that could be given to the shape of the lounge,and finding appropriate functional arrangements to modify this without effecting the use of the building, and finally by designing a new element, a lower extremely flat roof,we believe that the complex's overall conception has been improved. Some might say that this new roof is more Chinese-it is not up to us to judge-but it is clearly more modern in its shape and its construction. It will crown the theater in a way that is at once more humble and more distinctive. Its gilt underside will glow at night with the same radiant chandelier effect and mysterious quality that we have been seeking all along.

北侧透视 (修改)
PERSPECTIVE OF NORTH ELEVATION(MODIFIED)

南侧透视 (修改)
PERSPECTIVE OF SOUTH ELEVATION(MODIFIED)

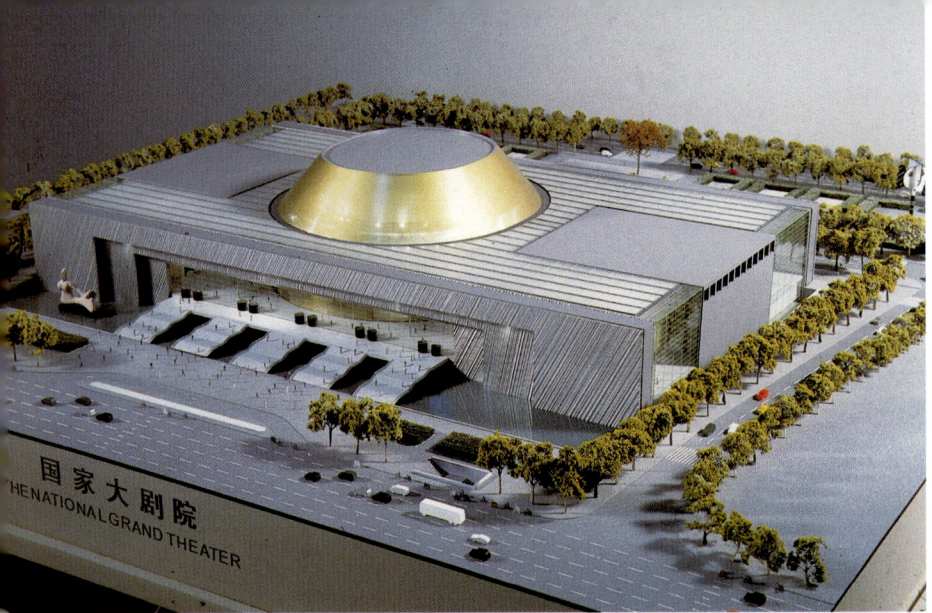

北侧模型透视
NORTH SIDE OF THE MODEL

南侧模型透视
SOUTH SIDE OF THE MODEL

戏剧院入口
THEATER ENTRANCE

歌剧厅和音乐厅之间的公众大厅
PUBLIC HALL BETWEEW THE OPERA HOUSE AMD THE COWCERT-HALL

歌剧厅入口
OPERA HOUSE ENTRANCE

南主大厅
VIEW ON THE SOUTHERN MAIN HALL

戏剧厅休息大厅
LOUNGE OF THE THEATER

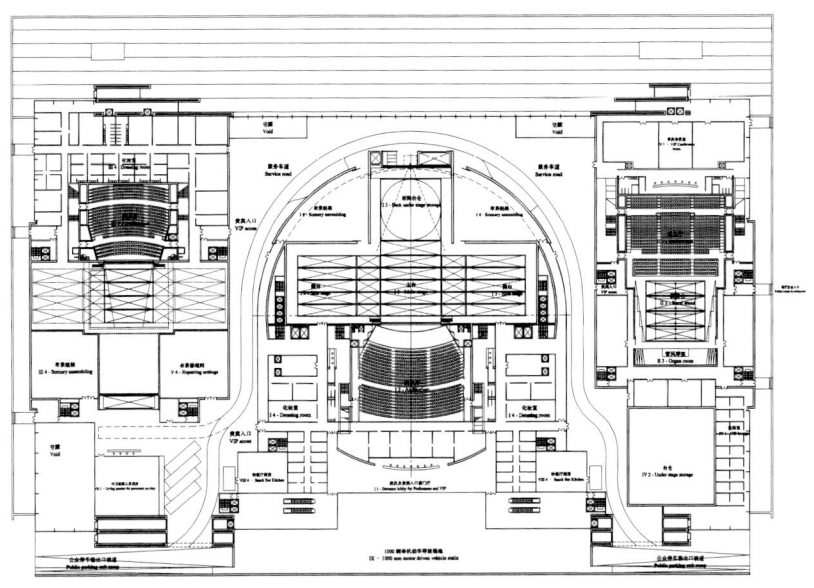

舞台层(-0.65)修改
STAGE LEVEL (-0.65)(MODIFIED)

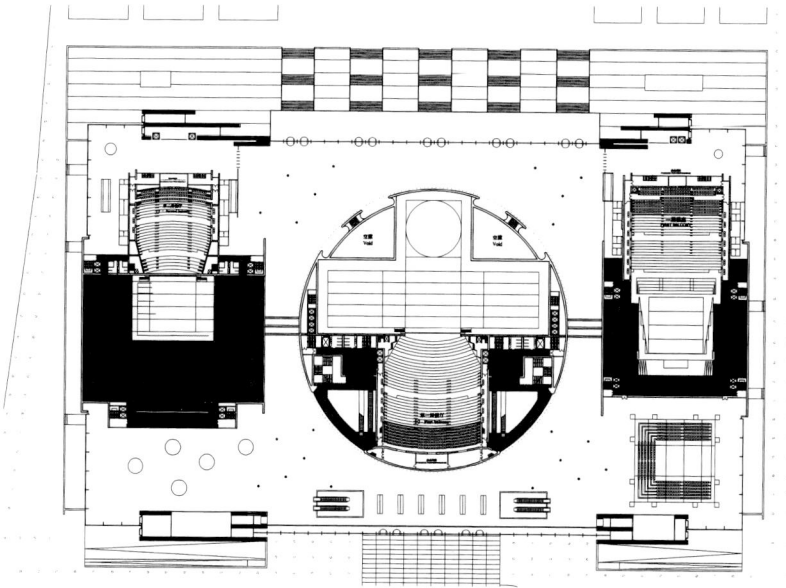

楼厅层(11.50)修改
BALCONIES LEVEL(11.50)(MODIFIED)

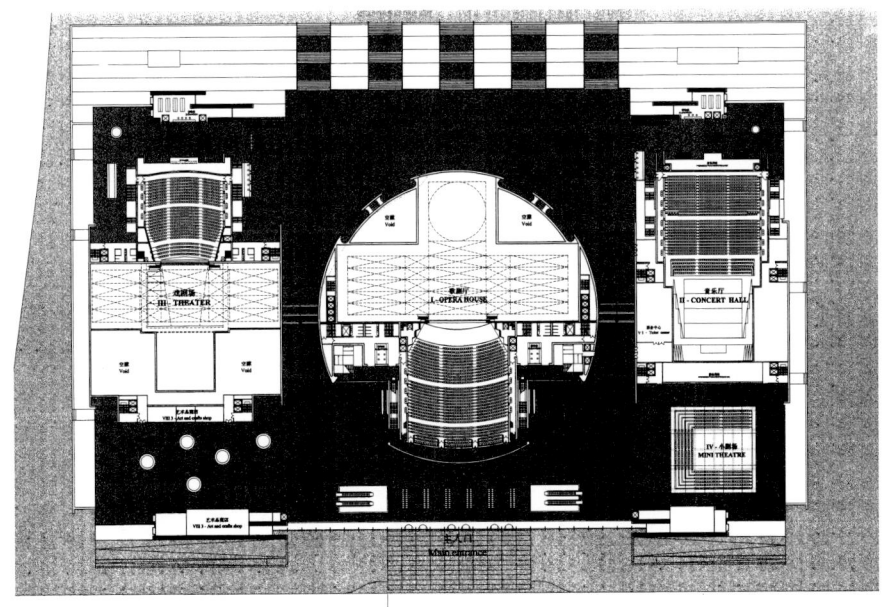

公众入口及池座层(4.35,5.75)修改
PUBLIC ENTRANCE AND STAUS LEVEL(4.35,5.75)(MODIFIED)

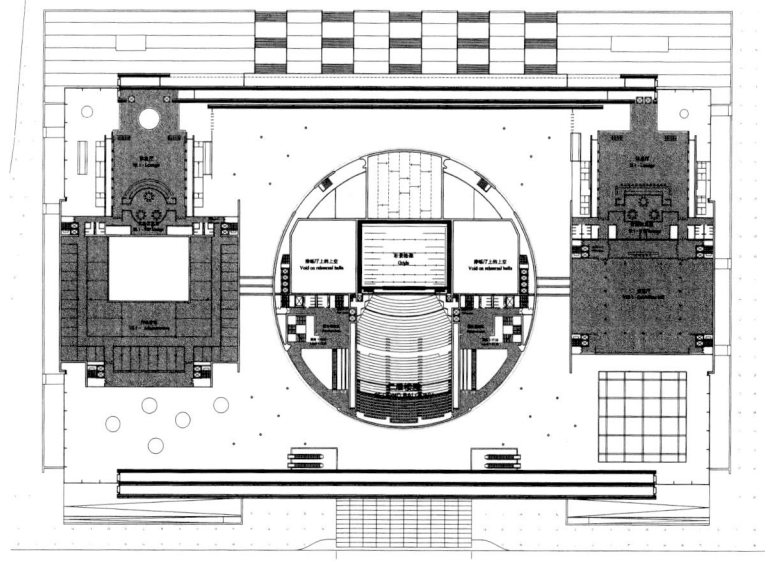

歌剧厅及音乐厅休息室层(17.25)修改
OPERA HOUSE AND CONCENT HALL LOUWGE LEVEL(17.25)(MODIFIED)

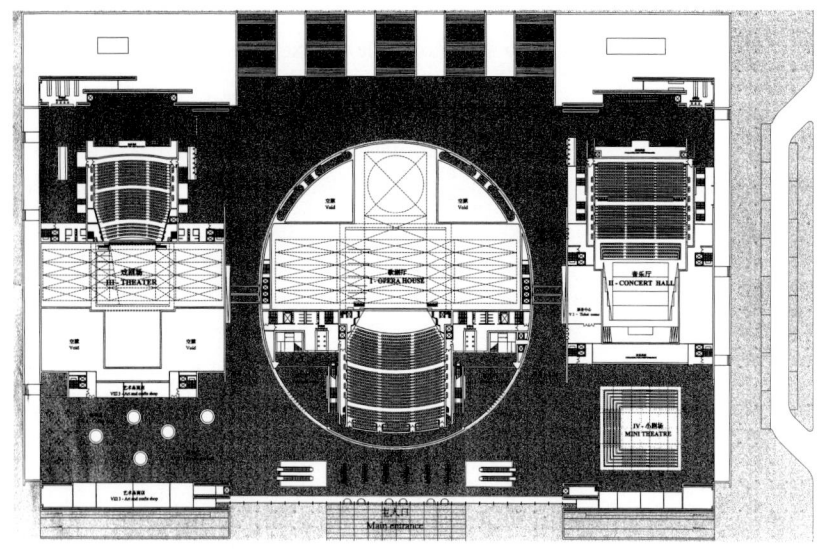

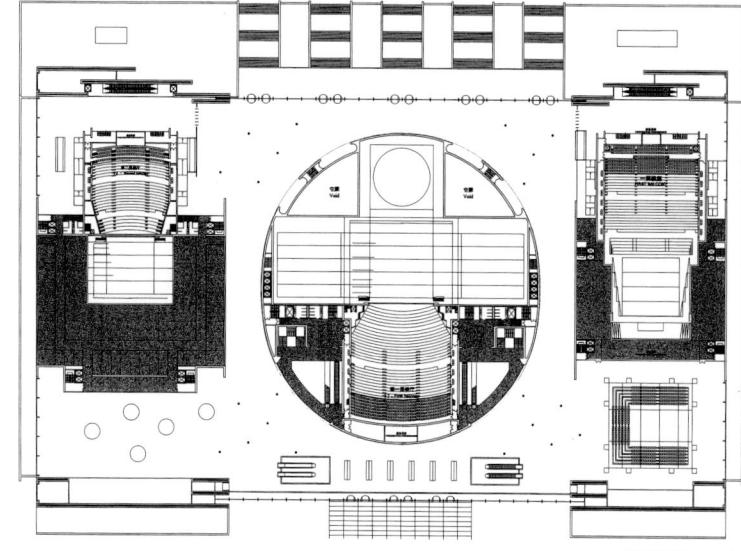

公众入口及池座层(4.35,5.75)
PUBLIC ENTRANCE AND STALLS LEVEL(4.35,5.75)

楼厅层(11.50)
BALCONIES LEVEL(11.50)

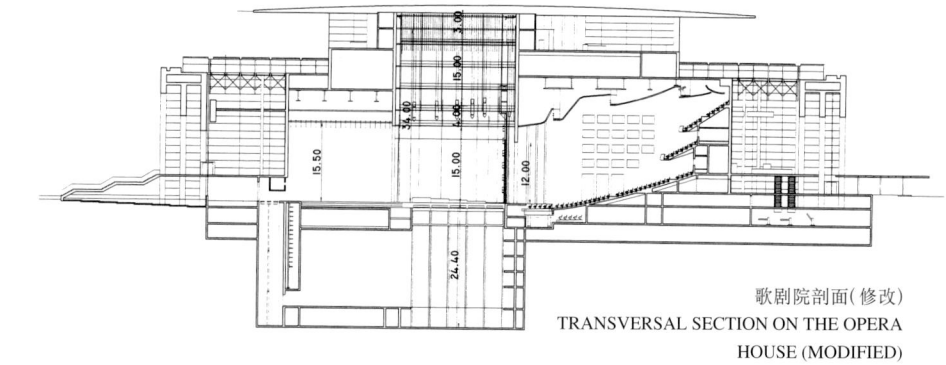

歌剧院剖面(修改)
TRANSVERSAL SECTION ON THE OPERA
HOUSE (MODIFIED)

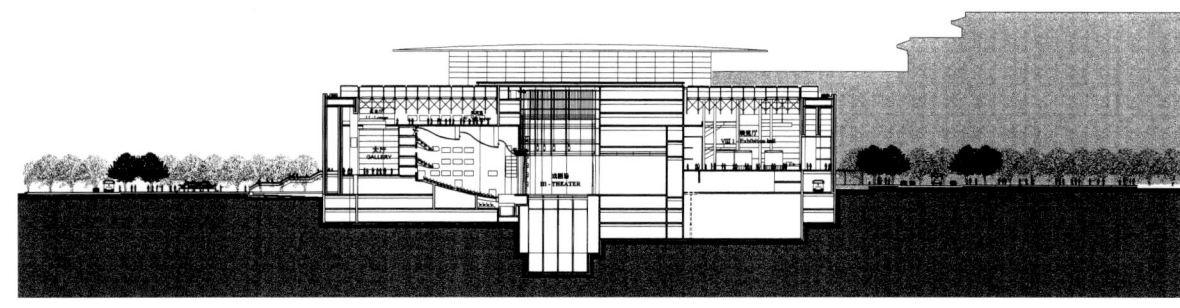

戏剧厅剖面(修改)
THEATER SECTION (MODIFIED)

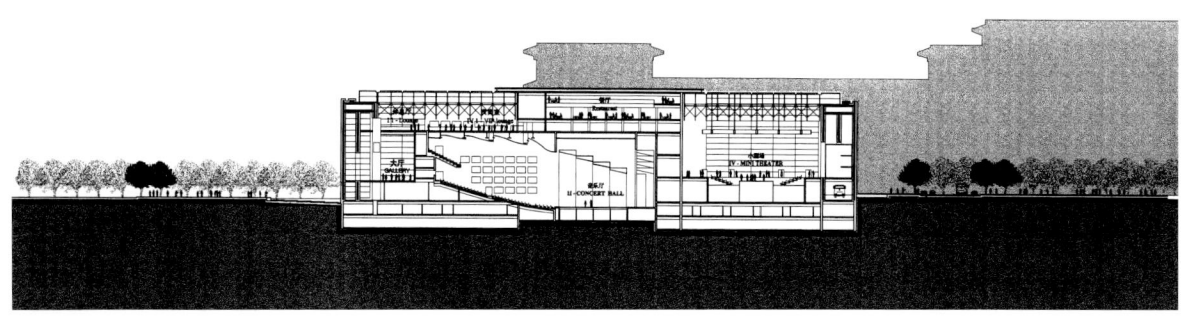

音乐厅及小戏剧厅剖面(修改)
SECTIONS OF THE CONCERT HALL AND
MINI-THEATER(MODIFIED)

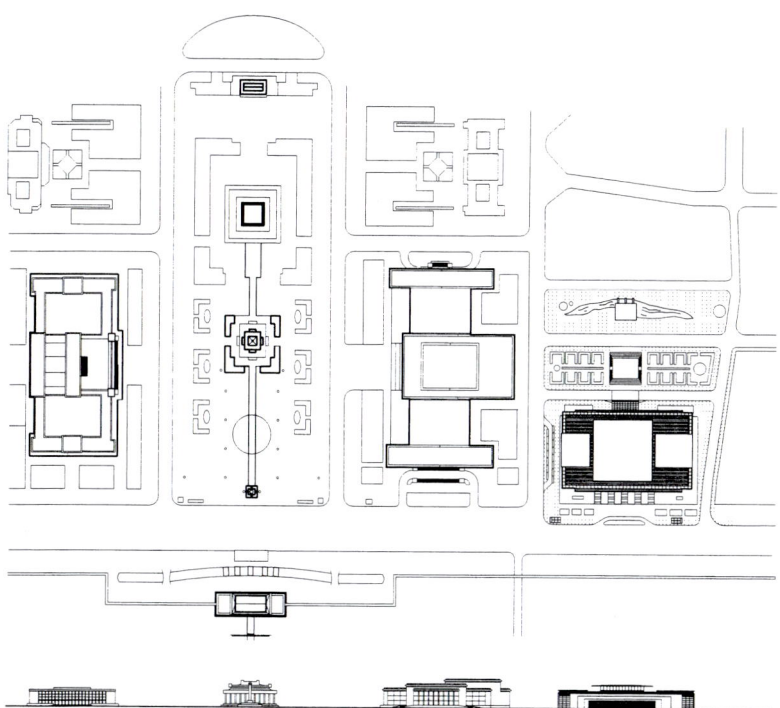

地域总平面(修改)
SURROUNDING ENVIRONMENT MAP (MODIFIED)

基地面积(m²) Site Area	总建筑面积(m²) Total Floor Area	建筑高度(m) Building Height	建筑覆盖率(%) Building Coverage Ratio
43200	地上：69000 地下：51000	27.50/38.30(局部)	75
容积率 Plot Ratio	绿化覆盖率(%) Green Coverage Ratio	停车数量 Number of Stalls	自行车停车数量 Number of Bicycle Stalls
2.8		地上：550 地下：20	1200

各观众厅数据
Data For Auditoria

		歌剧院 Opera House	音乐厅 Concert Hall	戏剧院 Theater	小戏场 Mini Theater
	观众厅面积(m²) Floor Area of Auditorium	2400	1600	1200	980
	观众厅体积(m³) Volume of Auditorium	19500	21500	9000	11000
座席数 Seating Capacity	总座席数 Seating Capacity(Seats)	2500	2000	1200	520
	其中：池座 Among Them:Auditorium	1630	1450	810	300/500
	楼座 Floor Seat	870	550	390	
	休息厅面积(m²) Lounge Floor Area	4000	3800	3200	1462
舞台尺寸 Stage Dimentions (m²)	主舞台 Main Stage	860	480	570	700
	左右侧台 Left and Right Side Stages	995	300	640	
	后舞台 Back Stage	525		320	
	台仓 Under Stage Storage	1400	200	570	960
	升降乐池 Elevating Orchestra Pit	120	320	75	
	最大视距(m) Max. Visual Distance	40	33	26	
	最大俯角(°) Max. Angle of Depression	28	22	30	

总平面
OVERALL SITE PLANE

塔瑞·法若建筑师
事务所（英国）
Terry Farrell &
Partners(UK)
北京市建筑设计
研究院（中国）
Beijing Institute of
Architectural Design &
Research (China)

经过第二阶段竞赛之后，此方案集中对评审委员所提供的意见重点加以修改及发挥。

●北京是中国历代的首都，中国国家大剧院之位置处于此文化及政治之中心—北京，其设计必须要和谐地配合整个城市设计。

●建筑物的布局安排及轮廓面貌与毗邻建筑物平衡并置，互相协调。

●将现代建筑设计及古代传统中国建筑设计达到一个真正的交流。

●形成与古代传统及现代技术的一条建筑桥梁。

●创造一种融汇开放性及包容性的建筑设计。

●利用技术性的巧妙布局安排来发挥设计效益。

●采用智能型设计来发挥能源效益。

设计目的在于将第二轮建议书的成功元素结合评委意见书的意见以形成此方案书的宗旨：在北京设计一座最优秀的建筑物，亦是世界殿堂级的设计……，并为中国建筑艺术创造出一个新方向。

The design development since the stage two competition design submission has concentrated on the issues raised by the competition jury panel as follows.

● Harmonisation of the design with its unique setting in the cultural and political heart of Beijing,the ancient capital of China.

● Development of the massing and profile of the building to make a balanced composition in harmony with its surroundings.

● Development of the design that achieves a genuine fusion of modern world architecture and the ancient traditions of Chinese architecture.

● Development of architecture that forms a bridge between ancient traditions and modern technology.

● Creation of an architecture that is open and welcoming.

● The refinement of the technical planning of the building to create an efficient economical design.

● The application of intelligent building technology to create an energy efficient design.

The objective has been to integrate the successful elements form the Stage 2 submission with the requirements from the Jury Report to create a proposal that "…will be the best architectural work in Beijing and one of the top works in the world… to set a new architectural direction for China".

北侧透视(修改)
PERSPECTIVE VIEW OF NORTH ELEVATION(MODIFIED)

南侧透视(修改)
PERSPECTIVE VIEW OF SOUTH ELEVATION(MODIFIED)

北侧模型透视
VIEW FROM NORTH SIDE OF THE MODEL

南侧模型透视
VIEW FROM SOUTH SIDE OF THE MODEL

大厅内部透视(修改)
INTERIOR OF GRAND HALL (MODIFIED)

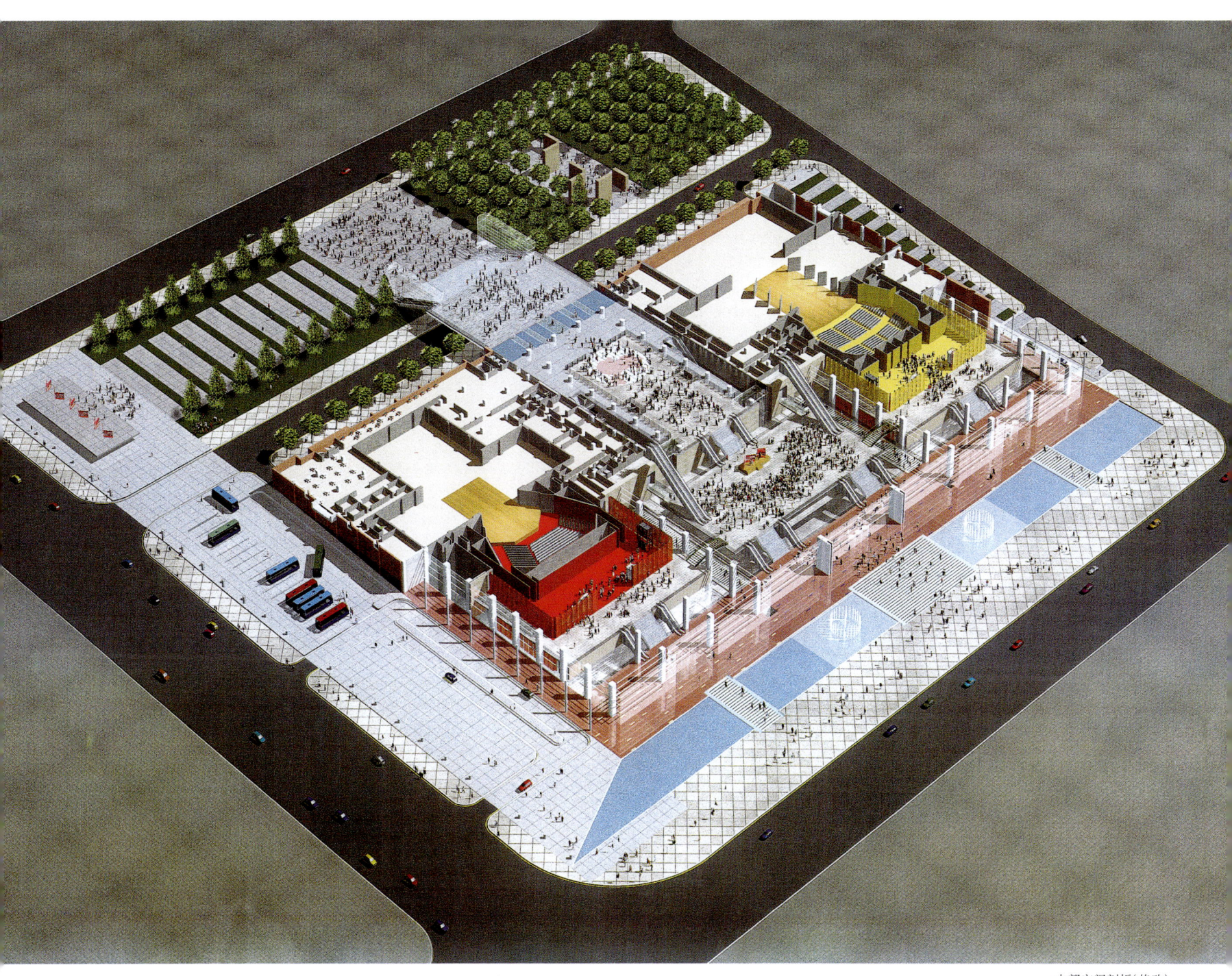

内部空间剖析(修改)
VIEW OF THE INTERIOR ARRANGEMENT(MODIFIED)

歌剧院观众厅
OPERA HOUSE AUDITORIUM

公共大厅透视
PUBLIC HALL

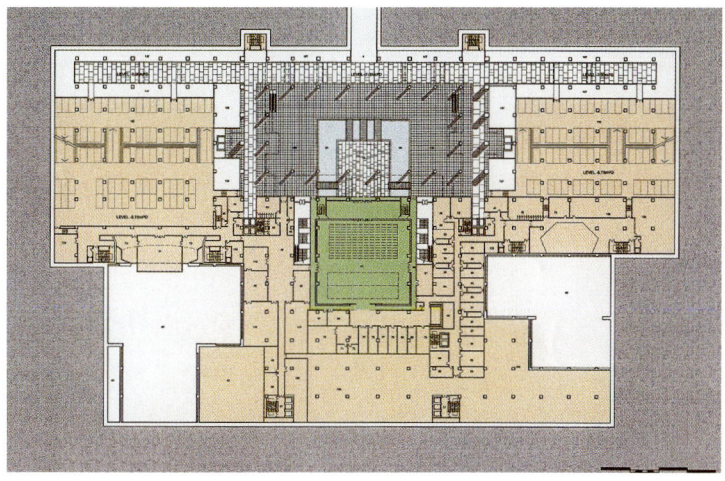

地下一层平面(修改)
FIRST BASEMENT PLAN (MODIFIED)

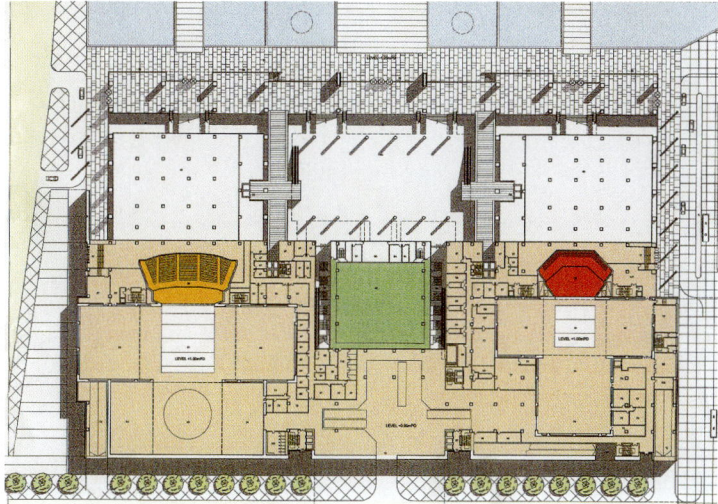

一层平面(修改)
GROUND FLOOR PLAN(MODIFIED)

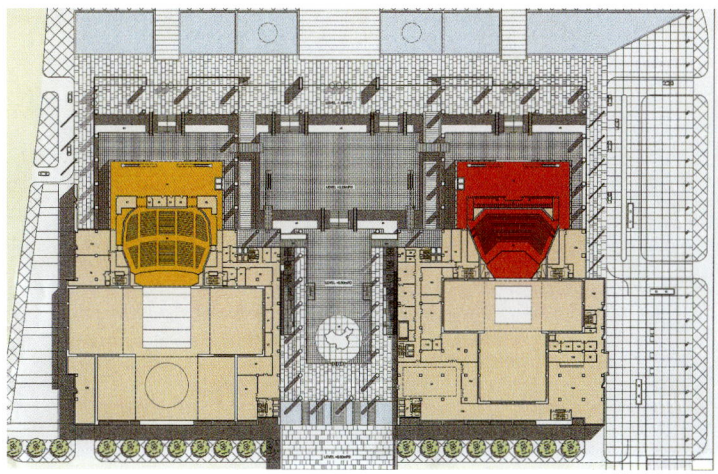

二层平面(修改)
2nd FLOOR PLAN(MODIFIED)

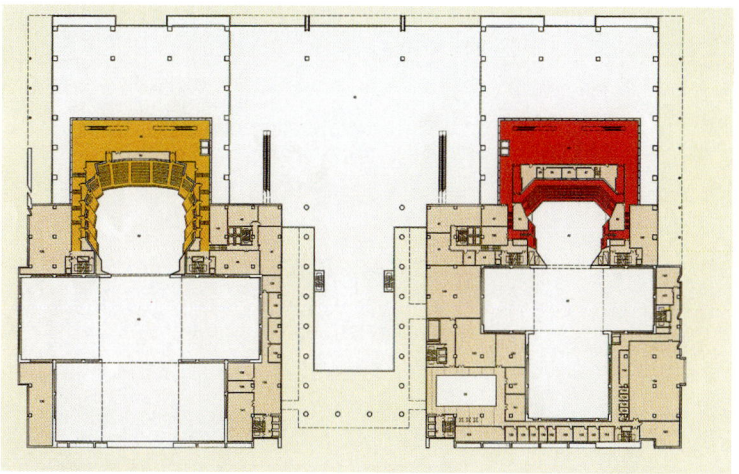

三层平面(修改)
3rd FLOOR PLAN(MODIFIED)

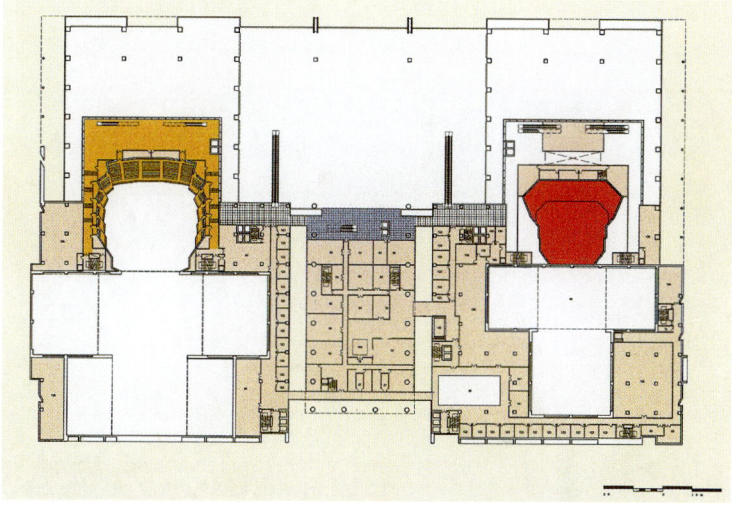

四层平面(修改)
4th FLOOR PLAN(MODIFIED)

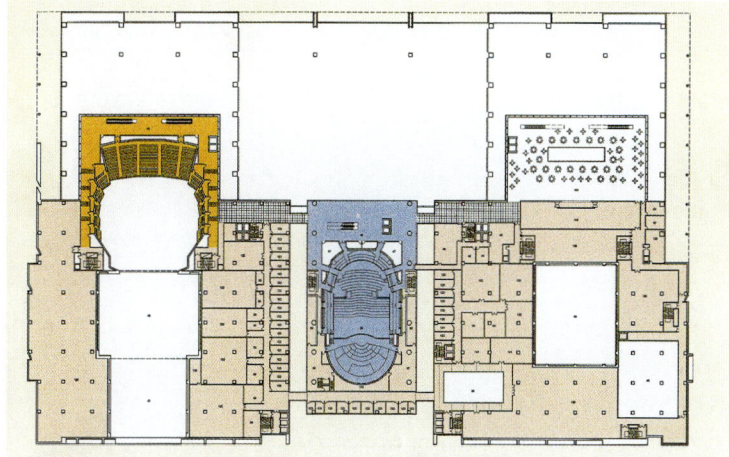

五层平面(修改)
5th FLOOR PLAN(MODIFIED)

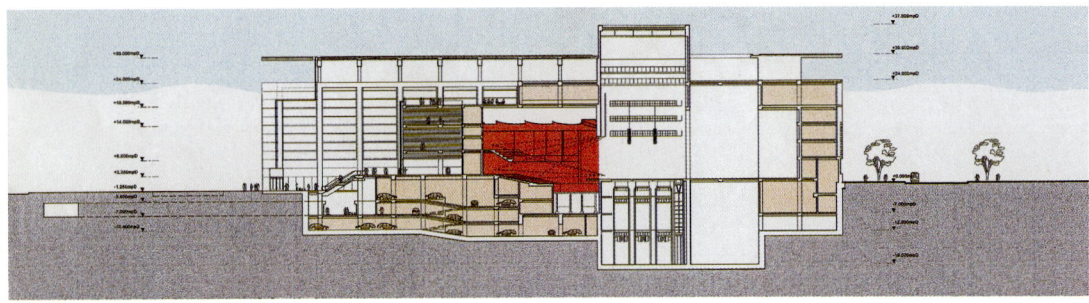

A-A 剖面(修改)
SECTION A-A(MODIFIED)

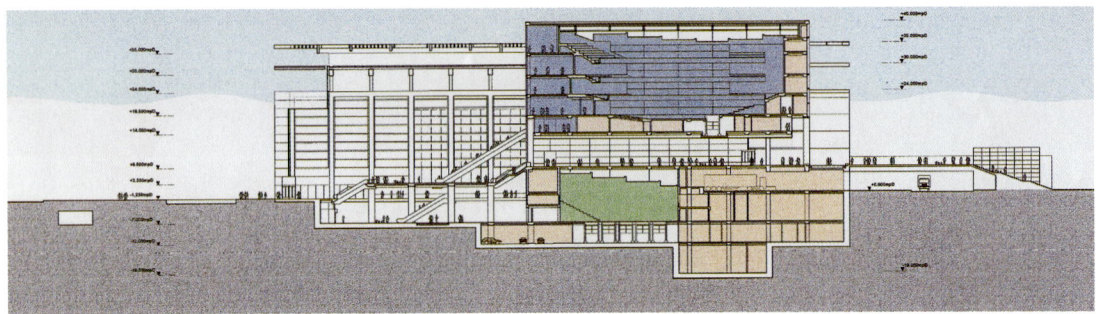

B-B 剖面(修改)
SECTION B-B(MODIFIED)

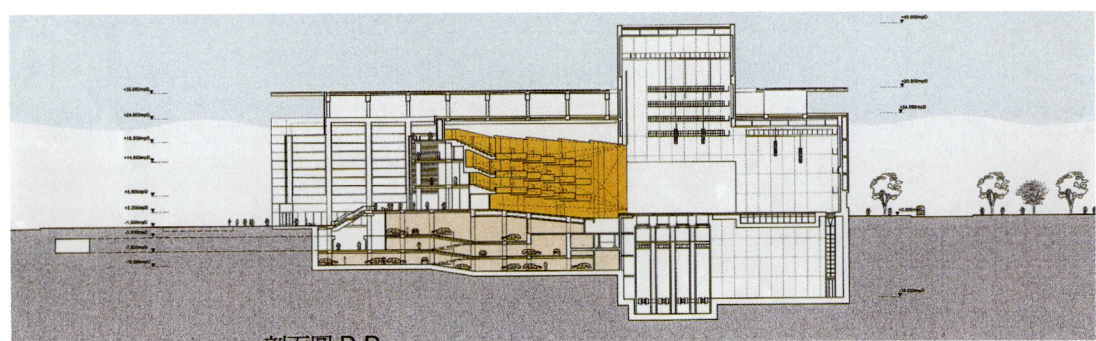

D-D 剖面(修改)
SECTION D-D(MODIFIED)

E-E 剖面(修改)
SECTION E-E(MODIFIED)

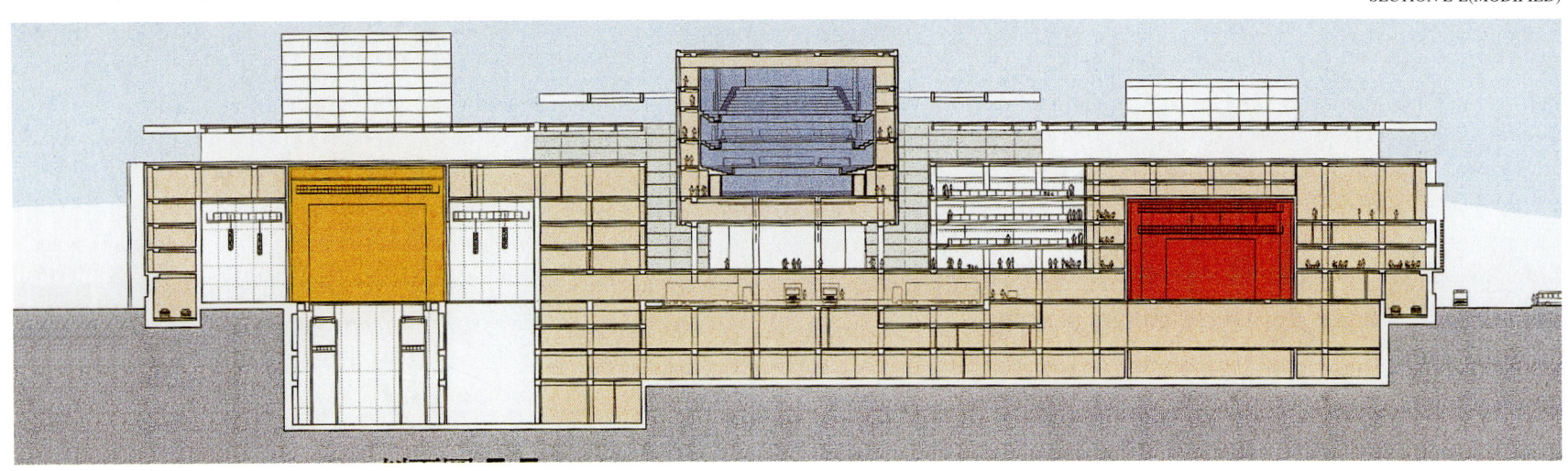

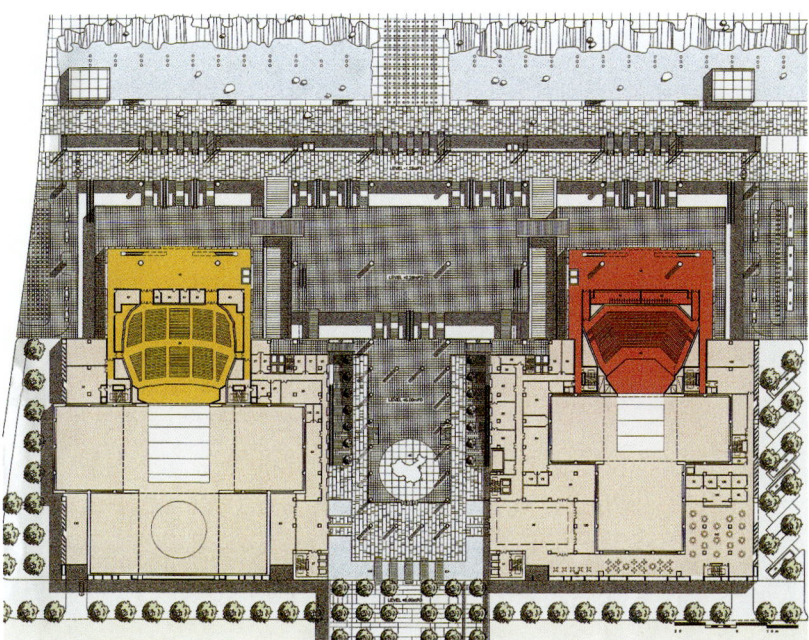

二层平面
2nd FLOOR PLAN

总平面(修改)
OVERALL SITE PLAN (MODIFIED)

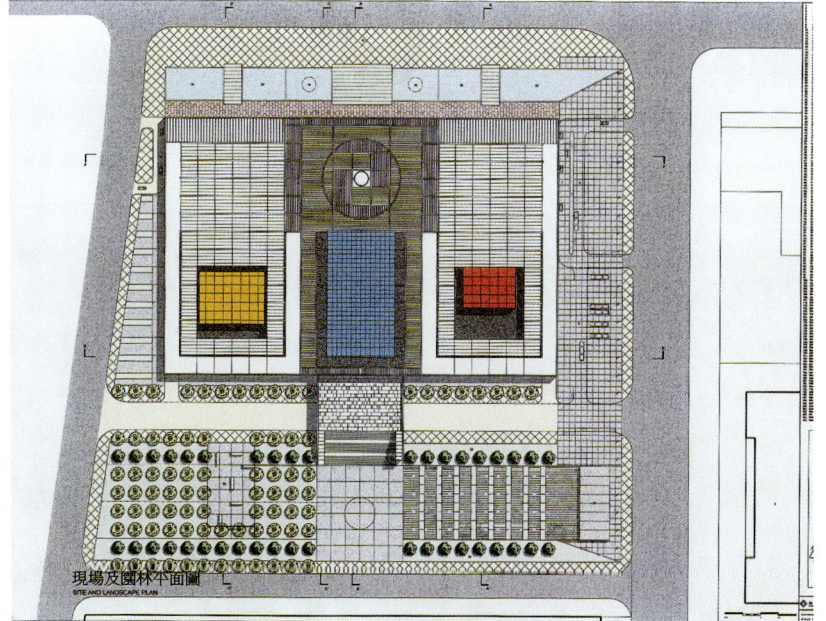

现场及园林平面图
SITE AND LANDSCAPE PLAN

技术经济指标
Technical Economic Index

基地面积(m²) Site Area	总建筑面积(m²) Total Floor Area	建筑高度(m) Building Height	建筑覆盖率(%) Building Coverage Ratio
38900	地上：696000 地下：59700	45/128(局部)	85
容积率 Plot Ratio	绿化覆盖率(%) Green Coverage Ratio	停车数量 Number of Stalls	自行车停车数量 Number of Bicycle Stalls
3.3	15	地上：0 地下：500	1000

各观众厅数据
Data For Auditoria

		歌剧院 Opera House	音乐厅 Concert Hall	戏剧院 Theater	小戏场 Mini Theater
	观众厅面积(m²) Floor Area of Auditorium	2170	1053	1027	942.5
	观众厅体积(m³) Volume of Auditorium	19000	22113	9310	9425
座席数 Seating Capacity	总座席数 Seating Capacity(Seats)	2500	2211	1200	500
	其中：池座 Among Them:Auditorium	1200	801	872	500
	楼座 Floor Seat	1300	1410	328	
	休息厅面积(m²) Lounge Floor Area	2100	1152	1008	420
舞台尺寸(m²) Stage Dimentions	主舞台 Main Stage	36 × 27	26 × 18	27 × 21	30 × 31
	左右侧台 Left and Right Side Stages	22 × 27		17 × 21	
	后舞台 Back Stage	26 × 25		27 × 26	
	台仓 Under Stage Storage	2406	200	1383	1092
	升降乐池 Elevating Orchestra Pit	26 × 6	24 × 16	21 × 11	
	最大视距(m) Max. Visual Distance	44	40	27	15
	最大俯角(°) Max. Angle of Depression	29.2	25.5	22.4	23.0

11

卡洛斯·奥特建筑师
事务所（加拿大）
Carlos Ott & Associates,
Architects(Canada)
建设部建筑
设计院（中国）
Architectural Design
Institute of Ministry of
Construction(China)

■ 传统
TRADITION

■ 现代性
MODERNITY

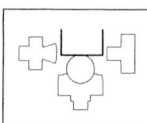 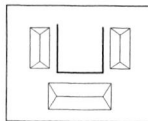

相似于"紫禁城"的三合院式布局
Similar to courtyards of THE FOR-
BIDDEN CITY,the new theatre is
formed by three pavillion

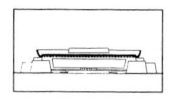 

传统的曲线屋顶
The traditional curvilinear roof

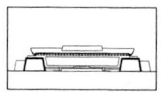 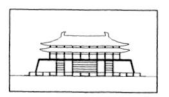

有收分的墙
Sloping walls

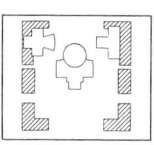 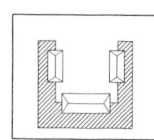

环绕建筑的护城河
A water moat surrounds
the buildings

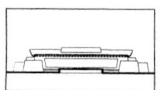 

中央大台阶作为主入口
Central stair as main access

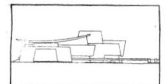 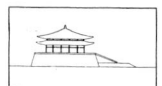

与基部相离的曲线屋顶
Curved roof separate from base

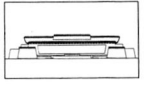 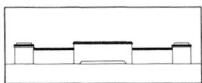

和天安门广场建筑群相似的
高度　Same height as
Tian'anmen Square buildings

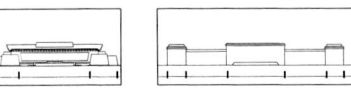

和天安门广场建筑群相似的
比例关系　Similar propotion
as Tian'anmen Square buildings

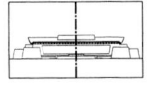 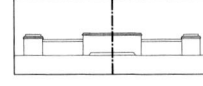 

对称
Symmetry

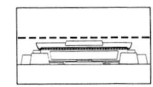 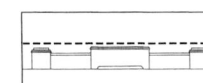

水平
Horizontality

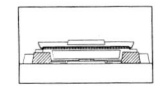 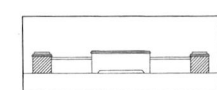

相似的材料和颜色
Similar colors and materials

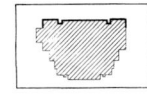 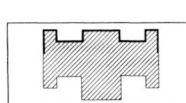

连续正交的立面
Articulate orthogonal facade

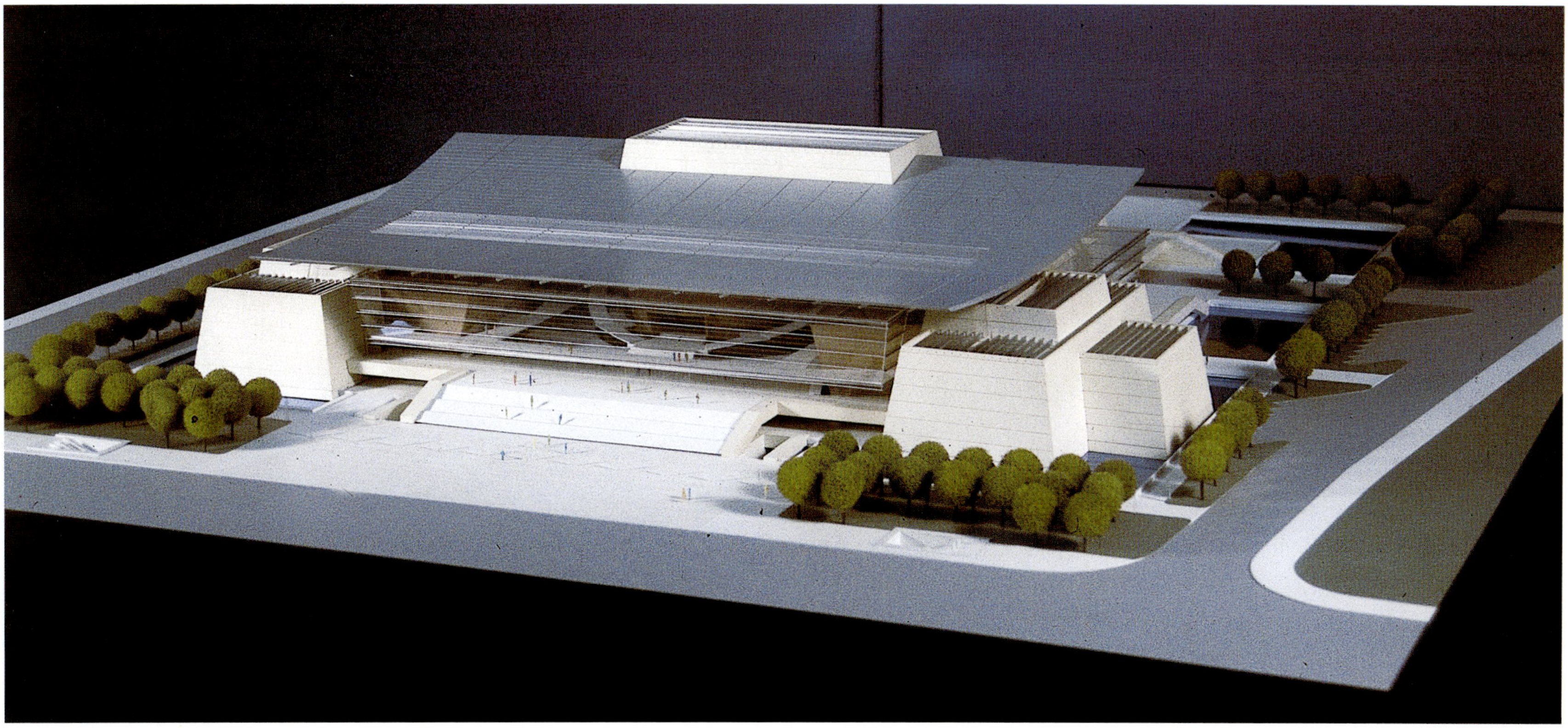

北侧模型透视(修改)
VIEW FROM NORTH SIDE OF THE MODEL(MODIFIED)

南侧模型透视(修改)
VIEW FROM SOUTH SIDE OF THE MODEL(MODIFIED)

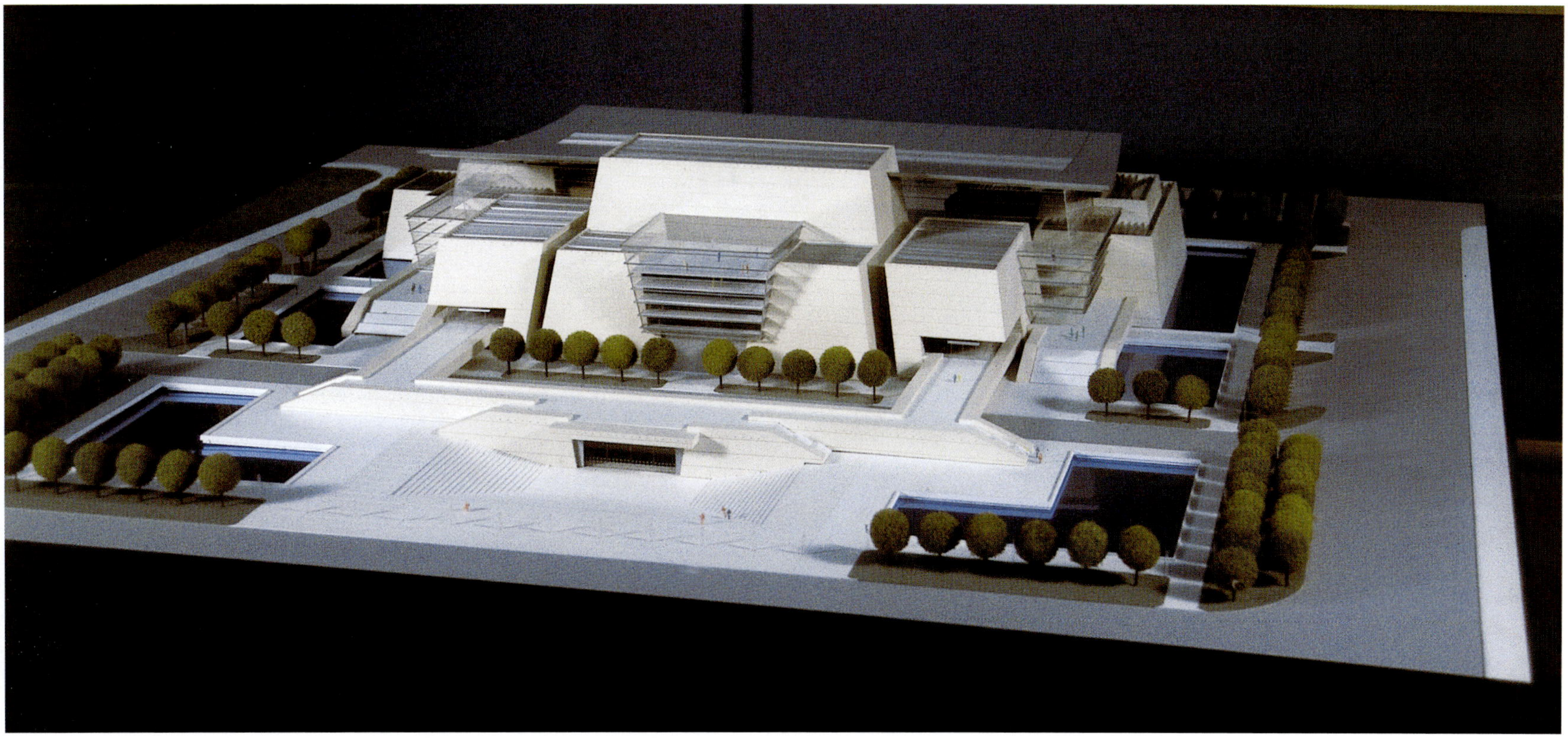

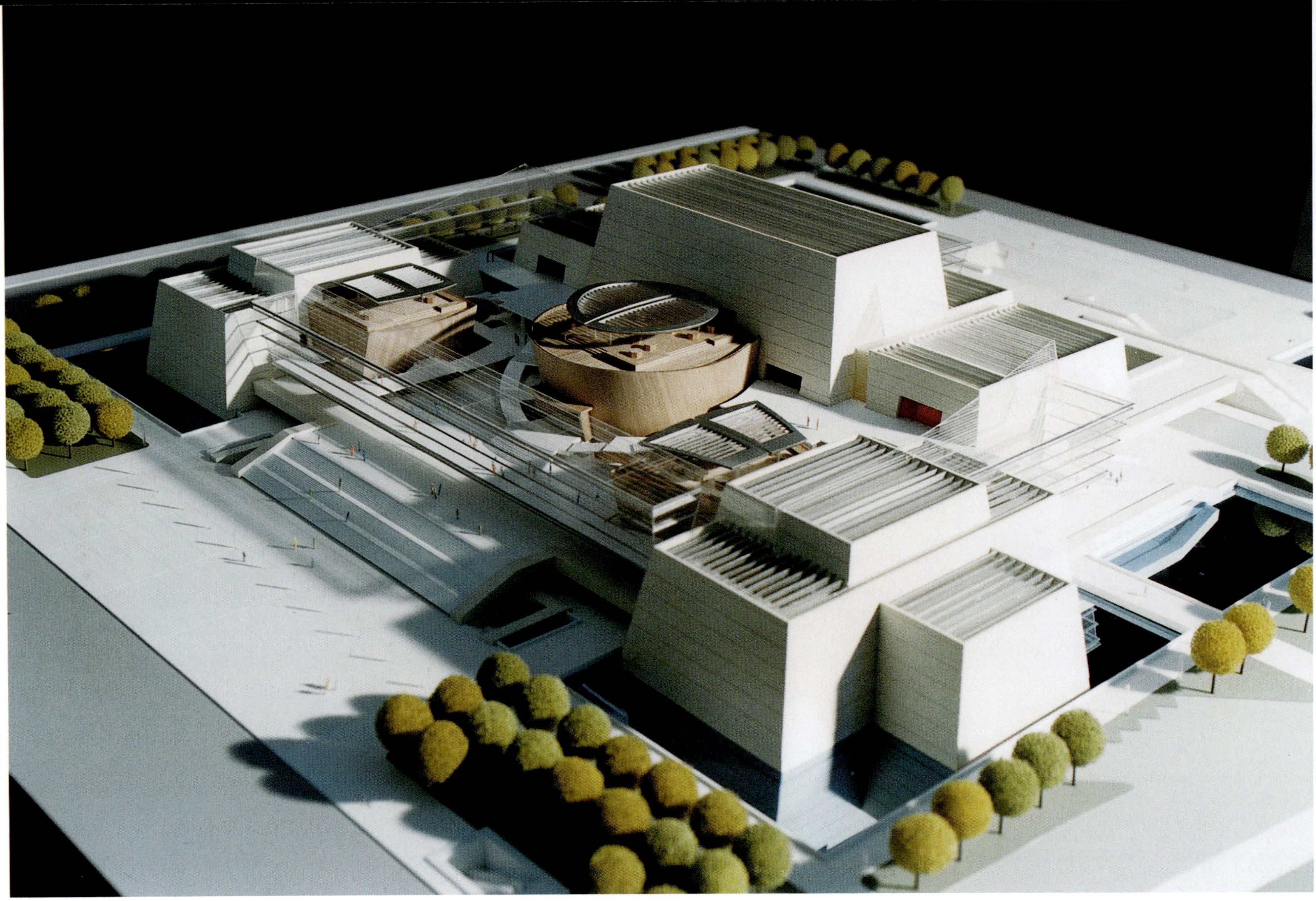

内部透视(修改)
INTERIOR PERSPECTIVE (MODIFIED)

入口透视(修改)
VIEW OF ENTRANCE (MODIFIED)

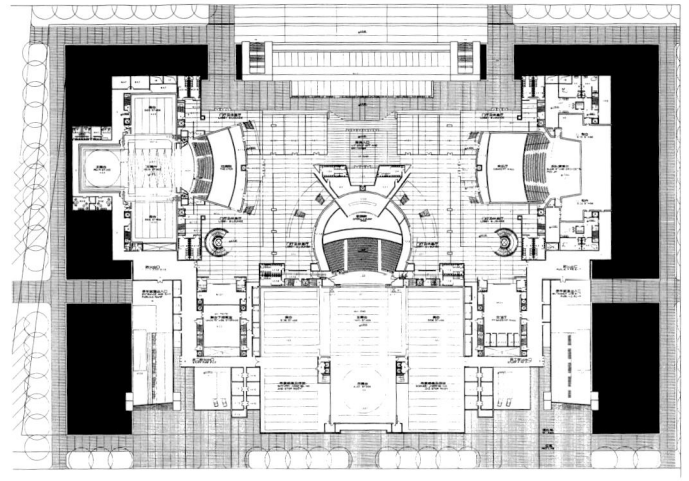

± 0.0m 平面(修改)
GROUND LEVEL(± 0.0m) (MODIFIED)

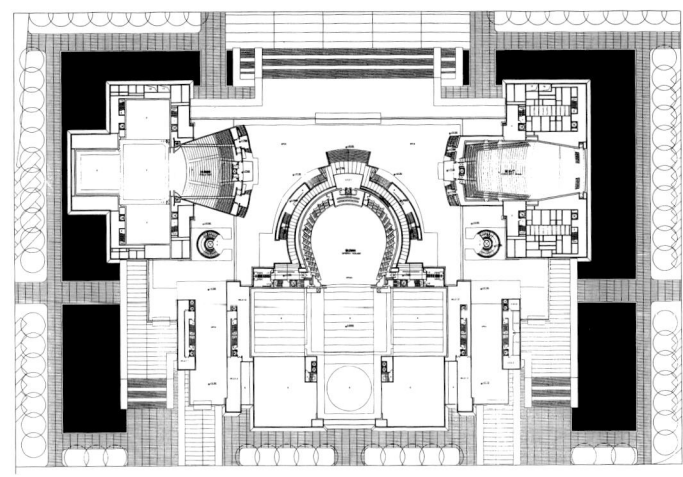

7.7m 层平面(修改)
GROUND LEVEL (+7.7m) (MODIFIED)

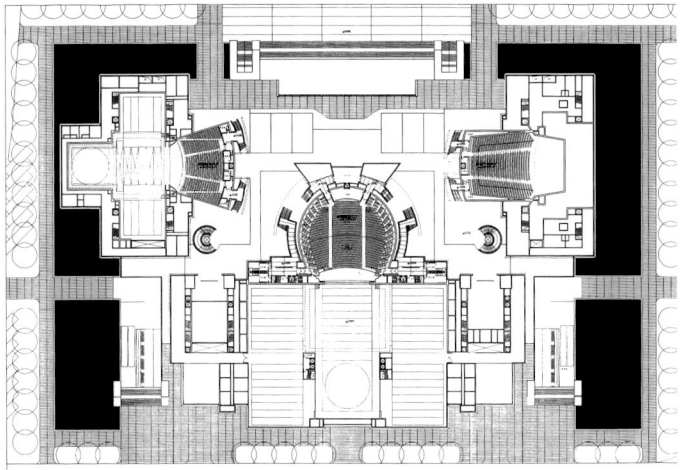

2.5m 层平面(修改)
GROUND LEVEL (+2.5m) (MODIFIED)

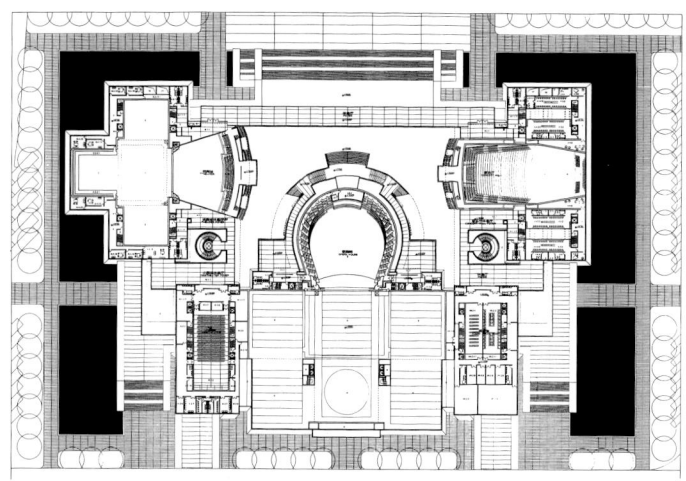

10.3m 层平面(修改)
GROUND LEVEL (+10.3m) (MODIFIED)

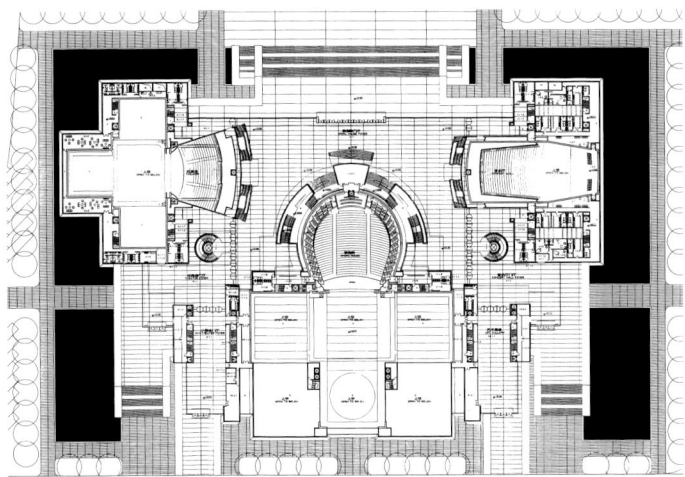

5.1m 层平面(修改)
GROUND LEVEL (+5.1m) (MODIFIED)

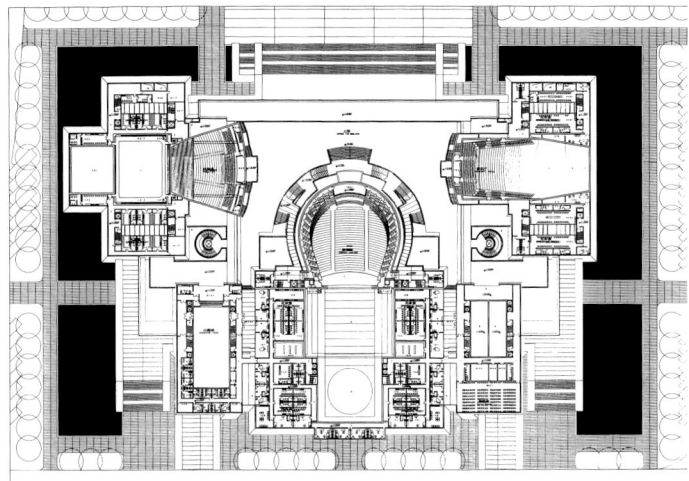

12.9m 层平面(修改)
GROUND LEVEL (+12.9m) (MODIFIED)

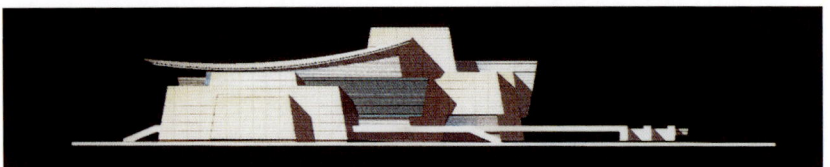

西立面(修改)
WEST ELEVATION (MODIFIED)

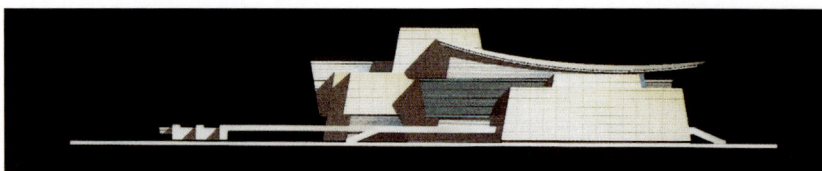

东立面(修改)
EAST ELEVATION (MODIFIED)

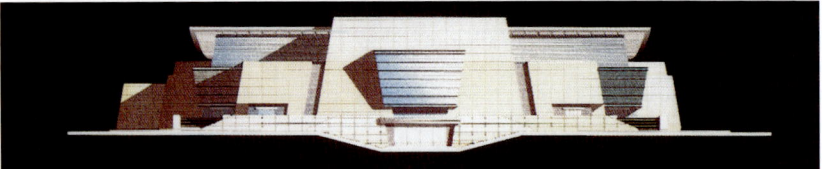

南立面(修改)
SOUTH ELEVATION (MODIFIED)

北立面(修改)
NORTH ELEVATION (MODIFIED)

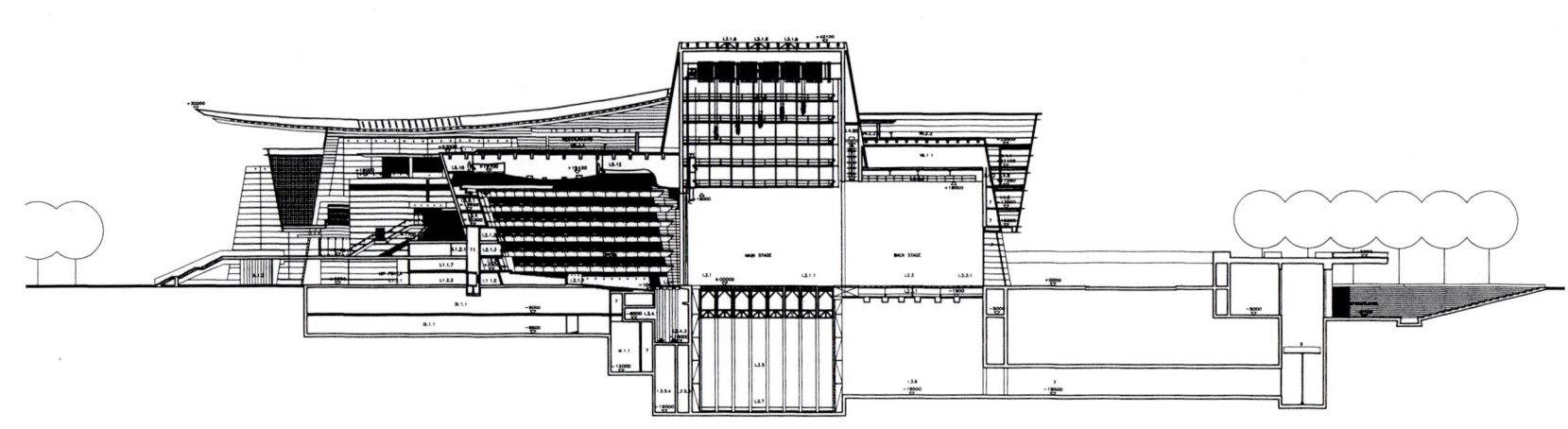

剖面(修改)
SECTION (MODIFIED)

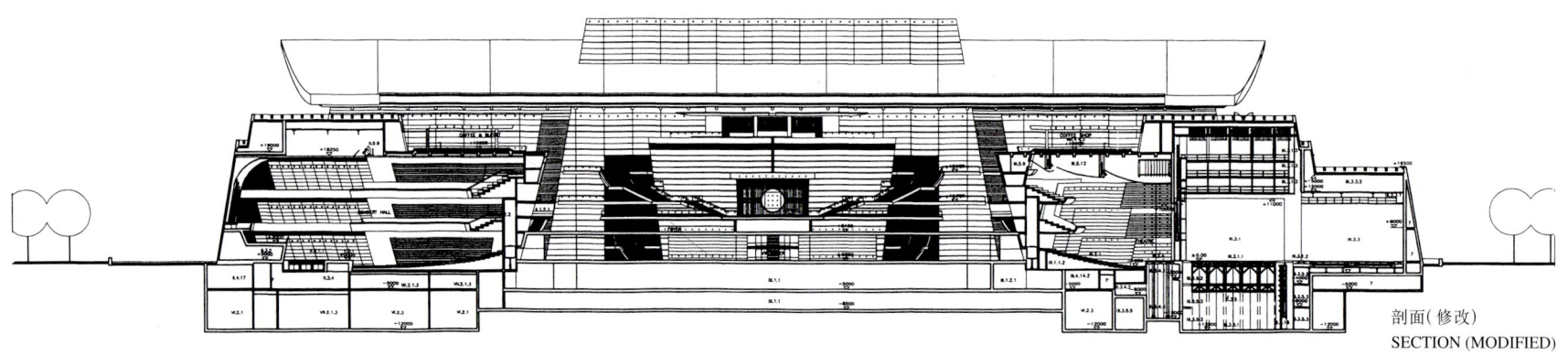

剖面(修改)
SECTION (MODIFIED)

北模型透视
VIEW FROM NORTH SIDE OF THE MODEL

南模型透视
VIEW FROM SOUTH SIDE OF THE MODEL

总平面(修改)
OVERALL SITE PLAN(MODIFIED)

技术经济指标
Technical Economic Index

基地面积(m²) Site Area	总建筑面积 (m²) Total Floor Area	建筑高度(m) Building Height	建筑覆盖率 (%) Building Coverage Ratio
38915	119835	30/42(局部)	70
容积率 Plot Ratio	绿化覆盖率 (%) Green Coverage Ratio	停车数量 Number of Stalls	自行车停车数量 Number of Bicycle Stalls
3.08	18.55	地上：50 地下：450	1060

各观众厅数据
Data For Auditoria

		歌剧院 Opera House	音乐厅 Concert Hall	戏剧院 Theater	小戏场 Mini Theater
	观众厅面积(m²) Floor Area of Auditorium	2794	1610	852	444/252
	观众厅体积(m³) Volume of Auditorium	22250	20200	10137	2615/2012
座席数 Seating Capacity	总座席数 Seating Capacity(Seats)	2514	1860	1146/953/814	453/341
	其中：池座 Among Them:Auditorium	1319	1052	552	225/162
	楼座 Floor Seat	2794	1710	850	444/252
	休息厅面积(m²) Lounge Floor Area	4882	2831	2530	991
舞台尺寸 (m²) Stage Dimentions	主舞台 Main Stage	700	438	566	165/111
	左右侧台 Left and Right Side Stages	730	438	548	
	后舞台 Back Stage	652		324	00/45
	台仓 Under Stage Storage	710	239	566	300/300
	升降乐池 Elevating Orchestra Pit	91/60	131/110	81/30	132/55
	最大视距(m) Max. Visual Distance	39	27.8	28.1	21.4/17.1
	最大俯角(°) Max. Angle of Depression	22	27	26	9/9

清华大学(中国)
Tsinghua University
(China)
法国巴黎机场公司
(法国)协作
Joint Collaborator:
ADP Aeroports de
Paris (France)

a.本方案在充分满足设计任务书基本要求的前提下，建筑体量相对缩小。东西向总长小于200m，南北向长度为136m，建筑主体檐口高度为28m及31.2m，舞台突出部分最高处为45m，均符合城市总体规划的要求，且与周边建筑高度相呼应。由于体量大小适宜，外形规整，有利于与大会堂纪念性尺度的协调。建筑周边用地较为宽敞，更有利于营造良好的室外绿化环境。

b.本方案大面积室外绿地与中南海、中山公园、劳动人民文化宫、天安门广场南部绿地有机地连接成完整的城市绿化体系，形成用绿化包围天安门广场及周边建筑群的格局。从宏观的城市规划角度来看，这样做能够形成北京市新的市中心绿地生态系统，恰似市中心的绿"肺"，有利于改善生态环境。本方案在对大剧院南侧60m宽的城市绿地设计中，不采用大片草坪，而是种植成片的常绿针叶树，更适应北方气候干燥的特点，树木下又宜于市民休憩与进行多种活动，具有浓厚的北京传统风味与特色。大剧院南侧出入口通过宽敞的过街天桥与南侧大片绿地中方形露天剧场连接起来，露天剧场的舞台地面以玻璃砖镶嵌成"曲水流觞"的传统形式，并装点以平地音乐喷泉(dry musical fountain)。对大片绿地进行了精心设计，取名"鲲鹏绿海"之美称。

c.本方案采用不对称的平面布局，歌剧院和戏剧场的舞台、后台集中布置在用地西侧，所有布景及后台运输出入口面向石碑胡同，既不影响主干道的交通，又在北、南、东方向自然地形成重要的主立面。这种不对称布局有利于突出人民大会堂，并烘托出天安门的主导地位与重要性。同时，大剧院的布局也有利于做到最大限度地南北贯通。

a.Overall building volume has been reduced while still satisfying the basic requirements of the design specifications in the commissioning letter. Total length is 200 meters in east-west direction and 136 meters in north-south direction. Building heights are 28 meters from ground to top of front gate,32.2 meters to roof, and 45 meters to the highest point of the stage tower, capable of meeting urban planning requirements of Beijing. The building dimensions appear to be the right size with a formal style,is in harmony with neighboring architectures like the Great Hall of the People. We intentionally leave plenty of open spaces near the National Grand Theater to build a green belt landscape to achieve an excellent, park-like outdoor environment.

b.The large, landscaped green belt around the National Grand Theater is naturally connected to the established green lands of Zhongnanhai, Zhong Shan Park,Cultural Palace, and south end of Tian'anmen Square to form a complete green system in the city center. This large, interconnected greenery land will surround the Tian'anmen Square and its nearby buildings. From the macroscopic view of urban planning, the large greenery land will form a new ecological system in Beijing's city center. It will function as the lung of the city to improve environment and ecosystem of Beijing. The large, 60 meter wide landscape in the south-end of the National Grand Theater is designed with evergreen pine trees rather than grass turf to be suitable to the dry climate in northern China. Characteristic to Beijing's tradition and cultural heritage, the pine trees and green landscape will form a park-like environment enjoyed by the city dwellers.

The south entrance of the National Grand Theater is connected to the square-shaped open theater in the center of the green belt through an elevated bridge. The floor of the open theater is decorated to Chinese classic graphic patterns with colored glass tiles, imitating strearms of clean water. A nearby dry music fountain adds artistic flavor to the "Ocean of Green"surroundings.

c.A non-symmetric building layout is adopted for this design proposal. The stages of the Opera House and the Play Theater are located in the west side of the building. With this layout, all scenery cargo shipments are directed to the rear entrances located in the Shibei Hutong,not only avoiding interference with traffic on the main roads, but also naturally forming main facades in the north, south, and east directions. This non-symmetric layout is beneficial to show a leadership presence of the Tian'anmen and the Great Hall of the People in nearby locations. This layout also allows maximum people traffic in the north-south direction within the huge atrium.

北侧透视(修改)
PERSPECTIVE OF NORTH ELEVATION(MODIFIED)

绿色中庭(修改)
GREEN ATRIUM (MODIFIED)

南侧透视(修改)
PERSPECTIVE OF SOUTH ELEVATION(MODIFIED)

入口局部(修改)
PARTIAL ENTRANCE(MODIFIED)

北侧模型透视(修改)
VIEW FROM NORTH SIDE OF THE MODEL(MODIFIED)

南侧模型透视(修改)
VIEW FROM SOUTH SIDE OF THE MODEL(MODIFIED)

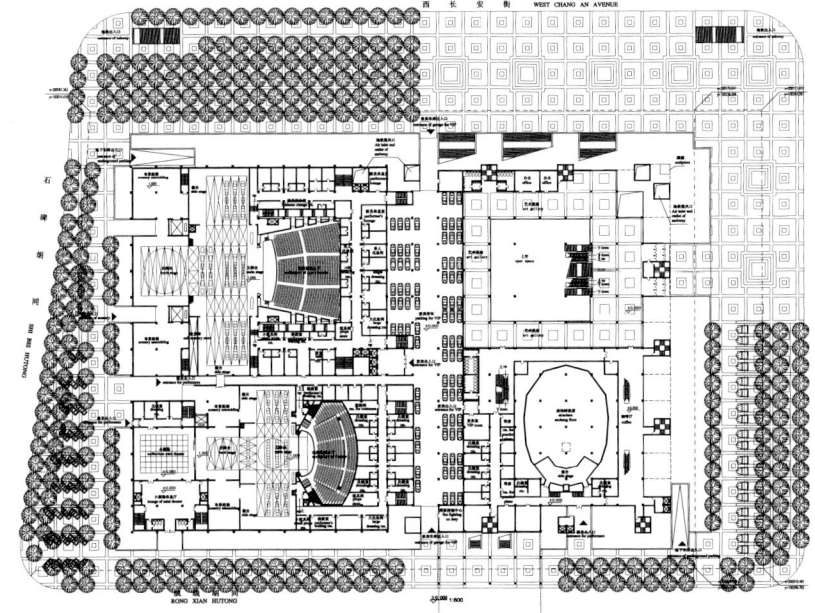

0.0m 层平面(修改)
GROUND LEVEL(± 0.0m)(MODIFIED)

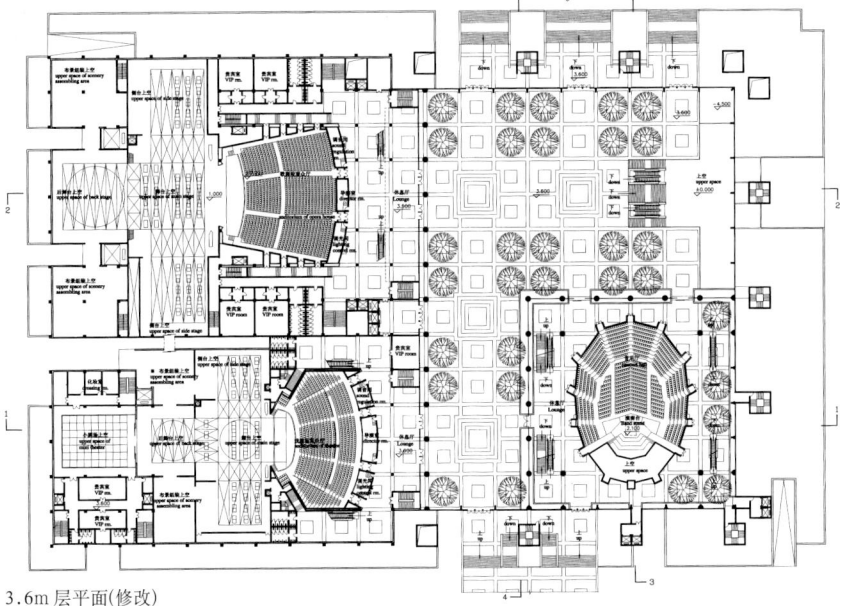

3.6m 层平面(修改)
GROUND LEVEL(± 3.6m)(MODIFIED)

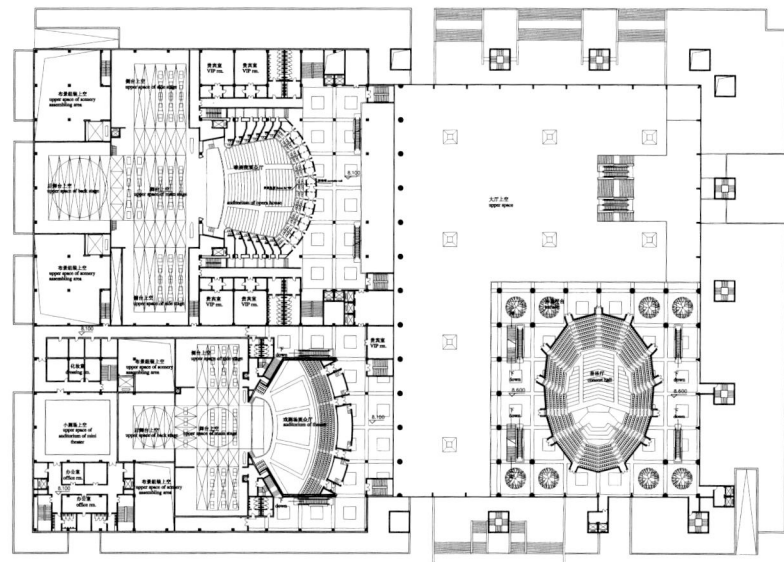

8.1m 层平面(修改)
GROUND LEVEL(+8.1m)(MODIFIED)

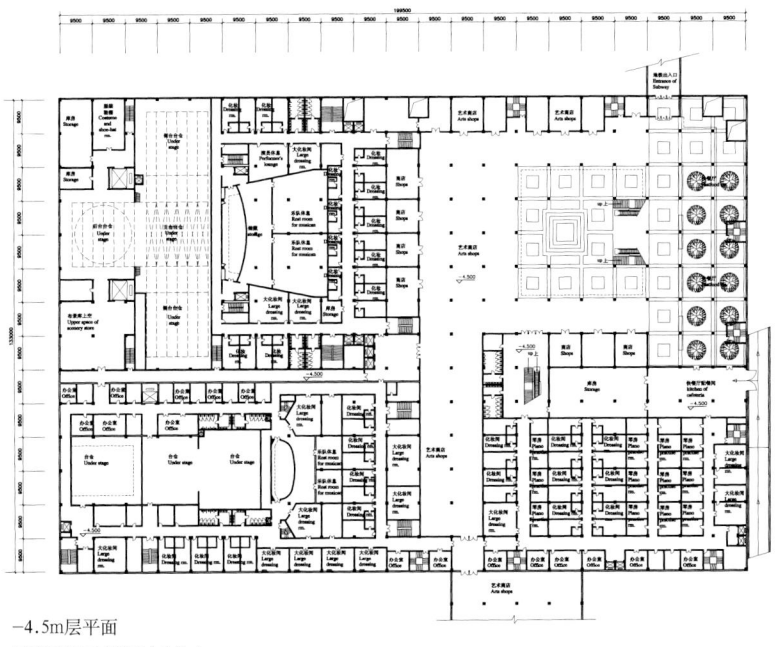

−4.5m层平面
GROUND LEVEL(-4.5m)

北立面(修改)
NORTH ELEVATION (MODIFIED)

南立面(修改)
SOUTH ELEVATION (MODIFIED)

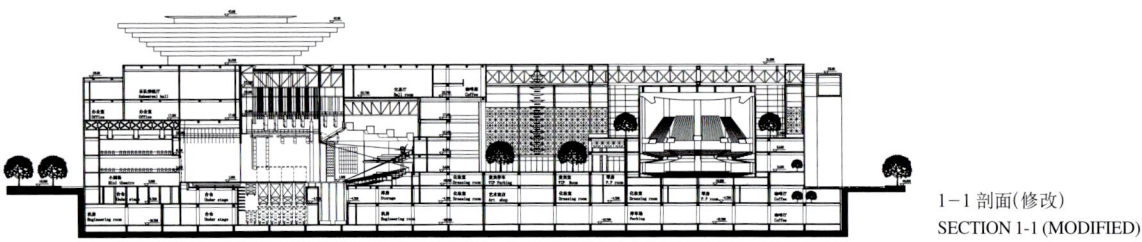

1—1 剖面(修改)
SECTION 1-1 (MODIFIED)

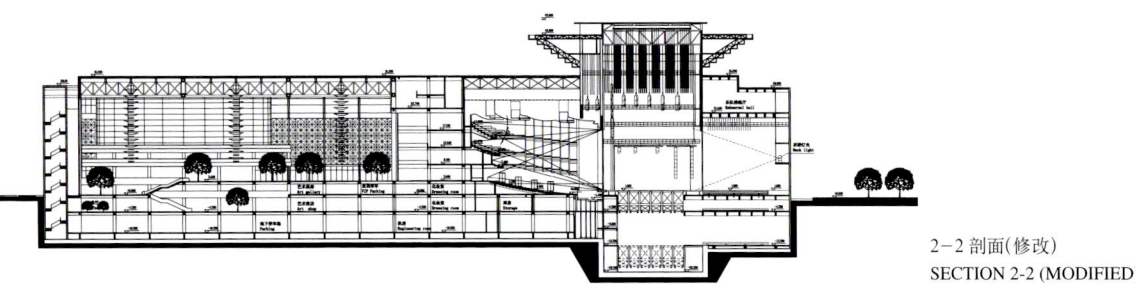

2—2 剖面(修改)
SECTION 2-2 (MODIFIED)

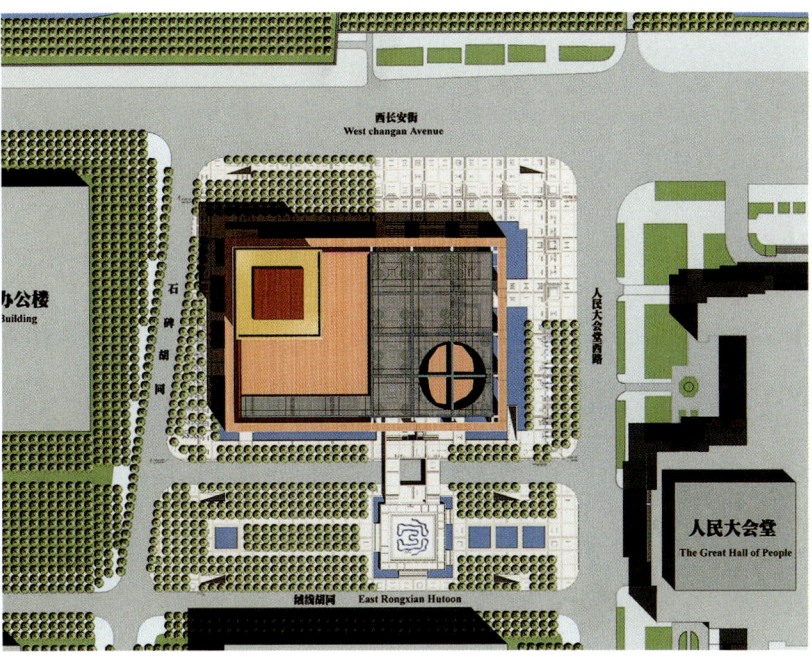

总平面(修改)
OVERALL SITE PLAN (MODIFIED)

环境透视(修改)
SURROUNDING VIEW (MODIFIED)

技术经济指标
Technical Economic Index

基地面积(m²) Site Area	总建筑面积(m²) Total Floor Area	建筑高度(m) Building Height	建筑覆盖率(%) Building Coverage Ratio
43300	地上：78570 地下：46643	28.0/31.2(局部)	60
容积率 Plot Ratio	绿化覆盖率(%) Green Coverage Ratio	停车数量 Number of Stalls	自行车停车数量 Number of Bicycle Stalls
2.91	22	地上：150 地下：320	1000

各观众厅数据
Data For Auditoria

		歌剧院 Opera House	音乐厅 Concert Hall	戏剧院 Theater	小戏场 Mini Theater
	观众厅面积(m²) Floor Area of Auditorium	2000	1200	1070	440
	观众厅体积(m³) Volume of Auditorium	19800	20700	5995	4000
座席数 Seating Capacity	总座席数 Seating Capacity(Seats)	2518	2006	1209	300-500
	其中：池座 Among Them:Auditorium	1385	1082	889	
	楼 座 Floor Seat	1133	924	320	
	休息厅面积(m²) Lounge Floor Area	6500		4000	
舞台尺寸(m²) Stage Dimentions	主舞台 Main Stage	33 × 26	16 × 12	22 × 21.2	14 × 18
	左右侧台 Left and Right Side Stages	21 × 22.2		13.6 × 19.2	
	后舞台 Back Stage	25.2 × 20.8		19 × 16.1	
	台 仓 Under Stage Storage	33 × 26 25.2 × 20.8		22 × 21.2 19 × 16.1	14 × 18
	升降乐池 Elevating Orchestra Pit	22 × 6		15 × 6	
	最大视距(m) Max. Visual Distance	36	24	24	
	最大俯角(°) Max. Angle of Depression	25	16	27	

第二轮竞赛第二、三次修改后的方案（4项）

Schemes of the 2nd and the 3rd Modification on
the 2nd Round Competition (four items)

法国巴黎机场公司(法国)
ADP Aeroports de Paris(France) (Implement Scheme P.53)
清华大学(中国) 协作
Joint Collaborator: Tsinghua University(China)

塔瑞 · 法若建筑师事务所(英国)
Terry Farrell & Partners(UK) (Candidate Plans For
 the Higher Level's Approual P.63)
北京市建筑设计研究院(中国)
Beijing Institute of Architectural Design &
Research(China)

卡洛斯 · 奥特建筑师事务所(加拿大)
Carlos Ott & Associates, Architects(Canada)
建设部建筑设计院(中国)
Architectural Design Institute of Ministry of
Construction(China)

清华大学(中国)
Tsinghua University(China)
法国巴黎机场公司(法国) 协作
Joint Collaborator:ADP Aeroports de Paris(France)

**卡洛斯·奥特建筑师
事务所（加拿大）**
Carlos Ott & Associates,
Architects(Canada)
**建设部建筑
设计院（中国）**
Architectural Design
Institute of Ministry of
Construction(China)

第二轮第二次修改
Schem of the 2nd Modi-
fication on the 2nd Round
Competition

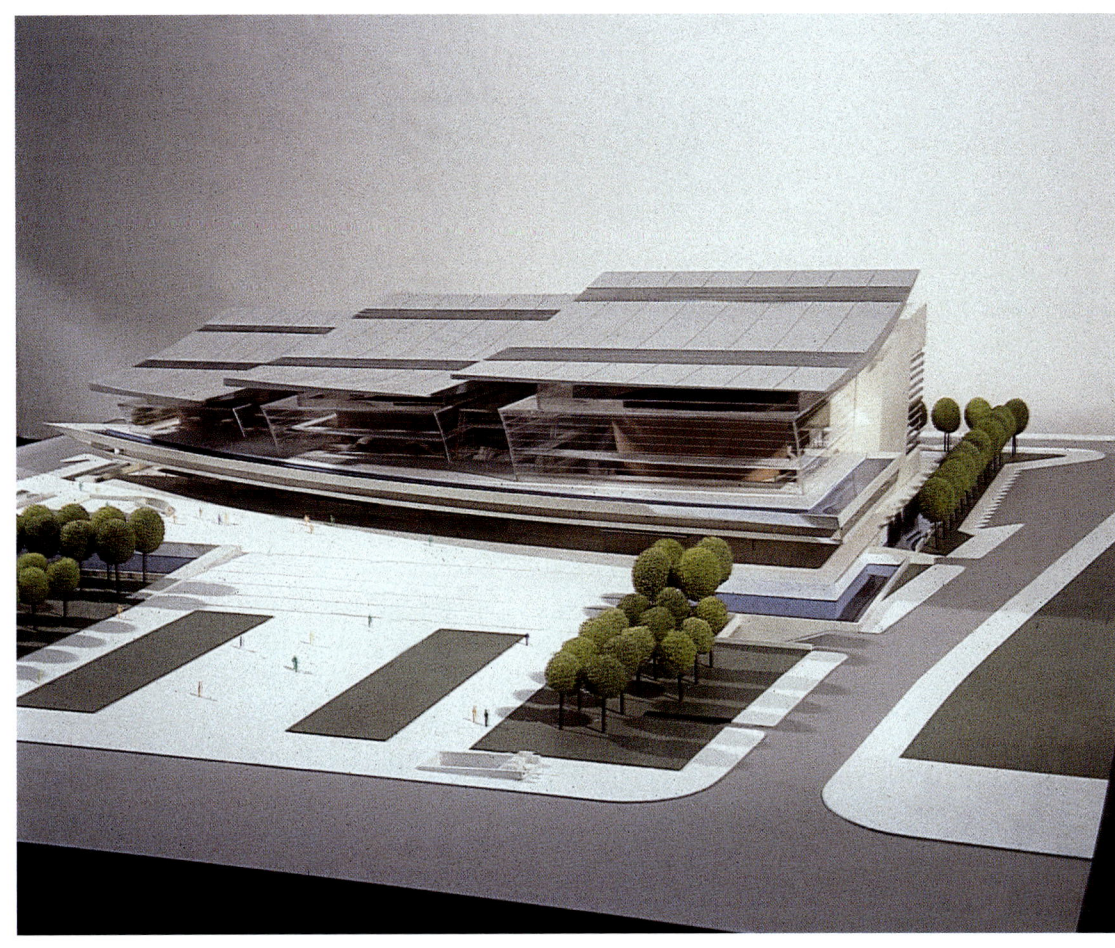

北侧模型透视
VIEW FROM NORTH SIDE OF
THE MODEL

南侧模型透视
VIEW FROM SOUTH SIDE OF
THE MODEL

北立面
NORTH ELEVATION

南立面
SOUTH ELEVATION

总平面
GENERAL PLAN

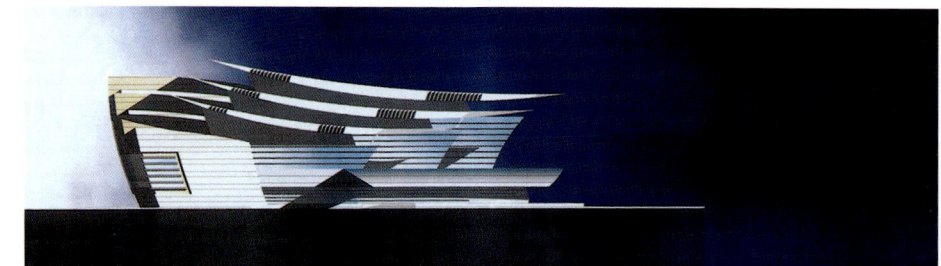

东立面
EAST ELEVATION

西立面
WEAST ELEVATION

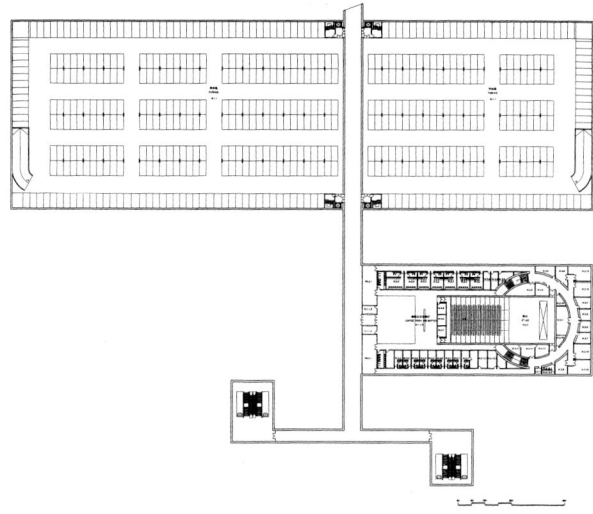

−8.0m平面
-8.0m LEVEL PLAN

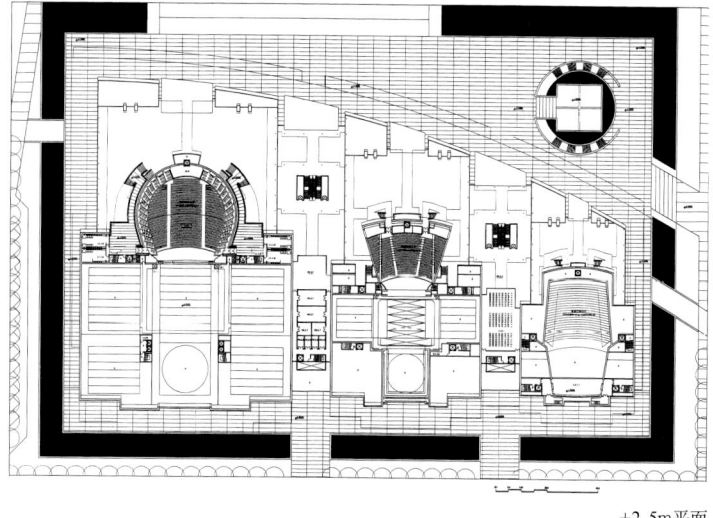

+2.5m平面
+2.5m LEVEL PLAN

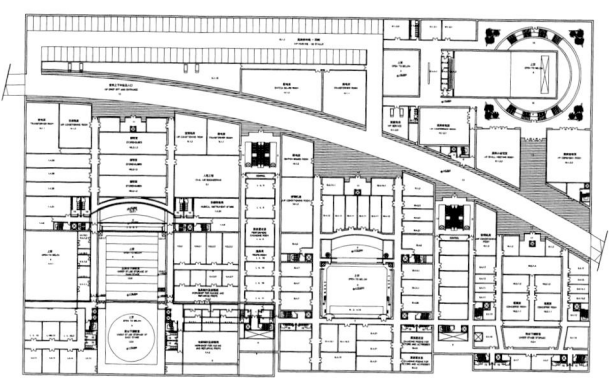

−4.0m平面
-4.0m LEVEL PLAN

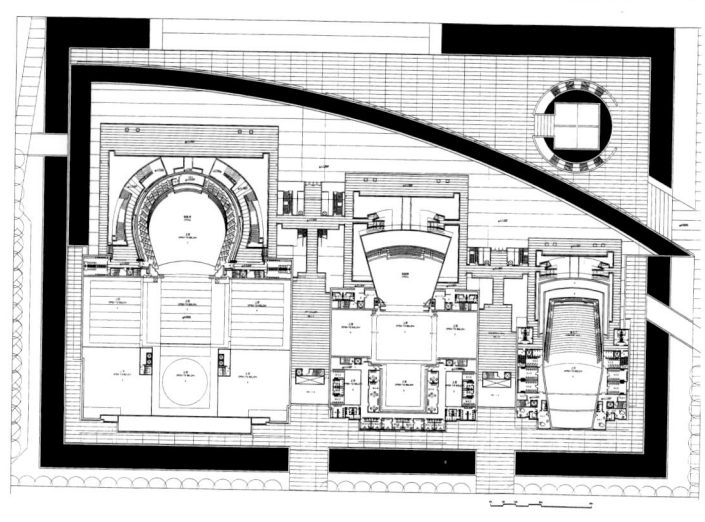

+7.5m平面
+7.5m LEVEL PLAN

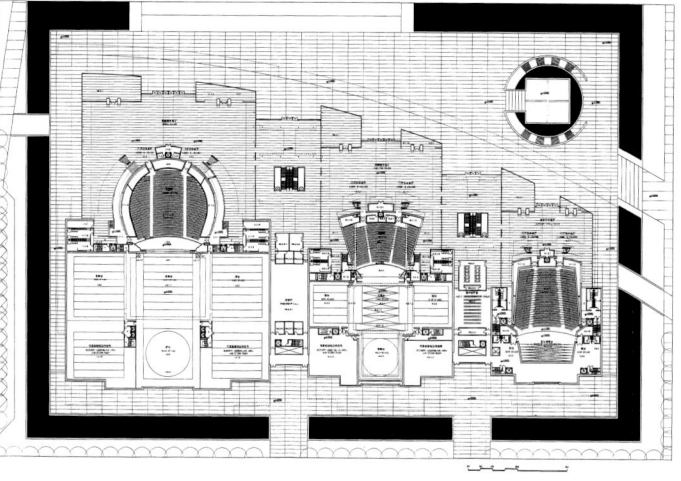

±0.0m平面
± 0.0m LEVEL PLAN

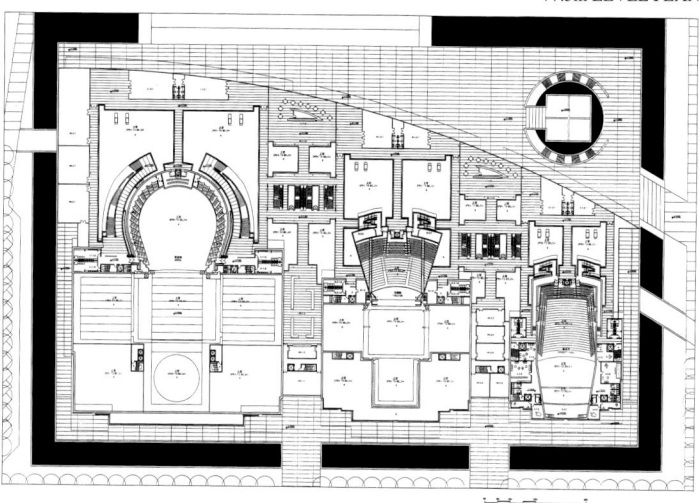

+11.25m平面
+11.25m LEVEL PLAN

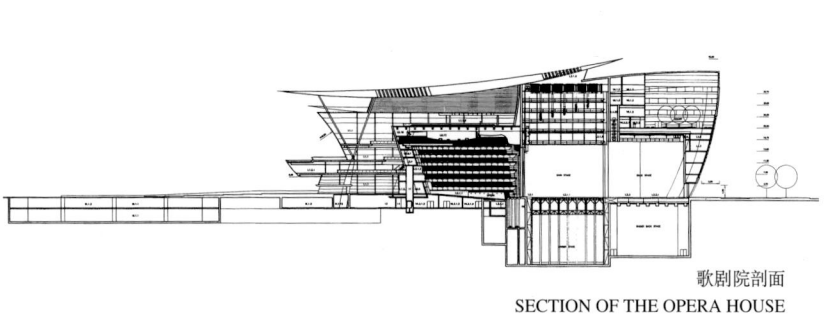

歌剧院剖面
SECTION OF THE OPERA HOUSE

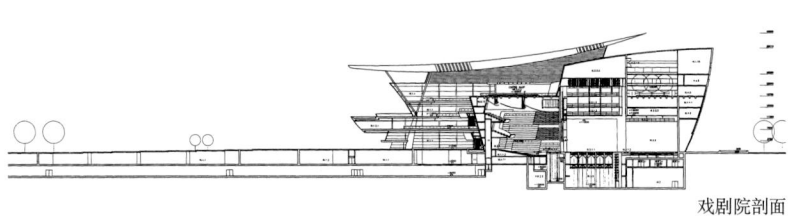

戏剧院剖面
SECTION OF THE THEATER

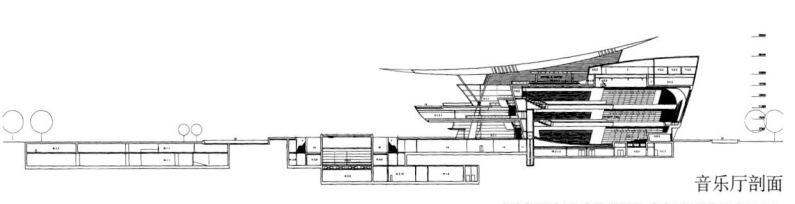

音乐厅剖面
SECTION OF THE CONCERT HALL

技术经济指标
Technical Economic Index

基地面积(m²) Site Area	总建筑面积 (m²) Total Floor Area	建筑高度(m) Building Height	建筑覆盖率（%） Building Coverage Ratio
37185	107965	33.8/45(局部)	45

容积率 Plot Ratio	绿化覆盖率(%) Green Coverage Ratio	停车数量 Number of Stalls	自行车停车数量 Number of Bicycle Stalls
2.9	3.98	地上：37 地下：1040 地下贵宾：104	1304

各观众厅数据
Data For Auditoria

		歌剧院 Opera House	音乐厅 Concert Hall	戏剧院 Theater	小戏场 Mini Theater
	观众厅面积(m²) Floor Area of Auditorium	2794	1610	852	699
	观众厅体积(m³) Volume of Auditorium	22250	20200	10137	5504/4234
座席数 Seating Capacity	总座席数 Seating Capacity(Seats)	2514	1860	1146/953/814	541
	其中：池座 Among Them:Auditorium	1319	1052	552	269
	楼 座 Floor Seat	2794	1710	850	699
	休息厅面积(m²) Lounge Floor Area	7319	2123	4350	556
舞台尺寸 Stage Dimentions (m²)	主舞台 Main Stage	700	438	566	296
	左右侧台 Left and Right Side Stages	730	438	548	0
	后舞台 Back Stage	652		324	0
	台 仓 Under Stage Storage	710	239	566	576
	升降乐池 Elevating Orchestra Pit	91-60 (可动)	131-110 (可动)	81-30 (固定)	60(可动)
	最大视距(m) Max. Visual Distance	39	27.8	28.1	21.4
	最大俯角（°） Max. Angle of Depression	22	27	26	9

清华大学（中国）
Tsinghua University
(China)

法国巴黎机场
公司协助（法国）
Joint Collaborator:
ADP Aeroports de
Paris(France)

第二轮第二次修改
Schem of the 2nd Modi-
fication on the 2nd Round
Competition

①本方案在认真研究建设地段比原用地红线向南退后70m及设计任务书基本要求不变的前提下，结合周边环境及建设用地的具体情况，采用了圆形的总体布局形态，圆形内半虚半实、阴阳互补、先抑后扬。建筑实体东西向总长为225m 南北向长度为137m，建筑主体檐口高度为28m，歌剧院舞台突出部分最高处为45m，均符合城市总体规划的要求，且与周边建筑高度和尺度相呼应。由于建筑体量大小适宜，外形规整、明晰，有利于与大会堂纪念性尺度的协调。建筑周边用地较为宽敞，向南退后70m后，大剧院北侧视野开阔，为室外环境气氛和中景的营造打下了基础。

②本方案向南退后70m后，北侧大面积室外绿地和水面的设计营造，与中南海、太庙、劳动人民文化宫、天安门广场南部绿地有机地连接成完整的城市绿化体系，形成用绿化包围天安门广场及周边建筑群的格局。从宏观的城市规划角度来看，这样做能够形成北京市新的市中心绿地生态系统，怡似市中心的绿"肺"，有利于改善生态环境。本方案在对大剧院北侧70m宽的城市绿地设计中，采用大片草坪与水面的结合，水面的形状呈半圆形，完善出圆形的整体布局形态，建筑的形态倒影在水面上，与水中的喷水列阵相映成辉。由于地段位置的变迁，大剧院用地南侧红线与大会堂东西向主轴线几乎重合，正圆形总体布局形态的规划设计，正是为了缓和与消解这一尖锐矛盾，并软化与大会堂及周边环境的关系，使大剧院在与周边建筑及环境和谐相处的同时，呈现出自身特色。

③通过对新的地段条件认真研究，本方案采用对称的平面布局，歌剧院剧中，音乐厅和戏剧场分居东西两侧。朝北的主立面是大剧院公共主要出入口方向，所有布景及后台运输出入口设置在用地南侧。

This design proposal is based on careful researches on the new and former design requirements. Under the change of the site set back 70m from former property line of the north side , the general layout is set up with a clear circular one, which is half void and half solid, just like the unity of the 'Yin' and the 'yang', down in front and up behind. Overall building volume has been carefully studied. Its total length is 225 meters in east-west direction and 137 meters in north-south direction. Building heights are 28 meter from ground to roof, and 45 meter to the highest point of the Opera House stage tower, capable of meeting urban planning requirements of Beijing. The building dimensions, appear to be the right size with a formal style, is in harmony with neighboring architectures like the Great Hall of People. We intentionally leave plenty of open space near the Naitonal Grand Theater to build a green belt landscape to achieve an excellent, park-like outdoor environment.

After receding 70m from the former north side property line, the large, landscaped green space at the north side of the Grand Theater is naturally connected to the established green area of Zhongnanhai, Temple for Worshiping Emperor's Ancestor, Cultural palace, and south end of Tiananmen Square to form a complete green system in the city center. This large, interconnected greenery space will surround the Tiananmen Square and its near-by buildings. From the macroscopic view of urban planning, the large greenery land will form a new ecological system in Beijing's city center. It will function as the lung of the city to improve environment and ecosystem of Beijing. The large, 70 meter wide landscape in the north side of the National Theater is manifested into broad grass turf and spacious pools to be suitable to the dry climate in northern China. The large pool is shaped as half -circular one, which forms the general circular layout. The architectural reflection and the artificial fountains in the pool make a pleasing scene. Due to the change of the site, the south property line of the Grand Theater is almost overlapping with the main east-west axis of the Great Hall. The design of general circular layout is to alleviate and clear up this contradiction, and to soften the relationship between the Grand Theater and its surrounding environment, like the Great Hall of people. Such design makes the Grand Theater in perfect harmony with its surroundings and also embody its own characteristic. A symmetric building layout is adopted for this design proposal under careful consideration of the new site . The Opera House is located at the middle of the layout. The Theater and the Concert Hall are set up in the west side and the east side of the building. With this layout, all scenery cargo shipments are directed to the rear entrances located in the south side, not only avoiding interference with traffic on the main roads, but also naturally forming main facades in the north direction. This layout allows maximum people traffic in the north direction and forms the north side as the main facade of the Grand Theater.

北侧透视图
PERSPECTIOE OF NORTH ELEVATION

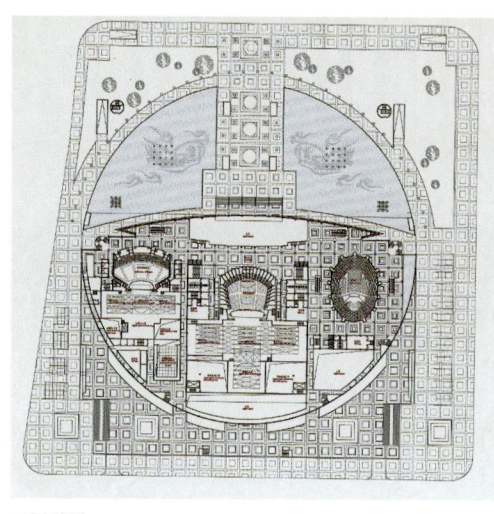

二层平面
2nd FLOOR PLAN

夜景透视图
NIGHT VIEW

地面层平面
GROUND FLOOR PLAN

首层平面
1st FLOOR PLAN

三层平面
3rd FLOOR PLAN

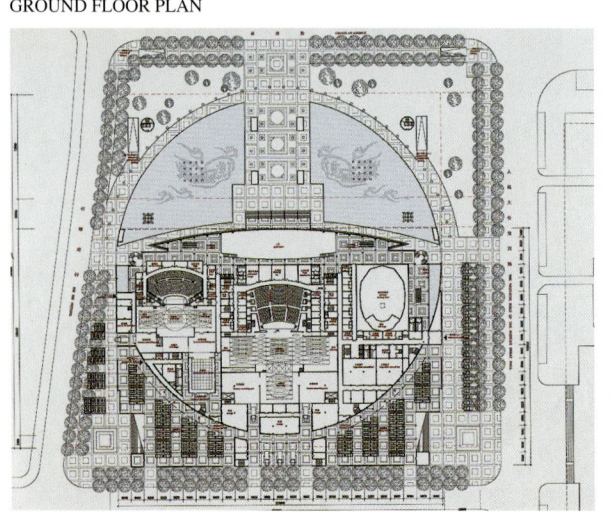

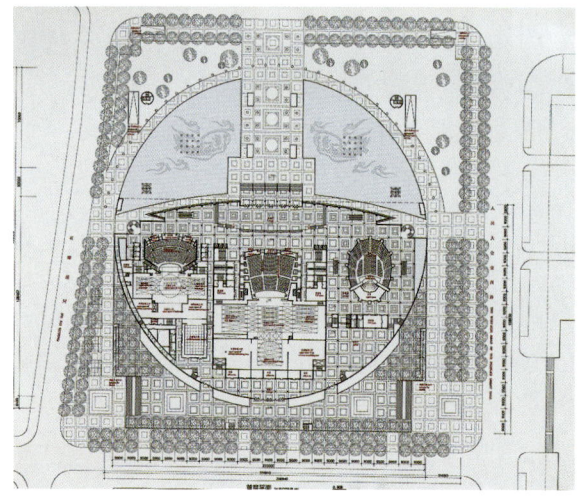

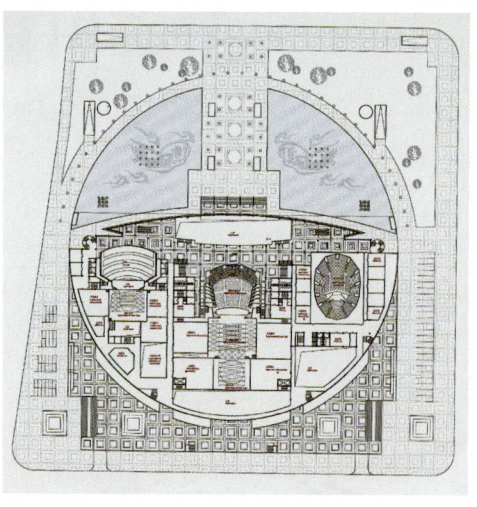

技术经济指标
Technical Economic Index

基地面积(m²) Site Area	总建筑面积(m²) Total Floor Area	建筑高度(m) Building Height	建筑覆盖率(%) Building Coverage Ratio
	105700	28/45	

容积率 Plot Ratio	绿化覆盖率(%) Green Coverage Ratio	停车数量 Number of Stalls	自行车停车数量 Number of Bicycle Stalls
		地上：大车27， 小车175 地下：636	1000

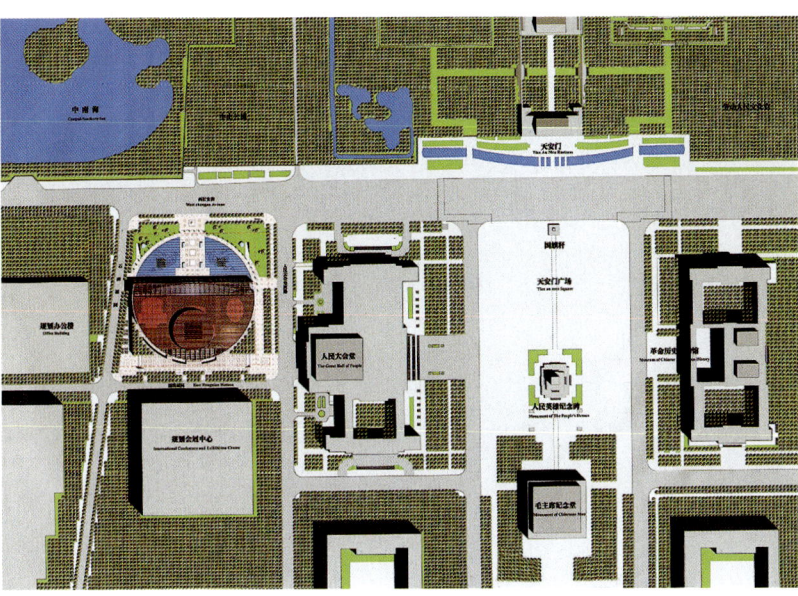

总平面
OVERALL SITE PLAN

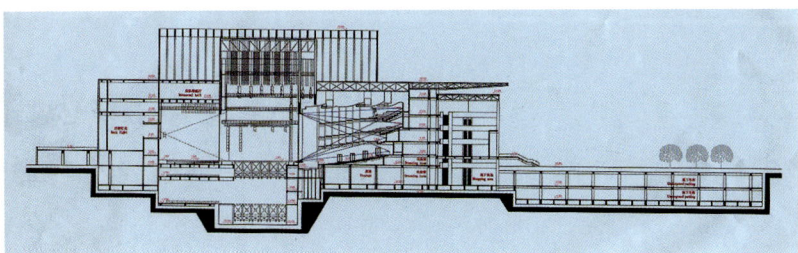

剖面
SECTION

各观众厅数据
Data For Auditoria

			歌剧院 Opera House	音乐厅 Concert Hall	戏剧院 Theater	小戏场 Mini Theater
		观众厅面积(m²) Floor Area of Auditorium	2000	1200	1070	432
		观众厅体积(m³) Volume of Auditorium	19800	20700	5995	4000
座 席 数 Seating Capacity		总座席数 Seating Capacity(Seats)	2518	2006	1209	300-500
		其中：池座 Among Them:Auditorium	1385	1082	889	可变换
		楼座 Floor Seat	1133	924	320	
		休息厅面积(m²) Lounge Floor Area				
舞 台 尺 寸 (m²) Stage Dimentions		主舞台 Main Stage	33×27 (宽×深)	16×12 (宽×深)	26×21 (宽×深)	14×20 (宽×深)
		左右侧台 Left and Right Side Stages	左：24×20 右：24×20		13.6×18	
		后舞台 Back Stage	27×20.8		18×11	
		台仓 Under Stage Storage	主：33×27 后：25×20.8		主：26×21 后：18×11	14×20
		升降乐池 Elevating Orchestra Pit	22×6 可升降部分		15×6 可升降部分	
		最大视距(m) Max. Visual Distance	36	24	24	
		最大俯角(°) Max. Angle of Depression	25	16	27	

图书在版编目（CIP）数据

中国国家大剧院建筑设计国际竞赛方案集／《中国国家
大剧院建筑设计国际竞赛方案集》编委会编. —北京：中
国建筑工业出版社,1999
　ISBN 7-112-03923-1

　Ⅰ.中…　Ⅱ.中…　Ⅲ.①剧院－建筑设计－中国－图集②
剧院－建筑方案－中国　Ⅳ.TU242.2

　中国版本图书馆 CIP 数据核字(1999)第 13654 号

责任编辑：郭洪兰
装帧设计：冯彝诤

中国国家大剧院建筑设计国际竞赛方案集
《中国国家大剧院建筑设计国际竞赛方案集》编委会编
＊
中国建筑工业出版社 出版、发行(北京西郊百万庄)
新华书店经销
北京广厦京港图文有限公司制作
北京华联公司印刷厂印刷
＊
开本：787 × 1092毫米 1/12　印张：27⅔
2000 年 6 月第一版　2006 年 1 月第二次印刷
印数：2001-5500 册　定价：266.00 元
ISBN 7-112-03923-1
　TU · 3054(9292)